Carotenoids

Carotenoids

Volume 5: Nutrition and Health

Edited by G. Britton
S. Liaaen-Jensen
H. Pfander

Birkhäuser Verlag
Basel · Boston · Berlin

Editors:

Dr. George Britton
53 Forest Road
Meols
Wirral
Merseyside
CH47 6AT
UK

Professor Dr. Dr. h.c. Synnøve Liaaen-Jensen
Organic Chemistry Laboratories
Department of Chemistry
Norwegian University of Science and
Technology (NTNU)
7491 Trondheim
Norway

Prof. Dr. Hanspeter Pfander
CaroteNature GmbH
Chief Operating Officer
Muristrasse 8e
3006 Bern
Switzerland

Library of Congress Control Number: 2008932322

Bibliographic information published by Die Deutsche Bibliothek
Die Deutsche Bibliothek lists this publication in the Deutsche Nationalbibliografie;
detailed bibliographic data is available in the internet at http://dnb.ddb.de

ISBN 978-3-7643-7500-3 Birkhäuser Verlag, Basel – Boston – Berlin

Part of Springer Science+Business Media
Printed on acid-free paper produced from chlorine-free pulp. TFC ∞
Printed in Germany

Cover design: Markus Etterich, Basel
Cover illustration: Katrin Uplegger, Birkhäuser, Basel

ISBN 978-3-7643-7500-3

e-ISBN 978-3-7643-7501-0

9 8 7 6 5 4 3 2 1

www.birkhauser.ch

Dedication

NORMAN I. KRINSKY

29th June 1928 – 28th November 2008

For a period of more than 50 years from his Ph.D work on 'Studies of Carotenoids and Vitamin A Complexes with Proteins in Plasma and Tissues', through his post-doctoral work on vision, in the Harvard laboratory of Nobel laureate George Wald, to his long and productive career at Tufts University, Boston, the name 'Norman Krinsky' has been synonymous with 'Carotenoids in Human Health and Nutrition'. A true giant in the field, Norman pioneered so much of current thinking and understanding of the nutritional value and health benefits of carotenoids, and this work is continued by the talented and dedicated group of colleagues that he assembled at Tufts. It is no coincidence that so many chapters in this Volume come from his associates, past and present, with him as a co-author on two of them.

A great teacher and communicator, Norman leaves us other legacies. The Gordon Research Conferences on Carotenoids, initiated by him in 1992, continue to go from strength to strength. He was co-chair of the 8th International Symposium on Carotenoids in Boston in 1987, and co-editor of the proceedings, published as Carotenoids: Chemistry and Biology (editors N. I. Krinsky, M. M. Mathews-Roth and R. F. Taylor, 1990), and his more recent book Carotenoids in Health and Disease (editors N. I. Krinsky, S. T. Mayne and H. Sies, 2004) is a most valuable collection of research reports on the subject.

Norman Krinsky's contribution to carotenoid science is immense but he is also remembered for his humanity and humour, his warmth, wit and wisdom. Always interested in other people and their work, he was ever ready to guide and advise and has been a great inspiration to so many. His perceptive comments and questions at conferences, always constructive, never destructive, are legendary.

The editors all have our own memories of Norman, who was a good friend of all of us. He will be very much missed; we will not see his like again.

That the field of Carotenoids in Human Nutrition and Health is as active and exciting as it is today is due in no small part to our friend Norman Krinsky. It is most fitting and a great honour for us to dedicate this volume to him and his memory.

Contents

Chapter 3: Carotenoids in Food

George Britton and Frederick Khachik

Chapter 4: Supplements

Alan Mortensen

Chapter 5: Microbial and Microalgal Carotenoids as Colourants and Supplements

Laurent Dufossé

Chapter 6: Genetic Manipulation of Carotenoid Content and Composition in Crop Plants

Paul D. Fraser and Peter M. Bramley

Chapter 7: Absorption, Transport, Distribution in Tissues and Bioavailability

Kirstie Canene-Adams and John W. Erdman Jr

Chapter 8: Carotenoids as Provitamin A

Guangwen Tang and Robert M. Russell

Chapter 9: Vitamin A and Vitamin A Deficiency

George Britton

Chapter 10: Epidemiology and Intervention Trials

Susan T. Mayne, Margaret E. Wright and Brenda Cartmel

Chapter 11: Modulation of Intracellular Signalling Pathways by Carotenoids

Paola Palozza, Simona Serini, Maria Ameruso and Sara Verdecchia

Chapter 12: Antioxidant/Pro-oxidant Actions of Carotenoids

Kyung-Jin Yeum, Giancarlo Aldini, Robert M. Russell and Norman I. Krinsky

Chapter 13: Carotenoids and Cancer

Cheryl L. Rock

Chapter 14: Carotenoids and Coronary Heart Disease

Elizabeth J. Johnson and Norman I. Krinsky

Chapter 15: The Eye

Wolfgang Schalch, John T. Landrum and Richard A. Bone

Chapter 16: Skin Photoprotection by Carotenoids

Regina Goralczyk and Karin Wertz

Chapter 17: The Immune System

Boon P. Chew and Jean Soon Park

Chapter 18: Biological Activities of Carotenoid Metabolites

Xiang-Dong Wang

Chapter 19: Editors' Assessment

George Britton, Synnøve Liaaen-Jensen and Hanspeter Pfander

Contents of *Carotenoids Volume 4: Natural Functions*

List of Contributors

Giancarlo Aldini
Department of Pharmaceutical Sciences
Faculty of Pharmacy
University of Milan
Via Mangiagalli 25
20133 Milan
Italy
(giancarlo.aldini@unimi.it)

Maria Ameruso
Institute of General Pathology
Catholic University
School of Medicine
L. go F. Vito 1
00168 Rome
Italy

Richard A. Bone
Department of Physics
Florida International University
Miami
FL 33199
U.S.A.
(bone@fiu.edu)

Peter M. Bramley
Centre for Systems and Synthetic Biology
School of Biological Sciences
Royal Holloway University of London
Egham Hill
Egham
Surrey
TW20 0EX
U.K.
(p.bramley@rhul.ac.uk)

George Britton
53 Forest Road
Meols
Wirral
CH47 6AT
U.K.
(george.britton19@gmail.com)

Kirstie Canene-Adams
Department of Pathology
Johns Hopkins School of Medicine
The Bunting-Blaustein Cancer Research Building
1650 Orleans Street
Baltimore
MD 21231-1000
U.S.A.
(kirstieadams26@yahoo.com)

Brenda Cartmel
Yale School of Public Health
60 College Street
P. O. Box 208034
New Haven
CT 06520-8034
U.S.A.
(brenda.cartmel@yale.edu)

Boon P. Chew
FSHN 110
School of Food Science
Washington State University
Pullman
WA 99164-6376
U.S.A.
(boonchew@wsu.edu)

Laurent Dufossé
Université de la Réunion
ESIDAI
LCSNSA
Parc Technologique
2 Rue Joseph Wetzell
F-97490 Sainte-Clotilde
La Réunion
France
(laurent.dufosse@univ-reunion.fr)

John W. Erdman Jr.
Department of Food Science and Human Nutrition
University of Illinois,
451 Bevier Hall
905 S. Goodwin Avenue
Urbana
IL 61801
U.S.A.
(jwerdman@illinois.edu)

Paul D. Fraser
Centre for Systems and Synthetic Biology
School of Biological Sciences
Royal Holloway University of London
Egham Hill
Egham
Surrey
TW20 0EX
U.K.
(p.fraser@rhul.ac.uk)

Regina Goralczyk
DSM Nutritional Products Ltd.
P.O. Box 2676
CH-4002 Basel
Switzerland
(regina.goralczyk@dsm.com)

Elizabeth J. Johnson
Carotenoids and Health Laboratory
Jean Mayer USDA Human Nutrition Center on Aging
Tufts University
711 Washington Street
Boston
MA 02111
U.S.A.
(elizabeth.johnson@tufts.edu)

Norman I. Krinsky
Jean Mayer USDA Human Nutrition Research Center on Aging
Department of Biochemistry
School of Medicine
Tufts University
711 Washington Street
Boston
MA 02111
U.S.A.

John T. Landrum
Department of Chemistry and Biochemistry
Florida International University
Miami
FL 33199
U.S.A.
(landrumj@fiu.edu)

Synnøve Liaaen-Jensen
Organic Chemistry Laboratories
Department of Chemistry
Norwegian University of Science and Technology (NTNU)
7491 Trondheim
Norway
(slje@chem.ntnu.no)

Susan T. Mayne
Yale School of Public Heath
60 College Street
P.O. Box 208034
New Haven
CT 06520-8034
U.S.A.
(susan.mayne@yale.edu)

Alan Mortensen
Product Development
Color Division
Chr. Hansen
Bøge Allé 10-12
DK-2970 Hørsholm
Denmark
(alanm@mailme.dk)

Paola Palozza
Institute of General Pathology
Catholic University
School of Medicine
L. go F. Vito 1
00168 Rome
Italy
(p.palozza@rm.unicatt.it)

Jean Soon Park
P & G Pet Care
Upstream R & D
6571 State Route 503 North
P.O. Box 189
Lewisburg
OH 45338
U.S.A.
(park.j.15@pg.com)

Hanspeter Pfander
Muristrasse 8E
CH-3006 Bern
Switzerland
(hanspeter.pfander@carotenature.com)

Cheryl L. Rock
Moores USCD Cancer Center
University of California San Diego
3855 Health Sciences Drive
La Jolla
CA 92093-0901
U.S.A.
(clrock@ucsd.edu)

Robert M. Russell
Office of Director
National Institutes of Health
Washington DC,
U.S.A.
(russellr2@od.nih.gov)

Wolfgang Schalch
DSM Nutritional Products Ltd.
Wurmisweg 576
CH-4303 Kaiseraugst
Switzerland
(wolfgang.schalch@dsm.com)

Simona Serini
Institute of General Pathology
Catholic University
School of Medicine
L. go F. Vito 1
00168 Rome
Italy

Guangwen Tang
Carotenoids and Health Laboratory
Jean Mayer USDA Human Nutrition
Research Center on Aging
Tufts University
711 Washington Street
Boston
MA 02111
U.S.A.
(guangwen.tang@tufts.edu)

Sara Verdecchia
Institute of General Pathology
Catholic University
School of Medicine
L. go F. Vito 1
00168 Rome
Italy

Xiang-Dong Wang
Nutrition and Cancer Biology Laboratory
Jean Mayer USDA Human Nutrition
Research Center on Aging
Tufts University
711 Washington Street
Boston
MA 02111
U.S.A.
(xiang-dong.wang@tufts.edu)

Karin Wertz
DSM Nutritional Products Ltd.
P.O. Box 2676
CH-4002 Basel
Switzerland
(karin.wertz@dsm.com)

Margaret E. Wright
Department of Pathology
College of Medicine
University of Illinois at Chicago
840 South Wood Street
Chicago
IL 60612
U.S.A.
(mewright@uic.edu)

Kyung-Jin Yeum
Jean Mayer USDA Human Nutrition
Research Center on Aging
Tufts University
711 Washington Street
Boston
MA 02111
U.S.A.
(kyungjin.yeum@tufts.edu)

Preface

More than twenty years after the idea of this *Carotenoids* book series was first discussed, we finally reach the end of the project with *Volume 5*, which covers the functions and actions of carotenoids in human nutrition and health. In 1971, in Isler's book *Carotenoids*, functions of carotenoids and vitamin A were covered in just two chapters. Now, thanks to technical developments and multidisciplinary approaches that make it possible to study functional processes in great detail, this subject is the most rapidly expanding area of carotenoid research, and occupies two full volumes, *Volumes 4: Natural Functions* and *5: Nutrition and Health.* Although *Volume 5* can be used as a single stand-alone volume, the two were planned as companion volumes to be used together. To understand the mechanisms of functions and actions of carotenoids, including how carotenoids may be involved in maintaining human health, requires understanding of the underlying principles, which are presented in the first part of *Volume 4*.

The general philosophy and strategy of the series, to have expert authors review and analyse critically a particular topic and present information and give guidance on practical strategies and procedures is maintained in *Volumes 4* and *5*. It is also the aim that these publications should be useful for both experienced carotenoid researchers and newcomers to the field.

The material presented in the earlier volumes of the series is relevant to studies of biological functions and actions. Biological studies must be supported by a rigorous analytical base and carotenoids must be identified unequivocally. It is a common view that carotenoids are difficult to work with. This may be daunting to newcomers to the field, especially if they do not have a strong background in chemistry and analysis. There are difficulties; carotenoids are less stable than most natural products, but ways to overcome the difficulties and to handle these challenging compounds are well established and are described and discussed in *Volume 1A* which, together with *Volume 1B*, gives a comprehensive treatment of the isolation, analysis and spectroscopic characterization of carotenoids as an essential foundation for all carotenoid work. This is complemented by the *Carotenoids Handbook* (2004), which was produced in association with this series and provides key analytical data for each of the 750 or so known naturally-occurring carotenoids.

Volume 2 describes methods for the chemical synthesis of carotenoids, including those that are needed as analytical standards and on a larger scale for biological trials. Functions and actions are inextricably linked with biosynthesis and metabolism, covered in *Volume 3*.

Note that the original editions of *Volumes 1A*, *1B* and *3* are now 'out of print' and not available from the usual booksellers. It is still possible to obtain reprinted paperback copies or CD versions. For information on this please see the website of the International Carotenoid Society (www.carotenoidsociety.org) or contact the editors by e-mail.

There are many other major publications in the carotenoid field which are still extremely valuable sources of information. The history of key publications up to around 1994 was outlined in the preface to the series, in *Volume 1A*. Since then there have been other progress reports, notably the published proceedings of the International Carotenoid Symposia in 1996, 1999, 2002, 2005 and 2008. References to specialized monographs and reviews on particular topics can be found in the relevant *Chapters*.

Volume 5 and its companion, *Volume 4*, are the last volumes in the *Carotenoids* series, and in many ways point the way to the future of carotenoid research. If the insight that these books provide stimulates chemists, physicists, biologists and the medical professions to understand and communicate with each other and thus serves as a catalyst for interdisciplinary studies that will bring advances and rewards in the future, then the editors will feel that their time and effort has been well spent.

Editors' Notes on the Use of this Book

The *Carotenoids* books are planned to be used together with the *Carotenoids Handbook*. Whenever a known natural carotenoid is mentioned, its number in the *Handbook* is given in bold print. Other compounds, including synthetic carotenoids and analogues that do not appear in the *Handbook*, are numbered separately in italics in sequence as they appear in the text for each *Chapter*, and their structural formulae are shown. The number is given at the first mention of a particular compound and may be repeated for clarity, for example when chemical comparisons are made.

The *Carotenoids* books form a coordinated series, so there is substantial cross-referencing between *Volumes 4* and *5*, and with earlier *Volumes* in the series. Earlier *Volumes* and *Chapters* therein are usually not included in reference lists.

Carotenoid nomenclature

The IUPAC semi-systematic names for all known naturally occurring carotenoids are given in the *Carotenoids Handbook*. Trivial names for many carotenoids are, however, well-established and convenient, and are generally used in biological publications, including β-carotene rather than β,β-carotene. These common, trivial names are used throughout these volumes.

The *E/Z* and *trans/cis* denominations for describing the stereochemistry about a double bond are not always equivalent. In most cases in this book, the *E/Z* system is used to designate geometrical isomers of carotenoids. The terms *cis* and *trans* are in common biological usage for retinoids and have been retained, *e.g.* (9-*cis*)-retinoic acid, and they are also used at times for general statements about carotenoids.

Naming of organisms

The correct classification and naming of living organisms is essential. The editors have not checked all these but have relied on the expertise of the authors to ensure that classification schemes and names in current usage are applied accurately, and for correlation between new and old names.

Abbreviations

The abbreviations listed are mainly ones that occur in more than one place in the book. Abbreviations defined at their only place of mention are not listed.

Terms, designations and abbreviations used in the context of biochemistry and cell biology, for signalling molecules, *etc.*, are those in common usage by journals and in advanced textbooks.

Indexing

For many purposes the *List of Contents* is sufficient to guide the reader to a particular topic. The subject *Index* at the end of the book complements this and lists key topics that occur, perhaps in different contexts, in different places in the book. No author index, index of compounds or index of organisms is given.

In memoriam

During the 20 years or so that we have been working on this project, the 'Carotenoid Club' has lost many members, including some of the great personalities and pioneers in various aspects of the carotenoid field. In just the last few months of the preparation of this *Volume*, we were saddened to learn of the passing of three great names in carotenoid research.

Trevor W. Goodwin, pioneer of carotenoid biochemistry, especially in plants, who died on 7th October 2008, at the age of 92, is well known for his many books, especially '*The Biochemistry of the Carotenoids, Volume 1: Plants* (1980) and *Volume 2: Animals* (1984). George Britton is particularly indebted to Trevor Goodwin for introducing him to the wonderful world of carotenoids and being his guide and mentor in the carotenoid field.

Although Hans-Dieter Martin, who died on 8th March 2009, aged 70, came to the carotenoid field relatively late, his work contributed much to carotenoid chemistry, especially to our knowledge and understanding of the properties that form the basis of many of the natural functions of carotenoids. We are pleased to have had his *Chapter* 'Aggregation and Interface Behaviour of Carotenoids' as a major contribution to *Volume 4*.

Norman Krinsky, who was at the forefront of many new ideas and developments in the field of carotenoids and human health, died on 28th November 2008, aged 80. He gave us much valuable advice during the planning of this series and especially *Volumes 4* and *5*, and we are delighted to have chapters co-authored by him and his colleagues from Tufts University, Boston, in *Volume 5*.

It is an honour and pleasure to pay tribute to them and other carotenoid friends and colleagues who have passed on during the time we have been working on the *Carotenoids* books, and to acknowledge the immense contributions they have made to the carotenoid field. They have greatly enhanced the 'carotenoid world' and provided inspiration for many of us.

Editors' Acknowledgements

We repeat our comment from the earlier *Volumes*. "Although we are privileged to be the editors of these books, their production and publication would not be possible without the efforts of many other people".

The dedicated work of the authors, their attention to requests and questions and their gracious acceptance of the drastic editing that was sometimes needed for coordination, to avoid duplication and to meet the stringent limitations of space, is gratefully acknowledged. The job of the editors is made so much easier when authors provide carefully prepared manuscripts in good time.

We thank Detlef Klüber and the editorial staff at Birkhäuser for their forbearance as deadlines slipped, and especially Kerstin Tüchert who was responsible for the 'hands-on' work to get the book into publication.

Discussions with carotenoid colleagues during the planning of these *Volumes* were very useful and much appreciated.

Finally, we again express our gratitude to DSM and BASF for the financial support without which this project would not have been possible.

Abbreviations

ADI	Acceptable daily intake
AIDS	Autoimmune deficiency syndrome
AMD	Age-related macular degeneration
AMI	Acute myocardial infarction
AMS	Accelerator mass spectrometry
ARE	Antioxidant response element
ATBC	α-Tocopherol, β-Carotene Cancer Prevention Study
AUC	Area under the curve
BCO1	β-Carotene oxygenase 1 (β-carotene 15,15′-oxygenase)
BCO2	β-Carotene oxygenase 2 (β-carotene 9,10-oxygenase)
BHT	Butylated hydroxytoluene (2,6-di-*t*-butyl-*p*-cresol)
BMI	Body mass index
CARET	Carotene and Retinol Efficacy Trial
CC	Column chromatography
CD	Circular dichroism
cdk	Cyclin-dependent kinase
CHD	Coronary heart disease
CI	Confidence interval
Cv	Cultivar
CVD	Cardiovascular disease
CWD	Cold-water dispersible
Cx43	Connexin 43
CYP	Cytochrome P450
DTH	Delayed-type hypersensitivity
EPIC	European Prospective Investigation into Cancer and Nutrition
EPP	Erythropoietic protoporphyria
FAO	Food and Agriculture Organization (United Nations)
FDA	Food and Drug Administration (U.S.A.)
GJC	Gap junction communication
G(L)C	Gas (liquid) chromatography
GM	Genetic modification/manipulation
HDL	High density lipoprotein

HIV	Human immunodeficiency virus
HPLC	High performance liquid chromatography
HPV	Human papilloma virus
HR	Hazard ratio
IARC	International Agency for Research on Cancer
Ig	Immunoglobulin
IGF	Insulin-like growth factor
IL	Interleukin
IU	International Unit
IVACG	International Vitamin A Consultative Group
JECFA	Joint FAO/WHO Expert Committee on Food Additives
kDa	kiloDalton
L	Litre
LC	Liquid chromatography
LDL	Low density lipoprotein
MAPK	Mitogen-activated protein kinase
MDA	Malondialdehyde
MMP	Matrix metalloprotease
MPOD	Macular pigment optical density
MS	Mass spectrometry
NHANES	National Health and Nutrition Examination Survey (U.S.A.)
NIST	National Institute of Standards and Technology (U.S.A.)
NK	Natural killer
NMR	Nuclear magnetic resonance
NP	Normal phase
ODS	Octadecylsilanyl
OR	Odds ratio
PCR	Polymerase chain reaction
PDA(D)	Photodiode array (detector)
PG	Prostaglandin
PHS	Physicians' Health Study (U.S.A.)
PKC	Protein kinase C
PPAR	Peroxisome-proliferator activated receptor
PPARE	PPAR response element
PPC	Post-prandial chylomicron
PUFA	Polyunsaturated fatty acid
RABP	Retinoic acid binding protein
RAE	Retinol activity equivalent
RAR	Retinoic acid receptor
RARE	Retinoic acid response element
RBP	Retinol-binding protein
RDA	Recommended daily allowance
RE	Retinol equivalent
ROS	Reactive oxygen species

RP	Reversed phase
RPE	Retinal pigment epithelium
RR	Risk ratio
RT-PCR	Real-time PCR
RXR	Retinoid X receptor
RXRE	Retinoid X response element
SMD	Standard mean deviation
SOD	Superoxide dismutase
TBARS	Thiobarbituric acid reactive substances
TBME	*t*-Butyl methyl ether
THF	Tetrahydrofuran
TLC	Thin-layer chromatography
TNF	Tumour necrosis factor
UNICEF	United Nations Children's Fund
UV	Ultraviolet
UVA	UV wavelength range 320-400 nm
UVB	UV wavelength range 290-320 nm
UVR	UV radiation
UV/Vis	UV/visible
VAD	Vitamin A deficiency
VC	Variability coefficient
VLDL	Very low density lipoprotyein
WHO	World Health Organization

Carotenoids
Volume 5: Nutrition and Health

Chapter 1

Editors' Introduction: A Healthy Debate

George Britton, Synnøve Liaaen-Jensen and Hanspeter Pfander

A. Introduction

Interest in carotenoids and human health goes back some 80 years, when the link between β-carotene (**3**) and vitamin A was first demonstrated and the dietary importance of β-carotene and some other carotenoids as provitamin A was established. This alone is sufficient to ensure that carotenoids will always have an important place and value in human nutrition. But there is more. This is now the era of 'functional foods', when a major goal is to identify roles of chemical components of foods as important micronutrients. Dietary intake can be manipulated by adopting a 'healthy diet', *i.e.* one rich in fruit and vegetables. Many supplements are now available to augment supplies when intake is limited or considered to be sub-optimal.

Carotenoids feature high on the list of food components that are of interest in relation to human health. The first great catalyst and stimulus for this was the publication in 1981 of a paper in *Nature* in which the authors addressed the question 'Can dietary β-carotene materially reduce human cancer rates?' [1]. Three years later another key paper [2], revealing that β-carotene could be a new kind of antioxidant, stimulated the imagination of many carotenoid researchers. Antioxidants are now big business and their importance in maintaining health and as major players in the fight against serious and chronic diseases such as cancer is widely accepted. Even after 25 years of intensive study, however, it is still not clear if carotenoids have an important place in the hierarchy of natural antioxidants *in vivo*.

In recent years, investigations have spread in directions as diverse as whole population studies (epidemiology), detailed investigation of effects on molecular processes and intricate mechanistic studies. The literature is vast and expanding rapidly. This, the final volume of the *Carotenoids* series, surveys the field of carotenoids in human nutrition and health. In the past 15 years or so, the topic has been covered in several books. Two of these, '*Carotenoids in*

Human Health' [3] and '*Carotenoids and Retinoids: Molecular Aspects and Health Issues*' [4] comprise collections of papers presented at meetings. A recent book '*Carotenoids in Health and Disease*' [5], covers the topic comprehensively with a series of expert research reports. A '*Handbook of Cancer Prevention, Volume 2: Carotenoids*' [6] summarizes the conclusions of an expert working group which evaluated the evidence, to provide an overview of the relationship between carotenoids and cancer.

B. *Volume 5*

1. Strategy

Carotenoids, Volume 5 was planned as a coordinated, integrated treatment providing up-to-date research surveys by leading authorities in the field, and incorporating some background material to help make the chapters accessible to carotenoid researchers who are not specialists on the particular topic. The practical approach that has been a feature of the series is maintained. Not only are experimental findings reported but the methods by which the data were obtained are explained and evaluated.

2. Relation to other volumes

Although *Volume 5* may be used as a single stand-alone volume, *Volume 4: Natural Functions* and *Volume 5: Human Nutrition and Health* were planned as companion volumes to be used together. To understand the mechanisms of functions and actions of carotenoids requires understanding of the underlying fundamental principles. The treatment of fundamental properties of carotenoids presented in the first part of *Volume 4* is intended also as a foundation for understanding how carotenoids may be involved in maintaining human health.

Each carotenoid has a precise three-dimensional shape which is vital for ensuring that the carotenoid fits into cellular, sub-cellular and molecular structures in the correct location and orientation to allow it to function efficiently. Absolute configuration, conformation and geometrical isomeric form are considered in *Chapters 2-4* of *Volume 4*. Geometrical isomeric form (*cis/trans* or *E/Z*) may be an important factor in the biological activity of carotenoids, especially in relation to bioavailability, transport and deposition in tissues.

The conjugated double-bond system of carotenoids determines the photochemical properties and chemical reactivity that form the basis of most of their functions. Light absorption is the basis of detection and analysis. Excitation, energy transfer and quenching (*Volume 4*, *Chapter 9*) are relevant to protective roles in the eye and skin. The susceptibility of the electron-rich polyene chain to attack and breakdown by electrophilic reagents and oxidizing free radicals is the basis for the behaviour of carotenoids as antioxidants or pro-oxidants (*Volume 4, Chapter 7*). This instability can have serious consequences for large-scale trials of carotenoids for biological activity. Samples used in such investigations must be free from peroxides and other degradation products, otherwise misleading results may be obtained and comparison between studies is difficult.

Interactions strongly influence the properties of a carotenoid *in situ* and are crucial to functioning. As discussed in *Volume 4, Chapter 5*, the physical and chemical properties of carotenoids are inevitably influenced by interactions with other molecules in the immediate vicinity, and by aggregation, especially in an aqueous environment and in membranes.

Other *Volumes* in the series are also relevant. Biological studies must be supported by a rigorous analytical base. Newcomers to the field, especially if they do not have a strong background in chemistry and analysis, often feel that carotenoids are difficult to work with. There are difficulties; carotenoids are less stable than most natural products, but ways to overcome the difficulties and to handle these challenging compounds are well established. These are described and discussed in *Volume 1A*: *Isolation and Analysis*, which, together with *Volume 1B*: *Spectroscopy* gives a comprehensive treatment of the isolation, analysis and spectroscopic characterization of carotenoids that is an essential foundation for all carotenoid work. Complementary to this, data on all known naturally occurring carotenoids are compiled in the *Carotenoids Handbook.*

Volume 2: *Synthesis* describes strategies and methods for the total synthesis of carotenoids, including synthesis in bulk for commercial applications and use in experimental trials. It also provides methods for the synthesis of isotopically labelled samples for use in the powerful methods now being applied to assay bioavailability and conversion into vitamin A. *Volume 3*: *Biosynthesis and Metabolism* provides information relevant to the occurrence and distribution of various carotenoids in food, improvement of crop plants for nutritional quality and the optimization of microbial production for colourants and supplements by conventional and genetic manipulation (GM) methods. The biochemistry of metabolic processes including conversion into vitamin A is also covered.

3. Content of *Volume 5*

a) Nutrition

i) Analysis. Building on the coverage in *Volumes 1A* and *1B*, some recommended procedures for the analysis of carotenoids in food and human blood and tissues, especially by HPLC, are described in *Chapter 2.*

ii) Sources of carotenoids. Humans do not biosynthesize carotenoids so all carotenoids that are found in human blood and tissues must be ingested, either in food or as supplements. *Chapter 3* summarizes the main features of the occurrence of carotenoids in food, especially in vegetables and fruits. It provides a guide to what foods are good dietary sources of of both the provitamin A β-carotene (**3**) and of other carotenoids, namely lycopene (**31**), lutein (**133**), zeaxanthin (**119**) and β-cryptoxanthin (**55**), that are now under investigation for possible health benefits against serious diseases. It does not include comprehensive tables of quantitative data, which can be accessed on-line.

Carotenoids are also used as colourants in manufactured food products, and carotenoids and carotenoid-rich extracts are available as supplements. An overview of the various products and formulations and the legislation that governs their use is given in *Chapter 4*. The major production of carotenoid additives and supplements is by chemical synthesis but biological production is becoming of increasing interest and concern. Production by microbial biotechnology is evaluated in *Chapter 5*. *Chapter 6* then describes the application of molecular genetics to increasing or modifying carotenoid production by plants, both to provide natural food with optimized carotenoid content and composition and for use as colourants and in supplements. The use of genetically modified (GM) crops, including 'Golden Rice' engineered to accumulate β-carotene, is discussed.

iii) Bioavailability and provitamin A. Ingested carotenoids must reach appropriate tissues in the body in order to have any biological effect. The story of this process, covering digestion, absorption and transport, and deposition in tissues is told in *Chapter 7*. Factors that determine the bioavailability of β-carotene and other carotenoids of interest are assessed, covering food structure, digestion, absorption, transport and regulatory processes. The localization and distribution of carotenoids in body tissues is described.

Full coverage of vitamin A and vitamin A deficiency could easily fill a whole volume. Carotenoids are the main provitamin source of vitamin A for many of the world's population, so some treatment of this important topic is essential in this book. *Chapter 8* considers the nutritional aspects of the conversion of β-carotene into vitamin A, the biochemistry and enzymology of which were treated in *Volume 4*, *Chapter 16*. In spite of the many sophisticated methods that are now available to determine numerical conversion factors as a guideline for dietary recommendations, the results are varied, and the topic remains controversial. To complement this survey, an outline of the global problem of vitamin A deficiency and ways of alleviating it, by supplementation, fortification and sustainable dietary improvement of provitamin A carotene intake, is presented in *Chapter 9*.

b) Carotenoids in health and disease

i) Experimental approaches. Three chapters describe experimental approaches and strategies for the study of other health effects of carotenoids. There have been extensive epidemiological surveys to identify associations between carotenoids, either from normal dietary intake or administered in intervention trials, and relative risk of serious conditions such as cancer and coronary heart disease. *Chapter 10* describes the various experimental designs, how data are obtained and evaluated, and the strengths and limitations of the various methods. Some large studies are described and the relative merits of different endpoints evaluated.

Effects of applied carotenoids on various cellular and molecular processes in cells cultured *in vitro* are evaluated in *Chapter 11*. Microarray technology makes it possible to detect simultaneously effects on the expression of any of thousands of genes. Many effects of carotenoids on important signalling mechanisms and fundamental processes such as the cell cycle and apoptosis have been reported. Their significance in relation to cancer and other diseases is discussed.

Oxidizing free radicals are implicated in the progression of many serious diseases, and the efficacy of antioxidants and other body defences in fighting against the effects of these oxidations is considered of particular importance. One of the most widely discussed actions of carotenoids is their ability to act as antioxidants or pro-oxidants, that can be demonstrated under appropriate conditions in many model systems. *Chapter 12* builds on the fundamental treatment of the properties of carotenoid radicals, carotenoid oxidation and reactions with oxidizing free radicals that was given in *Volume 4*, and discusses the possible relevance of carotenoids as antioxidants or pro-oxidants in biological systems and disease prevention.

ii) Carotenoids and major diseases. The remaining part looks at evidence for effects of carotenoids on different aspects of health. Results from all the above experimental approaches, and work that specifically addresses the particular condition, are integrated. *Chapter 13* evaluates findings on the association between several dietary carotenoids and risk of major cancers of various body tissues, to try to identify indications of any protective role of any carotenoid on any particular cancer. Similarly, *Chapter 14* addresses the relationship between dietary carotenoid and coronary heart disease, particularly whether an antioxidant effect of lycopene or other carotenoids may be a factor.

There is a clear involvement of carotenoids in eye health (*Chapter 15*), not only as precursors of the vitamin A on which eye health and vision depend. An essential role of lutein (**133**) and (*meso*)-zeaxanthin (**120**) in the macula lutea and protection against photodamage is well established. The relationship between low carotenoid concentration in the macula and age-related macular degeneration, a leading cause of blindness in the elderly, is well documented; dietary or supplementary intervention in people or populations at risk may prove to be beneficial.

The skin is also often exposed to high intensity light and UV irradiation that can lead to photodamage and cancer. Detailed studies of how carotenoids, particularly β-carotene, may protect the skin against photodamage are discussed in *Chapter 16*..

Carotenoids have been shown to influence many parameters of the immune system. Any promoting effect on the human immune response should have a generally beneficial effect on health, and may also be significant for any tumour-suppressing role. Evidence and possible mechanisms are discussed in *Chapter 17*.

The essential roles of vitamin A in vision and of (all-*trans*)- and (9-*cis*)-retinoic acid in regulating key aspects of growth, development and hormone response are well known (*Volume 4, Chapter 15*). Whether other health effects attributed to carotenoids may in fact be due to metabolites or other breakdown products of provitamin A and non-provitamin A carotenoids is discussed in *Chapter 18*.

When reading the extensive literature on carotenoids and human health, one thing that becomes clear is the need to consider critically the experimental design and the actual results obtained rather than simply relying on the abstract and the published interpretation of the data, often in isolation. The authors of each *Chapter* in this *Volume* have done this objectively. Finally, in *Chapter 19*, the editors take a broad view and, with no prejudgement, make their personal evaluation of the importance of carotenoids in human health.

C. Conclusions

In a field where indications, which may or may not seem promising, rather than firm conclusions, are the norm, most studies lead to more questions. To answer these questions, interaction is essential between biologists, chemists and physical scientists, and between biologists from different disciplines so that the application of advanced techniques of molecular biology and the use of microarrays complements other biochemical and epidemiological studies. All of this must be integrated with clinical observations and other medical aspects.

It is clear that carotenoids are not the 'magic bullet' that will rid the world of major scourges such as cancer and coronary heart disease, but even a small percentage reduction in risk of some major diseaes in some populations could be of great benefit to millions of individuals.

References

[1] R. Peto, R. Doll, J. D. Buckley and M. D. Sporn, *Nature*, **290**, 201 (1981).

[2] G. W. Burton and K. U. Ingold, *Science*, **224**, 569 (1984).

[3] L. M. Canfield, J. A. Olson and N. I. Krinsky (eds.), *Carotenoids in Human Health*, *Ann. NY Acad. Sci*, **691** (1993).

[4] L. Packer, U. Obermüller-Jevic, K. Kraemer and H. Sies (eds.), *Carotenoids and Retinoids: Molecular Aspects and Health Issues*, AOCS Press, Champaign, Illinois, USA (2004).

[5] N. I. Krinsky, S. T. Mayne and H. Sies (eds.), *Carotenoids in Health and Disease*, CRC Press, Boca Raton, USA (2004).

[6] IARC/WHO, *Handbook of Cancer Prevention, Volume 2: Carotenoids*, International Agency for Research on Cancer, Lyon, France, (1999).

Carotenoids
Volume 5: Nutrition and Health

Chapter 2

Analysis of Carotenoids in Nutritional Studies

Frederick Khachik

A. Introduction

The ability to establish a statistically sound relationship between dietary intake of carotenoids and the incidence of chronic disease requires detailed knowledge of the qualitative and quantitative distribution of these compounds in the food supply as well as in human blood, major organs, and tissues. The occurrence and distribution of nutritionally important carotenoids in foods, especially fruits and vegetables, is surveyed in *Chapter 3* and the distribution of carotenoids in human blood, organs and tissues is covered in *Chapter 7*. In the past two decades, technological advances in high-performance liquid chromatography (HPLC) have provided analysts with powerful, sensitive tools to separate carotenoids and low levels of their metabolites and analyse them quantitatively with great precision. For HPLC analysis to be of value, it must be based on rigorous identification of the compounds under study. Even in laboratories with good HPLC facilities, the traditional, classical methods thin-layer and column chromatography (TLC and CC) are still widely used, for rapid preliminary screening of extracts, for isolating and purifying carotenoids for further study, for comparison of samples with standards, and for monitoring the course of reactions. This *Chapter* evaluates HPLC methods, describes quantitative analysis of carotenoids in extracts from foods and human samples and discusses a systematic approach to separation and identification. The procedures described and illustrated are ones that have been used extensively in the author's laboratory for many years, but they are not the only ones. Many other procedures have been described and widely used by other researchers.

Principles of HPLC of carotenoids, including a general description of normal phase and reversed phase chromatography and examples of separations that can be achieved were presented in *Volume 1A, Chapter 6, Parts I* and *IV*. Some particular separations are described in the *Worked Examples* in *Volume 1A*. Some of the basic aspects are included again in this *Chapter* so that it can be used as a 'self-sufficient' review.

B. Isolation and Characterization

1. Strategy

Although the methods for extraction and isolation of carotenoids have not changed significantly in the past 20 years or so, the techniques for analysis and identification have improved greatly.

HPLC has been shown to be the most efficient technique for the routine analysis of carotenoids in complex mixtures and is the method of choice in most laboratories today. The reproducibility and high sensitivity provide reliable analytical data, and the reasonably short analysis time minimizes the isomerization and decomposition of these sensitive compounds. In a systematic approach for separation and identification of carotenoids from a given extract, the sample is first examined by reversed-phase (RP) and/or normal-phase (NP) HPLC with UV/Vis photodiode array (PDA) detection. The HPLC elution profile provides useful information about what classes of carotenoids are present in the sample. The PDA allows the UV/Vis spectrum of each component to be determined on line. If the HPLC system is also interfaced into a mass spectrometer, additional information about the mass of individual carotenoids can also be obtained. Recent advances in linking HPLC with NMR show great promise. Usually though, for rigorous identification and structure elucidation, the individual carotenoids must be isolated from the mixture and purified.

For a preliminary fractionation, the extract may be subjected to flash CC or preparative TLC. Each of the isolated carotenoid fractions is then examined by HPLC on an appropriate analytical column, to determine the purity of the individual fractions. In most cases, the fractions obtained by CC or TLC require further purification, *e.g.* by semi-preparative HPLC employing conditions that are slightly modified from those that were used for the original analytical separation. A partially purified carotenoid is often a mixture of geometrical (*E/Z*) isomers and may contain other closely related carotenoids, so two successive semi-preparative HPLC separations may be necessary to achieve purity.

The purified carotenoid can then be identified by UV-visible (UV/Vis) spectrophotometry, mass spectrometry (MS) and nuclear magnetic resonance (NMR) spectroscopy. In the case of optically active carotenoids, a combination of NMR and circular dichroism (CD) can often be employed to determine the absolute configuration. The application of these spectroscopic techniques for the analysis of carotenoids is described in *Volume 1B*. The *Carotenoids*

Handbook, which gives data on all the more than 700 known naturally occurring carotenoids, should also be consulted. When the carotenoids have been identified, quantitative analysis of individual carotenoids, including their geometrical isomers, can be undertaken by HPLC.

Numerous HPLC conditions for separation of these pigments have been developed by different workers. The choice of the column, *i.e.* column dimensions, adsorbent and particle size, the eluting solvents and their composition, and the column flow rate are critical factors in developing the optimum separation conditions. Principles and guidance on this are given in *Volume 1A, Chapter 6, Part IV*. These will be summarized in this *Chapter*, and methods used by the author for the HPLC analysis of extracts from fruits, vegetables and human samples will then be described.

2. Extraction

Procedures for extraction of carotenoids from different kinds of samples are described and evaluated in *Volume 1A, Chapter 5*. *Worked Examples* of the extraction of carotenoids from higher plants, blood serum *etc.* are also presented in that *Volume*. In general, the choice of the organic solvent to be used for extraction depends on the nature of the material to be extracted and the solubility properties of the major carotenoids expected to be present. A variety of organic solvents, including acetone, tetrahydrofuran (THF), diethyl ether, ethyl acetate, and mixtures of petroleum ether/methanol, or hexane/methanol have been used. For food samples that contain a fairly large amount of water, it is desirable to use an organic solvent that is miscible with water, to optimize the release of carotenoids from the matrix and prevent the formation of emulsions. Acetone and THF are the recommended solvents for this but, if THF is used, it is advisable to stabilize it with 0.1% of an antioxidant such as 2,6-di-*t*-butyl-4-methylphenol (butylated hydroxytoluene, BHT) to prevent the formation of peroxides which could lead to some degradation of the carotenoids. When dried or canned fruits and vegetables *e.g.* apricots, peaches, sweet potato and citrus fruits, are to be analysed, the extraction can be facilitated by keeping the food samples in water for several hours before extraction. As well as improving the efficiency of extraction, this removes sugars and other water-soluble substances, thereby reducing the risk of forming emulsions.

In a general homogenization procedure, the sample is mixed in a blender with the extracting solvent, in the presence of Na_2CO_3 or $MgCO_3$ to neutralize organic acids that are often present in foods, and the resulting mixture is homogenized at a moderate speed for about one hour. The extraction temperature should not be allowed to exceed ambient temperature; to accomplish this, the blender vessel is placed in an ice-bath. The solid residue is removed by filtration and homogenized again with fresh solvent until the filtrate is no longer coloured. In some cases, the homogenization of foods with an organic solvent can cause an emulsion that can block or slow the vacuum filtration. Filtering the extract through celite or diatomaceous earth can overcome this problem. Alternatively, celite may be added to the sample and organic solvent mixture before homogenization. The combined filtrate is concentrated, and it is then partitioned between a water-immiscible organic solvent and water. The most commonly used organic solvents for this are *t*-butyl methyl ether (TBME) and diethyl ether, but dichloromethane, ethyl acetate, or hexane/dichloromethane (3:1) can be used. The organic phase is washed with water to remove unwanted water-soluble substances. Approximately 0.1% of an amine such as N,N-diisopropylethylamine (DIPEA) or triethylamine may be added to the organic solvent to neutralize any traces of acids that may remain in the extract. This is particularly necessary when the organic solvent is dichloromethane because chlorinated solvents may contain traces of HCl. The organic layer is then removed, dried over Na_2SO_4 or $MgSO_4$, and redissolved in an appropriate volume of a solvent suitable for injection into HPLC (see Section **D**. 8). In most cases the extract should be centrifuged and filtered through a microfilter (0.45 μm) to remove small particles which might create difficulties with HPLC analysis.

When human blood or tissue carotenoid analysis is used as a biomarker of carotenoid status in epidemiological studies (*Chapter 10*) many thousands of samples may need to be analysed and usually only small amounts of material are available. Carefully controlled and reproducible extraction procedures are essential. Section **H** describes the HPLC analysis of human serum/plasma samples. The extraction procedure, with ethanol and diethyl ether as solvents, has been described in detail [1]. Another routine procedure for extracting carotenoids from small samples of blood plasma or serum (250 μL) with ethanol and hexane is given in detail in *Volume 1A, Worked Example 7.*

In all work with carotenoids, the working practices and precautions described in *Volume 1A, Chapter 5* must be observed, to minimize carotenoid isomerization and degradation. In particular, heating the carotenoid extracts above 40°C must be avoided, the extraction should preferably be conducted under subdued light, and the exposure of carotenoid-containing extracts and samples to air should be minimized.

3. Saponification

With few exceptions, the carotenoids present in food samples are stable under alkaline conditions, so saponification can be used to remove unwanted chlorophylls and lipids. For most fruits and vegetables, moderate saponification conditions are satisfactory. This involves treatment of the extracts with alcoholic potassium or sodium hydroxide (5-10% w/v in ethanol or methanol), under an inert atmosphere such as nitrogen or argon. When the solubility of carotenoids in the alcohol is poor, an alkali-stable co-solvent such as dichloromethane, THF, diethyl ether or TBME may be added. Acetone or ethyl acetate must not be used for this. Normally, the saponification is complete within a few hours at ambient temperature. For high-fat foods that contain a high concentration of lipids and sterol esters, however, saponification at higher temperature or the use of lipase enzymes that can digest the fat may be required.

When carotenoids sensitive to alkali may be present, *e.g.* astaxanthin (**404-406**), enzymic hydrolysis with enzymes such as lipase or cholesterol esterase may be satisfactory [2,3]. Even if the enzymic hydrolysis is not complete, sufficient lipid is usually removed to allow subsequent efficient HPLC separation and analysis.

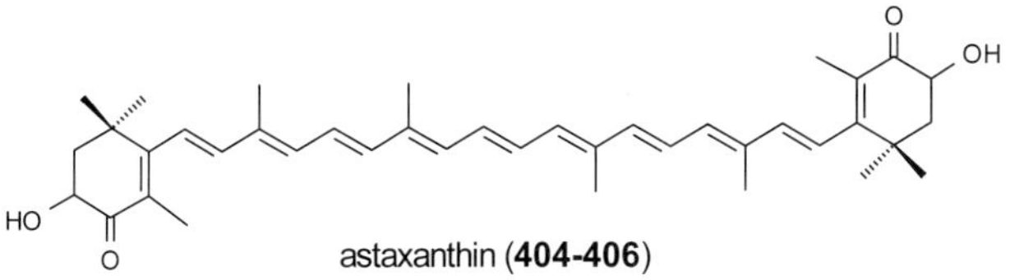

astaxanthin (**404-406**)

Saponification also hydrolyses carotenol fatty acyl esters and liberates the corresponding free carotenols. Unless the esters are of interest, saponification is recommended because it greatly simplifies the HPLC analysis. When the esters are of interest, however, the extract is not saponified but HPLC conditions are used that allow the separation and detection of the naturally occurring monoacyl and diacyl esters (see Sections **E**.4 and **G**.3). Comparison between the chromatographic profiles of a saponified and an unsaponified extract is useful in any analysis of food samples.

4. Fractionation of carotenoids by thin-layer and column chromatography

For any sample with a complex carotenoid profile, especially if it is of interest to isolate and identify a variety of minor components, preliminary fractionation of the extract by CC and/or TLC is often advantageous. Details of CC and TLC of carotenoids are given in *Volume 1A, Chapter 6, Parts II* and *III*. Carotenoids are most often separated by TLC on silica gel, but C_{18} reversed-phase plates have also been employed successfully to separate carotenoids and chlorophylls in the extracts of several green leafy vegetables [4]. Contaminating non-carotenoid impurities can be removed by rechromatography on a different adsorbent. As an example, the major carotenoids in the extracts of several varieties of squash have been separated by using a combination of C_{18} reversed phase and normal phase TLC [5]. An advantage of employing a C_{18} reversed phase system is the excellent separation of a range of different classes of carotenoids. Hydroxycarotenoids and their epoxides are best separated on silica gel whereas C_{18} reversed phase plates are most appropriate for the separation of carotenes and xanthophyll esters. For preparative TLC, high concentrations of carotenoids can be loaded onto plates of 1 mm or 2 mm thickness.

For large samples, flash column chromatography [6] is used for fractionation. The flash CC separation of carotenoids is normally carried out on silica gel (60-200 mesh) and the appropriate solvent compositions are determined by initial examination of the mixture by analytical TLC. Preparative CC or TLC of complex mixtures yields fractions from which individual carotenoids are isolated by further purification procedures.

C. Identification and Structure Elucidation

All analytical work requires that the components of interest are identified rigorously. When a new compound is encountered, its structure must be elucidated unequivocally. The sample must have a high level of purity, and is then characterized by the application of a range of the spectroscopic techniques described in *Volume 1B*. The strategy for characterization is discussed in *Volume 1B, Chapter 9*, and a general approach for structure determination of carotenoids in extracts from natural food sources is outlined below. The development of HPLC conditions under which the carotenoid constituents of an extract are well resolved is a very important initial step. The HPLC detection of carotenoids in a mixture is then optimized by monitoring each compound at its λ_{max}, and other HPLC monitoring wavelengths should also be optimized to detect other constituents that may co-elute with carotenoids of interest. In carotenoid work this is preferably accomplished by employing a UV/Vis photodiode array detector (PDAD) which records simultaneously absorbance data across a broad spectral range (*ca.* 300-700 nm). From the HPLC profile, supported by preliminary analytical TLC, a protocol can be developed for isolating carotenoids from the extract by various chromatographic techniques. When a carotenoid has been isolated that appears to be pure, as determined by analytical HPLC, it is then subjected to various spectroscopic techniques (MS, NMR, CD) to determine its identity. Ideally a sample of a few milligrams of an isolated carotenoid is required for complete structure elucidation. With the HPLC-MS and HPLC-NMR techniques that are now available, microgram quantities may be sufficient. For acceptable identification of a known carotenoid, the criteria listed in *Volume 1A* should be fulfilled: the UV/Vis λ_{max} and spectral fine structure must be in agreement with the chromophore suggested, chromatographic properties must be identical in two different systems and identity demonstrated by failure to separate when co-chromatographed or co-injected with a fully characterized synthetic or natural standard, and at least the molecular mass confirmed by mass spectrometry. The reactions of various functional groups, *e.g.* reduction of a carbonyl group with $NaBH_4$, can also provide valuable information about some structural features of a carotenoid (*Volume 1A*, *Chapter 4*).

D. HPLC of Carotenoids: General Aspects

HPLC of carotenoids has a long and interesting history, beginning in the early 1970s [7]. The first work used ‘home-packed’ columns with stationary phases such as MgO or $Ca(OH)_2$ that had been widely used as TLC and CC adsorbents. These stationary phases were not designed for application in HPLC. They did not have uniform particle size and shape or controlled pore size, and HPLC columns of these adsorbents were not available commercially. Reproducibility between different ‘home-packed’ columns and between batches of the same adsorbent was not good, and factors such as moisture content were difficult to control. Some successful procedures were developed, but these materials had serious limitations. Virtually all HPLC of carotenoids now uses conventional commercial NP and RP columns.

When embarking on HPLC analysis of a particular group of carotenoids, it is logical and sensible to make use of available expertise and search the literature for an established method. The choice can be bewildering, however; thousands of procedures have been published. Ask twenty carotenoid analysts to recommend an HPLC procedure and you will probably get twenty different answers. It is therefore essential to have a sound understanding of the principles of HPLC and of the properties and separation behaviour of the compounds of interest, in order to choose an appropriate procedure and to be able to adapt and modify this to optimize a particular separation. It is also important to consider what is the purpose of the analysis and what level of information is needed. Is the objective to obtain a complete profile of a complex extract and identify all the components present, or is it to have a rapid, reproducible routine analysis of one or a small number of main components? The HPLC procedure selected should be one that most effectively provides the desired level of information.

1. Special features of carotenoids and HPLC

The long conjugated polyene system of carotenoids makes the all-*trans* isomers rigid, linear molecules, a property that has an important influence on interactions with HPLC stationary phases. It also has other practical consequences because it gives the carotenoids the property of absorbing light in the visible region, 400-550 nm, where most other substances do not absorb. Carotenoids are thus especially well suited to detection by PDAD. Chromatograms can be monitored simultaneously at the λ_{max} of each component, and the absorption spectrum of each compound, a first criterion for identification, can be determined on line during the HPLC run. In the absence of a PDAD, fixed wavelength monitoring must be used. A wavelength around 450 nm is usually selected but allowance must be made for the different λ_{max} of different compounds (see Fig. 2, Section **E**.1). For a particular compound, the ratio of absorbance at the monitoring wavelength to that at λ_{max}, for which absorption coefficients are known, allows quantitative analysis of different components in a mixture.

Carotenoids generally are hydrophobic molecules. As with other classes of compound, the presence of polar functional groups (type, number, position) is a major determinant of behaviour on HPLC. Because of the distinctive rigid linear polyene structure, however, other, more subtle influences also come into play. The nature of the end groups (cyclic or acyclic, position of double bonds) alters the overall size and shape of the molecule. The degree of unsaturation is also an influence, and *E/Z* isomers, having different shapes, are well resolved in some systems. Finally, some optical isomers can be resolved on chiral phase columns.

Many of the features discussed in Sections **D-F** are illustrated in the chromatograms of carotenoids in various food and human samples presented later in Sections **G** and **H**. Cross-references to the appropriate Figures are given where relevant.

2. Strategy

For an unfamiliar or unknown extract it is recommended first to perform a rapid screening by preliminary TLC and RP-HPLC to get an indication of the complexity of the sample and of what kinds of carotenoids are present, especially the main components. This information is then used to plan the detailed analysis, considering the purpose of the investigation. If the main components are known, a routine procedure is selected that will give good resolution of these components. If the purpose is to have a full profile of the extract and identification of the components, then at least selected components must be isolated for characterization

When selecting an appropriate HPLC procedure, a good starting point is to use first a method that has been published for the separation of similar components, or to use a familiar general method that can be adapted as necessary.

When a favoured procedure has been optimized and standardized, it is a great advantage to use this as a routine method. Different analyses can then be compared and correlated. The procedures described in Sections **G** and **H** are ones that the author has found satisfactory for routine use for the analysis of many kinds of foods, especially different kinds of fruits and vegetables, and of human blood and tissue samples.

3. Choice of system: Normal phase or reversed phase?

In most cases, a stationary phase for which the carotenoids of interest have the greatest affinity gives longer retention times and more efficient separation. Apolar carotenoids, especially carotenes, have greatest affinity for apolar reversed-phase stationary phases. Xanthophylls have greatest affinity for polar normal-phase stationary phases.

NP-HPLC uses the conventional adsorption stationary phases alumina and especially silica, with a mobile phase of low polarity. Bonded alkylnitrile (CN) columns are also particularly useful for carotenoid work, especially for resolving mixtures of closely related xanthophylls. Most RP-HPLC applications use C_{18} or C_{30} reversed phase columns and a polar mobile phase.

4. Normal phase

a) Silica columns

The most commonly used stationary phase for NP-HPLC of carotenoids is silica, specially prepared with uniform particle size and shape and pore size. Mobile phases have been adapted from those used for silica TLC and usually consist of an apolar hydrocarbon solvent, most commonly hexane, to which is added a stronger, *i.e.* more polar, modifier such as TBME, acetone, isopropanol or methanol. For gradient elution, the proportion of polar modifier is increased according to a pre-determined program. When isocratic elution is used, the solvent composition is optimized to give the best resolution of the components of interest.

Polarity, due to the possession of polar functional groups, is the main determinant of interactions between carotenoids and the polar silanol Si-OH groups of the silica stationary phase. Polar groups or regions of the carotenoid molecule compete with polar solvent modifiers for adsorption sites on the stationary phase. The carotenes, with no polar substituents and overall low polarity, have little affinity for normal phase columns and are eluted almost immediately, so selectivity and resolution are not good. The general tendency, however, is for carotenes to be eluted in the order dicyclic → monocyclic → acyclic, and for compounds with a lower level of unsaturation to run earlier. Under very carefully controlled moisture-free conditions, some small separation of carotenes can be achieved, but NP-HPLC is not considered to be satisfactory for carotenes or other carotenoids of low polarity, such as ethers, carotene monoepoxides, carotenol esters and carotenediol diesters.

For carotenoids containing oxygen functions (xanthophylls), interactions between polar substituents and the polar stationary phase are the key determinants of chromatographic behaviour. The strongest interactions are with OH or COOH groups. Thus, for example, adsorption affinities and hence retention times increase in the order monohydroxy → dihydroxy → trihydroxy. Additional polar groups further increase the adsorption affinity. Thus xanthophyll epoxides are retained more strongly than the corresponding simple hydroxycarotenoids. Corresponding 5,6- and 5,8-epoxides are not well resolved. The effect of polar functional groups is modulated by other features and interactions, for example the positions of the functional groups, the carbon skeleton (cyclic or acyclic), ring type (*e.g.* β or ε), level of unsaturation, and geometrical isomeric form (*E*/*Z*).

Resolution of lutein (**133**), zeaxanthin (**119**), antheraxanthin (**231**), lutein 5,6-epoxide (**232**), violaxanthin (**259**) and neoxanthin (**234**) is very good.

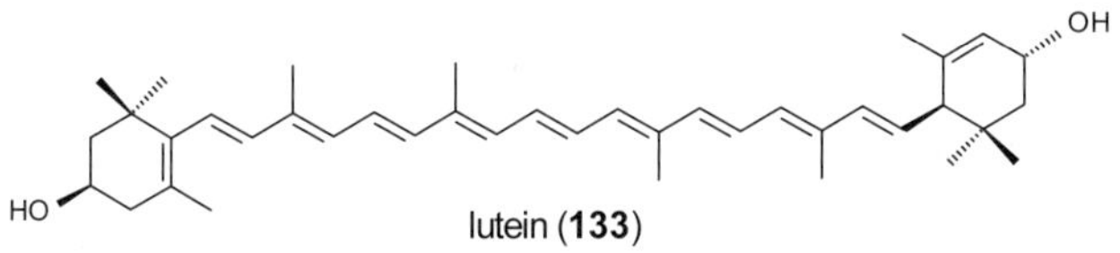

lutein (**133**)

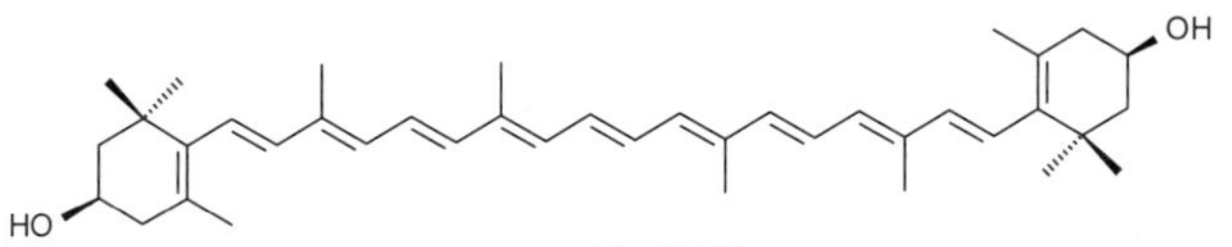

zeaxanthin (**119**)

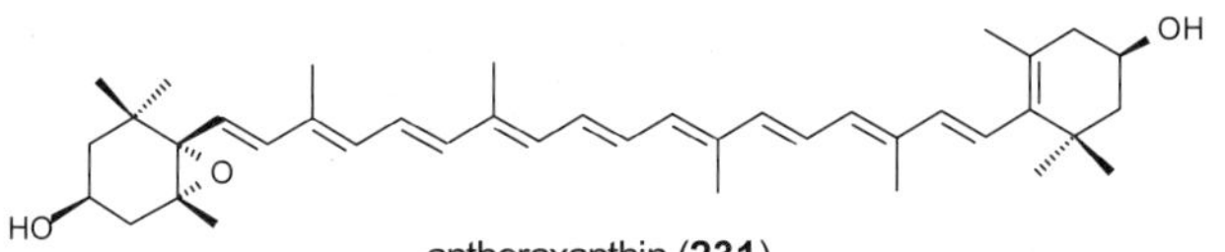

antheraxanthin (**231**)

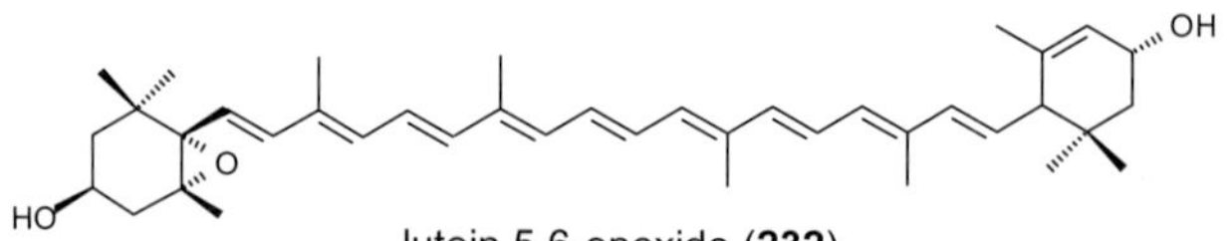

lutein 5,6-epoxide (**232**)

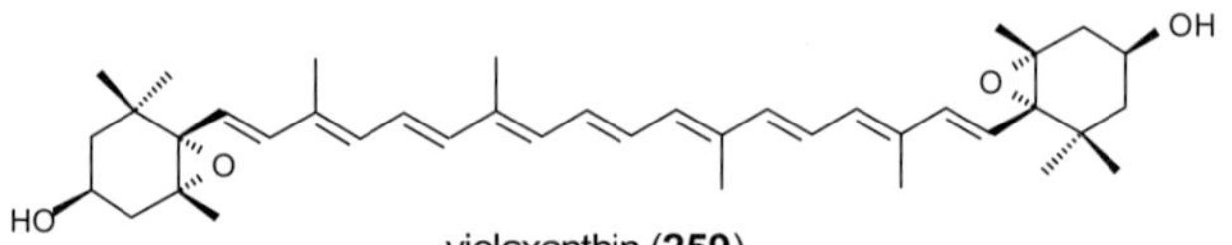

violaxanthin (**259**)

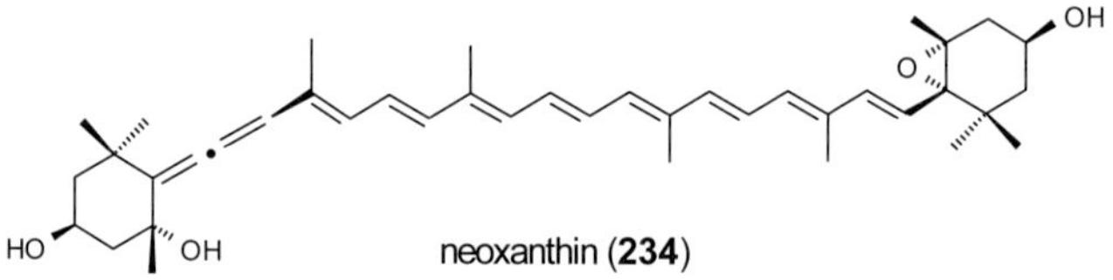

neoxanthin (**234**)

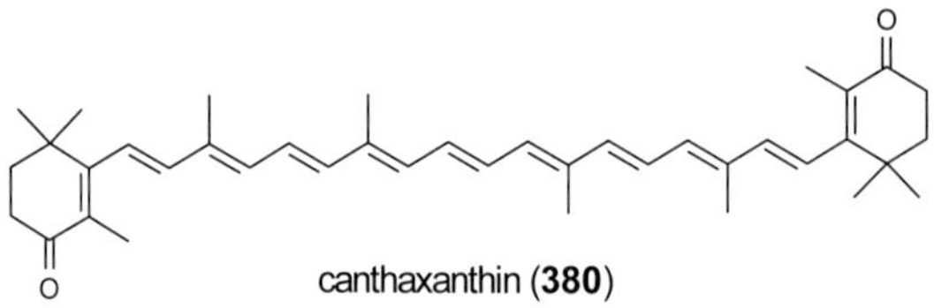

canthaxanthin (**380**)

Ketocarotenoids are less polar than carotenoids with the same number of hydroxy groups. Canthaxanthin (**380**) therefore is eluted before zeaxanthin (**119**) on NP-HPLC. The combination of hydroxy and keto group in the 3-hydroxy-4-oxo-β end group leads to behaviour that is unexpected but compatible with internal hydrogen bonding. Astaxanthin (3,3′-dihydroxy-β,β-carotene-4,4′-dione, **404-406**) normally runs ahead of the corresponding diol zeaxanthin. NP-HPLC of compounds such as astaxanthin is improved if the silica column is acidified with phosphoric acid. A rapid, reproducible method for quantitative analysis of astaxanthin, including resolution of *E/Z* isomers, on an acidified column, is described in *Volume 1A, Worked Example 9*.

Silica columns are appropriate for chromatography of complex extracts containing more polar components of a range of different polarities. The behaviour of solutes compares closely with their behaviour on silica TLC.

b) Silica-based bonded nitrile (CN) columns

First introduced some 30 years ago, the NP bonded nitrile column gives particularly impressive separation of xanthophylls [8]. The usual mobile phase consists of a mixture of hexane and dichloromethane containing a small amount of methanol and an organic amine. Not only the isomeric diols lutein (**133**) and zeaxanthin (**119**) but also diastereoisomers (lutein and 3′-epilutein, **137**) and *Z* isomers are resolved. Good separation of corresponding 5,6- and 5,8-epoxides of the diols is achieved and 8*R*/8*S* epimers of the furanoid oxides are resolved. A good example is illustrated in Fig. 4 in Section **G**.

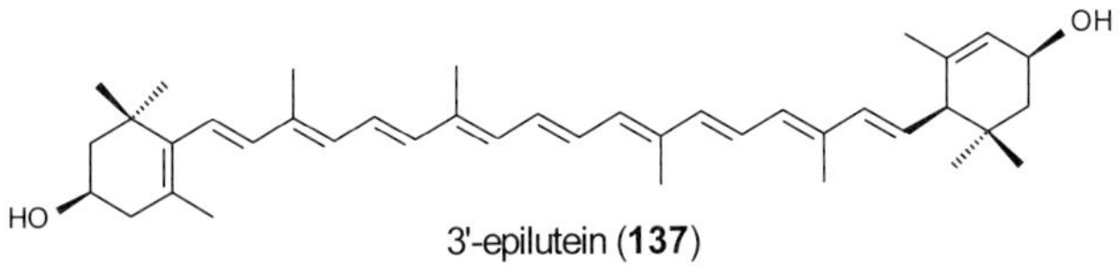

3'-epilutein (**137**)

HPLC on a CN column is not recommended as a general method for determining profiles of complex extracts, but is very powerful if fractions obtained by conventional CC or TLC are then used for the HPLC analysis.

5. Reversed phase

a) C_{18} columns

Features and principles of RP-HPLC of carotenoids, including extensive lists of published methods, are given in a review [9] and in *Volume 1A, Chapter 6, Part IV*. The most commonly used columns for RP-HPLC of carotenoids have an ODS (octadecylsilanyl) stationary phase, *i.e.* a silica support carrying bonded C_{18} alkyl chains. These columns are compatible with most solvents and are useful for the entire polarity range of carotenoids. There are many different forms of C_{18} columns, and the resolution of carotenoids is influenced by a number of factors, *e.g.* particle size and shape, pore diameter, surface coverage (carbon load), monomeric or polymeric synthesis, endcapping. Good and reproducible resolution is achieved with uniform spherical particles of 5 μm or 3 μm diameter. The effect of pore size is important but not well documented. For optimum resolution the carotenoid molecules (length *ca.* 35 Å) must penetrate the pores and interact with the C_{18} alkyl chains which are about half the length of the carotenoid molecule. If the pore size is too small the carotenoid may only enter the pores lengthwise. High surface coverage increases retention/resolution but is achieved by using silica particles with a small pore diameter. These factors must be balanced to obtain an optimum pore diameter/surface area combination.

Endcapping, *i.e.* methylation of silanol groups that remain unreacted after bonding the C_{18} chains, can have a substantial effect. Endcapping minimizes polar interactions and improves column reproducibility, but the presence of non-endcapped silanol groups can be beneficial for the separation of xanthophylls. The resolution of lutein and zeaxanthin is much better on a non-endcapped Zorbax-ODS column than on a similar but endcapped one.

Most C_{18} stationary phases are prepared by monomeric synthesis. Columns prepared by polymeric synthesis, however, are reported to have better selectivity for separation of geometrical isomers, though they do have a low carbon load because they are prepared from wide-pore, low surface-area silica.

The choice of stationary phase is therefore a crucial factor. When following a published procedure it is important to use the same kind of column, or results may not be compatible. Also once a procedure is established, the same kind of column should always be used.

The many mobile phases reported in published procedures are mainly variations on a general theme. Most use acetonitrile or methanol as the primary solvent. A stronger (less polar) solvent is added as a modifier, the most frequently used ones being dichloromethane, THF, TBME, ethyl acetate and acetone. If water is included, it should be at relatively low concentration. Acetonitrile is the most commonly used primary solvent; it gives slightly better selectivity for xanthophylls on a monomeric column [4]. A detailed evaluation has shown that higher recovery is obtained with methanol, though that with acetonitrile can be improved by adding an ammonium acetate buffer [10].

For carotenoids, C_{18} RP-HPLC is very versatile and has a wide range of applications. It is good for all classes of carotenoids and is especially good for resolving mixtures of carotenes and other compounds of low polarity (see Fig. 6). Separation is usually considered to be based on partition between the hydrophobic C_{18} chains of the stationary phase and the polar mobile phase. The polar xanthophylls partition most efficiently into the polar mobile phase, whereas the low-polarity carotenes are associated preferentially with the non-polar stationary phase and require a stronger, *i.e.* lower polarity, solvent for elution within a reasonable time. Compounds are expected to be eluted in order of decreasing polarity and the running order of carotenoids on RP-HPLC may be expected to be approximately the reverse of that on NP adsorbents. Although this is broadly correct, the relationship does not always hold. Interestingly, the order of elution, lutein before zeaxanthin, is the same for normal phase and most reversed phase systems. Other factors and interactions have to be taken into consideration, in particular, associations between bonded C_{18} alkyl chains of the stationary phase and the linear carotenoid molecule, especially with unsubstituted end groups.

Interactions with acyl groups of carotenol esters are also strong. On RP-HPLC, carotenol esters and carotenediol diesters that contain no other functional groups chromatograph together with or even slightly later than the carotenes (see Fig. 8).

Apocarotenoids occur quite commonly in food, especially in *Citrus* and other fruits (summarized in [11-13]). Their separation by HPLC is influenced mainly by chain length and the number and nature of functional groups. The HPLC of carotenoids in orange juice has been reviewed [14]. On RP columns, chain length is a major factor; short chain-length compounds are eluted before longer chain analogues. Retention times for apocarotenoids are usually much shorter than for C_{40} carotenoids with the same functional groups, so they can be confused with more polar structures. On NP-HPLC, both on silica and bonded CN phases, polar functional groups are the main influence; chain length has little effect.

b) C_{30} columns

A few years ago, a reversed phase column with a bonded C_{30} chain stationary phase, prepared by polymeric synthesis, was introduced and has become increasingly popular for the analysis of carotenoids [15]. The selectivity is high and very good resolution of carotenes can be achieved. Molecular shape seems to be a major factor in determining selectivity, and geometrical isomers are well resolved. The high degree of resolution makes the C_{30} column the one of choice for HPLC-MS of carotenoids. A disadvantage for routine quantitative analysis is the long run time, typically 60-100 minutes, as opposed to 10-25 minutes for a C_{18} column.

Careful identification of compounds is essential. It is not feasible to relate running order on a C_{30} column to that on a C_{18} column. The order of elution can change quite drastically between the two systems [16]. For example, the acyclic lycopene (**31**) usually runs faster than the cyclic β-carotene (**3**) and α-carotene (**7**) on a C_{18} column, but is eluted considerably later on a C_{30} column.

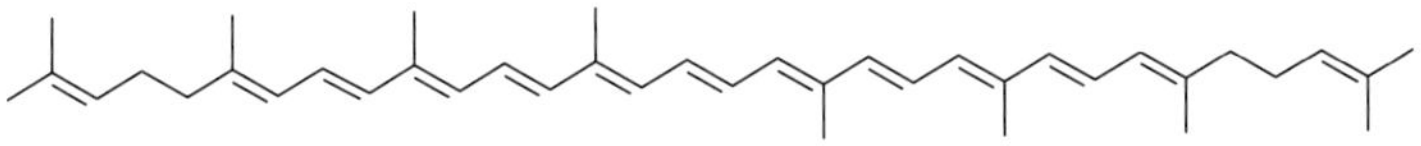

lycopene (**31**)

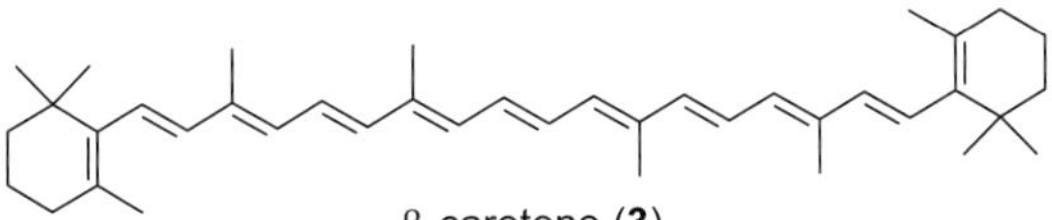

β-carotene (**3**)

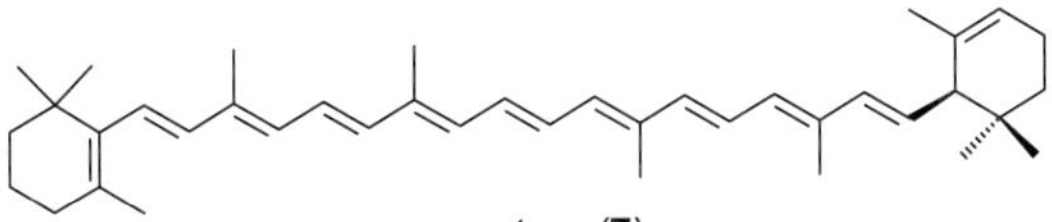

α-carotene (**7**)

Interestingly, the monocyclic monohydroxycarotenoid rubixanthin (β,ψ-caroten-3-ol, **72**), which runs in the same region as other monohydroxycarotenoids such as β-cryptoxanthin (**55**) on C_{18} columns, runs much later, even after β-carotene, on a C_{30} column [16]. This is attributed to interaction of the C_{30} chains of the stationary phase with the unsubstituted acyclic hydrocarbon half-molecule of rubixanthin. When a C_{30} column is used, therefore, it is essential to identify components in the chromatogram carefully and not rely on assumed extrapolation from behaviour on a C_{18} column.

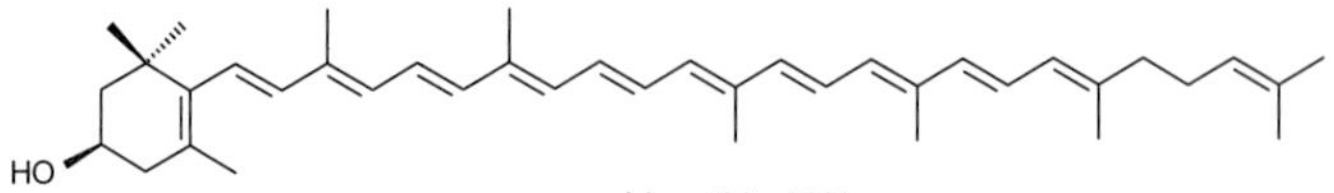

rubixanthin (**72**)

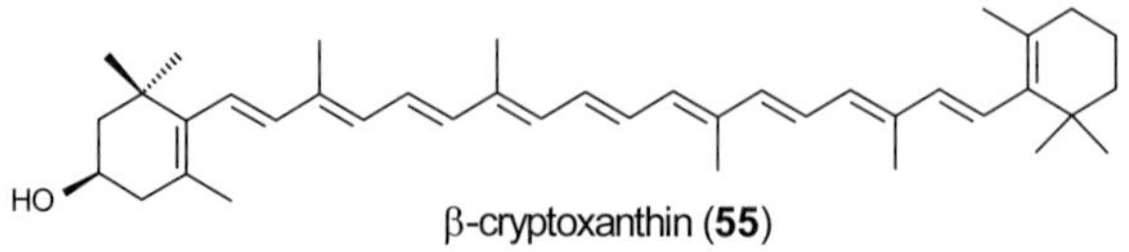

β-cryptoxanthin (**55**)

6. Temperature

Carotenoids are more sensitive to changes in HPLC conditions than many other classes of compounds. This may be due to the unusual linear shape and rigidity of the carotenoid molecules, which strongly influence the ability of carotenoids to enter the pores and interact with the alkyl chains of the stationary phase.

In most HPLC of carotenoids ambient temperature is used and small variations are neglected. Temperature can have a significant effect, however, especially on RP systems [9, 17]. In particular, temperature can strongly influence recognition of different shapes of molecules because the bonded alkyl chains of the RP stationary phase are more rigid at lower temperatures. Temperature control is, therefore, important to maintain day-to-day reproducibilty, especially in conditions where there are wide variations in ambient temperature. Temperature drift can cause substantial fluctuations in retention times of carotenoids.

It should also be noted that more volatile components of a mixed solvent may be lost by evaporation, so the composition of the solvent mixture, and consequenntly retention times, may change over the course of the day. A useful general guideline is to keep the system simple. For routine automated quantitative analysis an isocratic procedure is preferred because there is no equilibration time between runs.

7. Test chromatograms – standard mixture

A new column should always be tested and the performance of a well used column checked from time to time by injection of a standard mixture of easily available carotenoids. A mixture of carotenoid standards, namely α-carotene (**7**), β-carotene (**3**), nonapreno-β-carotene (*1*), lycopene (**31**), 8′-apo-β-caroten-8′-al (**482**), canthaxanthin (**380**) and zeaxanthin (**119**) covers the main polarity range, as illustrated in the RP chromatogram in Fig. 1a.

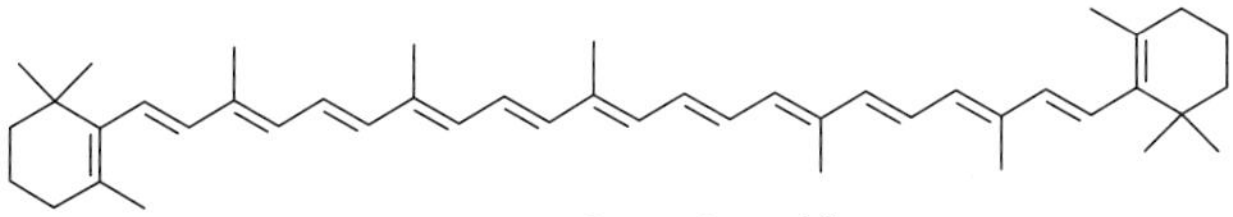

nonapreno-β-carotene (*1*)

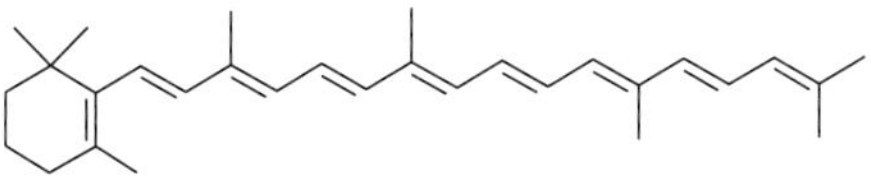

8'-apo-β-caroten-8'-al (**482**)

In the absence of these standards, a fresh leaf extract containing β-carotene (sometimes with α-carotene), lutein, violaxanthin and neoxanthin, together with chlorophylls *a* and *b* and the corresponding phaeophytins, may be used. Even with test chromatograms it is important to identify the components properly, at least by their UV/Vis spectra and not simply to rely on retention times or assumed relative positions in the chromatogram. Surprising changes in the order of elution can occur between systems. Under RP conditions the diol zeaxanthin would be expected to be eluted before the diketone canthaxanthin, but this order can be reversed on the same column with different mobile phases (*Volume 1A, Chapter 6, Part IV*).

8. Avoiding injection artefacts and peak distortion

The choice of injecting solvent is critical. It must be compatible with the mobile phase. The interaction between carotenoid solute molecules, injecting solvent and mobile phase can result in HPLC artefacts, which may seem to indicate that impurities such as *Z* isomers are present. If the injection solvent is much stronger than the mobile phase and nearly saturated with carotenoid, the carotenoids may precipitate on injection into the mobile phase, or may remain with the injection solvent as this passes through the column. The effect of this sample/solvent interaction on the C_{18} RP-HPLC of several carotenoids, injected in various solvents and chromatographed under various isocratic and gradient elution conditions has been studied extensively [18]. Depending on the solubility of the carotenoid in the mobile phase and the nature of the injecting solvent, double, even multiple and broad unresolved peaks can be generated reproducibly. This is especially pronounced if the sample is injected in dichloromethane, chloroform, benzene, toluene or THF.

For example, injection of (all-*E*)-β-carotene (**3**) in the HPLC solvent [acetonitrile (55%), methanol (25%), dichloromethane (20%)] results in a single symmetrical HPLC peak but, if dichloromethane is used as the injecting solvent, the resulting chromatogram shows an additional peak; both components are (all-*E*)-β-carotene and no *E/Z* isomerization had taken place [18]. Comparison of the chromatographic profiles of the mixture of carotenoid standards injected in acetone (Fig. 1a) and dichloromethane (Fig. 1b) clearly demonstrates that these HPLC artefacts are not unique to β-carotene.

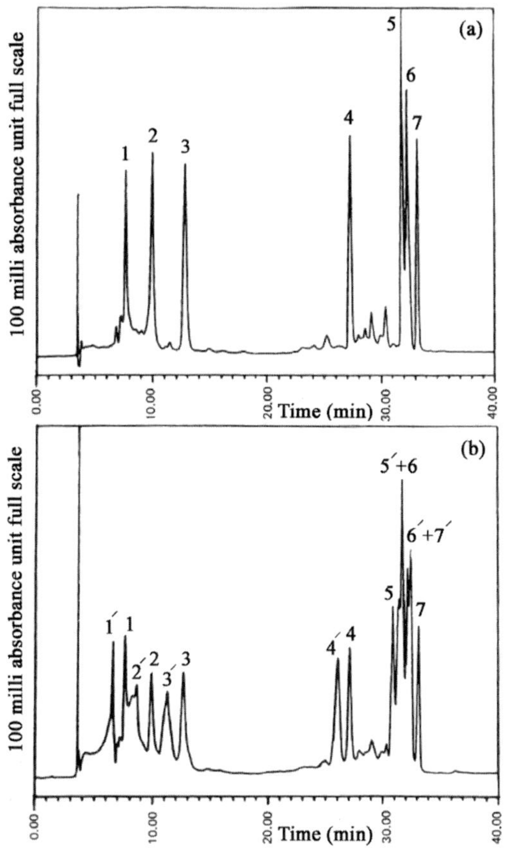

Fig. 1. (a) RP-HPLC profile of a mixture of carotenoid standards, injected in acetone. (b) Profile of the same mixture after injection in dichloromethane. HPLC conditions. Column: C_{18} Microsorb, 5 μm; 25 cm x 4.6 mm. Solvent A: acetonitrile (90%), methanol (10%). Solvent B: hexane (45%), dichloromethane (45%), methanol (10%), diisopropylethylamine (0.1%). Gradient: 0 - 10 min 95% A, 5% B, isocratic; 10 - 40 min linear gradient to 45% A, 55% B. Flow rate: 0.7 mL/min. Peak identifications. 1 (and 1′): zeaxanthin (**119**); 2 (and 2′): canthaxanthin (**380**): 3 (and 3′): 8′-apo-β-caroten-8′-al (**482**); 4 (and 4′): lycopene (**31**); 5 (and 5′): α-carotene (**7**); 6 (and 6′): β-carotene (**3**): 7 (and 7′): nonapreno-β-carotene (*1*). Reproduced from [18] with permission.

The generation of multiple HPLC peaks has been attributed to the relative solubility of the carotenoids in the injecting and eluting solvents as the sample bolus first interacts with the column [9,18]. As the injection volume is increased, the peak distortion becomes greater. Injection with more polar solvents such as acetone, methanol or acetonitrile generally produces a single symmetrical peak. Ideally, the sample should be injected in the initial HPLC eluent, thus minimizing unfavourable interactions between solute, injecting solvent and mobile phase. Stronger, miscible solvents can be used for injection if the volume is small (10 μL) and the concentrations of carotenoids are not greatly in excess of their solubility in the mobile phase. The solubility of some carotenoids in the mobile phase may be low, however, and it is then convenient to use a different solvent to ensure that the sample is fully dissolved. When the carotenoids are more conveniently solubilized in a solvent other than the HPLC mobile phase, the generation of HPLC artefacts may be eliminated by lowering the concentration of the sample or by reducing the injection volume. If the injection solvent is too weak, the carotenoid or extract may not dissolve completely. Agitation for 30-60 seconds in an ultrasonic bath is recommended to facilitate the dissolution.

E. Examples of Separations

1. Separation of carotenes

Reversed-phase HPLC is normally used for the separation of the hydrocarbon carotenes. Most C_{18} columns will give some separation but the selectivity varies considerably with the different kinds of stationary phases. The main structural features that determine separation are:

(i) Cyclic or acyclic end groups. On a C_{18} column, the usual order of elution is acyclic → monocyclic → dicyclic, *e.g.* lycopene (**31**) → γ-carotene (**12**) → β-carotene (**3**).

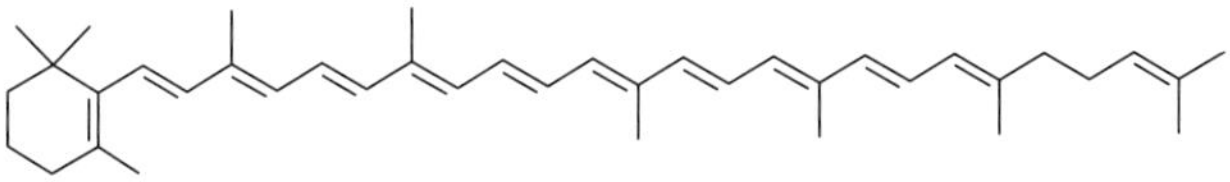

γ-carotene (**12**)

(ii) Double-bond position in cyclic end groups. For cyclic end groups the β ring has greater affinity for the stationary phase than does the ε ring, leading to the elution order ε-carotene (ε,ε-carotene, **20**) → α-carotene (β,ε-carotene, **7**) → β-carotene (β,β-carotene, **3**).

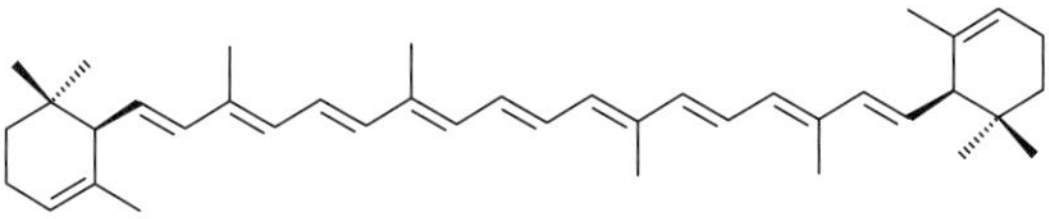

ε-carotene (**20**)

(iii) Degree of unsaturation. For carotenes with the same carbon skeleton, the more saturated compounds are eluted later. For example for the biosynthetic desaturation series the elution order is lycopene (**31**) → neurosporene (**34**) → ζ-carotene (**38**) → phytofluene (**42**) → phytoene (**44**). Note that, because of the different λ_{max} values for these compounds (*e.g.* lycopene 470 nm, β-carotene 450 nm, ζ-carotene 400 nm), different monitoring wavelengths have to be used, as shown in Fig. 2.

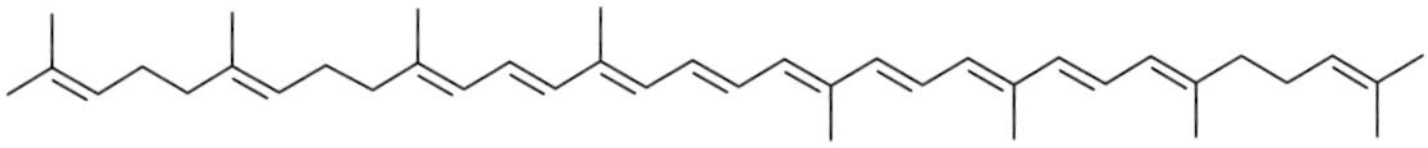

neurosporene (**34**)

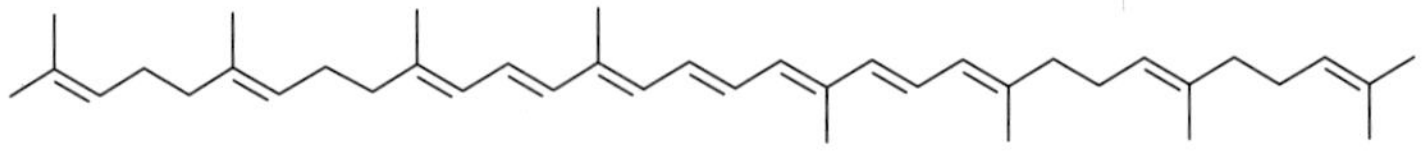

ζ-carotene (**38**)

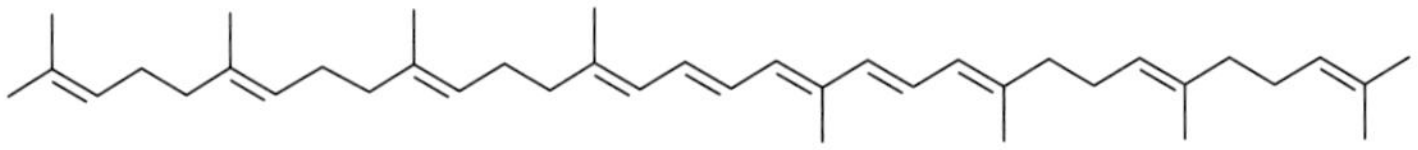

phytofluene (**42**)

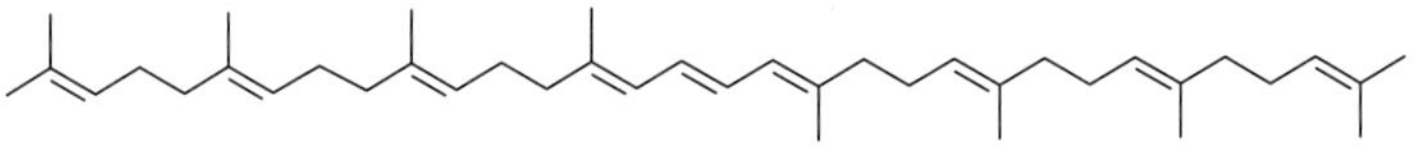

phytoene (**44**)

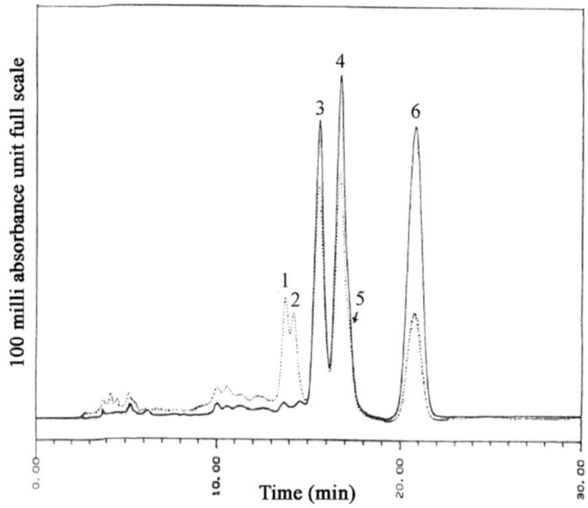

Fig. 2. RP-HPLC profile of carotenoids from canned pumpkin, illustrating the effect of monitoring at two different wavelengths, 402 nm (·····) and 475 nm (—). HPLC conditions. Column: C_{18} Microsorb, 5μm; 25 cm x 4.6 mm. Solvent: acetonitrile (55%), dichloromethane (23%, methanol (22%), containing diisopropylethylamine (0.1%). Isocratic, flow rate 0.7 mL/min. Peak identification. 1: (*Z*)-ζ-carotene; 2: (*E*)-ζ-carotene (**38**); 3: (*E*)-α-carotene (**7**); 4: (*E*)-β-carotene (**3**); 5: (*Z*)-β-carotene; 6: nonapreno-β-carotene (*I*) internal standard. Reproduced from [19] with permission.

For a natural extract, carotenes containing various combinations of these structural features may be present, all influencing retention in different ways. Also each component may be present in the all-*E* and various *Z* isomeric forms. Assignment of peaks in such a multi-component chromatogram is challenging and proper identification is essential.

2. Separation of xanthophylls

Good resolution of xanthophylls can be achieved on both NP (silica or bonded nitrile) and RP stationary phases (see Figs. 3 and 4). Much effort has been expended on developing systems for resolving the isomeric pair lutein and zeaxanthin. Most popular for routine use are RP systems, which can give good resolution not only of lutein (**133**) and zeaxanthin (**119**) but also of other xanthophylls including antheraxanthin (**231**), lutein 5,6-epoxide (**232**), violaxanthin (**259**) and neoxanthin (**234**). Resolution of these and related xanthophylls on NP columns, silica or bonded nitrile, is also very good. Interestingly, the order of elution, lutein before zeaxanthin, is the same for NP and most RP systems, though some RP conditions have been described in which zeaxanthin is eluted before lutein. Most of the published RP procedures have used fully endcapped stationary phases *e.g.* ODS-2, and not all give baseline separation of lutein and zeaxanthin. The best resolution and baseline separation have been achieved with non-endcapped reversed phases *e.g.* ODS-1.

3. *E/Z* Isomers

The first HPLC resolution of geometrical isomers of carotenoids was achieved on $Ca(OH)_2$ columns [20]. Now, some resolution of *E/Z* isomers can be achieved with almost any NP or RP system, and many efficient procedures are available, especially the use of normal phase bonded CN columns for xanthophylls and of reversed phase C_{18} columns for carotenes. Recently reversed-phase C_{30} columns have been shown to be ideal for separating *E/Z* isomers of carotenes [15,16].

The choice of system is directed by the level of analysis needed. Rarely is it necessary to identify and analyse all the geometrical isomers of a particular carotenoid. Most frequently, only a small number of the main and well characterized isomers are of interest, especially 9*Z* and 13*Z*, and an HPLC method is selected accordingly. A further challenge with carotenoids such as α-carotene (**7**) and lutein (**133**) that are not symmetrical, is the need to separate and distinguish the 9*Z* and 9′*Z*, and the 13*Z* and 13′*Z* forms.

A major concern is the identification and characterization of individual *Z* isomers, which readily undergo stereomutation (see *Volume 4, Chapter 3*). On both NP and RP columns, it is common for geometrical isomers to be eluted in the order di-*Z* → all-*E* → mono-*Z*, with the mono-*Z* isomers in the order 9*Z* → 13*Z* → 15*Z*, but this is not universal and components must always be properly identified. Elution order is not always predictable and should not be extrapolated from one system to another. Because of the different interactions, it is not possible to extrapolate the elution order on a C_{18} RP column to a C_{30} one. UV/Vis spectra (λ_{max}, *cis* peak) give an indication of possible identification, but proper characterization requires NMR analysis. There are few examples where individual geometrical isomers of a carotenoid have been isolated and fully characterized by NMR. Standards of particular *Z* isomers of some carotenoids are now available commercially. When following a published procedure, it is essential to check if identifications claimed have been proved rigorously.

a) Carotenes

The earliest separations of geometrical isomers of β-carotene (**3**) were achieved on columns of one particular form of $Ca(OH)_2$ [20]. A wide range of isomers were isolated and characterized. Good resolution of the main *Z* isomers can be achieved by C_{18} RP-HPLC, especially on non-endcapped columns such as Zorbax and ODS-2 (see Fig. 6). A rapid isocratic procedure for the routine analysis of geometrical isomers of β-carotene on a Vydac 218 TP54 column is described in *Volume 1A, Worked Example 8*. With careful control of column temperature at 30°C, clear resolution of the 13,15-di-*Z*, all-*E*, 9*Z*, 13*Z* and 15*Z* isomers is obtained in a 25 minute run time. Other C_{18} columns are usually less effective.

An early study described the separation of fifteen geometrical isomers of lycopene (**31**), all identified by NMR, on a NP silica column [21], whereas six components (not identified) were resolved on a RP column [22]. A particular challenge is the separation of the all-*E* and terminal 5*Z* isomers of lycopene and neurosporene (**34**), which can be achieved on a Spherisorb ODS-5 column [23] or a C_{30} column.

b) Xanthophylls

The procedures that give good separation of closely related xanthophylls also resolve the main *Z* isomers, which generally run later than the all-*E* compound in both NP and RP systems (see Figs. 3 and 4). In cases where the identification of the isomers has been confirmed, *e.g.* zeaxanthin (**119**) [24] the elution order of the main forms is all-*E* → 9*Z* → 13*Z* → 15*Z*, as with the carotenes. For other xanthophylls, and for other HPLC methods, the isomers must be fully identified. When the two end groups are not identical, as in lutein (**133**), the number of possible *Z* isomers is greater; 9*Z* and 13*Z* are not the same as 9′*Z* and 13′*Z*.

4. Acyl esters

In fruits and flowers it is common for the xanthophylls to be present largely in the form of esters with long-chain fatty acids, most frequently C_{12} to C_{18} saturated and unsaturated. Although esters with one particular fatty acid may predominate, it is usual to have a mixture of esters with different fatty acids. For a carotenediol, the two esterifying acids may be the same or different, which increases the number of possible molecular species, especially if the two carotenoid end groups are different. A complex pattern of carotenol monoesters and carotenediol diesters appears in the same chromatographic range as the carotenes (Figs. 6-8). For most routine analyses, when esters are not of interest *per se*, it is normal practice to saponify the extract to hydrolyse the esters and liberate the free carotenoids, resulting in much simpler chromatograms (Fig. 9). When it is of interest to identify the esters, a reversed-phase procedure is used; individual esters can be well resolved. Many methods have been published (see *Volume 1A, Chapter 6, Part IV*), most of which use a C_{18} column. One such procedure should be selected and conditions adapted and optimized for the particular collection of esters in the extract under study. Little separation of the different esters can be achieved with normal phase columns.

A number of features determine the separation of mixtures of esters on a C_{18} RP column. In general, esters of a more polar carotenoid are eluted first, *i.e.* esters of violaxanthin (**259**), which has two additional epoxide groups, run ahead of the corresponding esters of the simple diols lutein (**133**) and zeaxanthin (**119**) (Fig. 9). Esters of the isomeric diols lutein and zeaxanthin with the same fatty acid are resolved more efficiently than the free xanthophylls; the different shape of the 3-hydroxy-β and 3-hydroxy-ε end groups is accentuated by the bulky ester group. For esters of the same carotenoid, the chain length and degree of unsaturation of the esterifying fatty acid become the determining factors, because of interactions of the acyl chains with the C_{18} chains of the stationary phase.

The overall pattern can be complex, with peaks for different esters of one carotenoid overlapping with peaks for different esters of another carotenoid, *e.g.* the collection of lutein esters overlaps with the collection of zeaxanthin esters.

It appears that, for esters of the same carotenol the HPLC retention times of carotenol fatty acid esters on a RP column increase as the number of carbon atoms in the fatty acyl chain increases, but not in a uniform way. In gas-liquid chromatography (GLC), the correlation between the number of carbon atoms in saturated fatty acids and related esters has been used extensively to predict the elution sequence of these compounds [25]. Similar correlations appear to exist between the HPLC retention times of carotenol fatty acid esters and the number of carbon atoms in the fatty acid chains. For example, in the separation of the squash carotenoids in baby foods illustrated in Fig. 7, there is an increase of about 4 minutes in the HPLC retention times as the total number of carbon atoms in the fatty acyl side chains increases by four, but the tight mathematical relationship seen with the fatty acid GLC does not hold [26]. As the number of carbons in straight chain fatty acids is increased, some noticeable changes in the physical properties (melting point and solubility behaviour) are observed. These transitions occur at around the ambient temperature used for HPLC but not at the higher temperatures used in GLC. Because of these changes, the increase in HPLC retention times for the carotenol esters is not uniform. Unsaturation in the fatty acid chain generally decreases the retention time compared with that of the ester with a saturated fatty acid of the same chain length.

5. Optical isomers/enantiomers

Many natural carotenoids contain at least one chiral centre or axis. Two strategies may be used (*Volume 1A, Chapter 6, Part IV*) to resolve enantiomeric mixtures of a carotenoid, either direct resolution on a bonded chiral-phase column or reaction with a chiral reagent (*Volume 1A, Chapter 4*) followed by separation of the diastereoisomeric derivatives on an ordinary column. Resolution of the diastereoisomeric carbamates formed by reaction with (+)-(*S*)-1-(1-naphthyl)ethyl isocyanate or of the (–)-dicamphanate esters on normal phase silica or nitrile columns has been described for several xanthophylls [27,28]. The direct separation of (3*R*,3'*R*)-zeaxanthin (**119**) and (3*R*,3'*S*)-zeaxanthin (**120**) on a chiral column with bonded amylose tris-(3,5-dimethylphenylcarbamate) is illustrated in Fig. 13.

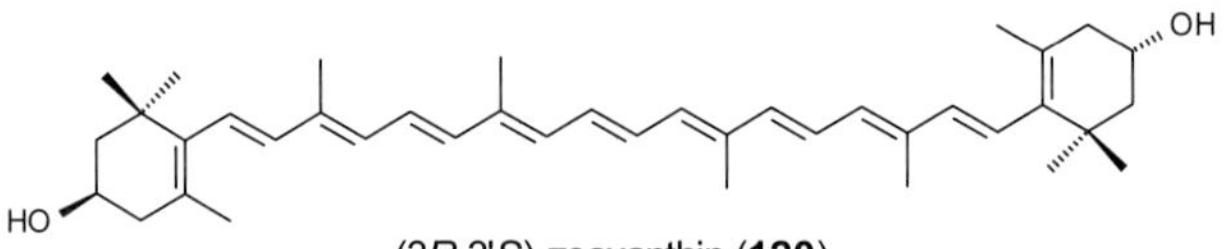

(3*R*,3'*S*)-zeaxanthin (**120**)

F. Quantitative Analysis of Carotenoids by HPLC

1. Selection of an internal standard

The preparation of carotenoid samples for HPLC analysis requires extensive extraction and work-up procedures that can be accompanied by various losses and analytical errors. For accurate quantitative analysis, therefore, the use of an internal standard is essential. Many examples of the separation and quantitative analysis of plant carotenoids have appeared in the literature, but few of these have employed an internal standard.

a) Requirements of an internal standard

The general requirements of an internal standard can be summarized as follows.

(i) It must not be present in the original sample and preferably should not be a naturally occurring carotenoid.

(ii) It must produce a well-resolved HPLC peak with no interference with the compounds of interest.

(iii) It must be eluted relatively close to the compounds of interest.

(iv) The solubility and chromatographic response of the internal standard in the mobile phase must be similar to those of the compounds of interest; this is normally the case when the internal standard is structurally similar or related to the compounds of interest.

(v) Its light absorption properties (λ_{max}) should be similar to those of compounds of interest and its absorption coefficient must be known accurately.

(vi) It must have similar stability to the compounds of interest and it must not react with sample components, column packing, or mobile phase.

(vii) It must be added at a concentration that will produce a peak area or peak height similar to those of the compounds of interest.

(viii) It is desirable for it to be available commercially in high purity.

(ix) For analyses of multi-component mixtures, more than one internal standard may be required to achieve highest precision.

b) Examples of internal standards

The readily available 8′-apo-β-caroten-8′-al (**482**) is used quite extensively in RP-HPLC, *e.g.* in the separation and quantitative analysis of carotenoids of citrus fruit [29]. It is generally suitable for quantitative determination of naturally occurring xanthophylls [4] but is less satisfactory for carotenes. Its reduction product 8′-apo-β-caroten-8′-ol (*2*) has some advantages as an internal standard for the analysis of xanthophylls; it is readily prepared and is more stable under saponification conditions.

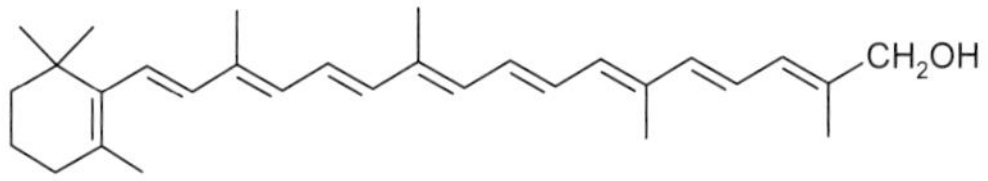

8'-apo-β-caroten-8'-ol (*2*)

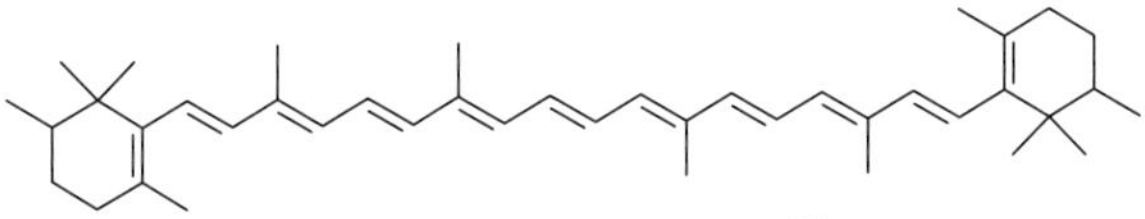

2,2'-dimethyl-β-carotene (*3*)

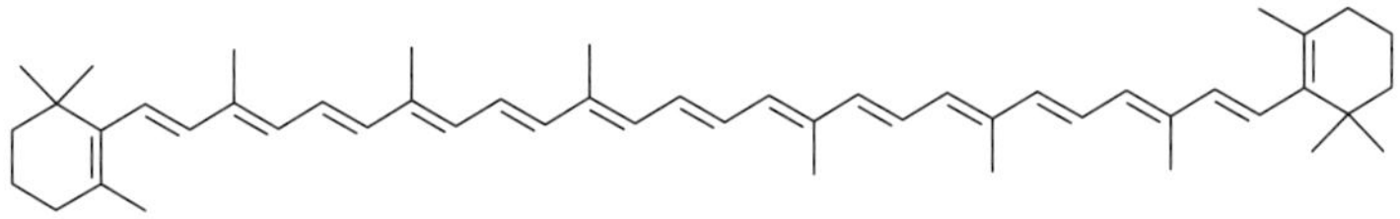

decapreno-β-carotene (*4*)

(2*R*,2′*R*)-2,2′-Dimethyl-β-carotene (*3*) has been used successfully as an internal standard in quantitative determination of β-carotene in human serum [30], but has the disadvantages that it is not available commercially, is difficult to prepare, and its stability is not good. Decapreno-β-carotene (*4*), the C_{50} analogue of β-carotene, has also been employed as an internal standard for quantitative analysis of carotenes (see Fig. 3) [4,31]. It can be synthesized [32] and has some good attributes, but its application is limited to carotenes. Its low solubility and ease of degradation are not ideal and its λ_{max} (502 nm) is far from the λ_{max} of natural carotenes. Nonapreno-β-carotene (*1*), the C_{45} analogue of β-carotene, fulfils many of the requirements of an internal standard and can be employed for quantitative analysis of carotenes, but other internal standards are required for extracts with a more complex chromatographic profile, including different classes of more polar carotenoids. Until an ideal general example becomes available it is often convenient to use a combination of two internal standards in the low and high polarity regions of the chromatogram (see Fig. 3).

c) Internal standard for carotenol esters

An unsaponified extract of fruit will often contain a range of xanthophyll esters, which run close to carotenes on RP-HPLC. For the specialized analysis of extracts of this kind an internal standard is needed that is eluted close to the natural esters and carotenes but does not overlap with them. An appropriate internal standard is a diacyl ester of a carotenediol that does not occur naturally but can be prepared from a readily accessible carotenoid. A suitable diol, isozeaxanthin (**129**), can easily be made by reduction of canthaxanthin (**380**) [33,34].

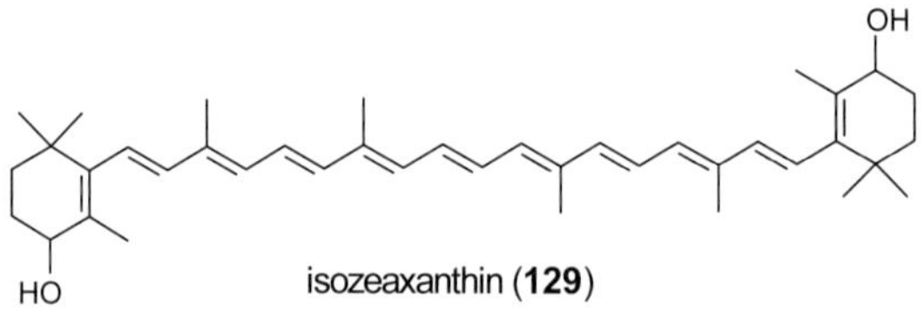

isozeaxanthin (**129**)

Esters of isozeaxanthin with fatty acids of different chain lengths can be prepared; isozeaxanthin dipelargonate (dinonanoate) has been found suitable for analysis of natural carotenoid esters (main components lutein diesters) in squash (Figs. 7 and 8) [35]. Esters of isozeaxanthin with fatty acids of other chain lengths can be prepared similarly and may be more appropriate for other applications.

2. Use of an internal standard

A known amount of the internal standard is added to the food or other sample before this is extracted, homogenized and partitioned, and the resulting extract is analysed by HPLC. The peak height, or preferably peak area, of this standard compound is then used as an internal marker against which the peak heights or areas for the sample components are compared. The assumption is made that any loss in carotenoid components as a result of sample preparation would be accompanied by the loss of an equivalent fraction of the internal standard. The accuracy of this approach therefore depends on the similarity in structure and properties between the internal standard and the carotenoids of interest.

3. Preparation of the internal standard calibration curves

Stock solutions of internal standard and reference samples of carotenoids of known concentrations are prepared, the HPLC response factor of each solution is determined at various concentrations, and calibration graphs are plotted. A known amount of the internal standard is then added to each solution of carotenoid standards at various concentrations, and the ratio of the peak area or peak height of each reference sample to that of the internal standard is plotted *versus* the concentration of each carotenoid. If calibration mixtures have been prepared properly, the calibration plot for each carotenoid should be linear and they should intercept at the origin. The concentrations of carotenoids in the extract are then determined by relating the area ratio of each carotenoid and the internal standard to those of the calibration curves. To generate accurate analytical data, the recovery of the internal standard after each extraction and the chromatographic reproducibility in preparation of the calibration curves must be monitored carefully. If a suitable single internal standard cannot be identified for quantitative analysis of all carotenoids within a complex HPLC profile, the external standard technique can be used. More information on the theory and practice of quantitative analysis of carotenoids by HPLC, by use of internal and external standard techniques, is given in *Volume 1A, Chapter 6, Part IV*.

G. HPLC of Carotenoids in Food

The distribution of carotenoids in food is surveyed in *Chapter 3*. Some examples of the HPLC analysis of different kinds of fruits and vegetables that generated this knowledge are described below. These examples also illustrate many of the principles and general features of carotenoid HPLC discussed in Sections **D-F**. Other examples of the application of HPLC to the analysis of carotenoids in foods are available elsewhere (see, for example, [36-38]).

1. Green vegetables and fruits

With very few exceptions, all green plant tissues, including green fruits and vegetables, contain in their chloroplasts the same collection of main pigments, namely chlorophylls *a* and *b*, β-carotene (**3**), lutein (**133**), violaxanthin (**259**) and neoxanthin (**234**) (*Chapter 3*), all of which have fundamental roles in photosynthesis (*Volume 4, Chapter 14*). Minor amounts of α-carotene (**7**), zeaxanthin (**119**), antheraxanthin (**231**), lutein epoxide (**232**) and β-cryptoxanthin (**55**) may also be detected. It is most common to use RP methods for routine HPLC screening of green plant and other samples across a range of polarities (*Volume 1A, Chapter 6, Part IV*). An example of such a procedure for analysis of a green vegetable (Brussels sprout) is illustrated in Fig. 3. In samples such as this, in which the leaves are not uniformly green but inner leaves are yellow, the content of carotenoids such as lutein epoxide

(**232**) and zeaxanthin (**119**), which are associated with etiolated leaves, is comparatively high. It should be noted that, in the initial qualitative screening of extracts, no internal standard should be employed. Indeed, the preliminary chromatography will guide the choice of an appropriate internal standard that is well resolved from components of the extract, for subsequent quantitative analysis.

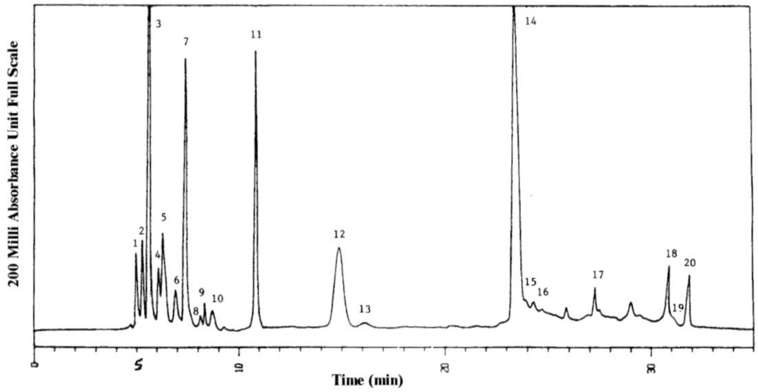

Fig. 3. RP-HPLC profile of a green vegetable (Brussels sprouts).
HPLC conditions. Column: C_{18} Microsorb, 5 μm; 25 cm x 4.6 mm. Solvent A: acetonitrile (90%), methanol (10%). Solvent B: hexane (45%), dichloromethane (45%), methanol (10%), diisopropylethylamine (0.1%). Gradient: 0 - 10 min 95% A, 5% B, isocratic; 10 - 40 min linear gradient to 45% A, 55% B. Flow rate: 0.7 mL/min.
Main peak identifications. 1, 2: (*E*) and (*Z*) neoxanthin (**234**); 3: violaxanthin (**259**); 5: lutein epoxide (**232**); 7: lutein (**133**) and zeaxanthin (**119**); 8-10: (*Z*)-lutein; 11: 8′-apo-β-caroten-8′-al (**482**), internal standard; 12-17: chlorophylls and phaeophytins; 18: β-carotene (**3**); 19: (*Z*)-β-carotene; 20: decaprenoxanthin (*4*), internal standard. Reproduced from [4] with permission.

It is usually convenient to saponify the extract; this destroys chlorophylls and simplifies the chromatographic profile. Separation is dependent on the kind of column used. Although some RP systems give clear resolution of lutein, zeaxanthin and their geometrical isomers, others do not. Consequently, food composition tables, in many cases, list only 'lutein + zeaxanthin' and do not provide data on the individual concentrations of lutein and zeaxanthin. Zeaxanthin is found in substantial concentrations in the human serum and retina, so its presence in foods should not be ignored [39-44]. To address this problem, either a RP procedure that does give baseline resolution of these two components should be used (examples are given in *Volume 1A, Chapter 6, Part IV*), or efficient separation of these xanthophylls and their geometrical isomers can be achieved by isocratic or gradient HPLC on a silica-based bonded nitrile column [45], as shown in Fig. 4 for a saponified extract from green beans. In this system, components with low polarity (carotenes *etc.*) have short retention times and are not well resolved, but very good separation of the polar carotenoids lutein, zeaxanthin, violaxanthin, lutein epoxide, neoxanthin and their *E/Z* isomers is achieved. When xanthophyll epoxides are not present, *e.g.* in wheat and pasta products, the analysis can be simplified and isocratic conditions used.

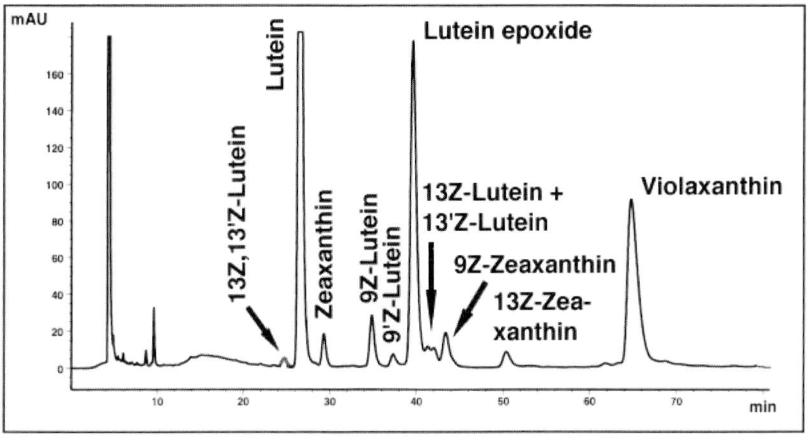

Fig. 4. NP-HPLC profile of an extract from green beans on a silica-based bonded nitrile column, optimized for resolution of xanthophylls and their *Z* isomers.
HPLC conditions. Column: silica-based bonded nitrile column, 5μm; 25 cm x 4.6 mm. Solvent A: hexane (75%), dichloromethane (25%), containing methanol (0.3%) and diisopropylethylamine (0.1%). Solvent B: hexane (75%), dichloromethane (25%), containing methanol (1.0%) and diisopropylethylamine (0.1%). Gradient: 0 - 25 min 95% A, 5% B, isocratic; 25 - 45 min linear gradient to 35% A, 65% B. Flow rate 0.7 mL/min. Reproduced from [45] with permission.

2. Yellow/red fruits and vegetables containing mainly carotenes

With fruits and vegetables of other colours, there is considerable diversity in carotenoid compositions (*Chapter 3*). The same colour can be produced by different carotenoids. A good example of this is provided by tomato (lycopene, **31**) and red pepper (capsanthin, **335**, and capsorubin, **413**). The similar red of strawberry is due not to carotenoids but to anthocyanins.

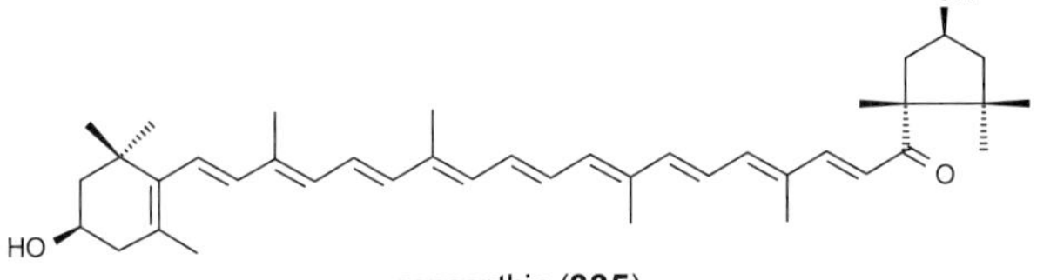

capsanthin (**335**)

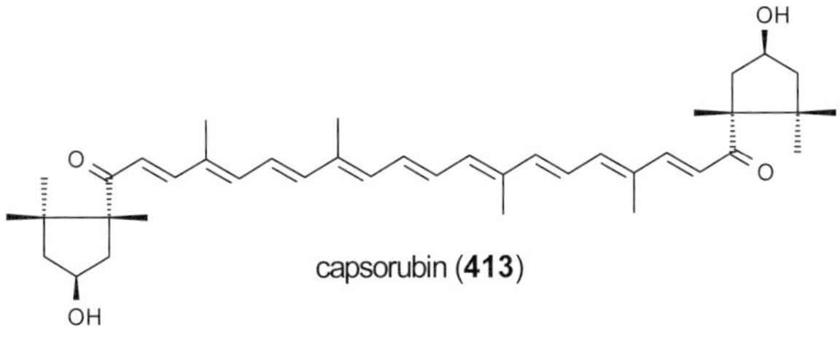

capsorubin (**413**)

The qualitative carotenoid composition of fruits or vegetables depends on what carotenoid biosynthesis genes are present and active. This has led to several different categories of carotenoid composition being identified (see *Chapter 3*). In addition to the carotenoids that accumulate in the ripe fruit, some chloroplast carotenoids may remain from the green pre-ripening stage. Many yellow/red fruits and vegetables contain mostly carotenes. Common yellow-orange examples are apricot, cantaloupe, carrot, pumpkin, and sweet potato [19,46]. Although extracts from these sources can be analysed readily by the RP gradient HPLC method described above for green vegetables (Fig. 3), when only carotenes are present it can be more effective to use isocratic HPLC conditions. For example, the carotenoid HPLC profiles of an extract from dried apricots on a Microsorb-C_{18} RP column with a high carbon loading (small pore size, 100 Å) and on a Vydac-C_{18} RP column with a low carbon loading (large pore size, 300 Å) are compared in traces A and B of Fig. 5. Both methods use a C_{18} RP column, but the conditions used and the separations achieved are different. For example, *E/Z* isomers of a carotene are resolved better on the Vydac (Trace B) than on the Microsorb column (Trace A) [46]. Other yellow fruits and vegetables may have a simpler HPLC profile. In carrots and sweet potatoes, for example, the only carotenoids present are α-carotene and β-carotene, and these can be resolved rapidly and effectively by an isocratic procedure.

Figure 5. Trace A

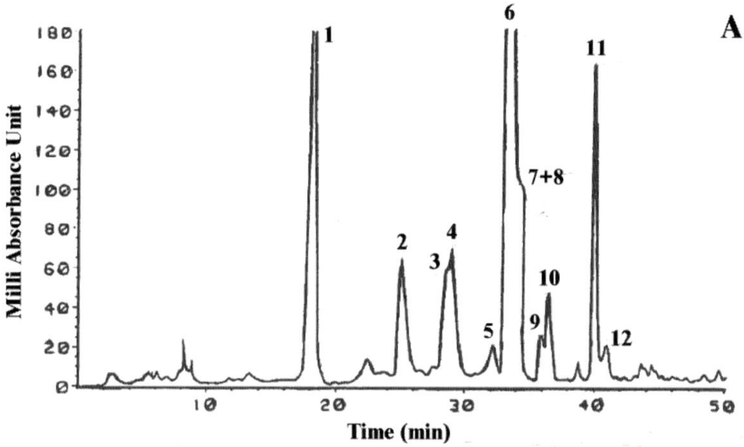

Figure 5. Trace B

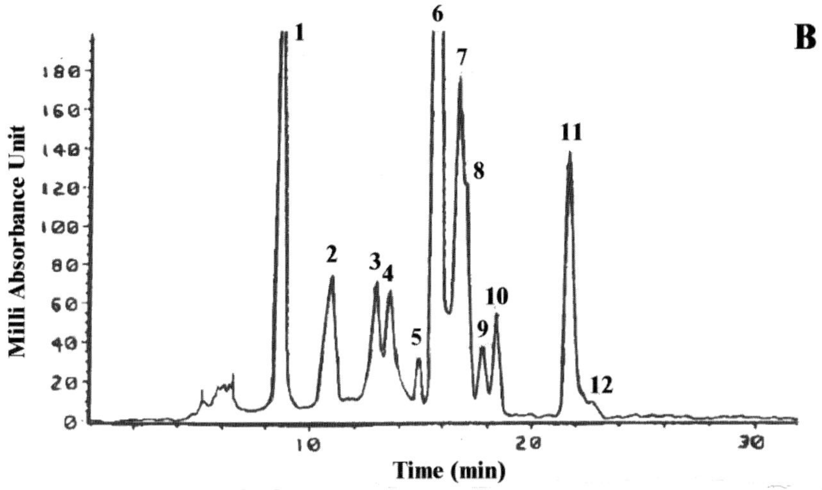

Fig. 5. RP-HPLC profile of the carotenes, including Z isomers, from dried apricot, comparing two different columns and conditions.
HPLC conditions. Trace A. Column: C_{18} Microsorb; conditions as in Fig. 3. Trace B. Column: C_{18} Vydac 201 TP54, 5µm, pore size 300 Å; 25 cm x 4.6 mm. Solvent: acetonitrile (85%), methanol (10%), dichloromethane (2.5%), hexane (2.5%), containing diisopropylethylamine (0.1%); isocratic, flow rate 0.7mL/min. Main peak identifications. 1: 8′-apo-β-caroten-8′-al (**482**), internal standard; 2: lycopene (**31**); 3: γ-carotene (**12**); 4: (*Z*)-γ-carotene; 5: ζ-carotene (**38**); 6: β-carotene (**3**); 7, 8: (*Z*)-β-carotene; 9, 10: (*E/Z*)-phytofluene (**42**); 11, 12: (*E/Z*)-phytoene (**44**). Reproduced from [46] with permission.

Various red fruits, *e.g.* tomato, pink grapefruit, and watermelon, are major dietary sources of the acyclic carotenoid lycopene (**31**) and the biosynthetic precursors ζ-carotene (**38**), phytofluene (**42**) and phytoene (**44**). They may also contain lesser amounts of neurosporene (**34**), γ-carotene (**12**), and β-carotene (**3**) [46-48]. The resolution of these hydrocarbons is illustrated by the HPLC profile of an extract from tomato paste (Fig. 6). It is known that, in tomato, the biosynthetic intermediates are present largely as the 15*Z* isomers, so peaks 17 and 15 probably represent (15*Z*)-phytoene and (15*Z*)-phytofluene, respectively, whereas peaks 18 and 16 are probably due to the corresponding all-*E* isomers.

Other tomato-based food products have a qualitatively similar HPLC profile, though the relative concentrations of individual carotenoids may vary.

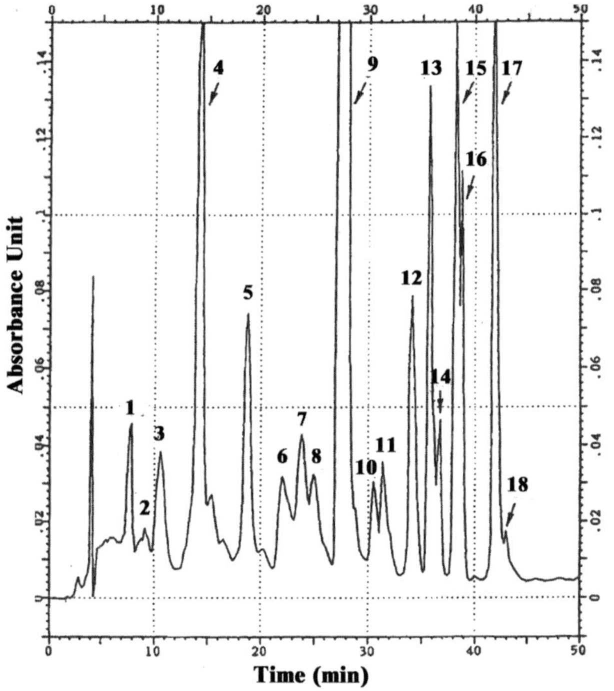

Fig. 6. C_{18} RP-HPLC profile of cyclic and acyclic carotenes from tomato paste.
HPLC conditions. As in Fig. 3. Main peak identifications. 1: lutein (**133**); 4: 8′-apo-β-caroten-8′-al (**482**), internal standard; 9: lycopene (**31**); 10: γ-carotene (**12**); 11: (*Z*)-γ-carotene; 12: ζ-carotene (**38**); 13: β-carotene (**3**); 14: (*Z*)-β-carotene; 15, 16: (*E/Z*)-phytofluene (**42**); 17, 18: (*E/Z*)-phytoene (**44**). Reproduced from [48] with permission.

3. Yellow/orange fruits and vegetables containing mainly xanthophylls and xanthophyll esters

These foods are sources of a variety of carotenoids that are absorbed, utilized, and metabolized by humans [1,41,42,49]. Good examples of foods that contain carotenol acyl esters are two strains of squash (*Cucurbita maxima*), the Northrup King and butternut varieties that are grown and processed by baby-food manufacturing companies. The carotenoid HPLC profiles of extracts from these squash varieties are shown in Figs. 7 and 8. The major carotenoids in the Northrup King variety, in the order of elution on a C_{18} RP column, are unesterified xanthophylls and their monoacyl and diacyl esters.

As shown in Fig. 8, the major carotenoids in the butternut squash are α-carotene, β-carotene, and carotenol diacyl esters. From extensive NMR studies of the isolated native lutein monomyristate and monopalmitate in the Northrup King squash and several synthetic compounds, it has been shown that the site of the ester moiety in these monoacyl esters of lutein is on the β rather than the ε end group [5].

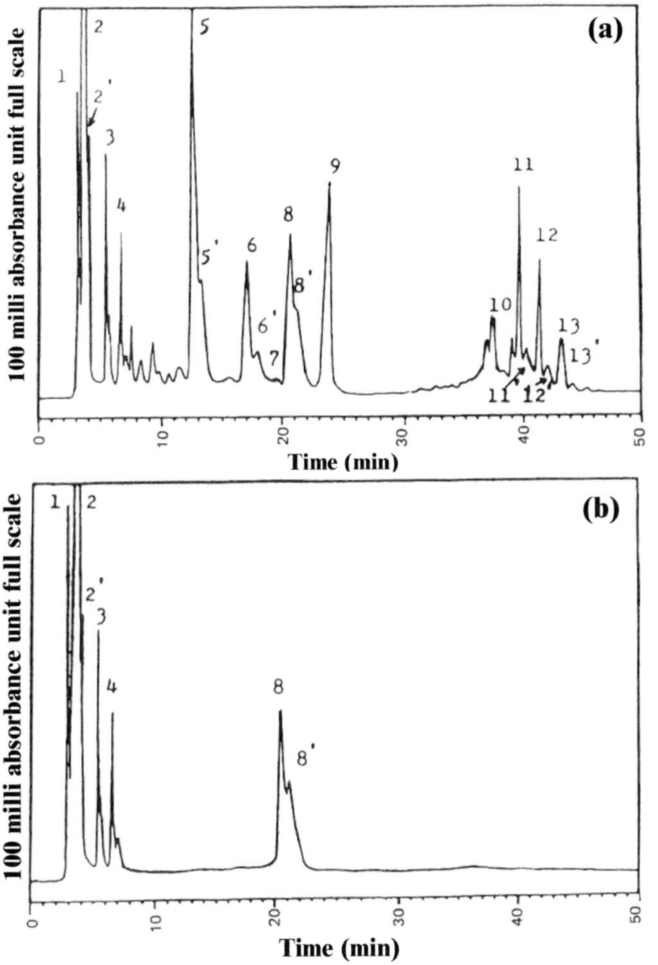

Fig. 7. RP-HPLC profile of baby-food squash (Northrup King), (a) before saponification, (b) after saponification. HPLC conditions. Column: C_{18} Microsorb, 5μm; 25 cm x 4.6 mm. Solvent A: acetonitrile (85%), methanol (15%). Solvent B: hexane (42.5%), dichloromethane (42.5%), methanol (15%), diisopropylethylamine (0.1%). Gradient: 0 - 23 min, 80% A, 20% B, isocratic; 23 - 33 min, linear gradient to 50% A, 50% B. Flow rate 0.7 mL/min. Main peak identifications. 2: lutein (**133**); 4: β-cryptoxanthin (**55**); 5: lutein monomyristate, 6: lutein monopalmitate; 7: α-carotene (**7**); 8: β-carotene (**3**); 9: isozeaxanthin (**129**) dipelargonate, internal standard; 10: lutein dilaurate; 11: lutein dimyristate; 12: lutein myristate palmitate; 13: lutein dipalmitate. Reproduced from [35] with permission.

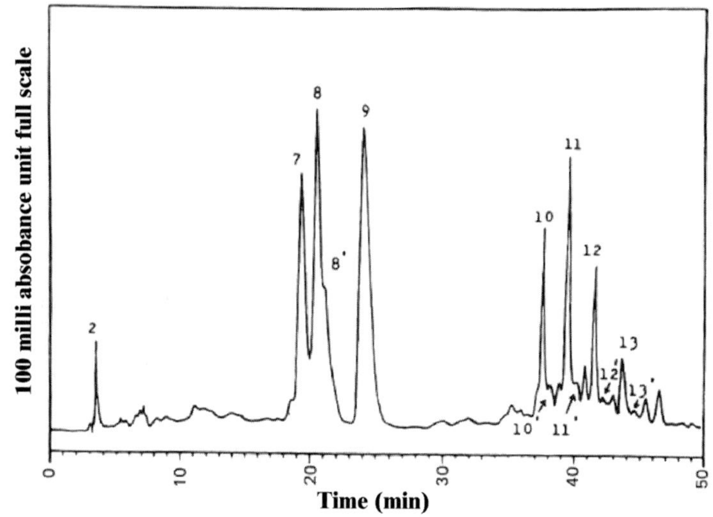

Fig. 8. RP-HPLC profile of baby-food squash (butternut). Chromatographic conditions and main peak identification as in Fig. 7. Reproduced from [35] with permission.

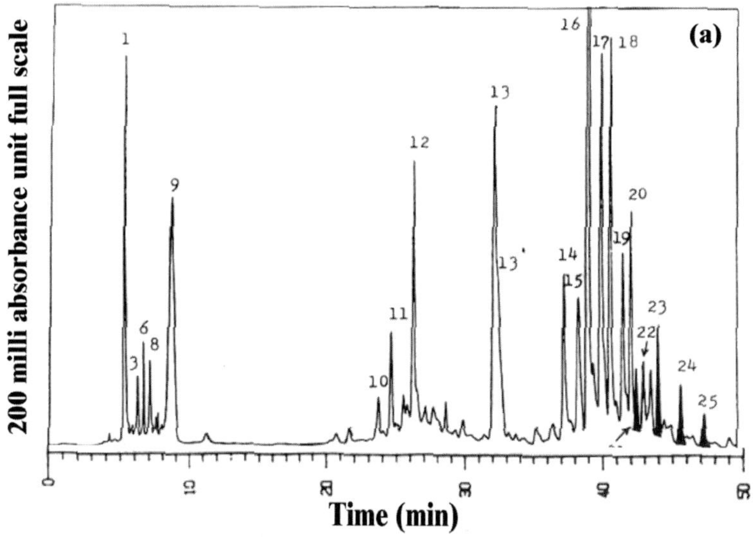

Fig. 9. RP-HPLC profile of acorn squash, illustrating the separation of esters of different carotenoids (violaxanthin and lutein) with the same fatty acids, and of the same carotenoid with fatty acids of increasing chain length. Chromatographic conditions as in Fig. 3. Main peak identifications. 1: violaxanthin (**259**); 8: lutein (**133**); 9: 8′-apo-β-caroten-8′-ol (*2*), internal standard; 10: violaxanthin monolaurate; 11: violaxanthin monomyristate; 12: violaxanthin monopalmitate; 13: β-carotene (**3**); 14: violaxanthin dilaurate; 16: violaxanthin dimyristate; 18: violaxanthin myristate palmitate; 20: violaxanthin dipalmitate; 21: lutein dilaurate; 23: lutein dimyristate; 24: lutein myristate palmitate; 25: lutein dipalmitate. Reproduced from [35] with permission.

A reasonable separation of carotenoid acyl esters is accomplished with the RP procedure shown in Figs. 7 and 8, but the separation is improved considerably when the different gradient, as used in Fig. 3, is applied (Fig. 9). Because of the complex carotenoid profile in squash, saponification of the extract can greatly simplify the chromatographic profile and is therefore a logical strategy in quantitative analysis of the carotenoids.

If there is a risk that saponification may result in destruction or structural transformation, the above HPLC methods that can separate carotenol fatty acid esters within a reasonable time can be used and the saponification step omitted. This can also provide valuable information on the identity and levels of the esters as they occur naturally in foods.

H. Analysis of Carotenoids in Human Serum, Milk, Major Organs, and Tissues

1. Human serum and milk

Carotenoids in human serum, milk, and tissues originate from the diet or supplements. Carotenes, monohydroxycarotenoids and dihydroxycarotenoids are found in human serum/ plasma and milk. Carotenol acyl esters are not detected; when ingested in the diet they are hydrolysed to their parent hydroxycarotenoid by pancreatic secretions. Carotenoid epoxides have not been detected in human serum/plasma or tissues. Typical HPLC profiles of an extract of serum from a lactating mother are shown in Figs. 10 (C_{18} RP column) and 11 (silica-based bonded nitrile NP column), respectively. The RP procedure (Fig. 10) will separate non-carotenoid compounds such as caffeine, vitamins A and E, and the antioxidant BHT that is added during the extraction. Carotenes and monohydroxycarotenoids are well separated. More polar carotenoids are not well resolved, but their complete separation can be achieved on a NP, silica-based bonded nitrile column (Fig. 11).

When geometrical isomers are included, as many as 35 carotenoids have been found in human serum. There are relatively simple procedures for routine quantitative analysis of the main carotenoids (see for example, *Volume 1A, Worked Example 7*) but complete separation and determination of all the carotenoids and their metabolites, including *Z* isomers, requires analysis by both RP and NP HPLC. The qualitative carotenoid profile of milk is quite similar to that of serum, but the carotenoids are present at much lower concentration [42]. Some HPLC peak overlap between carotenoids and non-carotenoid components, *e.g.* between (*Z*)-β-cryptoxanthin (**55**) and γ-tocopherol, ζ-carotene (**38**) and α-carotene (**7**), and 3′-epilutein (**137**) and caffeine is generally of no concern because the absorption maxima of the two overlapping components are sufficiently different. Serum and milk contain a high concentration of steryl esters which, because of similar retention times, can interfere with the analysis of phytoene (**44**) and phytofluene (**42**).

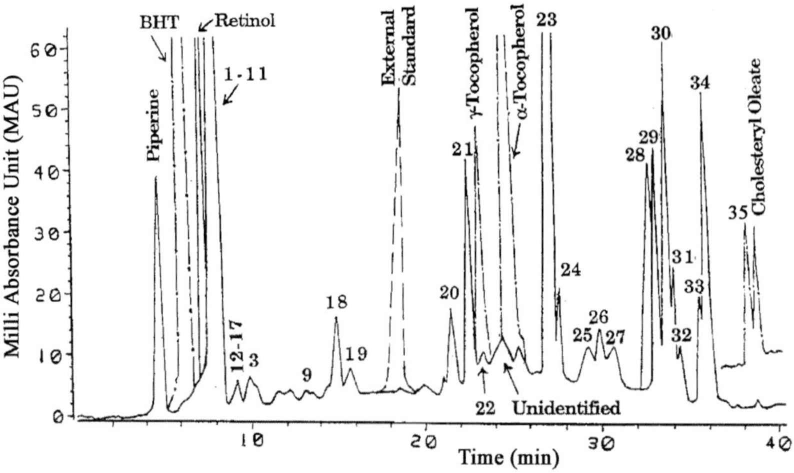

Fig. 10. Typical C_{18} RP-HPLC profile of carotenoids, retinol, tocopherols, and other non-carotenoid components from serum of a lactating mother, showing good resolution of carotenes. Chromatographic conditions as Fig. 3. Main peak identifications. 21: β-cryptoxanthin (**55**); 23: lycopene (**31**); 25: neurosporene (**34**); 27: γ-carotene (**12**); 28: ζ-carotene (**38**); 29: α-carotene (**7**); 30: β-carotene (**3**); 33,34: phytofluene (**42**); 35: phytoene (**44**); 18, 19, 39: carotenoid metabolites; 22, 24, 26, 31, 32: *Z* isomers. Reproduced from [42] with permission.

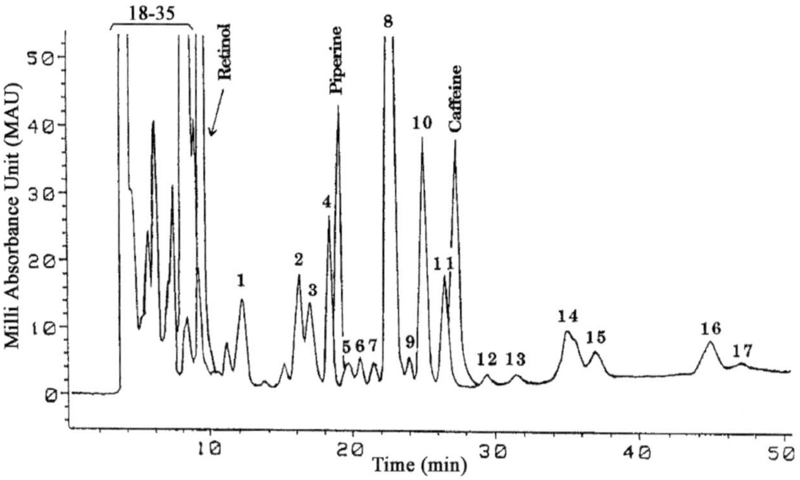

Fig. 11. Typical NP-HPLC profile of carotenoids, retinol, tocopherols, and other non-carotenoid components from serum of a lactating mother, showing good resolution of xanthophylls. HPLC conditions. Column: silica-based bonded nitrile column, 5 μm; 25 cm x 4.6 mm. Solvent: hexane (75%), dichloromethane (25%), containing methanol (0.4%) and diisopropylethylamine (0.1%); isocratic, flow rate 7 mL/min. Main peak identifications. 8: lutein (**133**); 10: zeaxanthin (**119**); 11: 3′-epilutein (**137**); 1-6, 9: metabolites of lycopene (**31**) and lutein; 7, 12-17: *Z* isomers of lutein and zeaxanthin. Reproduced from [42] with permission.

2. Major organs

Analysis of extracts from several human organs and tissues including liver, lung, breast, and cervix by these HPLC methods has revealed the presence of the same prominent carotenoids and their metabolites that are found in human serum [50] (see *Chapter 7*).

In the eye, carotenoids are present in the retina and other tissues (*Chapter 15*). Small amounts of lycopene (**31**) and its metabolites, other carotenes and β-cryptoxanthin (**55**) have beeen detected in some eye tissues but the characteristic carotenoids are the xanthophylls lutein (**133**) and zeaxanthin (**119**) and their metabolites. The efficient separation of these on a NP bonded nitrile column is illustrated in Fig. 12 [40].

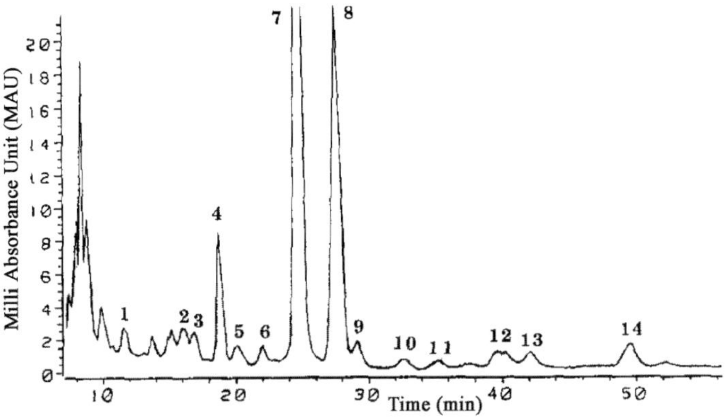

Fig. 12. Typical NP-HPLC profile of carotenoids from human retina. Chromatographic conditions as for Fig. 11. Main peak identifications. 7: lutein (**133**); 8: zeaxanthin (**119**); 9: 3′-epilutein (**137**); 1-6: metabolites of lutein; 12-14: *Z* isomers of lutein and zeaxanthin. Reproduced from [40] with permission.

Much of the zeaxanthin in the retina is the (3*R*,3′*S*) [(*meso*)] isomer (**120**), which cannot be distinguished from dietary (3*R*,3′*R*)-zeaxanthin (**119**) by the usual NP or RP procedures. The separation of these isomers requires the use of a chiral column with bonded amylose tris-(3,5-dimethylphenylcarbamate) [51]. In addition to resolving the stereoisomers of zeaxanthin, this column will also separate (all-*E*)-lycopene (**31**), its 5*Z* isomer, (3*R*,3′*R*,6′*R*)-lutein (**133**), and 3′-epilutein (**137**). The separation of standards is illustrated in Fig. 13. Analysis of eye and plasma samples, illustrated in Fig. 14, clearly shows that (3*R*,3′*S*)-zeaxanthin is present in the eye but not in plasma.

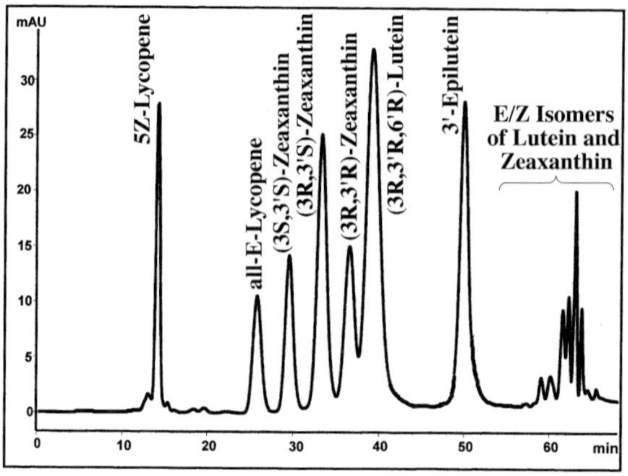

Fig. 13. HPLC separation of carotenoid standards on a chiral column, illustrating especially the separation of the optical isomers of zeaxanthin. HPLC conditions. Column: Chiralpak AD [amylose tris-(3,5-dimethylphenylcarbamate) coated on silica], 10 μm; 25 cm x 4.6 mm. Solvent A: hexane (95%), propan-2-ol (5%). Solvent B: hexane (85%), propan-2-ol (15%). Gradient: 0 - 10 min 90% A, 10% B (isocratic); 10 - 30 min linear gradient to 50% A, 50% B, then isocratic. Flow rate 0.7 mL/min. Reproduced from [51] with permission.

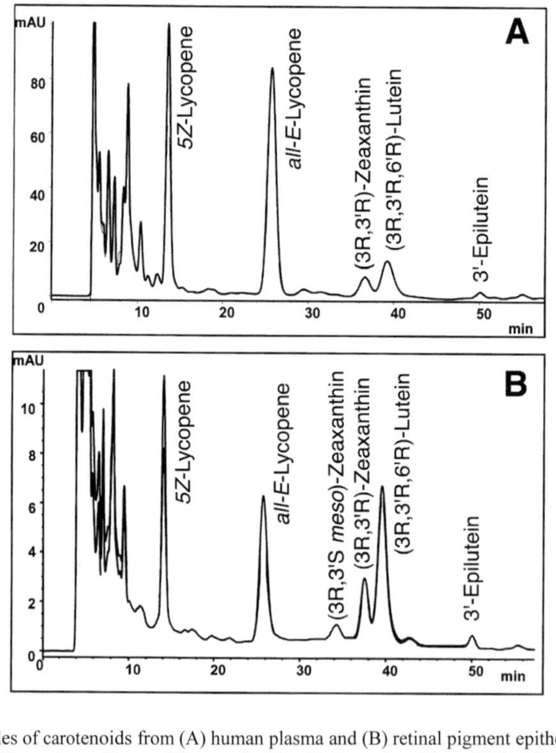

Fig. 14. HPLC profiles of carotenoids from (A) human plasma and (B) retinal pigment epithelium (RPE)-choroid on a chiral column. Chromatographic conditions as for Fig. 13. Reproduced from [51] with permission.

I. Conclusions

HPLC gives carotenoid researchers a powerful means of analysing carotenoid compositions and concentrations in foodstuffs and human samples. This brings with it a responsibility to avoid the risk of misleading information that can follow if identifications are not rigorous. Identification must be confirmed and not simply assumed. The photodiode array detector is a great benefit in this, allowing simultaneous monitoring at a range of selected wavelengths and providing UV/Vis spectra on-line for each component of a chromatogram as an aid to identification. Linked HPLC-MS and HPLC-NMR are now becoming more widely available but, without careful sample preparation and specialist and informed interpretation, serious mistakes can easily be made.

It is necessary to understand the properties of the carotenoids and the chromatographic principles and to balance realistically the great analytical precision of HPLC against the uncertainty arising from the wide and often uncontrollable sample variability. Caution and judgement must be applied to interpret what overall level of precision is justified.

When many thousands of human samples are being analysed as a biomarker of carotenoid status in epidemiological studies, rigorous quality control and careful standardization between laboratories are essential (*Chapter 10*). Non-invasive resonance Raman spectroscopic techniques show promise but lack the power to resolve individual carotenoids (*Chapter 10*).

References

[1] F. Khachik, G. R. Beecher, M. B. Goli, W. R. Lusby and C. E. Daitch, *Meth. Enzymol.*, **213**, 205 (1992).
[2] P. B. Jacobs, R. D. leBoeuf, S. A. McCommas and J. D. Tauber, *Comp. Biochem. Physiol.*, **72B**, 157 (1982).
[3] T. Matsuno, M. Katsuyama, T. Hirono, T. Maoka and T. Komori, *Bull. Jap. Soc. Sci. Fish.*, **52**, 115 (1986).
[4] F. Khachik, G. R. Beecher and N. F. Whittaker, *J. Agric. Food Chem.*, **34**, 603 (1986).
[5] F. Khachik, G. R. Beecher and W. R. Lusby, *J. Agric. Food Chem.*, **36**, 938 (1988).
[6] W. C. Still, M. Kahn and A. Mitra, *J. Org. Chem.*, **43**, 2923 (1978).
[7] I. Stewart and T. A. Wheaton, *J. Chromatogr.*, **55**, 325 (1971).
[8] P. Rüedi, *Pure Appl. Chem.*, **57**, 793 (1985).
[9] N. E. Craft, *Meth. Enzymol.*, **213**, 185 (1992).
[10] K. S. Epler, L. C. Sander, R. G. Ziegler, S. A. Wise and N. E. Craft, *J. Chromatogr.*, **595**, 89 (1992).
[11] T. W. Goodwin, *The Biochemistry of the Carotenoids, Vol. 1: Plants*, Chapman and Hall, London (1980).
[12] H. Kläui and J. C. Bauernfeind, in *Carotenoids as Colorants and Vitamin A Precursors* (ed. J. C. Bauernfeind), p. 48, Academic Press, New York (1987).
[13] J. Gross, *Pigments in Fruits*, Academic Press, Orlando (1987).
[14] A. J. Melendez, I. M. Vicario and F. J. Heredia, *J. Food Comp. Anal.*, **20**, 638 (2007).
[15] C. Emenhiser, N Simunovic, L. C. Sander and S. J. Schwartz, *J. Agric. Food Chem.*, **44**, 3887 (1996).
[16] A. Z. Mercadante, in *Food Colorants: Chemical and Functional Properties* (ed. C. Socaciu), p. 447, CRC Press, Boca Raton (2007).
[17] L. C. Sander and S. A. Wise, *Anal. Chem.*, **61**, 1749 (1989).

[18] F. Khachik, G. R. Beecher, J. T. Vanderslice and G. Furrow, *Anal. Chem.*, **60**, 807 (1988).

[19] F. Khachik and G. R. Beecher, *J. Agric. Food Chem.*, **35**, 732 (1987).

[20] K. Tsukida, K. Saiki, T. Takii and Y. Koyama, *J. Chromatogr.*, **245**, 359 (1982).

[21] U. Hengartner, K. Bernhard, K. Meyer, G. Englert and E. Glinz, *Helv. Chim. Acta*, **75**, 1848 (1992).

[22] F. W. Quackenbush, *J. Liquid Chromatogr.*, **10**, 643 (1987).

[23] A. Zumbrunn, P. Uebelhart and C. H. Eugster, *Helv. Chim. Acta*, **68**, 1540 (1985).

[24] G. Englert, K. Noack, E. A. Broger, E. Glinz, M. Vecchi and R. Zell, *Helv. Chim. Acta*, **74**, 969 (1991).

[25] G. R. Jamieson, in *Topics in Lipid Chemistry*, (ed. F. D. Gunstone), p. 107, Pergamon, New York (1970).

[26] F. Khachik and G. R. Beecher, *J. Chromatogr.*, **449**, 119 (1988).

[27] A. Rüttimann, K. Schiedt and M. Vecchi, *J. High Res. Chromatogr., Chromatogr. Commun.*, **6**, 612 (1983).

[28] M. Vecchi and R. K. Müller, *J. High Res. Chromatogr., Chromatogr. Commun.*, **2**, 195 (1979).

[29] G. Noga and F. Lenz, *Chromatographia*, **17**, 139 (1983).

[30] W. J. Driskell, M. M. Bashor and J. W. Neese, *Clin. Chem.*, **29**, 1042 (1983).

[31] F. Khachik and G. R. Beecher, *J. Chromatogr.*, **346**, 237 (1985).

[32] J. D. Surmatis and A. Ofner, *J. Org. Chem.*, **26**, 1171 (1961).

[33] F. J. Petracek and L. Zechmeister, *J. Am. Chem. Soc.*, **78**, 1427 (1956).

[34] R. Entschel and P. Karrer, *Helv. Chim. Acta*, **41**, 402 (1958).

[35] F. Khachik and G. R. Beecher, *J. Agric. Food Chem.*, **36**, 929 (1988).

[36] J. L. Bureau and R. J. Bushway, *J. Food Sci.*, **51**, 128 (1986).

[37] D. J. Hart and K. J. Scott, *Food Chem.*, **54**, 101 (1995).

[38] H. Müller, *Z. Lebensm. Forsch. A*, **204**, 88 (1997).

[39] R. A. Bone, J. T. Landrum, G. W. Hime, A. Cains and J. Zamor, *Invest. Ophthalmol. Vis. Sci.*, **34**, 2033 (1993).

[40] F. Khachik, P. Bernstein and D. L. Garland, *Invest. Ophthalmol. Vis. Sci.*, **38**, 1802 (1997).

[41] F. Khachik, G. R. Beecher and M. B. Goli, *Anal. Chem.*, **64**, 2111 (1992).

[42] F. Khachik, C. J. Spangler, J. C. Smith Jr., L. M. Canfield, A. Steck and H. Pfander, *Anal. Chem.*, **69**, 1873 (1997).

[43] G. J. Handelman, D. M. Snodderly, A. J. Adler, M. D. Russett and E. A. Dratz, *Meth. Enzymol.*, **213**, 220 (1992).

[44] P. S. Bernstein, F. Khachik, L. S. Carvalho, G. J. Muir, D. Y. Zhao and N. B. Katz, *Exp. Eye Res.*, **72**, 215 (2001).

[45] J. H. Humphries and F. Khachik, *J. Agric. Food Chem.*, **51**, 1322 (2003).

[46] F. Khachik, G. R. Beecher, and W. R. Lusby, *J. Agric. Food Chem.*, **37**, 1465 (1989).

[47] F. Khachik, M. B. Goli, G. R. Beecher, J. Holden, W. R. Lusby, M. D. Tenorio and M. R. Barrera, *J. Agric. Food Chem.*, **40**, 390 (1992).

[48] L. H. Tonucci, J. M. Holden, G. R. Beecher, F. Khachik, C. S. Davis and G. Mulokozi, *J. Agric. Food Chem.*, **43**, 579 (1995).

[49] F. Khachik, Z. Nir, R. L. Ausich, A. Steck, and H. Pfander, in *Food Factors for Cancer Prevention* (ed. H. Ohigashi, T. Osawa, J. Terao, S. Watanabe, and T. Yoshikawa), p. 204, Springer-Verlag, Tokyo (1997).

[50] F. Khachik, F. B. Askin, and K. Lai, in *Phytochemicals, a New Paradigm* (ed. W. R. Bidlack, S. T. Omaye, M. S. Meskin, and D. Jahner), p. 77, Technomic Publishing, Lancaster, PA (1998).

[51] F. Khachik, F. F. Moura, D. Y. Zhao, C. P. Aebischer and P. S. Bernstein, *J. Invest. Ophthalmol. Vis. Sci.*, **43**, 3383 (2002).

Carotenoids
Volume 5: Nutrition and Health

Chapter 3

Carotenoids in Food

George Britton and Frederick Khachik

A. Introduction

No members of the animal kingdom, including humans, can synthesize carotenoids. Even those animals (birds, fish, invertebrates) that use carotenoids for colouration must obtain them from the diet. Although humans, being mammals, are normally not coloured by carotenoids, analysis of human blood and tissues reveals a significant content of carotenoids which, as discussed later in this book, are associated with good health and reduced risk of diseases. Although some carotenoids are added to manufactured foods as colourants, or are taken as supplements (*Chapter 4*), most ingested carotenoid is obtained direct from natural food, especially vegetables and fruit.

In richer countries, where food is plentiful, much publicity is given to the possible benefits of a carotenoid-rich diet to maintain health and reduce risks of serious age-related degenerative diseases and conditions such as cancer, coronary heart disease and macular degeneration, as discussed in later *Chapters* in this *Volume*. Attention is focused on encouraging the consumption of 'healthy foods' or 'functional foods' which provide high intake of the carotenoids of interest. The target is to have dietary sources that provide a high concentration of those carotenoids, notably β-carotene (**3**), lycopene (**31**), lutein (**133**), zeaxanthin (**119**) and β-cryptoxanthin (**55**), which have been investigated most for an association with beneficial effects.

A large proportion of the world's population live in poverty and don't have the luxury of living long enough to develop these diseases. For people who live in poorer countries, the priority need for carotenoids is different but acute. The essential nutrient vitamin A is a metabolite of the provitamins β-carotene and some related carotenoids, notably α-carotene (**7**) and β-cryptoxanthin (**55**), and these carotenoids provide most of the vitamin A for many populations in the world. In countries where vitamin A deficiency is a real or potential problem, the availability and provision of food containing sufficient amounts of provitamin A carotenoids, especially β-carotene, can be a matter of life or death (see *Chapter 9*).

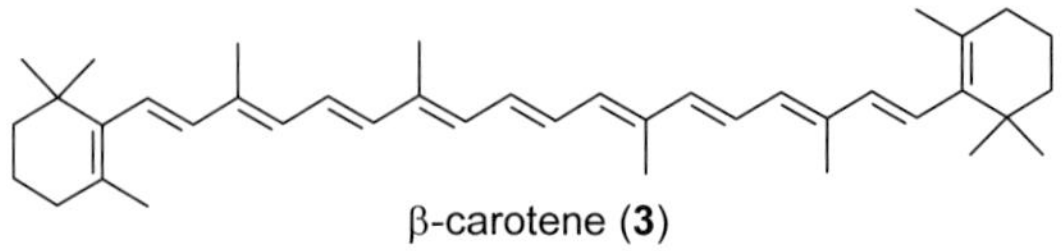

β-carotene (**3**)

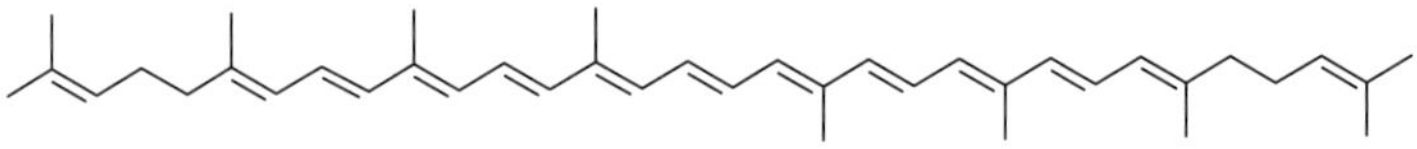

lycopene (**31**)

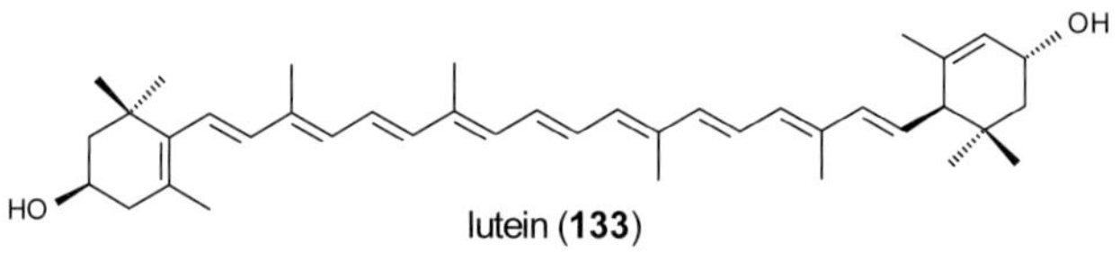

lutein (**133**)

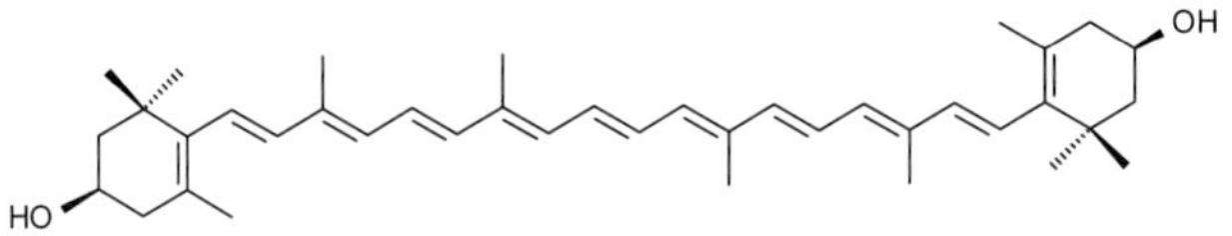

zeaxanthin (**119**)

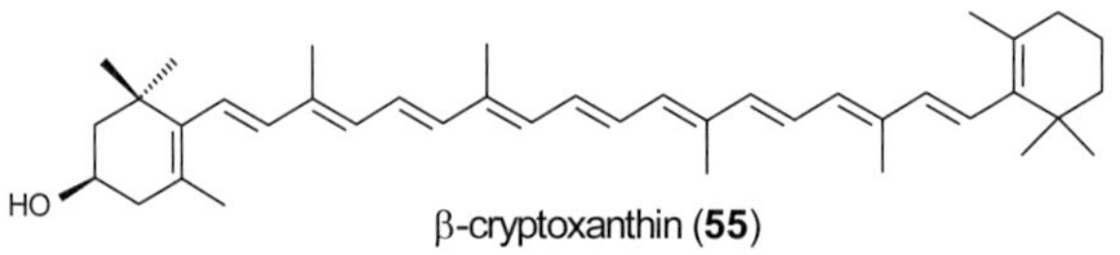

β-cryptoxanthin (**55**)

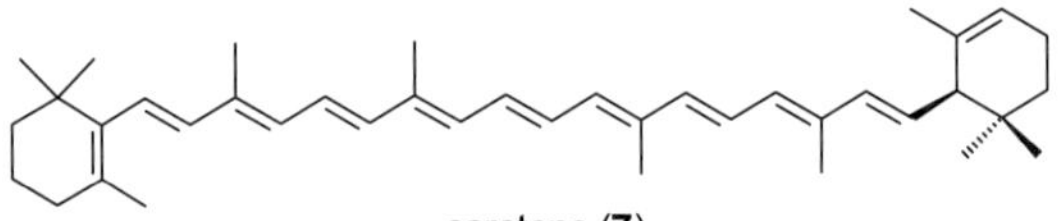

α-carotene (**7**)

In the context of both rich and poorer countries, knowledge of carotenoid content and composition is therefore essential in order that guidance can be given on what food sources can provide adequate supplies of desired carotenoids.

Over many years, thousands of papers have been published describing carotenoid content and composition of particular species and varieties under different conditions. The literature is flooded with numbers, reporting precise carotenoid concentrations, obtained by different analytical methods, especially HPLC (*Chapter 2*). The great variation in results can be extremely confusing. A major aim of this *Chapter* is to plot a way through this confusion and give some realistic evaluation and guidance.

The most important sources of carotenoids in the human diet are vegetables and fruit. The overall contribution of animal-derived food products is not large, but it must not be overlooked. Dairy products, eggs and some fish and seafood can have a significant carotenoid content. Also synthetic and natural carotenoids and carotenoid-rich extracts are widely used as natural colourants in manufactured food products such as cakes, confectionery, ice-creams and drinks.

Cultivation practices and methods of cooking and processing food vary widely around the world, and can have a profound effect on the stability and therefore the content of carotenoids.

B. Distribution of Carotenoids in Vegetables and Fruits

There can be some confusion over the description of a food as a fruit or a vegetable. Anatomical accuracy and culinary usage often do not coincide. Tomato and pepper, for example, are clearly fruits but are generally used as vegetables. In some reviews, the term 'fruit vegetables' has been used for such examples [1]. Also some foods eaten as vegetables are actually flowers (broccoli, cauliflower) or seeds and seed-bearing structures (peas, beans).

The ability to synthesize and accumulate carotenoids is determined genetically, but actual carotenoid compositions and contents are also highly dependent on environmental and cultivation conditions (Section **C**).

1. Green vegetables and fruits

All green plant tissues, not only leaves and stems but also green fruit and pods and seeds of legumes such as beans and peas, are green because of the presence of chlorophyll in the photosynthetic structures, the chloroplasts (see *Volume 4, Chapter 14*). In chloroplasts the chlorophyll-containing pigment-protein complexes of photosystems 1 and 2 also contain carotenoids. The carotenoid composition of plant chloroplasts is remarkably constant with, as the main components, β-carotene (25-30% of the total), lutein (40-50%), violaxanthin (**259**) (15%) and neoxanthin (**234**) (15%).

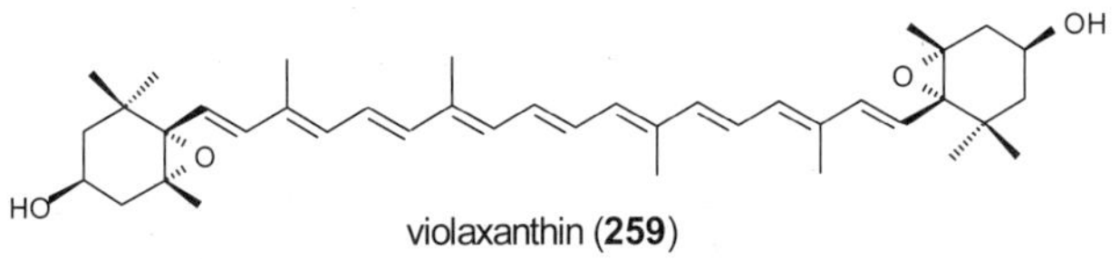

violaxanthin (**259**)

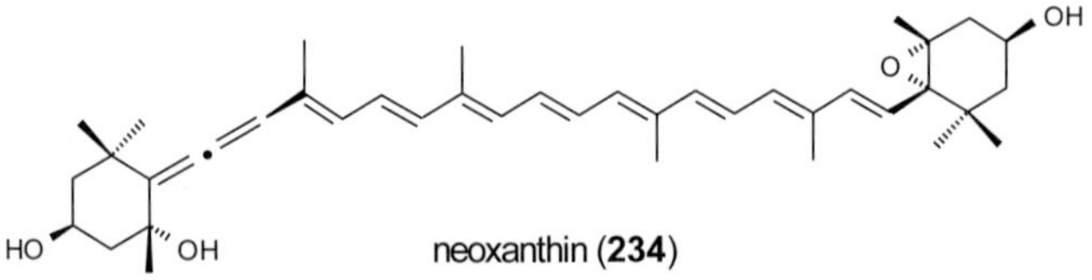

neoxanthin (**234**)

Small amounts of other carotenoids may be detectable, namely α-carotene, zeaxanthin, antheraxanthin (**231**) and lutein 5,6-epoxide (**232**). Very rarely, the only frequently consumed example being lettuce, some of the lutein may be replaced by lactucaxanthin (**150**).

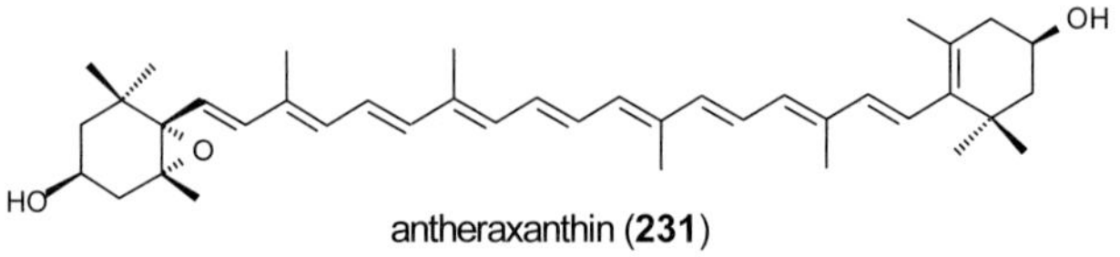

antheraxanthin (**231**)

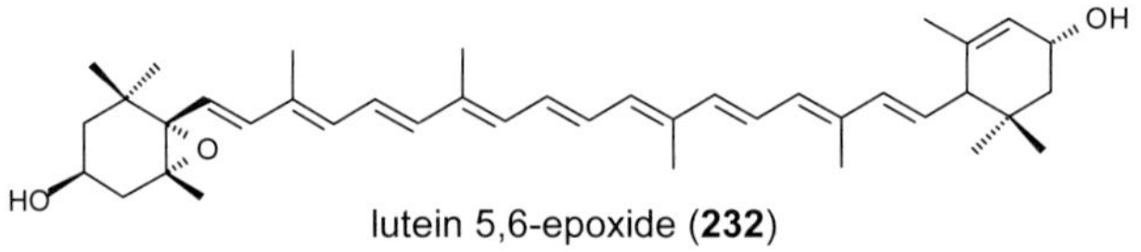

lutein 5,6-epoxide (**232**)

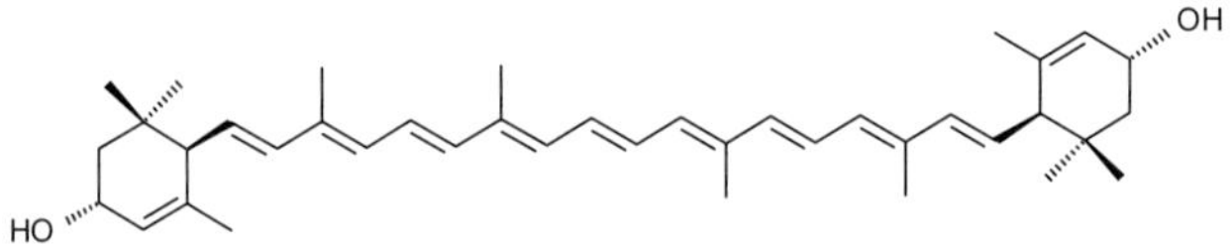

lactucaxanthin (**150**)

Although the carotenoid composition is almost constant, the quantitative carotenoid contents vary widely. There is, though, a clear correlation; darker green indicates a high population of chloroplasts and therefore a high concentration of carotenoids. In vegetables such as lettuce and members of the cabbage family which consist of more-or-less tightly packed leaves, the darkest green and hence the highest carotenoid concentration is in the outer leaves. Inner leaves that are not exposed to light may be pale green or almost white, with very little carotenoid, or may be yellow (etiolated) and have a different carotenoid composition, usually having little or no β-carotene and an altered xanthophyll composition. This variation must be taken into account when the carotenoid profile and content of a particular vegetable that is being consumed is estimated.

The reversed-phase HPLC profile of an extract from Brussels sprouts was illustrated in *Chapter 2*, *Fig. 3*. Other green fruits and vegetables have similar HPLC profiles [2] and some, *e.g.* green beans and lima beans show the additional presence of considerable amounts of α-carotene [2].

The qualitative distribution of the major carotenoids in some of the most commonly consumed green fruits and vegetables has been published in a review [3]. In all cases, these carotenoids are accompanied by considerable amounts of their geometrical isomers. For symmetrical carotenoids, the most common geometrical isomers in foods are 9*Z* and 13*Z* and, to a lesser extent, the 15*Z* isomer. With the unsymmetrical carotenoids, 9*Z*, 9'*Z*, 13*Z*, 13'*Z*, and 15*Z* isomers may all be present in variable concentrations. The occurrence of di-*Z* isomers of carotenoids in foods is rare.

2. Yellow, orange and red fruits and vegetables

Many of the richest sources of carotenoids are not green. Yellow, orange and red plant tissues, including fruits, flowers, roots and seeds, may contain high concentrations. It is important to realise, however, that these colours are not always due to carotenoids; anthocyanins, betalains and quinones provide other striking examples.

a) Fruits

Fruit represent a major dietary source of carotenoids and have been studied extensively. Although some fruits contain insignificant amounts of carotenoids or small amounts of the carotenoids that are usually found in chloroplasts, others contain larger amounts of different carotenoids. Some distinctive patterns have been recognized [4,5] that appear in a range of fruits: (i) large amounts of the acyclic carotene lycopene, as in tomatoes (red colour), (ii) large amounts of β-carotene and/or its hydroxy derivatives β-cryptoxanthin and zeaxanthin (orange colour), (iii) as (ii) but with also α-carotene and/or its hydroxy derivatives, especially lutein (yellow-orange), (iv) large amounts of carotenoid epoxides (yellow), and (v) carotenoids that appear to be unique to or characteristic of that species (yellow, orange or red), *e.g.* capsanthin (**335**) and capsorubin (**413**) in red peppers (*Capsicum annuum*).

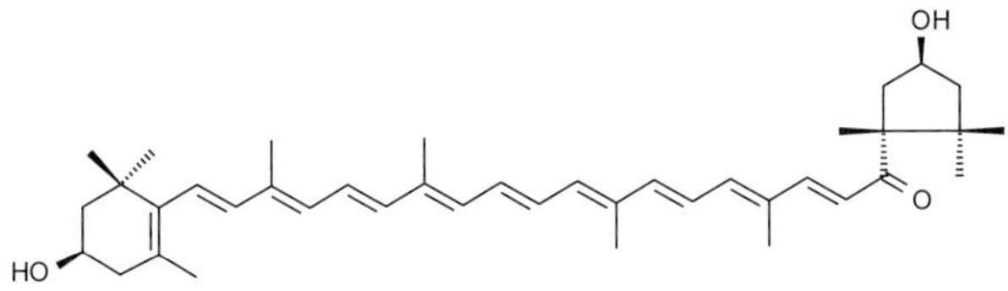

capsanthin (**335**)

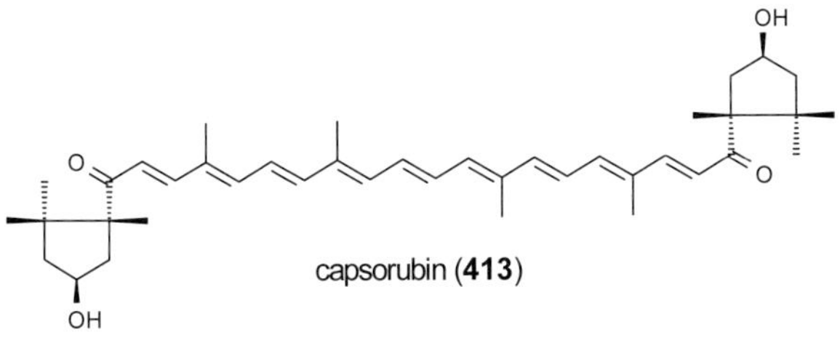

capsorubin (**413**)

Now that the genes of carotenoid biosynthesis are known and understood, these observations can be rationalized (see *Volume 3, Chapters 2* and *3*). Green, unripe fruits contain chloroplasts in which the usual collection of chloroplast carotenoids is found. As the fruits mature, these chloroplast pigments may remain or may be degraded. In many cases, however, familiar colour changes take place as the fruits ripen and develop chromoplasts, sub-cellular organelles that replace chloroplasts and may be derived from them. The carotenoids are biosynthesized and accumulate in the chromoplasts. The biosynthesis is controlled by a set of genes which are activated as a key feature of the ripening process. The carotenoid composition of the ripe fruit is determined by which ripening-specific genes are present and activated. So if the phytoene synthase and desaturase genes are active, lycopene will be produced (category i), if in addition the β-cyclase and hydroxylase genes are active, the dicyclic β-carotene and its hydroxy derivatives or the moncyclic γ-carotene (**12**) and its hydroxy derivative rubixanthin (**72**) will accumulate (category ii). When the ε-cyclase and ε-hydroxylase are also present, α-carotene and lutein are produced (category iii). Similarly, an active epoxidase gene gives category iv. Additional genes may also be present, leading to other end-products (category v).

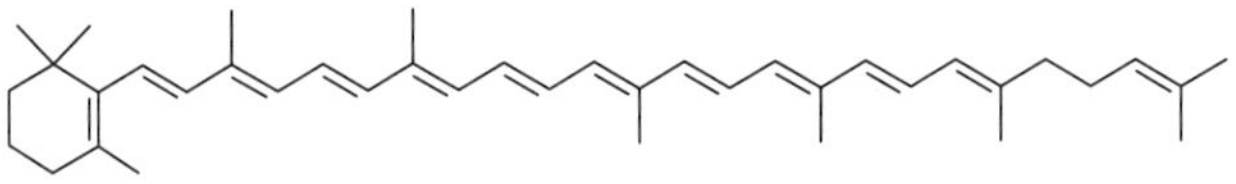

γ-carotene (**12**)

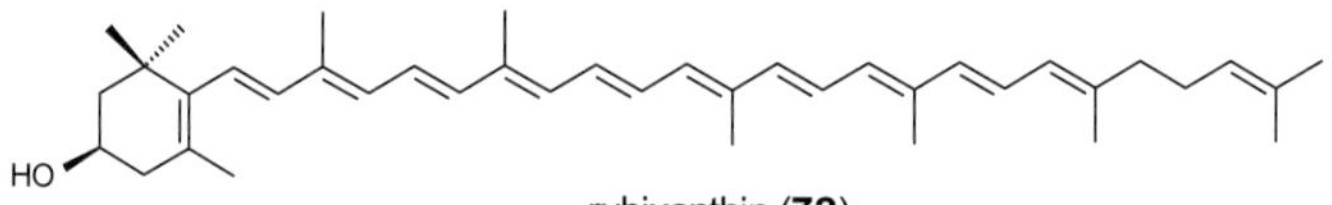

rubixanthin (**72**)

Application of this knowledge makes possible the genetic modification of carotenoid profiles or content in various crop plants (for details see *Chapter 6*).

Some yellow/red fruits and vegetables contain mostly carotenes and only small amounts of xanthophylls [6,7]. With some of these, *e.g.* apricot, the carotenoid profile, illustrated in *Chapter 2, Fig. 5*, is complicated, consisting not only of α-carotene and β-carotene but also containing a range of biosynthetic intermediates and *Z* isomers.

Other examples, *e.g.* the root vegetables carrot and sweet potato, have a simpler HPLC profile with only α-carotene and β-carotene and small amounts of biosynthetic intermediates present. The qualitative distribution of the major carotenoids in commonly consumed yellow fruits and vegetables has been published [3].

Some red fruits, *e.g.* tomato, are major dietary sources of lycopene and the biosynthetic intermediates phytoene (**44**), phytofluene (**42**) and ζ-carotene (**38**), and to a lesser extent also contain neurosporene (**34**), β-carotene, and γ-carotene [7-9]. The lycopene content of tomatoes can reach 5-10 mg/100g. This high content is maintained in tomato-based food products such as ketchup, soup and sauces [3,8-10].

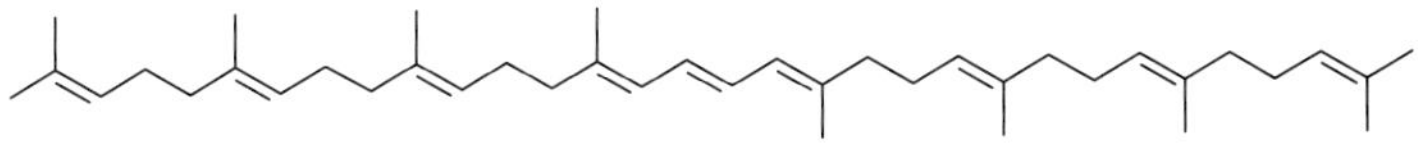

phytoene (**44**)

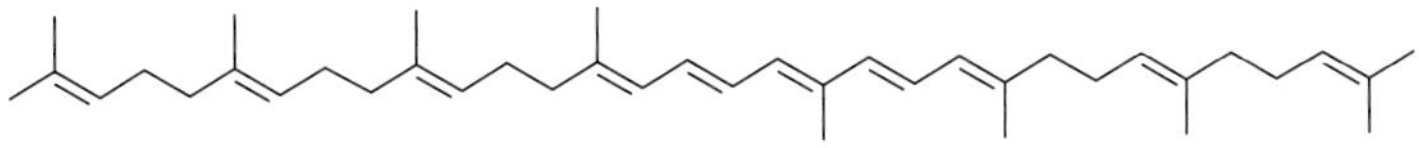

phytofluene (**42**)

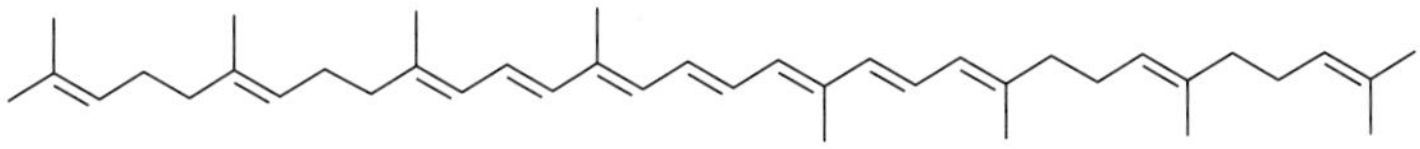

ζ-carotene (**38**)

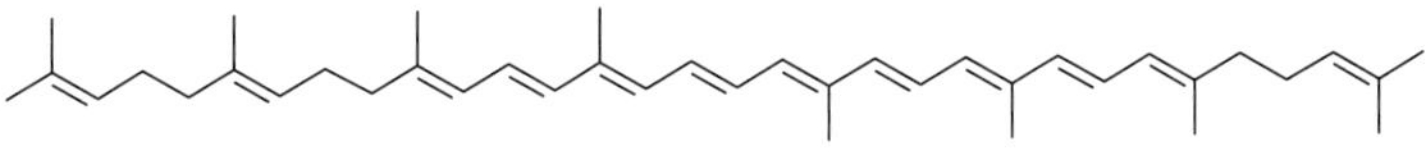

neurosporene (**34**)

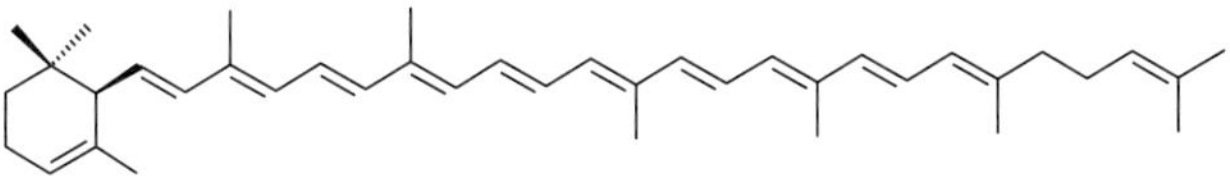

δ-carotene (**21**)

Other commonly consumed fruits that contain lycopene are pink grapefruit, water melon, papaya and, in low concentration, apricots (fresh, canned, dried) [7]. A typical carotenoid HPLC profile of an extract from tomato paste is shown in *Chapter 2, Fig. 6*.

Many strains and mutants of tomato have been produced with quite different carotenoid profiles, *e.g.* 'high-beta' and 'high-delta' strains in which the lycopene is replaced by high concentrations of β-carotene or δ-carotene (**21**), respectively. The genetic modification of carotenoid content and composition in tomatoes is discussed in *Chapter 6*.

Yellow/orange fruits and vegetables, *e.g.* mango, papaya, peaches, prunes, acorn and winter squash, and oranges [3,7,10-12], may contain, in addition to carotenes and often as the main pigments, xanthophylls and xanthophyll epoxides, esterified with straight chain fatty acids such as lauric, myristic, and palmitic acids [3,7,10-12]. Strains of squash (*Cucurbita maxima*) that contain a high concentration of esters of lutein and other xanthophylls have been studied extensively. There are significant differences in qualitative distribution of carotenoids and their esters in different cultivars. As shown in *Chapter 2, Fig. 8*, the major carotenoids in the 'butternut' squash are α-carotene, β-carotene, and lutein diacyl esters but not monoesters, whereas the 'Northrup King' squash contains mainly lutein and its monoesters and α-carotene is absent (*Chapter 2, Fig. 7*). In acorn squash, violaxanthin and its monoesters and diesters are also present.

Small amounts of lutein dehydration products are detected in squash but are rarely found elsewhere [13]. Their general occurrence in human plasma is attributed to metabolism of dietary lutein.

Citrus fruits, *e.g.* oranges, tangerines, grapefruit and lemons, and their juices, are widely consumed. They have been studied extensively and many different varieties, strains and hybrids analysed [14]. The carotenoid compositions are very variable and can be complex. β-Carotene, β-cryptoxanthin and violaxanthin are characteristic and apocarotenoids are common, sometimes as the main pigments. The carotenoids are present not only in the brightly coloured peel but also in internal tissues and juice. The carotenoid compositions of the different parts can differ considerably. The carotenoid content and composition of the juices depend on which parts of the fruit have been used in the processing.

b) Roots

In roots such as carrots and sweet potatoes that contain a high amount of carotene, the pigments are also synthesized and accumulate in chromoplasts. In carrots, the concentrations accumulated and the ratio of α-carotene to β-carotene vary considerably between strains [15]. α-Carotene can range from 5% to 50% of the total carotene. Varieties with a deeper orange-red hue have a higher proportion of β-carotene. The concentration in outer tissues (phloem) is generally greater than that in the inner core (xylem). Some red varieties also contain lycopene, which can be the main pigment. A yellow variety has a considerable concentration of lutein instead of β-carotene. Sweet potatoes also accumulate β-carotene, usually with little α-carotene. Common potatoes, even yellow varieties, contain only low levels of the common chloroplast xanthophylls. Genetically modified potatoes, engineered to accumulate carotene, have been produced (see *Chapter 6*). Other yellow root vegetables may contain carotenoids in low concentration, including, in swede (*Brassica rutabaga*) some lycopene [16].

c) Seeds

The yellow-orange colour of the outer coat of sweetcorn (maize, *Zea mays*) is due primarily to lutein, β-carotene, zeaxanthin and cryptoxanthins [17].

An extensive programme has led to the development of a genetically modified 'Golden' rice [18], that accumulates β-carotene in the butter-coloured endosperm (see *Chapter 6*).

Wheat and pasta products have been analysed [19]. The only carotenoids of note present were lutein and zeaxanthin, with lutein predominating. The concentrations were generally very low (ng/g) though somewhat higher (μg/g) in an Australian green-harvested wheat. The carotenoid content of wheat pasta is rather higher because of lutein and zeaxanthin from eggs used in the processing. Pasta made from durum wheat (*Triticum durum*) also contains more lutein from the durum flour. There is very little carotenoid in other cereals and flours.

Green seeds of legumes *e.g.* peas, contain β-carotene and chloroplast xanthophylls [4].

The seed coat of the shrub *Bixa orellana* accumulates an extremely high concentration of the apocarotenoid bixin (**533**). This product is widely used as a food colourant (annatto) but it is not known if it is of any consequence for human health.

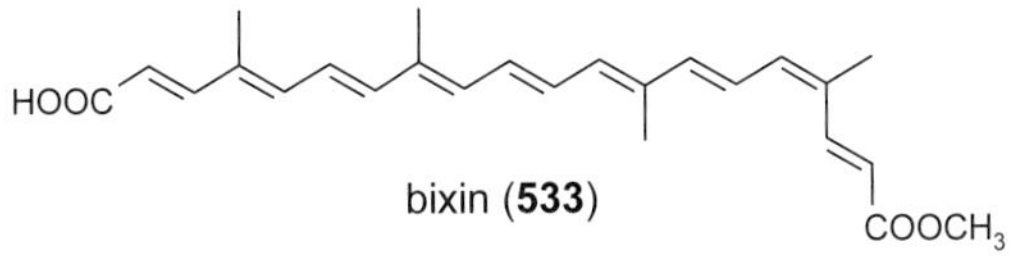

bixin (**533**)

d) Flowers

Flowers are not widely consumed, the best known examples being cauliflower and broccoli (*Brassica oleracea*, cv. group Botrytis and Italica, respectively). Familiar white cauliflowers contain little or no carotenoid, but an orange strain, accumulating β-carotene, first found in a field of white cauliflowers, is now available for growing commercially [20]. The flower heads of broccoli and calabrese are harvested before the petals open. In older or stored examples the yellow florets, rich in lutein esters, may start to show.

e) Oils

Fruit of some oil palms synthesize and accumulate a high concentration of α-carotene and β-carotene which are retained in the oil from the pressed fruit. Red palm oil from *Eleais guineensis* is processed on a large scale and refined in various forms for use as a cooking oil and ingredient in manufactured foods [21]. Some other palm fruits, *e.g.* the South American 'Buriti' (*Mauritia vinifera*) also have a very high carotenoid content [22].

Canola (rapeseed, *Brassica napus*) oil normally contains little or no carotenoid but the ability to produce carotenoids can be introduced by GM methods [23].

The oil of the South-East Asian 'Gac' fruit (*Momordica cochinchinnensis*) contains a high concentration of β-carotene [24] but has so far found only local use.

3. Animal-derived food products

Animals do not biosynthesize carotenoids but many can accumulate, sometimes in quite high concentration, carotenoids that they ingest. If such animal tissues or products are eaten as part of the human diet, they provide an additional source of carotenoids. There are well known examples of this.

a) Eggs

The yellow colour of egg yolk is due to carotenoids. The colour hue and intensity depend on the poultry feed used (see *Volume 4, Chapter 13*). Marigold flowers or lutein esters produced from them are widely used in chicken feed, so the eggs provide a good source of lutein. The more orange-yellow yolks from corn-fed hens also contain zeaxanthin. Some synthetic apocarotenoids, such as 8′-apo-β-caroten-8′-oic acid (**486**) ethyl ester are also used as additives to give a more orange hue. These apocarotenoids have provitamin A activity.

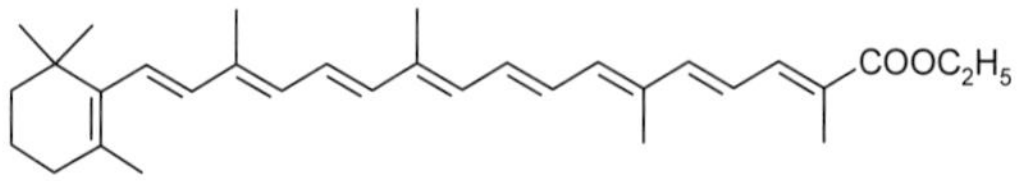

8'-apo-β-caroten-8'-oic acid (**486**) ethyl ester

b) Dairy produce

Cattle specifically absorb carotenes and not xanthophylls. This carotene may colour the fat yellow, and is also present in milk fat. Milk, cream, butter and cheese are therefore likely to contain some β-carotene, though the concentration is usually not great [14]. The presence of carotene often varies with season; it is highest in the early summer when the animals are grazing the best quality pastures.

c) Seafood

Pink-fleshed fish, notably salmon and trout, accumulate in the muscle high concentrations of astaxanthin (**404-406**) or canthaxanthin (**380**) that they obtain from their natural food or which is added to their feed (see *Volume 4, Chapter 12*). Invertebrate seafood, such as shrimp, lobster and other crustaceans and molluscs can contain quite high concentrations of carotenoids. In crustaceans this is often astaxanthin present as carotenoprotein complexes; the red carotenoid is released by cooking. The highest concentration of carotenoids is usually in the shell or integument which is commonly discarded before eating. Some products, including eggs (roe), can provide significant amounts in the diet, however. There is a great structural diversity of carotenoids in seafood; in most cases the possible biological activity of these has not been tested. The intake of these foods in a normal diet is generally small, and not of great significance. These carotenoids are generally not detected in human blood.

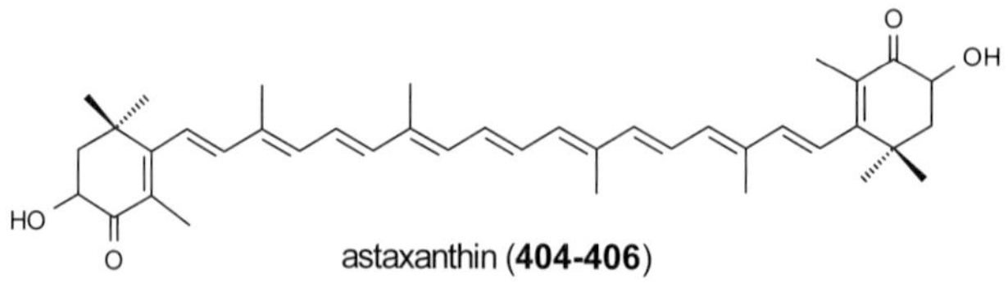

astaxanthin (**404-406**)

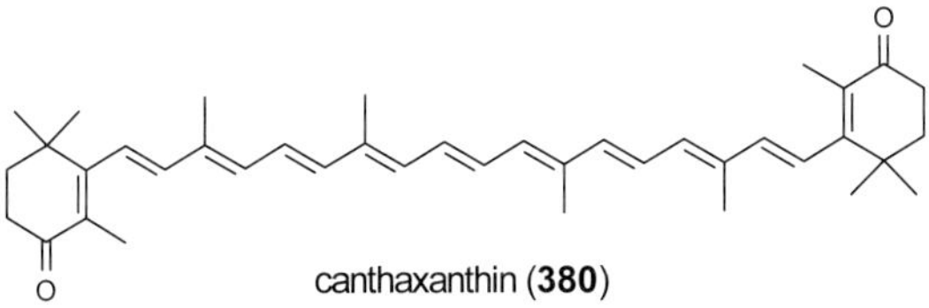
canthaxanthin (**380**)

4. Good sources

Many food composition tables are available in the literature [1,4,14,25-28] and references to carotenoid compositions of food in various parts of the world are given in reviews [24,29,30]. Extensive compilations of tabulated data can be found on the internet [31,32]. Evaluation of the data leads to the conclusions summarized in Table 1 about good sources of β-carotene and other carotenoids of nutritional interest.

Knowledge of carotenoid content is only part of the story, however. The efficiency with which the food is digested and the carotenoid released, solubilized, absorbed, transported and metabolized, *i.e.* 'bioavailability', is another key factor (see *Chapters 7* and *8*). The balance between content and bioavailability must always be considered.

Table 1. Good food sources of the nutritionally important carotenoids β-carotene, β-cryptoxanthin, lutein, lycopene and zeaxanthin, giving an indication of the likely carotenoid content. Precise values are not given but the content is indicated as a range, as follows. Low: 0 - 0.1 mg/100 g; Moderate: 0.1 - 0.5 mg/100 g; High: 0.5 - 2 mg/100 g; Very high: >2 mg/100 g.

Common name	Latin name	Content
β-Carotene		
Apricot	*Prunus armeniaca*	High - very high
Broccoli	*Brassica oleracea (Italiaca)*	Very high
Brussels sprouts	*Brassica oleracea (Gemmifera)*	High
'Buriti'	*Mauritia vinifera*	Very high
Butter		Low
Carrot	*Daucus carota*	Very high
'Gac' oil	*Momordica cochinchinnensis*	Very high
Grapefruit	*Citrus paradisi*	Low - moderate
Green leafy vegetables		Moderate - very high
Guava	*Psidium guajava*	Moderate
Kale	*Brassica oleracea (Acephala)*	Very high
'Karat' banana	*Musa troglodytarum*	High
Lettuce	*Lactuca sativa*	Moderate - high
Loquat	*Eriobotrya japonica*	Moderate
Mango	*Mangifera indica*	High - very high

β-Carotene (continued)		
Orange and juice	*Citrus* spp. and hybrids	Low - moderate
Papaya	*Carica papaya*	Moderate
Pea	*Pisum sativum*	Moderate
Peach	*Prunus persica*	High
Pepper (red, orange, green)	*Capsicum annuum*	High
Red palm oil	*Elaeis guineensis*	Very high
Spinach	*Spinacia oleracea*	Very high
Squash/pumpkin	*Cucurbita* spp.	Low - high
Sweet potato	*Ipomoea batatas*	Very high
Tangerine	*Citrus* spp. and hybrids	Low - moderate
Tomato	*Lycopersicon esculentum*	Moderate
Tomato, 'high-beta'		Very high
Tree tomato	*Cyphomandra betacea*	Moderate
West Indian cherry	*Malpighia glabra*	High
β-Cryptoxanthin		
Loquat	*Eriobotrya japonica*	Low - moderate
Papaya	*Carica papaya*	Moderate - high
Pepper (red, orange)	*Capsicum annuum*	Moderate
Persimmon	*Diospyros kaki*	High
Pitanga	*Eugenia uniflora*	High
Squash/pumpkin	*Cucurbita maxima*	Moderate - high
Tangerine	*Citrus* spp. and hybrids	Moderate - high
Tree tomato	*Cyphomandra betacea*	Moderate - high
West Indian cherry	*Malpighia glabra*	Low
Lutein		
Broccoli	*Brassica oleracea (Italica)*	Very high
Egg yolk		Moderate - high
Green leafy vegetables		Very high
Pepper (yellow, green)	*Capsicum annuum*	Very high
Squash/pumpkin	*Cucurbita* spp.	Moderate - very high
Lycopene		
Apricot	*Prunus armeniaca*	Low
Carrot (red)	*Daucus carota*	High
Grapefruit (red)	*Citrus paradisi*	Moderate - high
Guava	*Psidium guajava*	High
Papaya	*Carica papaya*	Moderate - high

Lycopene (continued)		
Persimmon	*Diospyros kaki*	Low - high
Tomato	*Lycopersicon esculentum*	Very high
Water melon	*Citrullus lanatus*	High - very high
Zeaxanthin		
'Buriti'	*Mauritia vinifera*	High
Chinese wolfberry 'Gou Qi Xi''	*Lycium chinensis*	Very high
Pepper (orange, red)	*Capsicum annuum*	Very high
Persimmon	*Diospyros kaki*	Moderate
Squash/pumpkin	*Cucurbita* spp.	Moderate
Sweetcorn	*Zea mays*	Moderate

5. Additives, colourants

β-Carotene and other synthetic or natural carotenoids or carotenoid-rich extracts are widely used as additives to colour processed food, drinks, confectionery, icecream *etc.* They are normally present in quite small amounts but in some cases the concentration can be significant. The concentration of β-carotene in some orange-flavoured drinks can be high enough to cause carotenodermia in people who drink large amounts.

C. Effects of Environmental Conditions and Cultivation Practice

Over more than 50 years there have been many studies of effects of conditions on carotenoid (often only β-carotene) content of green leaves, many with grasses and other forage plants. Analytical studies are reported from many parts of the world with many different species. The main findings have been summarized and discussed [4]. The results are very variable and sometimes conflicting. Because of variability of experimental design, analytical methods and species used, the precise numerical figures often reported and numerical comparisons between studies are of little real value. It is safe to say, however, that, as well as strongly affecting crop yield/productivity overall, *e.g.* the number and size of leaves, environmental conditions and cultivation practice also influence carotenoid content of leaves, since this is related to photosynthetic efficiency and density of chloroplasts.

In general, it can be concluded that optimal conditions that produce strong growth and healthy plants are consistent with good carotenoid content. This means soil that is of good quality and structure to allow strong root formation, and is well supplied with water, minerals and nutrients. Also, light is a significant factor. There may be differences in carotenoid composition and content between leaves or plants in sun and shade conditions, and excessive light can cause a reduction in photosynthetic efficiency, *via* photoinhibition and photodamage. Light quality, *i.e.* intensity at different wavelengths, varies with altitude and can also be influential. For most plants there are optimal day and night temperatures. Heat stress, light stress and drought stress, and stress by pollution or salt are detrimental to carotenoid content, as they are to plant growth and health in general.

The age and maturity of plant tissues at harvest is a significant factor. The time of day at harvest can have a profound effect on water content, leading to apparent variations in carotenoid concentration based on fresh weight. Treatment post-harvest is also important, especially the conditions and time of transport, and storage in the market and in the home.

Information on particular crops can be obtained from extensive reviews, e.g. [4] and references therein. A good illustration is provided by the baby-food squash ‘Northrup King’ grown in the U.S.A. When this is grown in Michigan the concentration of carotenoids is much lower than in the same cultivar grown in North Carolina, presumably because of environmental factors, though the qualitative composition is similar [11].

D. Effects of Storage, Processing and Cooking

1. Stability and loss or retention of carotenoids

Carotenoids *in situ* in vegetables and fruit are usually more stable than when they are isolated, because of the protective effect of the special conditions within the tissues due to molecular interactions with proteins *etc*., molecular aggregation and crystallization, and the presence of natural antioxidants, including antioxidant enzymes, such as superoxide dismutase (SOD). Any disruption of the tissues, such as may occur during processing or cooking, or during natural aging, may lessen this protection, leaving the carotenoids exposed to damaging factors and susceptible to change. When fruits and vegetables are cut, chopped, shredded or pulped, this increases exposure to oxygen and may remove the physical barriers that normally keep apart the carotenoids and oxidizing enzymes such as lipoxygenase.

Knowledge of the properties of carotenoids suggests that when foods are being stored, processed or cooked the greatest losses and changes are caused by prolonged exposure to air, strong light, high temperature or acid. To minimize destructive effects, prolonged heating and exposure to strong light and air should be avoided.

Transportation, storage and processing of foods must be optimized to prevent or reduce loss of quality and to preserve nutritional benefits. Some losses are unavoidable, *e.g.* removal of undesirable though carotenoid-rich peel or skin. Carotenoids may also be lost or altered during processing and storage, by enzymic or non-enzymic oxidation and by geometrical isomerization, rearrangement or other reactions. These factors can be addressed and monitored during industrial processing, but not during home preparation, where losses can be considerable and more difficult to control.

The losses or changes may be balanced by the improved bioavailability when the food structure has been weakened (*Chapter 7*).

The stability of carotenoids varies between different foods, even when processed and stored in the same way, so conditions that maximize carotenoid retention have to be determined for each individual case.

Analytical results may seem to indicate that carotenoid concentration increases during cooking or thermal processing. These erroneous results are likely to be analytical artefacts resulting from, for example, unaccounted loss of water or leaching of soluble solids during processing, so that calculation of carotenoid content on a fresh weight or a dry weight basis is not comparable to that of the fresh raw material. In fruits and root crops, carotenoid biosynthesis may continue after harvest and the carotenoid content actually increase, as long as the tissues are not damaged. Green vegetables, however, commonly lose carotenoid on storage, especially under conditions, such as high temperature, which favour wilting.

2. Storage, cooking and processing

There have been many experimental studies of effects of storage, processing and cooking of many different foods. Details of particular examples are given in a number of specialized surveys [1,14,25,33], which include extensive lists of references to the original literature. The main general findings and conclusions are summarized below.

a) Transport and storage

Many carotenoid-containing foods are seasonal. To make the products available all the year round, they must be harvested at the peak time and then stored or processed under conditions that preserve them and their carotenoid content. Harvested crops are often transported to the market, sometimed over long distances.

Sun-drying is a cheap and easy traditional method of food preservation in poor regions but the exposure to air and sunlight is particularly destructive of carotenoids. Protection from direct sunlight helps to lessen the losses.

Freezing (the more rapid the better) and storage frozen generally preserve the carotenoids but subsequent slow thawing, especially of unblanched products, can be detrimental.

The short heat treatment of blanching may cause some losses but it inactivates oxidative enzymes and thus prevents further greater losses later.

Packaging with exclusion of oxygen (vacuum or inert atmosphere) and storing at low temperature and protected from light preserves the carotenoid content. The presence of natural antioxidants, the addition of antioxidants, or treatment with bisulphite as a preservative may also reduce the extent of degradation.

b) Cooking and processing

Not surprisingly, more drastic methods of cooking, *e.g.* for longer times or at higher temperatures, lead to greater losses. Prolonged boiling or deep-frying lead to the greatest changes. Baking and pickling are also detrimental.

With other methods, *e.g.* boiling for a short time (blanching), stir-frying, and microwave cooking, changes are small. Loss of carotenoids is usually small but heat provides energy for geometrical isomerization so the proportion of *Z* isomers increases. Carotenoid 5,6-epoxides are not stable and undergo rapid isomerization to the furanoid 5,8-epoxide form.

For any processing method, loss of carotenoid increases with longer processing time, higher temperature and cutting or pureeing. Exposure of green vegetables to severe heat treatment over extended periods, *i.e.* boiling for one hour, has been shown to result in complete destruction of carotenoid epoxides [2].

There are many publications in which losses of carotenoids are reported, usually for one or a small number of examples. Comparisons between studies are difficult and data may be somewhat conflicting. As examples, a study of the effects of cooking and processing on a number of yellow-orange vegetables, *e.g.* carrot, sweet potato and pumpkin, has demonstrated that the destruction of α-carotene and β-carotene as a result of heat treatment is about 8-10% [6]. Comparison of the extracts from several green vegetables (kale, Brussels sprouts, broccoli, cabbage, spinach), raw and cooked, [2] shows that some of their major chlorophyll and carotenoid constituents undergo some structural transformation. About 60% of the xanthophyll in Brussels sprouts is destroyed as a result of cooking [2]. In cooked kale the figure is about 68%. Lutein and zeaxanthin survive, though with some loss, whereas the epoxide violaxanthin is mostly destroyed or converted into the 5,8-epoxide auroxanthin (**267**); only 10% and 12% of the violaxanthin survives in cooked Brussels sprouts and kale, respectively. The same study reported no significant changes in the ratio of the *E* and *Z* isomers of neoxanthin, lutein epoxide and lutein.

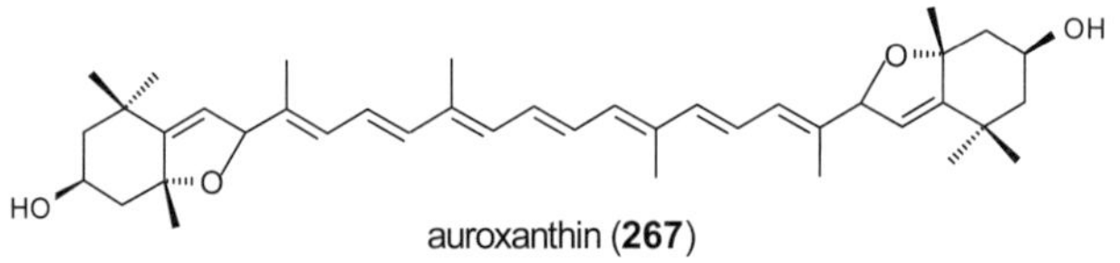

auroxanthin (**267**)

The epoxycarotenoids are sensitive to heat, light, and trace amounts of acids, and readily undergo rearrangement, stereoisomerization, and degradation. The cooking process also degrades the chlorophylls. Losses of β-carotene are small (about 15%) and the heat treatment does not cause extensive stereoisomerization [2]. In several varieties of squash the esters have been shown to be more stable than the free carotenols, and the diesters more so than the monoesters [11]. Monoesters of violaxanthin were completely destroyed by cooking, but the diesters of violaxanthin survived to some extent. No significant change in the ratio of *E/Z* isomers of violaxanthin diesters nor rearrangement of the esters to the 5,8-epoxide form as a result of cooking was indicated, in contrast to the unesterified violaxanthin. Lutein monoesters and diesters were also stable.

The quantitative loss and rearrangement of carotenoid 5,6-epoxides in cooked foods should not be of great concern since these compounds and their byproducts have not been detected in human serum or plasma [34,35].

As a result of various food preparation techniques, food carotenoids may undergo three main types of reactions, namely oxidation, rearrangement and dehydration. These reactions, which are described in the next Section, are dependent on the nature of the carotenoids, the food matrix, and the method of preparation.

3. Causes and mechanisms

The main changes that occur to carotenoid composition and content in foods during processing and cooking are oxidative breakdown and geometrical isomerization. Mechanisms of the isomerization and oxidative breakdown of carotenoids, in solution and in model systems, have been treated in *Volume 4, Chapters 3* and *7,* respectively. The same mechanisms apply to carotenoids in food, but the processes are modified by the special conditions *in situ*. There are many reports to show that carotenoids in foods vary in their susceptibility to degradation.

Oxidation, either enzymic or non-enzymic, is the main cause of destruction of carotenoids. Geometrical isomerization, which occurs particularly during heat treatment, increases the proportion of *Z* isomers and may alter the biological activity, but the total carotenoid content is not greatly changed. Conditions encountered during processing and storage may result in greater exposure to air, and so can induce greater losses than are caused by cooking. Also chopping or grinding can bring carotenoids into contact with degradative enzymes.

a) Oxidation

i) Enzymic. The main risk of enzyme-catalysed oxidative breakdown occurs during slicing, chopping or pulping of the fresh plant material, or when the food is allowed to wilt or become over-ripe, or in the early days of storage of minimally processed foods and unblanched frozen foods. The main breakdown is attributed to lipoxygenase enzymes. Oxidation of unsaturated fatty acids by these enzymes may be accompanied by bleaching (oxidative destruction) of carotenoids [36,37]. In the fresh healthy plant tissues lipoxygenase and carotenoids are in different locations. Only when the tissues are disrupted mechanically or as the tissues break down naturally can the enzyme come into contact with its carotenoid co-substrate. β-Carotene is usually most susceptible; with some leaves, up to 30% may be destroyed in seconds.

ii) Non-enzymic. Exposure to oxygen in the air during drying and processing leads to the generation of peroxides and oxidizing free radicals and can cause serious losses of carotenoids. Conditions of sun-drying in air are particularly damaging. Even under less drastic conditions, however, once the generation of oxidizing species has been initiated, the process can continue to progress during storage, even at freezer temperatures, leading to increasing losses with time.

Oxidative degradation of carotenoids produces apocarotenals. The amount of these detected is small compared with the amount of carotenoid lost. The large amounts of apocarotenoids found in some *Citrus* fruit and juices are likely to be formed by specific, controlled enzymic reactions. During the storage and processing of tomatoes, oxidation of lycopene also produces small amounts of lycopene 1,2-epoxide (**217**) and 5,6-epoxide (**222**), the latter leading to the 2,6-cyclolycopenediols (**168.1**) that may be detected in serum [9,35].

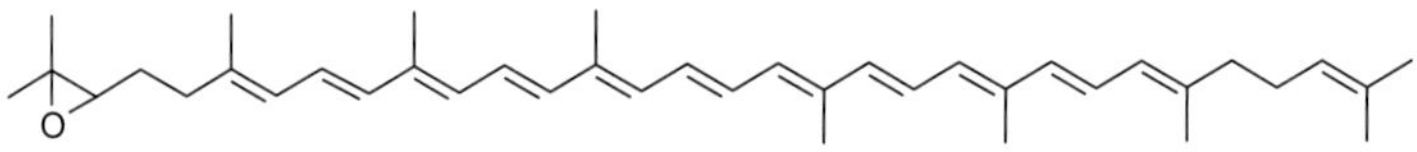

lycopene 1,2-epoxide (**217**)

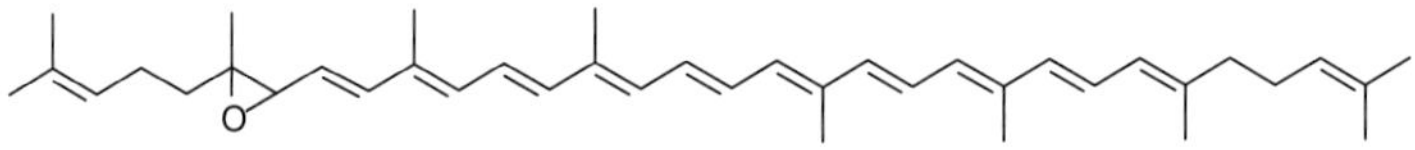

lycopene 5,6-epoxide (**222**)

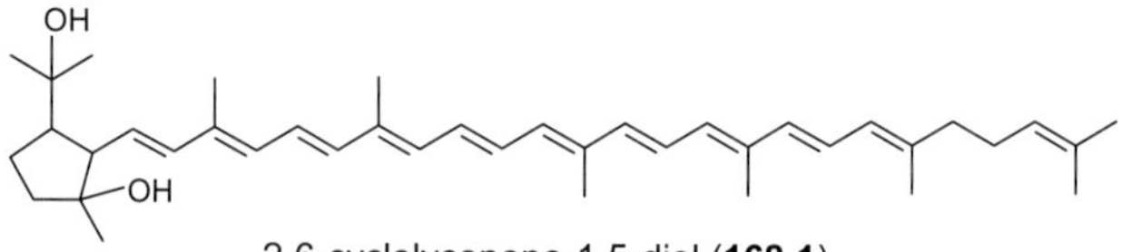

2,6-cyclolycopene-1,5-diol (**168.1**)

b) Geometrical isomerization

Geometrical isomerization is promoted by heat treatment and exposure to light, and may also result from exposure to acids. Increases of up to 40% in the proportion of *Z* isomers have been reported following heat treatment in the canning of several fruits and vegetables. The main *Z* isomer components found depend on the treatment and conditions. Isomerization of the Δ^{13} double bond has the lowest activation energy so the 13*Z* isomer is predicted to form most readily and to predominate if thermodynamic equilibrium has not been reached (*Volume 4, Chapter 3*). This has been shown to be the case in most heat-processed yellow, orange and red fruits and vegetables [38]. If thermodynamic equilibrium has been reached, it is predicted that the 9*Z* isomer should predominate. This has been reported in processed green vegetables, though chlorophyll-sensitized photoisomerization may be a factor in this. Lycopene in tomatoes is largely in a microcrystalline form and is comparatively resistant to isomerization [39]. In heat-processed tomatoes only about 5% of the lycopene is in the form of *Z* isomers, a similar figure to that in the raw fruit. A major product of the geometrical isomerization of lycopene is the 5*Z* isomer, which was overlooked prior to HPLC studies [40].

c) Other changes

i) Rearrangement of 5,6-epoxides. Carotene and xanthophyll 5,6-epoxides are readily transformed into the corresponding 5,8-epoxides, *e.g.* violaxanthin (**259**) into auroxanthin (**267**). The change in absorption spectrum that accompanies this isomerization leads to loss of colour intensity. Carotenoid epoxides are not detected in the human body so these changes are unlikely to have any nutritional consequences.

ii) Dehydration. Carotenoids with an allylic hydroxy group are susceptible to dehydration under acid conditions. A good example of this is lutein (**133**). Its dehydration products, especially anhydrolutein II (2′,3′-didehydro-β,ε-caroten-3-ol, **59.1**), are detected in the HPLC profile of squash [12].

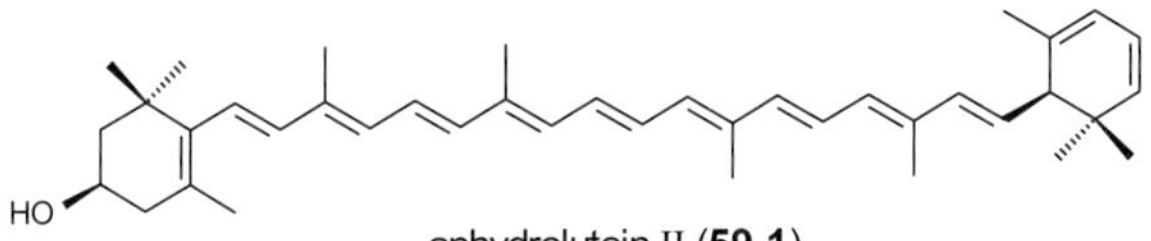

anhydrolutein II (**59.1**)

E. Conclusions and Recommendations

In both rich and poorer countries the aim is the same, namely to identify and increase the availability of the foods and products that give the highest amounts of the desired carotenoids in a form that is used efficiently, and to avoid losses of these carotenoids during storage, drying, processing and cooking.

1. Analytical data

a) HPLC

Modern HPLC analysis (*Chapter 2*) generates precise quantitative data about the sample being analysed. Based on this, extensive tables are available, listing concentrations of β-carotene and other carotenoids in a wide range of fruits, vegetables and other foods, including different varieties and strains. These tables provide a wealth of valuable information but a cautious and realistic approach is needed for the use and evaluation of the numerical data. The precise analytical figures refer to the particular sample that was actually analysed, and that sample was grown, harvested and subsequently stored and processed in a particular way. This history is usually not reported. As explained earlier in this *Chapter*, so many factors can affect carotenoid content and composition, *e.g.* environmental conditions, cultivation practice, method and time of harvest, age and state of maturity of the sample at the time of harvest, length and conditions of storage. There is also great variation in the analytical methods and calculations used. Water content can vary widely and is usually not controlled, leading to considerable uncertainties when carotenoid content is referred to fresh weight. Even figures for different fruits taken from the same plant at the same time and extracted and analysed under identical conditions can show considerable variation. So, too much reliance should not be be placed on the precise figures listed. Presenting a range of values is more realistic. If a table gives a value of 148 μg/g fresh weight for β-carotene in a particular fruit or vegetable, an apparently similar sample of the same variety analysed under the same conditions must not be expected to give the same precise figure. Discrepancies for the same material grown and analysed in different parts of the world may be large. The figures are indicative, however, and in this way very useful. If a figure of 148 μg/g is listed in a table, it is realistic to expect that for any sample the content is likely to be in the range 100-200 μg/g. It is also safe to assume that a fruit reported to contain 148 μg/g should contain in the order of ten times as much carotenoid as a different example reported to contain 15 μg/g. The composition tables when used in a realistic way, therefore provide useful guidelines.

b) Visual assessment

The importance of visual assessment should not be overlooked. The human eye is a sensitive instrument and, when it is used in an informed way, direct observation can give a reliable assessment of colour (hue and intensity) allowing broad judgement of carotenoid composition and content from which possible good sources of carotenoids can be identified.

One can get much guidance simply by observation, before consulting tables or performing analysis. Thus, as a simple but valuable guideline, dark green leaves and vegetables have a higher chloroplast density and hence β-carotene and lutein content than paler green ones. Yellow-orange-red fruits and roots may be coloured by carotenoids, but alternatively the colour may be due to other pigments. This can usually be ascertained by checking the tables for that or a similar species. Carotenoids are not water-soluble; the other pigments, especially anthocyanins, are. This can be tested quickly and easily. As with green leaves, the strongest colour indicates the highest pigment concentration. Observation of the colour/hue is also useful. A yellow-orange colour indicates that a source may contain α-carotene and β-carotene and/or their hydroxy derivatives lutein and zeaxanthin. A red source may contain lycopene. This can be supported by the UV/Vis absorption spectrum of the total extract. Any source that looks promising is then analysed by HPLC (*Chapter 2*).

It is also easy to see if the colour varies between tissues, leaves or at different depths within the tissue, as a good indication of which samples should be analysed in detail.

c) Instrumental

Instrumental evaluation of colour intensity and hue as CIELAB coordinates by spectroradiometry or tristimulus colorimetry can be informative [41]. This determines three colour coordinates, namely L^* (luminosity or lightness), a^* (positive values indicate redness, negative values greenness) and b^* (positive values indicate yellowness, negative values blueness). For the yellow-orange carotenoids, a^* and b^* are both positive. From these measurements, two parameters can be calculated, namely c_{ab} (chroma) and h_{ab} (hue). This rapid method has been validated by correlation with HPLC results [42] and used, for example, to characterize various orange juices [43] and estimate their provitamin A value [44].

2. Some general conclusions

The number of carotenoids for which associations with health and biological activity have been studied is small. Generally, these are the only ones that are included in the food composition tables. The number of carotenoids eaten in a varied diet is much larger. The possibility that, in the future, other carotenoids may become of interest in relation to human health should not be overlooked; data may not be available on their occurrence and content.

In regions where vitamin A deficiency is still a real or potential problem, the requirement is specific: to obtain sufficient β-carotene and other provitamin A carotenoids from whatever sources are available and to minimize destructive effects during storage, processing and cooking. This is discussed in *Chapter 9*.

For tropical regions, local sources may not be covered in the tables. Unusual carotenoid-rich local sources strongly merit further study. Known examples are 'buriti' and 'gac' fruit, but exploration will surely reveal other interesting ones.

References

[1] D. B. Rodriguez-Amaya, *A Guide to Carotenoid Analysis in Foods*, ILSI, Washington DC (1999).

[2] F. Khachik, G. R. Beecher and N. F. Whittaker, *J. Agric. Food Chem.*, **34**, 603 (1986).

[3] F. Khachik, G. R. Beecher, M. B. Goli and W. R. Lusby, *Pure Appl. Chem.*, **63**, 71 (1991).

[4] T. W. Goodwin, *The Biochemistry of the Carotenoids, Vol. 1: Plants*, Chapman and Hall, London (1980).

[5] T. W. Goodwin and L. J. Goad, in *The Biochemistry of Fruits and their Products, Vol. 1* (ed. A. C. Hulme), p. 305, Academic Press, London and New York (1970).

[6] F. Khachik and G. R. Beecher, *J. Agric. Food Chem.*, **35**, 732 (1987).

[7] F. Khachik, G. R. Beecher and W. R. Lusby, *J. Agric. Food Chem.*, **37**, 1465 (1989).

[8] F. Khachik, M. B. Goli, G. R. Beecher, J. Holden, W. R. Lusby, M. D. Tenorio and M. R. Barrera, *J. Agric. Food Chem.*, **40**, 390 (1992).

[9] L. H. Tonucci, J. M. Holden, G. R. Beecher, F. Khachik, C. S. Davis and G. Mulokozi, *J. Agric. Food Chem.*, **43**, 579 (1995).

[10] F. Khachik, G. R. Beecher, M. B. Goli and W. R. Lusby, *Meth. Enzymol.*, **213**, 347 (1992).

[11] F. Khachik and G. R. Beecher, *J. Agric. Food Chem.*, **36**, 929 (1988).

[12] F. Khachik, G. R. Beecher and W. R. Lusby, *J. Agric. Food Chem.*, **36**, 938 (1988).
[13] J. Deli, Z. Matus, P. Molnár, G. Toth, G. Szalontai, A. Steck and H. Pfander, *Chimia*, **48**, 102, (1994).
[14] H. Kläui and J. C. Bauernfeind, in *Carotenoids as Colorants and Vitamin A Precursors* (ed. J. C. Bauernfeind), p. 48, Academic Press, New York (1981).
[15] L. Laferriere and W. H. Gabelman, *Proc. Am. Soc. Hort. Sci.*, **93**, 408 (1968).
[16] A. E. Joyce, *Nature*, **173**, 311 (1954).
[17] F. W. Quackenbush, J. G. Firsch, A. M. Brunson and L. R. House, *Cereal Chem.*, **40**, 250 (1963).
[18] I. Potrykus, *Plant Physiol.*, **125**, 157 (2001).
[19] J. H. Humphries and F. Khachik, *J. Agric. Food Chem.*, **51**, 1322 (2003).
[20] M. H. Dickson, C. Y. Lee and A. E. Blamble, *Hort. Sci.*, **23**, 778 (1988).
[21] A. S. H. Ong and S. H. Goh, *Food Nutr. Bull.*, **23**, 11 (2002).
[22] H. T. Godoy and D. B. Rodriguez-Amaya, *Arq. Biol. Tecnol.*, **38**, 109 (1995).
[23] C. K. Shewmaker, J. A. Sheehy, M. Daley, S. Colburn and D. Y. Ke, *Plant J.*, **20**, 401 (1999).
[24] L. T. Vuong, *Food Nutr. Bull.*, **21**, 173 (2000).
[25] D. B. Rodriguez-Amaya, in *Shelf-life Studies of Foods and Beverages: Chemical, Physical and Nutritional Aspects* (ed. G. Charalambous), p. 591, Elsevier, Amsterdam (1993).
[26] J. Gross, *Pigments in Fruits*, Academic Press, London (1987).
[27] J. Gross, *Pigments in Vegetables: Chlorophylls and Carotenoids*, Avi:Van Nostrand Reinhold, New York (1991).
[28] K. L. Simpsom and S. C. S. Tsou, in *Vitamin A Deficiency and its Control* (ed. J. C. Bauernfeind), p. 461, Academic Press, Orlando (1986).
[29] A. Sommer and K. P. West Jr., *Vitamin A Deficiency. Health, Survival and Vision*, Chapter 13, Oxford University Press, New York and Oxford (1996).
[30] http://www.nal.usda.gov/fnic/foodcomp/Data/SR18/nutrlist/sr18w338.pdf
[31] http://www.ars.usda.gov/Services/docs.htm?docid=9673
[32] D. B. Rodriguez-Amaya, *Sight and Life Newsletter 3/2002*, 25 (2002).
[33] D. B. Rodriguez-Amaya, *Carotenoids and Food Preparation: The Retention of Provitamin A Carotenoids in Prepared, Processed and Stored Foods*, OMNI, Arlington (1997).
[34] F. Khachik, G. R. Beecher and M. B. Goli, *Anal. Chem.*, **64**, 2111 (1992).
[35] F. Khachik, C. J. Spangler, J. C. Smith Jr., L. M. Canfield, A. Steck and H. Pfander, *Anal. Chem.*, **69**, 1873 (1997).
[36] S. Aziz, Z. Wu and D. S. Robinson, *Food Chem.*, **64**, 227 (1999).
[37] Y. Wache, A. Bosser-DeRatuld, J.-C. Lhuguenot and J.-M. Belin, *J. Agric. Food Chem.*, **51**, 1984 (2003).
[38] W. J. Lessin, G. L. Catigani and S. J. Schwartz, *J. Agric. Food Chem.*, **45**, 3728 (1997).
[39] G. Britton, L. Gambelli, P. Dunphy, P. Pudney and M. Gidley, in *Functionalities of Pigments in Food* (ed. J. A. Empis), p. 151, Sociedade Portuguesa de Quimica, Lisbon (2002).
[40] S. J. Schwartz, in *Pigments in Food: A Challenge to Life Science* (ed. R. Carle, A. Schieber and F. S. Stintzing), p. 114, Shaker Verlag, Aachen (2006).
[41] F. J. Francis and F. M. Clydesdale, *Food Colorimetry: Theory and Applications*, AVI Publ. Co., Westport, CT (1975).
[42] A. J. Melendez-Martinez, G. Britton, I. M. Vicario and F. J. Heredia, *Food Chem.*, **101**, 1145 (2007).
[43] A. J. Melendez-Martinez, I. M. Vicario and F. J. Heredia, *J. Sci. Food Agric.*, **85**, 894 (2005).
[44] A. J. Melendez-Martinez, I. M. Vicario and F. J. Heredia, *J. Agric. Food Chem.*, **55**, 2808 (2007).

Carotenoids
Volume 5: Nutrition and Health

Chapter 4

Supplements

Alan Mortensen

A. Introduction

Dietary supplements are used either to increase the intake of dietary nutrients, such as vitamins, or to provide nutrients that are not usually found in foods, *e.g.* in the form of herbal extracts. Supplements in the form of vitamin pills have been known for decades. Supplements of non-essential nutrients are also widely available. The use of dietary supplements has become so widely accepted that they can be found in supermarkets alongside basic food items. Carotenoids are ubiquitous in a diet rich in fruits and vegetables. Thus, supplements containing carotenoids are intended either to boost carotenoid intake in individuals already well supplied with carotenoids from their diet, or to provide carotenoids to those whose diet contains only low amounts of them.

1. Market

Estimating the size of the global market for carotenoid supplements is not as straightforward as it may seem, because the same carotenoids are often used as food colourants. Also, it is difficult to obtain detailed information about carotenoid sales. In 1999, the total carotenoid market was estimated to be worth US$748 million [1] (excluding paprika), of which supplements accounted for 15.3%. As shown in Table 1, in 2004 the value of the total carotenoid market had increased by almost US$140 million [2], the supplements now accounting for 29.1%. As also shown in Table 1, the carotenoid market was predicted to increase to US$1,023 million by 2009, and supplements to constitute 31.7% of that market [2]. The latest figures show that the carotenoid market in 2007 was worth US$766 million, *i.e.*

with no increase over the 2004 figure, and is expected to increase to US$919 million by 2015 [3]. The reason for this lower value of the market than previously predicted is that increased competition has led to lower prices [3].

Table 1. Global market for carotenoids in 2004 and estimated global market for 2009 (US$ Million) [2].

	Supplements		Food		Cosmetics		Feed		Total	
	2004	**2009**	**2004**	**2009**	**2004**	**2009**	**2004**	**2009**	**2004**	**2009**
β-Carotene (**3**)	125.0	128.0	98.0	103.0	6.0	7.0	13.0	15.0	242.0	253.0
Astaxanthin (**404-6**)	3.5	4.8	0	0	0	0	230.5	252.2	234.0	257.0
Canthaxanthin (**380**)	3.0	3.0	7.0	8.0	0	0	138.0	145.0	148.0	156.0
Lutein (**133**)	54.0	85.0	18.0	20.0	0	0	67.0	82.0	139.0	187.0
Lycopene (**31**)	50.5	68.7	3.5	9.3	0	3.0	0	0	54.0	81.0
Annatto[1]	0	0	33.6	39.0	0	0	0	0	33.6	39.0
Zeaxanthin (**119**)	22.0	35.0	0	0	0	0	0	0	22.0	35.0
Apo-carotenal[2]	0	0	0	0	0	0	8.7	9.1	8.7	9.1
Apo-carotenoate[3]	0	0	0.4	0.5	0	0	5.2	5.4	5.6	5.9
Total	258.0	324.5	160.5	179.8	6.0	10.0	462.4	508.7	886.9	1023.0

[1] Main carotenoid bixin (**533**)

[2] 8′-Apo-β-caroten-8′-al (**482**)

[3] 8′-Apo-β-caroten-8′-oic acid (**486**) ethyl ester

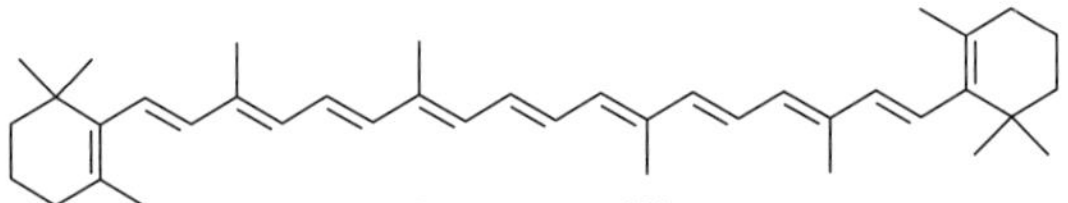

β-carotene (**3**)

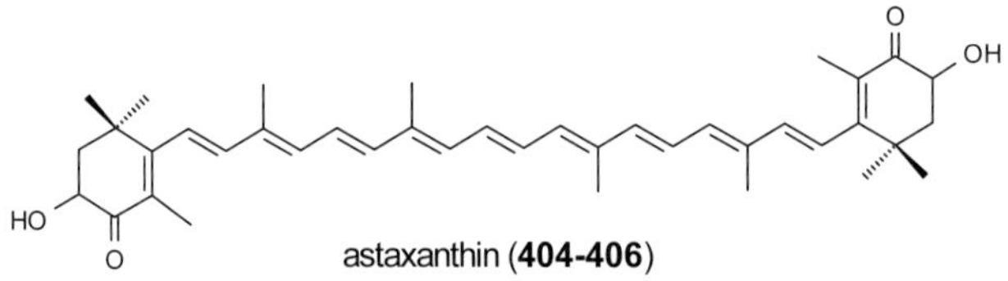

astaxanthin (**404-406**)

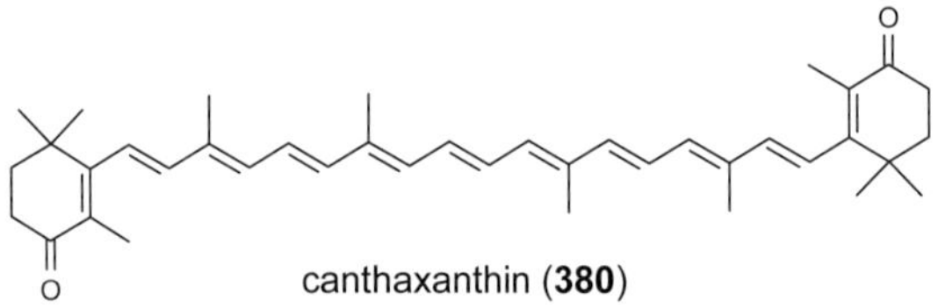

canthaxanthin (**380**)

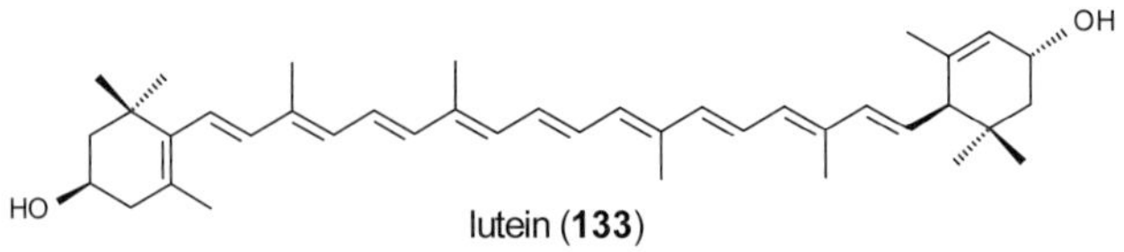

lutein (**133**)

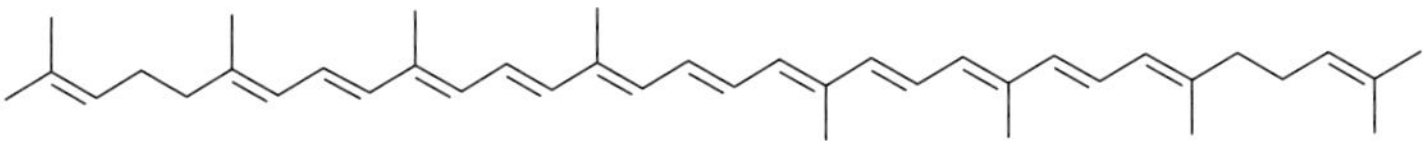

lycopene (**31**)

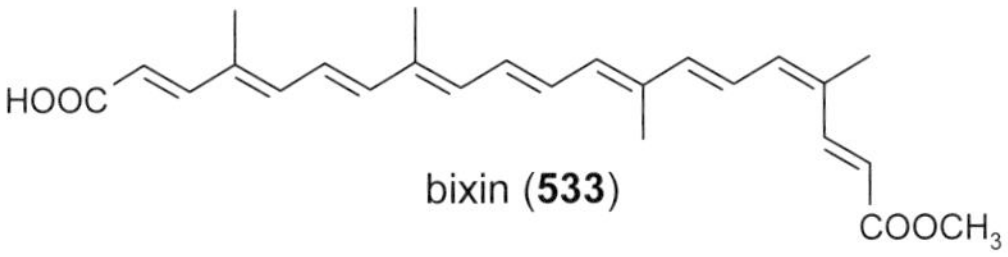

bixin (**533**)

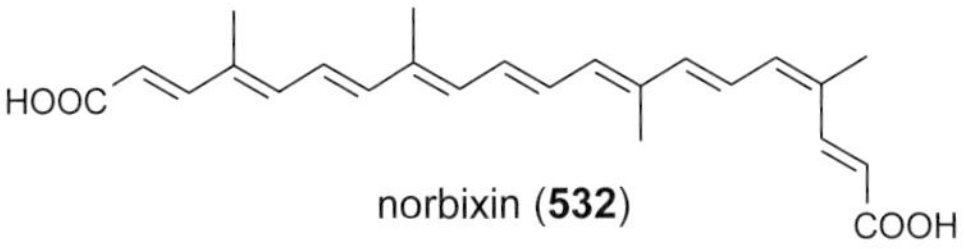

norbixin (**532**)

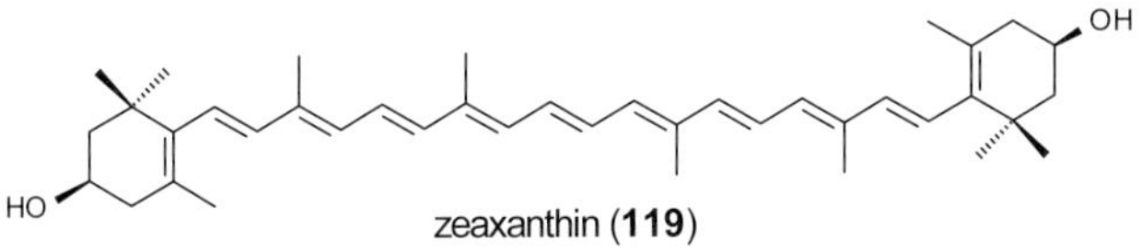

zeaxanthin (**119**)

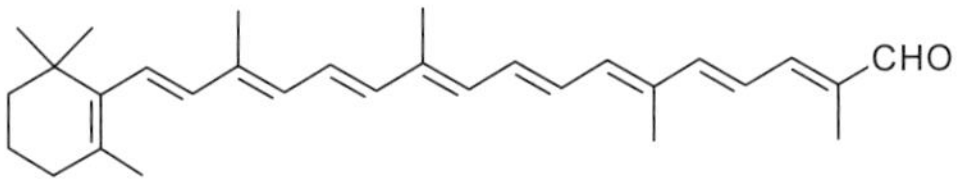

8'-apo-β-caroten-8'-al (**482**)

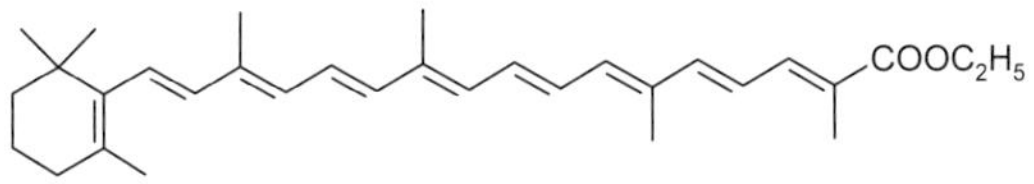

8'-apo-β-caroten-8'-oic acid (**486**) ethyl ester

It is not only predictions of the value of the total carotenoid market that should be regarded with caution, but also those for individual carotenoids. The prediction in the first report [1] for total sale of carotenoids in 2005 was fairly close to the actual value in 2004 given in the second report [2], but the predictions for particular markets and individual carotenoids were not accurate. This can be explained partly by increased competition in the food colourant market (leading to lower prices) and partly by the expanding demand for supplements.

It is mainly lutein (**133**) and zeaxanthin (**119**) that are predicted to drive the expected growth of the carotenoid supplements market between 2004 and 2009. β-Carotene (**3**), mainly synthetic, but also natural, will continue to be the best-selling carotenoid, though its share of the market is predicted to decrease. Lycopene (**31**) will also see a significant increase. These are currently the four most important carotenoids used in supplements. Astaxanthin (**404-406**) and canthaxanthin (**380**) have a large share of the total carotenoid market due to their use in feed, but have found little use in supplements. Astaxanthin supplements are now being promoted strongly, however, especially in Japan.

Considering the increasing publicity given to the link between diet and disease, it is no surprise that a large increase in the carotenoid supplements market is forecast. Furthermore, the number of older people in the Western world is increasing and carotenoids are often sold as possible preventive agents for conditions such as age-related macular degeneration and cancer that affect mainly the elderly.

2. Legal

Dietary supplements are subject to food legislation, although they are not perceived as foods in a traditional sense, given that their function is not to provide energy or otherwise form a basic part of the normal diet. Neither are dietary supplements medicines, as they are not intended for treatment of illnesses. Rather they are, as the name implies, means to supplement the diet with micronutrients, in the same way as vitamins, minerals *etc.* In the European Union (E.U.), dietary supplements (called food supplements) are defined as "concentrated sources of nutrients or other substances with a nutritional or physiological effect, alone or in combination, marketed in dose form" [4]. In the U.S., dietary supplements are covered by the Dietary Supplement Health and Education Act of 1994 (DSHEA) [5] and are defined as "a product containing one or more of the following dietary ingredients: a vitamin, a mineral, a herb or other botanical, an amino acid, or a dietary substance for use by man to supplement the diet by increasing the total dietary intake" [5]. The U.S. legislation does not require that the supplement is in dose form, but if it is not in dose form, it should not be represented as conventional food.

It is clear that carotenoids fulfil both the European and the American definitions of dietary supplements. The E.U. legislation [4] currently only covers vitamins and minerals, meaning that dietary supplements containing nutrients other than vitamins and minerals, *e.g.* most carotenoids, are regulated by national legislation. β-Carotene, however, as a source of vitamin A, is covered by the E.U. Directive. It is the intention to amend the existing legislation to cover all nutrients, so that carotenoids other than β-carotene should also become subject to E.U.-wide regulations [4].

Fortified foods, including energy bars and sports or energy drinks, occupy a position between traditional foods and dietary supplements, as they may contain higher levels of nutrients than traditional food, but not the concentrated levels as in dietary supplements. Besides, these categories of products also provide energy, in contrast to dietary supplements which do not provide any appreciable amount of energy. The use of carotenoids for fortification of foods is subject to national legislation, meaning that carotenoids may be approved for fortification in some countries but not in others. Finally, carotenoids can be added to foods as colourants [6], subject to the legislation governing their use, *i.e.* which foods may be coloured, which colourants are allowed and at what level. In this case, carotenoids should be labelled as colourants and not as nutrients.

Carotenoids "which have not hitherto been used for human consumption to a significant degree within the Community" or new processes for manufacturing existing carotenoids, require approval in the E.U. under the Novel Foods Regulation [7]. Under this scheme, the safety of the carotenoid is to be assessed and the potential intake evaluated. Up to July 2008, six applications had been made [8] concerning the use of carotenoids in dietary supplements and other foods, namely for lycopene from *Blakeslea trispora* (two), synthetic lycopene, tomato oleoresin (two) and synthetic zeaxanthin. Only one of the applications (lycopene from *Blakeslea trispora*) had been approved [9]; the rest are still being evaluated.

An important aspect of the legal issues is how the dietary supplements are marketed. One way of promoting a product is through claims. In the E.U., the use of health claims was harmonized in 2006 [10]. This regulation covers the following nutrients: protein, carbohydrate, fat, fibre, sodium, vitamins and minerals, and thus does not cover carotenoids, except for β-carotene which is a recognized source of vitamin A. Thus, health claims regarding carotenoids in dietary supplements are still regulated by national authorities. Some countries are very strict about which claims are allowed, whereas others allow a wide variety of claims. In general, claims that a product may prevent, treat or cure an illness are not allowed, since dietary supplements are regarded as foods and not as medicine. Thus, claims such as "lutein prevents age-related macular degeneration" are not allowed. Claims should also be truthful and not misleading. In the U.S., three types of claims are allowed: health claims, nutrient content claims and structure/function claims. Nutrient content claims are claims about the content of the active ingredient(s), either in the form of quantitative figures or giving qualitative information, *e.g.* "high in ..." or "low in ...". Structure/function claims are claims that describe the effect of the product on (parts of) the body, *e.g.* "lutein helps maintain healthy eyesight"; mention of a specific disease is not allowed. Health claims link a dietary ingredient with a reduced risk of disease. In contrast to structure/function claims, health claims must be approved by the U.S. Food and Drug Administration (FDA). Up to now, FDA has not allowed health claims for any carotenoid, and has in fact rejected health claims linking lycopene with reduction of cancer risk [11] and lutein esters with reducing susceptibility to age-related macular degeneration and cataract formation [12,13].

B. Carotenoids in Supplements

1. Which carotenoids?

As can be seen from Table 1, the four most important carotenoids marketed as dietary supplements are β-carotene, lutein, lycopene and zeaxanthin. β-Carotene, lutein and lycopene are also the most abundant carotenoids in human serum [14] and, therefore, the ones most extensively researched, and for which most information is available about possible beneficial health effects. Astaxanthin, canthaxanthin and paprika are only of minor importance in relation to dietary supplements, though paprika contains both β-carotene and zeaxanthin in addition to its main characteristic carotenoids, capsanthin (**335**) and capsorubin (**413**).

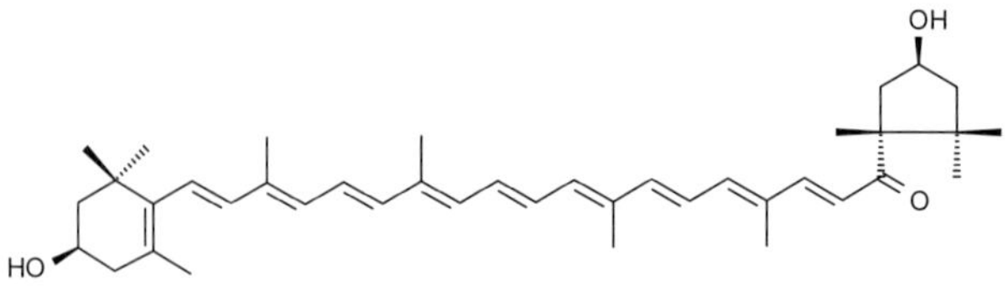

capsanthin (**335**)

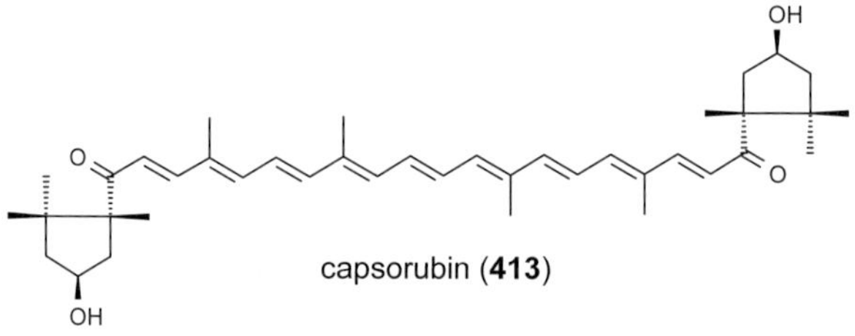

capsorubin (**413**)

Carotenoids used in supplements may be either of synthetic or natural origin. Thus, most of the β-carotene used in supplements is synthetic, but natural sources such as an alga (*Dunaliella salina*), a fungus (*Blakeslea trispora*), red palm oil (*Elaeis guineensis*) or carrot oil (*Daucus carota*) are also used. Lycopene is also either of synthetic or natural origin [from tomatoes (*Lycopersicon esculentum*) or *Blakeslea trispora*]. Lutein is extracted from marigold flowers (*Tagetes erecta*), whereas zeaxanthin is primarily made synthetically but may also be extracted from Chinese wolfberries (*Lycium barbarum* or *L. chinense*). Synthetic carotenoids are highly pure compounds that do not contain other carotenoids. In contrast, carotenoids extracted from natural sources often contain one major carotenoid, some closely related carotenoids, and some biosynthetic precursors. Thus, a tomato extract contains β-carotene, phytoene (**44**) and phytofluene (**42**) and others, alongside the major carotenoid lycopene [15]. Lutein from *Tagetes erecta* contains a few percent zeaxanthin [16]. Among the products with β-carotene as the major carotenoid, *Dunaliella salina* extracts contain some α-carotene (**7**) and a small amount of xanthophylls [17,18], extracts of *Blakeslea trispora* contain a little mutatochrome (**239**), β-zeacarotene (**13**) and γ-carotene (**12**) [19], and palm oil extracts contain a large amount of α-carotene (around 40%) together with smaller amounts of biosynthetic precursors [20]. Carotenoids extracted from natural sources often also contain other lipid-soluble material, including sterols, triacylglycerols, tocopherols *etc.*

Synthetic all-*E* carotenoids contain only a small amount of *Z* isomers. The *Z* isomer content of carotenoids in natural extracts is often substantial, however, *e.g.* β-carotene from *Dunaliella* may contain up to about 40% of the 9*Z* isomer (*Chapter 5*).

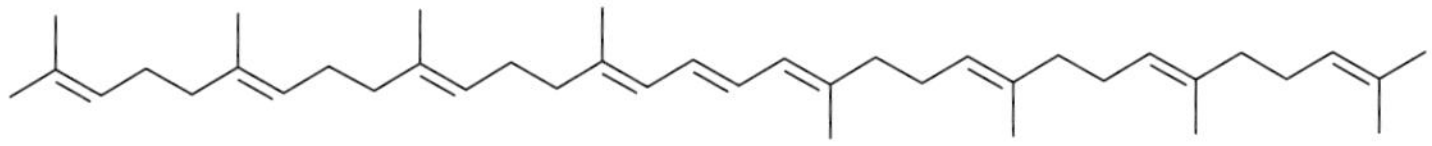

phytoene (**44**)

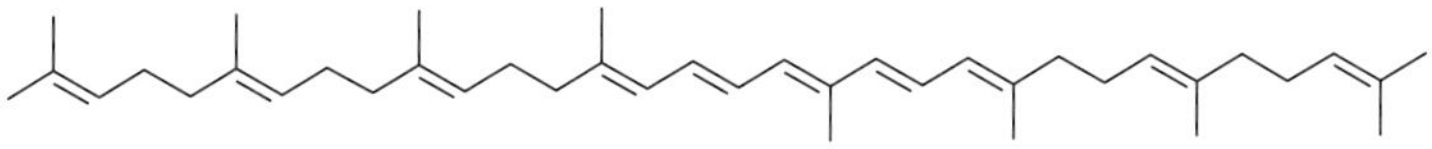

phytofluene (**42**)

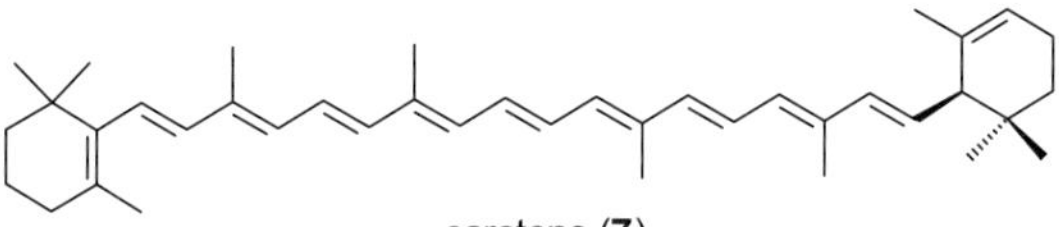

α-carotene (**7**)

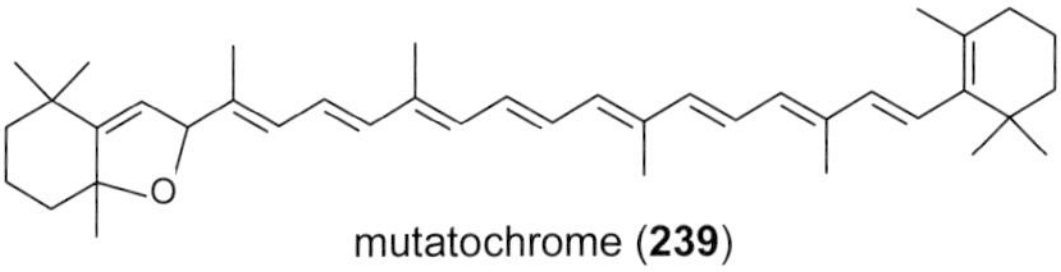

mutatochrome (**239**)

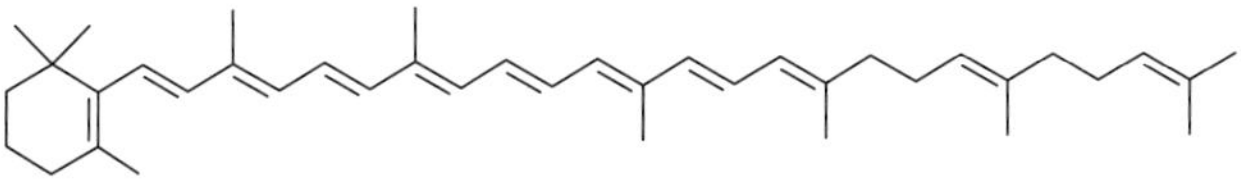

β-zeacarotene (**13**)

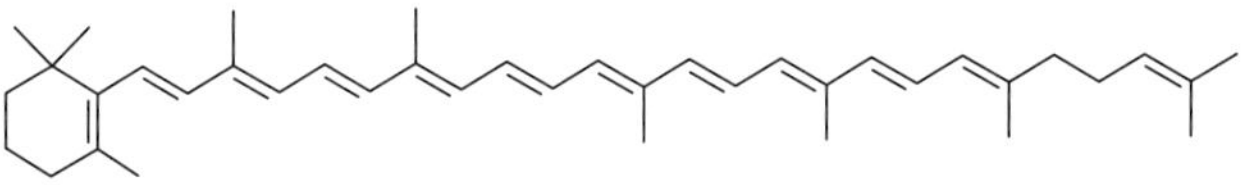

γ-carotene (**12**)

2. Formulations

Dietary supplements are available in many different forms and shapes, *e.g.* liquids, tablets, sachets, and capsules. These require different formulations.

a) Oil suspensions and oleoresins

The most basic formulation is an oil suspension or oleoresin. Micronized synthetic carotenoid crystals are usually suspended in vegetable oil as this increases the stability of the carotenoid compared to the pure crystalline material. The concentration of carotenoids is typically 20-30%, and the product is a viscous, yet still pourable liquid. When carotenoids are extracted from a natural source and the solvent is evaporated, the residue is called an oleoresin. The oleoresin contains the carotenoid(s) together with other oil-soluble material like triacyl-glycerols, sterols, wax *etc*. Often, wax and gums are removed because they would increase the viscosity and decrease the stability of emulsions. The carotenoid concentrations of oleoresins are typically lower than those in oil suspensions: tomato oleoresin commonly contains 6-15% lycopene, and oleoresin from *Tagetes erecta* contains 10-25% lutein. Oleoresins are also more viscous than oil suspensions. Thus, a 6% tomato oleoresin is much more viscous than a 20% oil suspension of synthetic lycopene; a 20-25% lutein extract from marigold is solid and a 10% oleoresin is a non-pourable sticky material. Natural carotenoids may also be manufactured as oil suspensions. Palm oil carotenes and carotenoids from *Dunaliella salina* and *Blakeslea trispora* are traded as 20-30% suspensions in vegetable oil. Lutein is present in marigolds as acyl esters, and this is the form found in oleoresins. The esters may also be hydrolysed and the free lutein suspended in vegetable oil to give a less viscous product than the oleoresin.

Oil suspensions and oleoresins may be used in softgel capsules, which consist of the active material covered by a soft gelatin shell. This is the most commonly used form of carotenoid supplements. There are also examples of carotenoid oil suspensions and oleoresins sold as hardgel capsules (a two-piece capsule made of a harder, yet still flexible gelatin) though hardgel capsules are traditionally used for powders. Capsules made from plant-based carbo-hydrates (often cellulose or starch), and therefore suitable for vegetarians, are also used for carotenoid supplements.

b) Water-miscible formulations

The carotenoid oil suspension or oleoresin is used as raw material for making various formulations. Pure crystalline carotenoid may also be used to achieve a higher concentration. Carotenoids can be made water-dispersible as emulsions. Two types of emulsifiers are commonly used: polymeric hydrocolloids and fatty acid esters. The fatty acid esters are made with polar alcohols, *e.g.* polyoxyethylene sorbitan (polysorbate). Among hydrocolloids, most commonly gum arabic and gelatin are used as emulsifiers. There are only a few examples of liquid carotenoid dietary supplements, and the emulsions find greater use as colourants and in fortification of beverages. However, emulsions of this type form the starting point for making water-dispersible solid formulations of carotenoids used in dietary supplements.

A carotenoid emulsion made with hydrocolloid may be spray-dried, either on its own or with a carrier (typically maltodextrin), to give a powder. This powder is water-dispersible and may be used in the same applications as the emulsion itself. However, the typical usage of carotenoid powders in dietary supplements is in the form of sachets or hardgel capsules; these may also be filled with granules, *i.e.* particles larger than powder particles. These powders are not suitable for making tablets, as the high pressures used in this process lead to leaking of the carotenoid which, therefore, becomes more susceptible to degradation by light and oxygen.

The preferred choice of carotenoid formulation for making tablets is beadlets, small spherical particles consisting of the carotenoid encapsulated in a gelatin-sucrose matrix. Beadlets are made by spray-cooling the 'solution' of carotenoid in oil, gelatin, sucrose and water, and covering the particles with starch to prevent them from sticking together. Synthetic carotenoids may be made into beadlets, with or without an oil phase, by dissolving the carotenoid in an alcohol or ketone solvent, at elevated temperature, and then precipitating the carotenoid in the presence of gelatin [21]. Finally, the beadlets are dried by spray-drying or in a fluidized bed. Beadlets can withstand high pressures and thus are ideal for tablet-making. They can also be used in hardgel capsules. Another advantage of beadlets is that the gelatin-sucrose matrix forms a barrier against oxygen, thus conferring high stability on the encapsulated carotenoid. Recently, vegetarian beadlets without gelatin have been introduced by several companies. Instead of gelatin, different types of plant-based carbohydrates may be used. The carbohydrate most often used is alginate; there are other possibilities but few details have been divulged.

Formulation of a carotenoid as described above may alter the isomeric composition. Thus, synthetic β-carotene suspended in oil is almost exclusively (all-*E*)-β-carotene, but heat treatment, *e.g.* during emulsion making, may cause formation of *Z* isomers (*cis* isomers).

Some dietary supplements containing carotenoids may be sold as a single carotenoid, *e.g.* lutein or lycopene, but more often the supplements contain a mixture of carotenoids or of carotenoids together with other active ingredients, *e.g.* as part of a multi-vitamin and multi-mineral preparation. The typical carotenoid content of dietary supplements ranges from a few mg up to around 20 mg per dose. Provitamin A supplements typically contain 10,000 IU (6 mg) or 25,000 IU (15 mg) β-carotene. This range of dosing is similar to the daily intake obtained through the diet.

3. Analysis

A manufacturer of dietary supplements must be able to substantiate any claim made for nutrition content. National authorities may not have the resources to analyse the contents of dietary supplements, so the consumer is left with the information provided by the manufacturer, without an official approval or analysis provided by an independent laboratory. The actual content of α-carotene, β-carotene, lutein and/or zeaxanthin in commercially available dietary supplements has been examined in a number of studies [22-24], which all showed that the actual content of carotenoids may be a long way from what is claimed.

In the U.S., the FDA has awarded a contract to validate methods for analysis of selected ingredients in dietary supplements. An RP-HPLC method for analysis of β-carotene in dietary supplements has been adopted as First Action Official Method (2005.07) [25,26]. The method employs a C_{18} column or, for products with high amounts of α-carotene, *i.e.* extracted from carrots or palm fruits, a C_{30} column. Official Methods for lutein and lycopene are under development.

C. Health Issues

1. Selling points

In order for a dietary supplement to be attractive to the customer, the active ingredient(s) must be associated by the consumer with a beneficial physiological or psychological effect. Thus, the efficacy of dietary supplements is often substantiated or supported by some form of documentation, be it scientific studies, anecdotes or otherwise. Scientific evidence of an effect of a particular compound, extract or herb is, of course, a strong selling point, but this requires costly clinical intervention studies. Epidemiological studies of dietary habits (see *Chapter 10*) have shown that a diet with plenty of carotenoid-rich vegetables and fruits carries a positive effect on health. Such studies can, at best, only provide an indication of which dietary components, *e.g.* carotenoids, may be beneficial. Based on these epidemiological studies, recommendations of national health and food authorities to eat five or six portions a day of fruits and vegetables have been used to market carotenoid supplements to those who do not meet these recommendations ('bridging the gap' by supplementing to reach or exceed a level of intake that would be achieved by eating five or six portions of fruit and vegetables every day).

One of the key attributes of β-carotene is its provitamin A activity (*Chapters 8* and *9*). It is a well-established fact that β-carotene and a few other carotenoids, *e.g.* α-carotene and β-cryptoxanthin (**55**), have vitamin A activity. Labels of dietary supplements containing β-carotene will often state the vitamin A content in International Units [IU; equivalent to 0.3 μg of (all-*E*)-retinol]. Carotenoids in dietary supplements are often sold under the heading 'antioxidants'. Carotenoids have been shown to be antioxidants *in vitro*, but in some cases they show pro-oxidant activity. Their role in the complex redox process in the human body has not been established (*Chapter 12*).

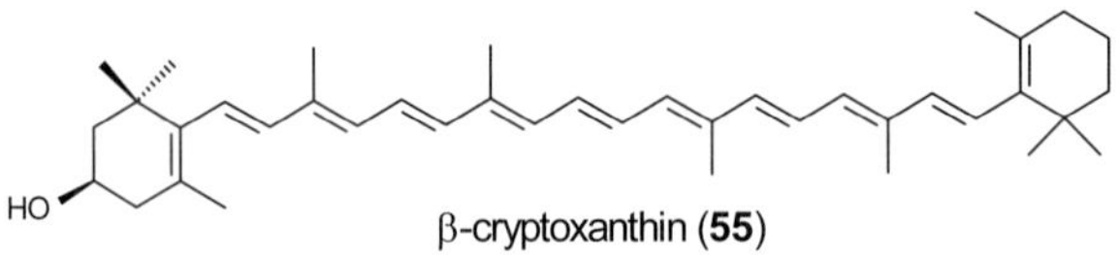

β-cryptoxanthin (**55**)

Carotenoids are also sold as tanning agents, though the use of canthaxanthin for this purpose in no longer permitted. It is well known that high intake of carotenoids may lead to their accumulation in the skin, causing yellow colouration, a reversible phenomenon known as carotenodermia. At the same time, the accumulated carotenoids provide some protection against sunburn (*Chapter 16*), though the effect is not comparable to that of traditional sunscreens (most carotenoids show only weak UV absorption).

Probably the most promising areas for carotenoid supplements are protection against cancer (*Chapter 13*) and eye disease, particularly age-related macular degeneration and cataract formation (*Chapter 15*). This may seem odd, as dietary supplements may not be sold as remedies to prevent, treat or cure an illness (see above), but many scientific studies of carotenoids in disease prevention are currently focused on these two conditions and are attracting much publicity.

An important feature of supplements in general is that the active principles should be naturally occurring substances. In the case of carotenoids, this is clearly the case, and there is less concern over whether the supplement itself is derived from a natural source or provided as a pure, synthetic, 'nature-identical' compound. The carotenoid most used in dietary supplements is synthetic β-carotene.

2. Bioavailability

In order for a carotenoid to exert its function on a part of the human body, the carotenoid must first be absorbed through the gastro-intestinal tract. An important aspect of carotenoid supplements is thus that the carotenoid should be in a form that may readily be absorbed. The topic of bioavailability is discussed in detail in *Chapter 7.*

It has been found that some fat is needed to facilitate absorption of carotenoids. Since the dietary sources of carotenoids (fruits and vegetables) are often low in fat, and the food matrix, in which the carotenoid is incorporated, may be difficult to digest, it is no surprise that the bioavailability of carotenoids from unprocessed foods is often low. In dietary supplements, the carotenoids are freed from their matrix and some oil is present, so a much higher bioavailability of carotenoids from supplements can be expected.

An important question concerning supplements is: Which carotenoid formulation gives the highest bioavailability? A study has shown that β-carotene from beadlets may have up to 50% greater bioavailability than that from softgel capsules containing an oil suspension [27]. It was speculated that the beadlet, being water-miscible, improved the incorporation of β-carotene into mixed micelles [27]. It could be that other components of the beadlet aid in the uptake of β-carotene. However, it could also simply be a matter of size. The size of the carotenoid crystals in beadlets is of the order of 0.1 μm, whereas carotenoid crystals in oil suspensions/oleoresins are much larger, of the order of 1-5 μm (after grinding) [21]. Thus, the much smaller carotenoid particles in beadlets can be expected to dissolve more readily in the fat typically taken together with the supplement; it has been shown that small carotenoid crystals (0.16 μm) had higher bioavailability than larger crystals (0.55 μm) when both were formulated as beadlets [21]. Therefore, even if sufficient fat is present to solubilize the carotenoid, there may not be enough time to dissolve the larger crystals from the oil suspension/oleoresin in the gastro-intestinal tract. In another study, the bioavailability of lycopene was better from softgel capsules containing tomato oleoresin than from ones containing synthetic lycopene beadlets or a spray-dried powder containing lycopene from tomatoes [28]. The supplements were taken after breakfast, but no details of the fat content of the breakfast were given; it may be that the amount of fat was a limiting factor for uptake, and the small amount of lipid in the oleoresin led to the higher bioavailability. Inclusion of unspecified surface-active agents decreased the bioavailability of lycopene [28]. In another study, though, polysorbate 80, a surfactant, was reported to increase the bioavailability of astaxanthin [29]. The difference between the two studies was that the lycopene was in an oleoresin and astaxanthin was in the form of an algal meal, and the polysorbate may have enhanced the release of astaxanthin from this algal meal.

A number of factors influence the bioavailability of carotenoids; some of these may be of relevance to the formulation of dietary supplements. Medium-chain triacylglycerols may lead to lower absorption of β-carotene than long-chain ones [30]. Vitamins C and E may enhance the bioavailability of lutein, though large variations between individuals were found, so the apparent increase seen was not statistically significant [31]. Vitamin C did, however, cause faster absorption of lutein, [31]. Lysophosphatidylcholine and phosphatidylcholine may increase the bioavailability of carotenoids from mixed micelle formulations [32].

Lutein diesters as a powder (no details on the formulation of the powder were given) were shown to have greater bioavailability than free lutein as an oil suspension [33]. This was probably due to the different formulations, since free lutein completely dissolved in oil had higher bioavailability than either of these products [33]. An effect of esterification on bioavailability was demonstrated, however, in another study, which showed that zeaxanthin dipalmitate had higher bioavailability than free zeaxanthin, when both were completely dissolved in oil [34]. In the case of β-cryptoxanthin though, the ester and the free form had equal bioavailability [35].

A question often raised is whether synthetic or natural carotenoids have better bioavailability. According to all chemical and physical principles, the two should be identical. The argument may be confused and complicated because it does not take into account other factors such as formulation, or whether the carotenoid is esterified. There are reports that esterification of xanthophylls may improve the bioavailability [34]. The demonstration that natural zeaxanthin (**119**) diester, from paprika, for instance, may have better bioavailability than synthetic zeaxanthin is not a direct comparison of natural *versus* synthetic but is also comparing ester and free. Second, one carotenoid may influence the absorption of other carotenoids. Natural carotenoid preparations typically contain a mixture of carotenoids (including *Z* isomers), in contrast to the single carotenoid found in synthetic preparations. Finally, plants may contain components that may either enhance, *e.g.* phospholipids [32], or lower, *e.g.* plant sterols and stanols [36], the bioavailability of carotenoids, and these components may be present in the oleoresins.

There are two studies that have directly compared the bioavailability of synthetic *versus* natural carotenoids, formulated in the same way, so as to avoid any effects of different matrices. In one study, the bioavailability of palm oil carotenes and synthetic carotenoids, both as 30% oil suspensions, was found to be equivalent [37]. Thus, α-carotene or other components of the purified palm oil carotenes did not diminish the uptake of β-carotene. In another study, beadlets containing either synthetic lycopene or tomato oleoresin provided the same increase in lycopene serum levels [38].

3. Recommendations

Carotenoids are not classified as essential nutrients, so values for recommended daily intake have not been established. β-Carotene and other provitamin A carotenoids are an important source of vitamin A, but there are no recommendations about how much of the daily requirement for vitamin A should come from retinol and how much from carotenoids.

In 2000, the Scientific Committee on Food in the E.U. found that there was insufficent scientific evidence to establish a tolerable upper intake level of β-carotene [39]. However, this committee did withdraw the Group Acceptable Daily Intake (ADI) of 0-5 mg/kg body weight for β-carotene, mixed carotenes, 8'-apo-β-caroten-8'-al (**482**) and 8'-apo-β-caroten-8'-oic acid (**486**) ethyl ester, but did not establish new ADIs for these carotenoids [40]. Also, the Institute of Medicine in the U.S. did not set a tolerable upper intake level for β-carotene or total carotenoids [14]. In 2003, the Expert Group on Vitamins and Minerals in the U.K. established a safe upper level of 7 mg for daily intake of β-carotene from dietary supplements [41].

Following the publication of the results of the much cited Alpha-Tocopherol, Beta-Carotene (ATBC) Cancer Prevention Trial [42], various governmental and independent organizations have examined the available evidence linking carotenoids with disease prevention and have begun to make recommendations for the intake of carotenoids in supplement form. The International Agency for Research on Cancer (IARC) in 1998 concluded [43] that there is:

(i) evidence suggesting a lack of cancer-preventive activity in humans for β-carotene when it is used as a supplement at high doses,

(ii) inadequate evidence with regard to the cancer-preventive activity of β-carotene at the usual dietary levels,

(iii) inadequate evidence with respect to the possible cancer-preventive activity of other individual carotenoids,

and declared that "supplemental β-carotene, canthaxanthin, α-carotene, lutein, and lycopene should not be recommended for cancer prevention in the general population". Based on the totality of evidence, the Institute of Medicine concluded that "β-carotene supplements are not advisable for the general population" [14], but could be used in populations with inadequate vitamin A intake or patients suffering from erythropoietic protoporphyria. The American Heart Association found that "the scientific data do not justify the use of antioxidant vitamin supplements for CVD (cardiovascular disease) risk reduction" [44] (antioxidant vitamins in this context are vitamin C, vitamin E and β-carotene). The strongest assertion comes from the U.S. Preventive Services Task Force, which specifically "recommends against the use of β-carotene supplements, either alone or in combination, for the prevention of cancer or cardiovascular disease" [45], because "β-carotene supplements are unlikely to provide important benefits and might cause harm in some groups" (*i.e.* smokers).

It should be stressed that these recommendations follow from the results of clinical trials in which the subjects were given pharmacological doses of β-carotene (around ten times more than the average dietary consumption), and the bioavailability of the supplemental β-carotene given was greater than that of β-carotene from fruits and vegetables.

The scientific studies have focused on β-carotene, lutein and lycopene because these are the most abundant carotenoids in the diet and in the body. About 750 carotenoids are listed in the *Carotenoids Handbook*, but only β-carotene has been studied to an extent that recommendations can be made. The pertinent questions are, therefore, not simply whether supplementation with carotenoids should be encouraged, but also whether supplements of those carotenoids already abundant in the diet, *e.g.* lutein and lycopene, should be used or whether some of the natural carotenoids that are less abundant or not present in a normal diet may improve health, so that their use as supplements could be advantageous.

Considering that medical associations and national authorities advise against the general use of dietary supplements containing β-carotene and that the FDA has denied the use of health claims linking lycopene with protection against cancer or lutein esters with reduction in age-related macular degeneration and cataract formation, one might think that the future for carotenoid supplements looks bleak. In the short-to-medium term, this is probably not so, as Table 1 indicates. What will happen in the long term is impossible to predict. The market for dietary supplements is volatile and will depend on the accumulation of more scientific evidence from well-designed trials and experiments. What is popular today may be forgotten tomorrow. Humans will, however, always ingest carotenoids as part of their diet, whether they are naturally present in the food, added as colourants or taken as supplements.

References

[1] U. März, *GA-110 – The Global Market for Carotenoids*, Business Communications Co. (2000).

[2] U. März, *GA-110R – The Global Market for Carotenoids*, Business Communications Co. (2005).

[3] U. März, *FOD025C – The Global Market for Carotenoids*, BCC Research (2008).

[4] Regulation (EC) No 1925/2006 of the European Parliament and of the Council, *Off. J. Eur. Commun.*, **404**, 26 (2006).

[5] *Dietary Supplement Health and Education Act of 1994*, http://www.fda.gov/opacom/laws/dshea.html

[6] A. Mortensen, *Pure Appl. Chem.*, **78**, 1477 (2006).

[7] Regulation (EC) No 258/97 of the European Parliament and of the Council, *Off. J. Eur. Commun.*, **L 043**, 1 (1997).

[8] http://ec.europa.eu/food/food/biotechnology/novelfood/index_en.htm

[9] COMMISSION DECISION of 23 October 2006, *Off. J. Eur. Commun.*, **L 296**, 13 (2006).

[10] Regulation (EC) No 1924/2006 of the European Parliament and of the Council, *Off. J. Eur. Commun.*, **L 404**, 9 (2006).

[11] *Qualified Health Claims: Letter of Partial Denial - "Tomatoes and Prostate, Ovarian, Gastric and Pancreatic Cancers (American Longevity Petition)" (Docket No. 2004Q-0201)*, http://www.cfsan.fda.gov/~dms/qhclyco.html

[12] *Qualified Health Claims: Letter of Denial – "Xangold® Lutein Esters, Lutein, or Zeaxanthin and Reduced Risk of Age-related Macular Degeneration or Cataract Formation" (Docket No. 2004Q-0180)*, http://www.cfsan.fda.gov/~dms/qhclutei.html

[13] P. R. Trumbo and K. C. Ellwood, *Am. J. Clin. Nutr.*, **84**, 971 (2006).

[14] Institute of Medicine, *Dietary Reference Intake of Vitamin C, Vitamin E, Selenium, and Carotenoids*, National Academy Press, Washington D.C. (2000).

[15] B. Olmedilla, F. Granado, S. Southon, A. J. A. Wright, I. Blanco, E. Gil-Martinez, H. van den Berg, D. Thurnham, B. Corridan, M. Chopra and I. Hininger, *Clin. Sci.*, **102**, 447 (2002).

[16] W. L. Hadden, R. H. Watkins, L. W. Levy, E. Regalado, D. M. Rivadeneira, R. B. van Breemen and S. J. Schwartz, *J. Agric. Food Chem.*, **47**, 4189 (1999).

[17] Commission Directive 95/45/EC, *Off. J. Eur. Commun.*, **L 226**, 1 (1995).

[18] Commission Directive 2001/50/EC, *Off. J. Eur. Commun.*, **L 190**, 41 (2001).

[19] J. Gerritsen and F. Crum., *Soft Drinks Int.*, 25 (October 2002).

[20] A. Mortensen, *Food Res. Int.*, **38**, 847 (2005).

[21] D. Horn, *Angew. Makromol. Chem.*, **166/167**, 139 (1989).

[22] P. R. Sundaresan, *J. AOAC Int.*, **85**, 1127 (2002).

[23] R. Aman, S. Bayha, R. Carle and A. Schieber, *J. Agric. Food Chem.*, **52**, 6086 (2004).

[24] D. E. Breithaupt and J. Schlatterer, *Eur. Food Res. Technol.*, **220**, 648 (2005).

[25] J. Schierle, B. Pietsch, A. Ceresa, C. Fizet and E. H. Waysek, *J. AOAC Int.*, **87**, 1070 (2004).

[26] J. Szpylka and J. W. DeVries, *J. AOAC Int.*, **88**, 1279 (2005).

[27] C. J. Fuller, D. N. Butterfoss and M. L. Failla, *Nutr. Res.*, **21**, 1209 (2001).

[28] V. Böhm, *J. Food Sci.*, **67**, 1910 (2002).

[29] J. M. Odeberg, Å. Lignell, A. Pettersson and P. Höglund, *Eur. J. Pharm. Sci.*, **19**, 299 (2003).

[30] P. Borel, V. Tyssandier, N. Mekki, P. Grolier, Y. Rochette, M. C. Alexandre-Gouabau, D. Lairon and V. Azaïs-Braesco, *J. Nutr.*, **128**, 1361 (1998).

[31] S. A. Tanumihardjo, J. Li and M. P. Dosti, *J. Am. Diet. Assoc.*, **105**, 114 (2005).

[32] R. Lakshminarayana, M. Raju, T. P. Krishnakantha and V. Baskaran, *Mol. Cell. Biochem.*, **281**, 103 (2006).

[33] P. E. Bowen, S. M. Herbst-Espinosa, E. A. Hussain and M. Stacewicz-Sapuntzakis, *J. Nutr.*, **132**, 3668 (2002).

[34] D. E. Breithaupt, P. Weller, M. Wolters and A. Hahn, *Br. J. Nutr.*, **91**, 707 (2004).

[35] D. E. Breithaupt, P. Weller, M. Wolters and A. Hahn, *Br. J. Nutr.*, **90**, 795 (2003).

[36] M. B. Katan, S. M. Grundy, P. Jones, M. Law, T. Miettinen and R. Paoleti, *Mayo Clin. Proc.*, **78**, 965 (2003).

[37] K. H. van het Hof, C. Gärtner, A. Wiersma, L. B. M. Tijburg and J. A. Weststrate, *J. Agric. Food Chem.*, **47**, 1582 (1999).

[38] P. P. Hoppe, K. Krämer, H. van den Berg, G. Steenge and T. van Vliet, *Eur. J. Nutr.*, **42**, 272 (2003).

[39] Scientific Committee on Food, *Opinion of the Scientific Committee on Food on the Tolerable Upper Intake Level of Beta-Carotene, SCF/CS/NUT/UPPLEV/37 Final (2000)*, http://ec.europa.eu/food/fs/sc/scf/out80b_en.pdf

[40] Scientific Committee on Food, *Opinion of the Scientific Committee on Food on the Safety of Use of Beta-Carotene from all Dietary Sources, SCF/CS/ADD/COL/159 Final (2000)*, http://ec.europa.eu/food/fs/sc/scf/out71_en.pdf

[41] Expert Group on Vitamins and Minerals, *Safe Upper Levels for Vitamins and Minerals*, (2003), http://www.food.gov.uk/multimedia/pdfs/vitamin2003.pdf

[42] The Alpha-Tocopherol, Beta-Carotene Cancer Prevention Study Group, *New Engl. J. Med.*, **330**, 1029 (1994).

[43] H. Vainio and M. Rautalahti, *Cancer Epidemiol. Biomark. Prev.*, **7**, 725 (1998).

[44] P. M. Kris-Etherton, A. H. Lichtenstein, B. V. Howard, D. Steinberg and J. L. Witztum, *Circulation*, **110**, 637 (2004).

[45] U.S. Preventive Services Task Force, *Ann. Intern. Med.*, **139**, 51 (2003).

Carotenoids
Volume 5: Nutrition and Health

Chapter 5

Microbial and Microalgal Carotenoids as Colourants and Supplements

Laurent Dufossé

A. Introduction

General aspects of the production and use of carotenoids as colourants and supplements were discussed in *Chapter 4*. For several decades, these carotenoids have been produced commercially by chemical synthesis or as plant extracts or oleoresins, *e.g.* of tomato and red pepper. Some unicellular green algae, under appropriate conditions, become red due to the accumulation of high concentrations of 'secondary' carotenoids. Two examples, *Dunaliella* spp. and *Haematococcus pluvialis*, are cultured extensively as sources of β-carotene (**3**) and (3*S*,3'*S*)-astaxanthin (**406**), respectively.

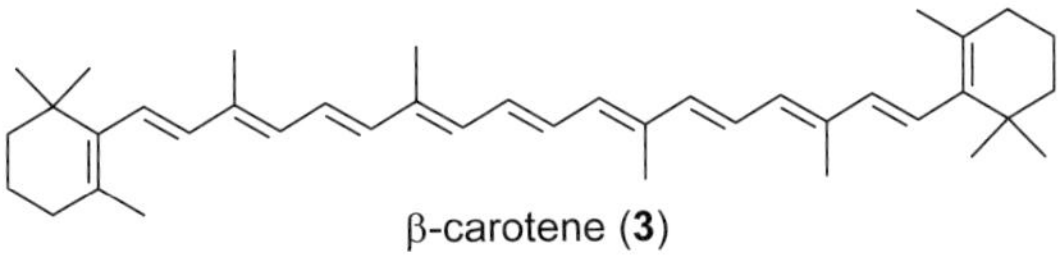

β-carotene (**3**)

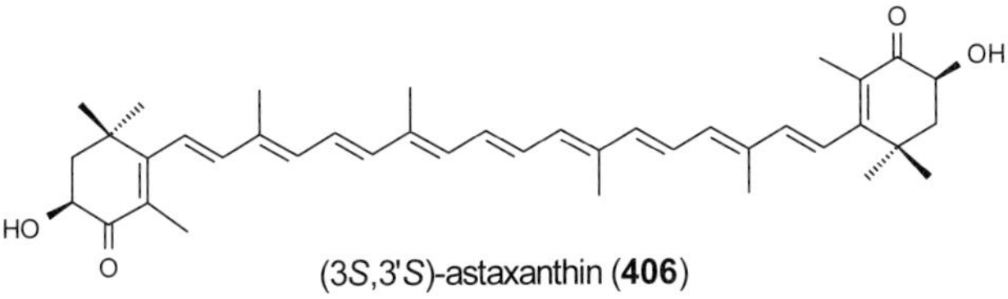

(3*S*,3'*S*)-astaxanthin (**406**)

Non-photosynthetic microorganisms, *i.e.* bacteria, yeasts and moulds, may also be strongly pigmented by carotenoids, so commercial production by these organisms is an attractive prospect. Penetration into the food industry by fermentation-derived ingredients is increasing year after year, examples being thickening or gelling agents (xanthan, curdlan, gellan), flavour enhancers (yeast hydrolysate, monosodium glutamate), flavour compounds (γ-decalactone, diacetyl, methyl ketones), and acidulants (lactic acid, citric acid). Fermentation processes for pigment production on a commercial scale were developed later but some are now in use in the food industry, such as production of β-carotene from the fungus *Blakeslea trispora*, in Europe, and the non-carotenoid heterocyclic pigments from *Monascus*, in Asia [1-3]. Efforts have been made to reduce the production costs so that pigments produced by fermentation can be competitive with synthetic pigments or with those extracted from natural sources. There is scope for innovations to improve the economics of carotenoid production by isolating new microorganisms, creating better ones, or improving the processes.

The microbial carotenoid products may be used as colour additives for food and feed, and are now under consideration for use as health supplements.

B. Carotenoid Production by Microorganisms and Microalgae

Commercial processes are already in operation or under development for the production of carotenoids by microalgae, moulds, yeasts and bacteria. The production of β-carotene by microorganisms, as well as by chemical synthesis or from plant extracts, is well developed, and the microbial production of several other carotenoids, notably lycopene (**31**), astaxanthin (**404-406**), zeaxanthin (**119**) and canthaxanthin (**380**), is also of interest. There is no microbial source that can compete with marigold flowers as a source of lutein (**133**).

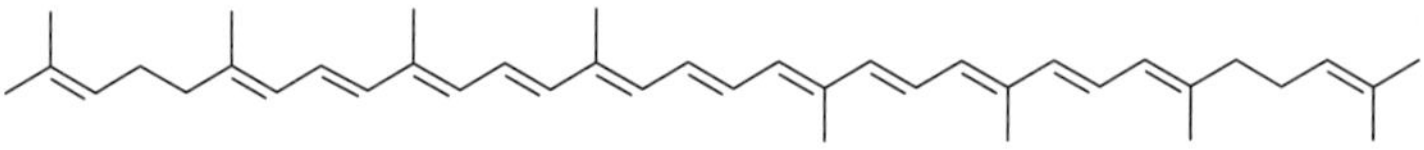

lycopene (**31**)

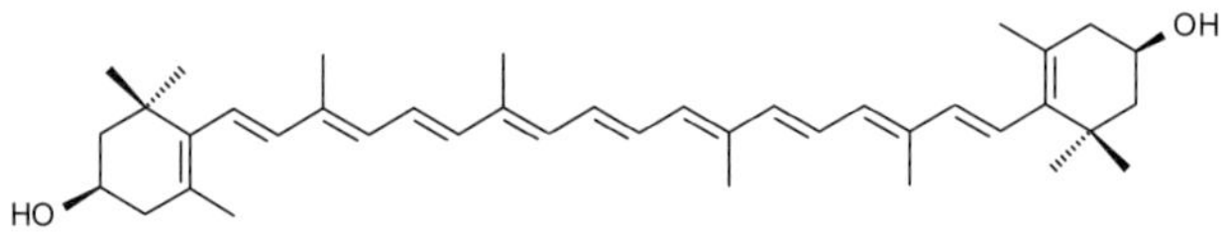

zeaxanthin (**119**)

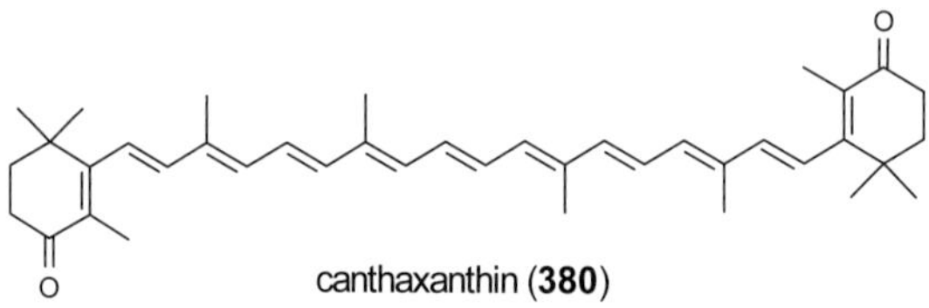

canthaxanthin (**380**)

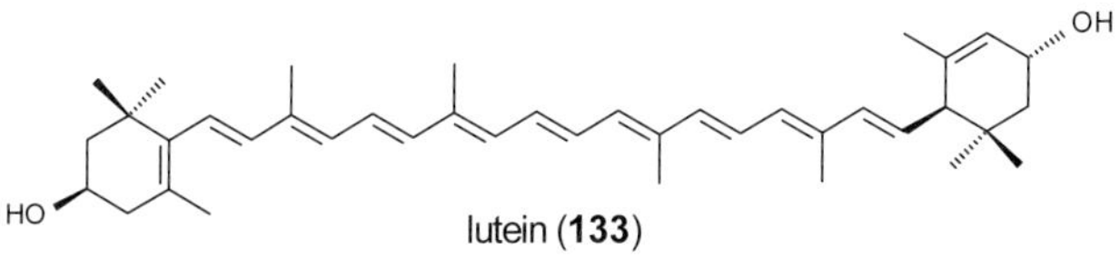

lutein (**133**)

1. β-Carotene

β-Carotene is produced on a large scale by chemical synthesis, and also from plant sources such as red palm oil, in addition to production by fermentation and from microalgae. The various preparations differ in the composition of geometrical isomers and in the presence of α-carotene (**7**) and other carotenoids, particularly biosynthetic intermediates (Table 1).

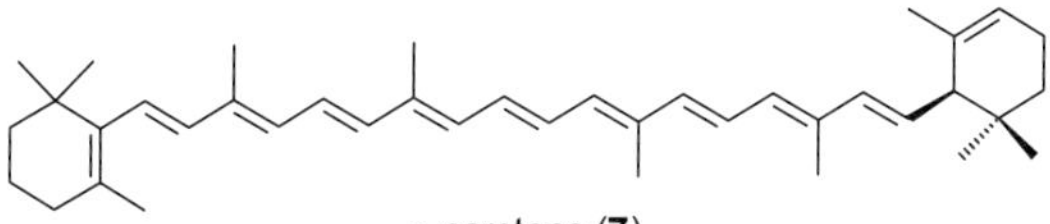

α-carotene (**7**)

Table 1. Percentage composition of 'β-carotene' from various sources.

Source	**(all-*E*)-β-carotene**	**(*Z*)-β-carotene**	**α-carotene**	**others**
Fungus *(Blakeslea)*	94	3.5	0	2.5
Chemical synthesis	98	2	0	0
Alga *(Dunaliella)*	67.4	32.6	0	0
Palm oil	34	27	30	9

a) *Dunaliella* species

Although the cyanobacterium (blue-green alga) *Spirulina* is able to accumulate β-carotene at up to 0.8-1.0 % w/w, *Dunaliella* species (*D. salina* and *D. bardawil*) produce the highest yield of β-carotene among the algae.

Dunaliella is a halotolerant, unicellular, motile green alga belonging to the family Chlorophyceae [4]. It is devoid of a rigid cell wall and contains a single, large, cup-shaped chloroplast which contains the characteristic carotenoid complement of green algae, similar to that of higher plant chloroplasts. In response to stress conditions such as high light intensity [5], it accumulates a massive amount of β-carotene [6]. The alga can yield valuable products, notably glycerol and β-carotene, and is also a rich source of protein that has good utilization value, and of essential fatty acids. *Dunaliella* biomass has GRAS (Generally Recognized As Safe) status and can be used directly as food or feed. As a supplement, *Dunaliella* has been reported to exhibit various biological effects, such as antihypertensive, bronchodilator, analgesic, muscle relaxant, and anti-oedema activity [7].

Dunaliella grows in high salt concentration (1.5 ± 0.1 M NaCl), and requires bicarbonate as a source of carbon, and other nutrients such as nitrate, sulphate and phosphate. The initial photosynthetic (vegetative) growth phase requires 12-14 days in nitrate-rich medium. The subsequent carotenogenesis phase requires nitrate depletion and maintenance of salinity. This technology is best suited for coastal areas where sea water is rich in salt and other nutrients. For carotenogenesis, nutrient, salt or light stress is essential; generally the vegetative phase requires 5-10 klux whereas the light should be around 25-30 klux for β-carotene accumulation.

Dunaliella salina can be cultivated easily and quickly compared to plants and, under ideal conditions, can produce a very high quantity of β-carotene compared to other sources (3-5%, w/w on a dry mass basis, 400 mg per square metre of cultivation area) [8]. At high light intensity, there can be a large proportion of *Z* isomers, with up to 50% of (9*Z*)-β-carotene. Cells are harvested by flocculation followed by filtration; the product can be directly utilized as feed or in food formulations, or it can be extracted for pigments.

For various food formulations and applications the carotene can be extracted either in edible oils or food grade organic solvents. Most of the pharmaceutical formulations are made with either olive oil or soybean oil. In the natural extracts, β-carotene is generally accompanied by small amounts of the residual chloroplast carotenoids and is marketed under the title 'Carotenoids Mix'.

The major sites of commercial production are Australia, China, India, Israel, Japan and the U.S., with smaller-scale production in several other countries with suitable environmental conditions. *Dunaliella* β-carotene is widely distributed today in many different markets under three different categories, namely β-carotene extracts; *Dunaliella* powder for human use; dried *Dunaliella* for feed use. Extracted purified β-carotene, sold mostly in vegetable oil in bulk concentrations from 1% to 20%, is used to colour various food products or, in soft gel capsules, for use as a supplement, usually 5 mg β-carotene per capsule.

b) *Blakeslea trispora*

Blakeslea trispora is a commensal mould associated with tropical plants. The fungus exists in (+) and (–) mating types; the (+) type synthesizes trisporic acid, which is both a metabolite of β-carotene and a hormonal stimulator of its biosynthesis. On mating the two types in a specific ratio, the (–) type is stimulated by trisporic acid to synthesize large amounts of β-carotene.

The production process proceeds essentially in two stages. Glucose and corn steep liquor can be used as carbon and nitrogen sources. Whey, a byproduct of cheese manufacture, has also been considered [9], with strains adapted to metabolize lactose. In the initial fermentation process, seed cultures are produced from the original strain cultures and subsequently used in an aerobic submerged batch fermentation to produce a biomass rich in β-carotene. In the second stage, the recovery process, the biomass is isolated and converted into a form suitable for isolating the β-carotene, which is extracted with ethyl acetate, suitably purified and concentrated, and the β-carotene is crystallized [10]. The final product is either used as crystalline β-carotene (purity >96%) or is formulated as a 30% suspension of micronized crystals in vegetable oil. The production process is subject to Good Manufacturing Practices (GMP) procedures, and adequate control of hygiene and raw materials. The biomass and the final crystalline product comply with an adequate chemical and microbiological specification and the final crystalline product also complies with the JECFA (**J**oint FAO/WHO **E**xpert **C**ommittee on **F**ood **A**dditives) and E.U. specifications as set out in Directive 95/45/EC for colouring materials in food.

The first β-carotene product from *B. trispora* was launched in 1995. The mould has shown no pathogenicity or toxicity, in standard pathogenicity tests in mice, by analysis of extracts of several fermentation mashes for fungal toxins, and by enzyme immunoassays of the final product, the β-carotene crystals, for four mycotoxins. HPLC analysis, stability tests and microbiological tests showed that the β-carotene obtained by co-fermentation of *Blakeslea trispora* complies with the E.C. specification for β-carotene (E 160 aii), listed in Directive 95/45/EC, including the proportions of *Z* and *E* isomers, and is free of mycotoxins or other toxic metabolites and free of genotoxic activity. In a 28-day feeding study in rats with the β-carotene manufactured in the E.U. no adverse findings were noted at a dose of 5% in the diet, the highest dose level used. The E.U. Scientific Committee considered that "β-carotene produced by co-fermentation of *Blakeslea trispora* is equivalent to the chemically synthesized material used as food colorant and is therefore acceptable for use as a colouring agent for foodstuffs" [11].

There are now other industrial productions of β-carotene from *B. trispora* in Russia, Ukraine, and Spain [12]. The process has been developed to yield up to 30 mg of β-carotene per g dry mass or about 3 g per litre of culture. *Blakeslea trispora* is now also used for the production of lycopene (Section **B**.2.a).

c) *Phycomyces blakesleeanus*

Another mould, *Phycomyces blakesleeanus*, is also a potential source of various chemicals including β-carotene [13]. The carotene content of the wild type grown under standard conditions is modest, about 0.05 mg per g dry mass, but some mutants accumulate up to 10 mg/g [14]. As with *Blakeslea trispora*, sexual stimulation of carotene biosynthesis is essential, and can increase yields to 35 mg/g [15]. The most productive strains of *Phycomyces* achieve their full carotenogenic potential on solid substrates or in liquid media without agitation. *Blakeslea trispora* is more appropriate for production in usual fermentors [16].

d) *Mucor circinelloides*

Mucor circinelloides wild type is yellow because it accumulates β-carotene as the main carotenoid. The basic features of carotenoid biosynthesis, including photoinduction by blue light [17], are similar to those in *Phycomyces* and *Mucor* [18]. *M. circinelloides* is a dimorphic fungus that grows either as yeast cells or in a mycelium form, and research is now focused on yeast-like mutants that could be useful in a biotechnological production [12].

2. Lycopene

Lycopene is produced on a large scale by chemical synthesis, and from tomato extracts, in addition to production by fermentation. As with β-carotene, the various preparations differ in the composition of geometrical isomers (Table 2).

Table 2. Percentage of geometrical isomers in 'lycopene' from various sources.

Source	(all-*E*)	(5Z)	(9Z)	(13Z)	Others
Chemical synthesis	>70	<25	<1	<1	<3
Tomato	94-96	3-5	0-1	1	<1
Blakeslea trispora	≥ 90		(mixed *Z* isomers) 1-5		

Lycopene is an intermediate in the biosynthesis of all dicyclic carotenoids, including β-carotene. In principle, therefore, blocking the cyclization reaction and the cyclase enzyme by mutation or inhibition will lead to the accumulation of lycopene. This strategy is employed for the commercial production of lycopene.

a) *Blakeslea trispora*

A commercial process for lycopene (**31**) production by *Blakeslea trispora* is now established. Imidazole or pyridine is added to the culture broth to inhibit the enzyme lycopene cyclase [19]. The product, predominantly (all-*E*)-lycopene, is formulated into a 20% or 5% suspension in sunflower oil, together with α-tocopherol at 1% of the lycopene level. Also available is an α-tocopherol-containing 10% or 20% lycopene cold-water-dispersible (CWD) product. Lycopene oil suspension is intended for use as a food ingredient and in dietary supplements. The proposed level of use for lycopene in food supplements is 20 mg per day.

Approval for the use of lycopene from *B. trispora* was sought under regulation (EC) No 258/97 of the European Parliament and the Council concerning novel foods and novel food ingredients [20]. The European Food Safety Authority was also asked to evaluate this product for use as a food colour. The conclusions were that the lycopene from *B. trispora* is considered to be nutritionally equivalent to lycopene in a natural diet, but further safety trials are necessary. Whilst the toxicity data on lycopene from *B. trispora* and on lycopene from tomatoes do not give indications for concern, nevertheless these data are limited and do not allow an ADI to be established. The main concern is that the proposed use levels of lycopene from *B. trispora* as a food ingredient may result in a substantial increase in the daily intake of lycopene compared to the intakes solely from natural dietary sources. The use of lycopene as a health supplement was not considered.

b) *Fusarium sporotrichioides*

The fungus *Fusarium sporotrichioides* has been genetically modified to manufacture lycopene from the cheap corn-fibre material, the 'leftovers' of making ethanol [21]. By use of sequential, directional cloning of multiple DNA sequences, the isoprenoid pathway of the fungus was redirected toward the synthesis of carotenoids *via* carotenoid biosynthesis genes introduced from the bacterium *Erwinia uredovora*. Cultures in laboratory flasks produced 0.5 mg lycopene per g dry mass within six days and improvements are predicted [22].

3. Astaxanthin

The application of astaxanthin in aquaculture feed to impart the desired colour to fish and crustaceans is described in *Volume 4, Chapter 12*. There are also reports of beneficial actions of astaxanthin for human health, so its use in supplements is of interest. The biotechnological production of astaxanthin from microalgae, yeasts and bacteria is the subject of intensive investigation, though synthetic astaxanthin remains the market leader.

a) *Haematococcus pluvialis*

Haematococcus pluvialis is a green alga known for its ability to accumulate (3*S*,3'*S*)-astaxanthin (**406**), up to 0.2 to 2.0% (on a dry mass basis). The alga can grow both under autotrophic and heterotrophic conditions. Astaxanthin from *Haematococcus* is under consideration for US Food and Drug Administration clearance and several European countries have approved its marketing as a dietary supplement ingredient for human consumption.

The production of astaxanthin by *Haematococcus pluvialis* is attractive [23], but has fewer advantages than the *Dunaliella* β-carotene process. *H. pluvialis* is a freshwater alga so open-air culture leads to contamination by undesirable species. Outdoor cultivation of *Haematococcus* is a challenge and requires curtailment of contamination and control of environmental conditions such as light and temperature. This organism grows at 20-28°C, below 15 klux light intensity and at pH 6.8-7.4, so contamination by bacteria, fungi and protozoa, is a serious problem. Under high light intensity the cell growth is significantly affected. Recently, however, processes involving completely closed photobioreactors with artificial light or a combination of closed photobioreactors and open culture ponds are being used for *Haematococcus* cultivation [24].

Unlike *Dunaliella*, *Haematococcus* changes from a motile, flagellated cell to a non-motile, thick-walled aplanospore during the growth cycle [25,26]; the astaxanthin is contained in the aplanospore. This means that the physical properties (density, settling rate, cell fragility) and nutrient requirements of the cells change during the culture process, and this alters the optimum conditions for growth and carotenoid accumulation during the growth cycle [27]. The content of astaxanthin in the aplanospores is about 1-2% of dry mass but their thick wall requires physical breakage before the astaxanthin can either be extracted or be available to organisms consuming the alga [28].

The development of a commercially viable algal astaxanthin process requires the development of an effective closed culture system and the selection (either from Nature or by mutagenesis) of strains of *Haematococcus* with higher astaxanthin content and an ability to tolerate higher temperatures than the wild strains. Successful commercial production is now operating in India, Japan and the U.S.

Astaxanthin is recognized by U.S. FDA under title 21 Part 73 (under List of Colour Additives Exempted from Certification) Subpart A - Foods (Sec.73.35 Astaxanthin). Formulations containing astaxanthin are: soft gelatin capsules containing 100 mg equivalent of total carotenoids; skin-care cream containing astaxanthin as one of the ingredients; food and feed formulations for shrimp and fish.

b) *Xanthophyllomyces dendrorhous* (formerly *Phaffia rhodozyma*)

Among the few astaxanthin-producing microorganisms, *Xanthophyllomyces dendrorhous* is one of the best candidates for commercial production of astaxanthin [29] though, in this case, the product is the (3*R*,3'*R*)-isomer (**404**).

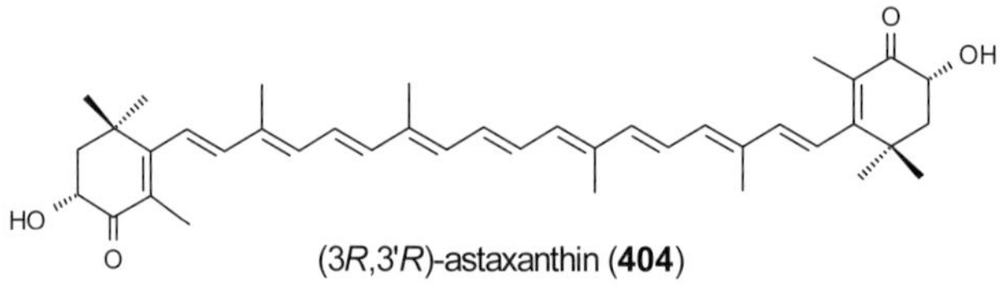

(3*R*,3'*R*)-astaxanthin (**404**)

The effects of different nutrients on *Xanthophyllomyces dendrorhous* have generally been studied in media containing complex sources of nutrients such as peptone, malt and yeast extracts. By-products from agriculture were also tested, such as molasses [30], enzymic wood hydrolysates [31], corn wet-milling co-products [32], bagasse or raw sugarcane juice [33], date juice [34] and grape juice [35]. However, in order to elucidate the nature of nutritional effects as far as possible, chemically defined or synthetic media have been used [36-38]. In one major study [38], the optimal conditions stimulating the highest astaxanthin production were found to be: temperature 19.7°C; carbon concentration 11.25 g/L; pH 6.0; inoculum 5%; nitrogen concentration 0.5 g/L. Under these conditions the astaxanthin content was 8.1 mg/L.

Fermentation strategy also has an impact on growth and carotenoid production of *Xanthophyllomyces dendrorhous* [39], as shown by studies with fed-batch cultures (*e.g.* limiting substrate is fed without diluting the culture) [40] or pH-stat cultures (*i.e.* a system in which the feed is provided depending on the pH) [41]. The highest biomass obtained was 17.4 g/L. Another starting point in optimization experiments is the generation of mutants [42], but metabolic engineering of the astaxanthin biosynthetic pathway is now attractive [43].

A major drawback in the use of *Xanthophyllomyces dendrorhous* is that, for efficient intestinal absorption of the pigment, disruption of the cell wall of the yeast biomass is required before addition to an animal diet. Several chemical, physical, autolytic, and enzymic methods for cell-wall disruption have been described, inluding a two-stage batch fermentation technique [44]. The first stage was for 'red yeast' cultivation. The second stage was the mixed fermentation of the yeast and *Bacillus circulans*, a bacterium with a high cell-wall lytic activity.

The case of *Xanthophyllomyces dendrorhous* (*Phaffia rhodozyma*) is peculiar; hundreds of scientific papers and patents deal with astaxanthin production by this yeast [45,46] but the process has not yet become economically efficient. New patents are filed almost each year, with improved astaxanthin yield; yields up to 3mg/g dry matter have been achieved [47].

c) *Agrobacterium aurantiacum* and other bacteria

Astaxanthin is one of ten carotenoids present in *Agrobacterium aurantiacum* [48]. The biosynthetic pathway, the influence of growth conditions on carotenoid production and the occurrence of astaxanthin glucoside have been described [49,50], but commercial processes have not yet been developed.

Numerous screenings have been conducted in the search for new bacterial sources of astaxanthin, and positive targets were isolated such as *Paracoccus carotinifaciens* [51] and a *Halobacterium* species [52]. The latter is particularly interesting because: (i) the extreme NaCl concentrations (about 20%) used in the growth medium prevent contamination with other organisms so no particular care has to be taken with sterilization; (ii) NaCl concentrations under 15% induce bacterial lysis, so that no special cell breakage technique is necessary, and pigments may be extracted directly with sunflower oil instead of organic solvents. This would eliminate possible toxicity problems due to trace amounts of acetone or hexane and facilitate pigment assimilation by animals. No commercial processes have yet been developed, however.

4. Zeaxanthin

Zeaxanthin (**119**) can be used, for example, as an additive in feeds for poultry to intensify the yellow colour of the skin or to accentuate the colour of the yolk of their eggs [53]. It is also suitable for use as a colourant, for example in the cosmetics and food industries, and as a health supplement in relation to the maintenance of eye health (see *Chapter 15*).

In the mid-1960s, several marine bacteria that produce zeaxanthin were isolated. Cultures of a *Flavobacterium* sp. (ATCC 21588, classified under the accepted taxonomic standards of that time) [54] in a defined nutrient medium containing glucose or sucrose as carbon source, were able to produce up to 190 mg of zeaxanthin per litre, with a concentration of 16 mg/g dried cell mass. One species currently under investigation in many studies [55-57] is *Sphingobacterium* (formerly *Flavobacterium*) *multivorum* (ATCC 55238). This was recently shown to utilize the deoxyxylulose phosphate or methylerythritol phosphate pathway [58,59]. A strain was constructed for over-production of zeaxanthin in industrial quantities [60].

Another zeaxanthin-producing '*Flavobacterium*' was recently reclassified as a *Paracoccus* species, *P. zeaxanthinifaciens* [61]; earlier findings that isoprenoid biosynthesis occurs exclusively *via* the mevalonate pathway were confirmed [62-65]. A second strain, isolated in a mat from an atoll of French Polynesia, produces also exopolysaccharides [66]. Another member of the Sphingobacteraceae, *Nubsella zeaxanthinfaciens*, was isolated recently from fresh water [67].

Chemical synthesis remains the method of choice for production of zeaxanthin, however.

5. Canthaxanthin

Canthaxanthin (**380**) has been used in aquafeed for many years to impart the desired flesh colour to farmed salmonid fish, especially trout (*Volume 4*, *Chapter 12*). Because extreme overdosage with canthaxanthin can lead to the deposition of minute crystals in the human eye (*Chapter 15*), canthaxanthin is not likely to be accepted as a health supplement and there is some pressure to limit its use in aquafeeds.

Some bacteria have potential for commercial canthaxanthin production. A strain of a *Bradyrhizobium* sp. was described as a canthaxanthin producer [68] and the carotenoid gene cluster was fully sequenced [69]. A second organism under scrutiny for canthaxanthin production is the extreme halophile *Haloferax alexandrinus*, a member of the family Halobacteriaceae (Archaea). Most members of the Halobacteriaceae are red due to the presence of C_{50}-carotenoids [70]. Some species, however, have been reported to produce C_{40}-carotenoids, including ketocarotenoids, as minor components. Recently, the biotechnological potential of these members of the Archaea has increased because of their unique features, which facilitate many industrial procedures. For example, no sterilization is required, because of the extremely high NaCl concentration used in the growth medium (contamination by other organisms is avoided). In addition, no cell-disrupting devices are required, as cells lyse spontaneously in fresh water [71]. A 1-litre-scale cultivation of the cells in flask cultures (6 days) under non-aseptic conditions produced 3 g dry mass, containing 6 mg total carotenoid and 2 mg canthaxanthin [72]. Further experiments in a batch fermenter also demonstrated increases in the biomass concentration and carotenoid production.

A third example is *Gordonia jacobea* (CECT 5282), a Gram-positive, catalase negative, G+C 61% bacterium which was isolated in routine air sampling during screening for microorganisms that produce pink colonies [73], with canthaxanthin as the main pigment [74]. The low carotenoid content (0.2 mg/g dry mass) does not support an industrial application but, after several rounds of mutations, a hyper-pigmented mutant (MV-26) was isolated which accumulated six times more canthaxanthin than the wild-type strain and, by varying the culture medium, canthaxanthin concentrations between 1 and 13.4 mg/L were achieved. Mutants of this species have potential advantages from the industrial point of view: (i) the optimal temperature for growth and carotenogenesis, 30°C, is usual in fermentors; (ii) glucose, an inexpensive carbon source, gives optimal growth and pigmentation; and (iii) >90% of the total pigments can be extracted directly with ethanol, a non-toxic solvent allowed for human and animal feed [75].

6. Torulene and torularhodin

Yeasts of the genus *Rhodotorula* synthesize carotenoids, mainly torularhodin (**428**) and torulene (**11**) accompanied by very small amounts of β-carotene. Most of the research has focused on the species *Rhodotorula glutinis* [76], though other species such as *R. gracilis*, *R. rubra* [77], and *R. graminis* [78] have been studied. These yeasts have potential as feed products rather than as health supplements.

Optimization studies [79,80] have mainly resulted in an increased yield of torulene and torularhodin, which are of minor interest, though some did succeed in increasing the β-carotene content up to about 70 mg/L.

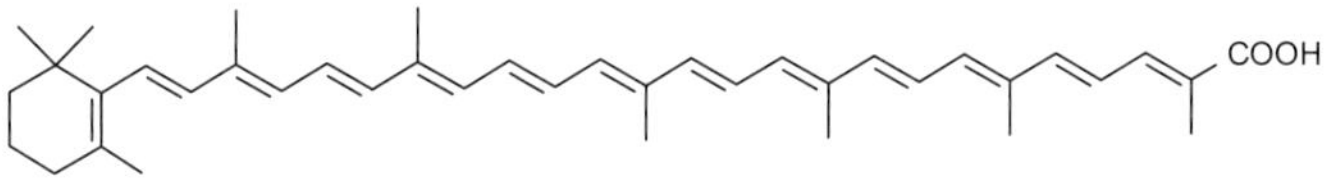

torularhodin (**428**)

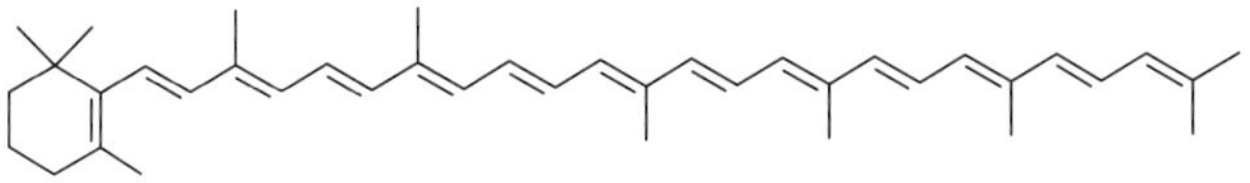

torulene (**11**)

C. Prospects for Carotenoid Production by Genetically Modified Microorganisms

1. *Escherichia coli* and other hosts

Metabolic engineering is defined as the use of recombinant DNA techniques for the deliberate modification of metabolic networks in living cells to produce desirable chemicals with superior yield and productivity. The traditional assumption was that the most productive hosts would be microbes that naturally synthesize the desired chemicals, but microorganisms that have the ability to produce precursors of the desired chemicals with superior yield and productivity are also considered as suitable hosts [81].

As a starting point a large number (>200) of genes and gene clusters coding for the enzymes of carotenoid biosynthesis have been isolated from various carotenogenic microorganisms, and the functions of the genes have been elucidated (*Volume 3, Chapter 3*).

In bacteria such as *Escherichia coli*, which cannot synthesize carotenoids naturally, carotenoid biosynthesis *de novo* has been achieved by the introduction of carotenogenic genes. *E. coli* does possess the ability to synthesize other isoprenoid compounds such as dolichols (sugar carrier lipids) and the respiratory quinones. It is thus feasible to direct the carbon flux for the biosynthesis of these isoprenoid compounds partially to the pathway for carotenoid production by the introduction of the carotenogenic genes. For example, plasmids carrying *crt* genes for the synthesis of lycopene, β-carotene and zeaxanthin have been constructed and expressed in *E. coli*. Transformants accumulated lycopene, β-carotene, and zeaxanthin, at 0.2-1.3 mg/g dry mass, in the stationary phase. With a few exceptions, such as the zeaxanthin C(5,6) epoxidase gene, almost all cloned carotenoid biosynthetic genes are functionally expressed in *E. coli*. The use of shot-gun library clones constructed with *E. coli* chromosomal DNA [82] has revealed that genes not directly involved in the carotenoid biosynthesis pathway are important, such as *appY*, which encodes transcriptional regulators related to anaerobic energy metabolism and can increase the lycopene production to 4.7 mg/g dry cell mass.

A most important challenge for biotechnology is to identify rate-limiting steps or to eliminate regulatory mechanisms in order to enhance further the production of valuable carotenoids [83]. Sufficient amounts of endogenous precursors (*i.e.* substrates for the reactions involved) must be available; by control of the pyruvate/glyceraldehyde 3-phosphate ratio, a yield of 25 mg lycopene/g dry mass has been reported [84]. A balanced system of carotenogenic enzymes should be expressed, to enable efficient conversion of precursors without the formation of pools of intermediate metabolites. The correct plasmid combination is important to minimize the accumulation of intermediates and to increase the yield of the end product. Finally, the host organism should exhibit an active central terpenoid pathway and possess a high storage capacity for carotenoids [85].

As well as *E. coli*, the edible yeasts *Candida utilis* [86] and *Saccharomyces cerevisiae* [87] acquire the ability to produce carotenoids when the required carotenogenic genes are introduced.

2. Directed evolution and combinatorial biosynthesis

Directed evolution involves the use of rapid molecular manipulations to mutate the target DNA fragment, followed by a selection or screening process to isolate desirable mutants. By various directed evolution protocols, several enzymes have been improved or optimized for a specific condition. Directed evolution was applied to geranylgeranyl diphosphate (GGDP) synthase (a rate-controlling enzyme) from *Archaeoglobus fulgidus* to enhance the production of carotenoids in metabolically engineered *E. coli* [88]. The production of lycopene was increased by about 2-fold.

A second example deals with the membrane-associated phytoene synthase which appears to be the major point of control over product diversity. By engineering the phytoene synthase to accept longer diphosphate substrates, variants were produced that can make previously unknown C_{35}-, C_{45}- and C_{50}-carotenoid backbones from the appropriate isoprenyl diphosphate precursors [89,90]. Once a carotenoid backbone structure is created, downstream enzymes, either natural or engineered, such as desaturases, cyclases, hydroxylases, and cleavage enzymes, can accept the new substrate, and whole series of novel C_{35}-, C_{45}- and C_{50}-carotenoid analogues can be produced.

A different approach is to combine available biosynthetic genes [91] and evolve new enzyme functions through random mutagenesis, recombination (DNA-shuffling) and selection. Prerequisites for this approach are that the enzymes from different species can function cooperatively in a heterologous host and display enough promiscuity regarding the structure of their substrates. The success of functional colour complementation in transgenic *E. coli* for cloning a number of carotenoid biosynthesis genes demonstrates that enzymes from phylogenetically distant species can assemble into a functional membrane-bound multi-enzyme complex through which carotenoid biosynthesis presumably takes place [92].

A related strategy [93] which can be used to produce novel carotenoids is to combine carotenogenic genes from different bacteria that alone normally produce different end products and to express them in a simple *E. coli* host that carries the biosynthetic machinery for phytoene production [94].

Much is now technically feasible, but there are still many problems to be overcome, especially in relation to control of the end product so that the desired target carotenoid is produced rather than a complex mixture, and to the ability of the host organism to accumulate the carotenoid in high concentration.

D. Concluding Comments

Nature is rich in colour, and carotenoid-producing microorganisms (fungi, yeasts, bacteria) are quite common.

The success of any pigment product manufactured by fermentation depends upon its acceptability in the market place, regulatory approval, and the size of the capital investment required to bring the product to market. A few years ago, doubts were expressed about the successful commercialization of carotenoids produced by fermentation because of the high capital investment needed for fermentation facilities and the extensive and lengthy toxicity studies required by regulatory agencies. Also, public perception of GM organisms is an important factor in the acceptance of biotechnology-derived products.

Now, however, some carotenoids produced by fermentation are on the market. This, and the successful marketing of algal-derived or vegetable-extracted carotenoids, both as food colours and as nutritional supplements, reflects the importance of 'niche markets' in which consumers are willing to pay a premium for 'all-natural ingredients'. Carotenoids play an exceptional role in the fast-growing 'over-the-counter medicine' and 'nutraceutical' sector.

Among carotenoids under investigation for colouring or for biological properties, only a small number of the 700 or so carotenoids listed in the '*Carotenoids Handbook*' are currently available from natural extracts or by chemical synthesis [95]. With imagination, biotechnology could be a solution for providing additional pigments for the market.

References

[1] U. Wissgott and K. Bortlik, *Trends Food Sci. Technol.*, **7**, 298 (1996).
[2] P. O'Carroll, *The World of Ingredients*, **3-4**, 39 (1999).
[3] A. Downham and P. Collins, *Int. J. Food Sci. Technol.*, **35**, 5 (2000).
[4] M. Avron and A. Ben-Amotz, *Dunaliella: Physiology, Biochemistry and Biotechnology*, CRC Press, London (1992).
[5] A. Ben-Amotz, A. Kartz and M. Avron, *J. Phycol.*, **18**, 529 (1983).
[6] Z. W. Ye, J. G. Jiang and G. H. Wu, *Progr. Prosp. Biotechnol. Adv.*, **26**, 352 (2008).
[7] R. Villar, M. R. Laguna, J. M. Callega and I. Cadavid, *Planta Med.*, **58**, 405 (1992).
[8] L. S. Jahnke, *J. Photochem. Photobiol. B*, **48**, 68 (1999).
[9] L. E. Lampila, S. E. Wallen, L. B. Bullerman and S. R. Lowry, *Lebensm. Wiss. Technol.*, **18**, 366 (1985).
[10] E. Papaioannou, T. Roukas and M. Liakopoulou-Kyriakides, *Prep. Biochem. Biotechnol.*, **38**, 246 (2008).
[11] European Commission, *Opinion of the Scientific Committee on Food on β-Carotene from* Blakeslea trispora, *SCF/CS/ADD/COL 158*, adopted on 22 June 2000 and corrected on 7 September 2000.
[12] E. A. Iturriaga, T. Papp, J. Breum, J. Arnau and A. P. Eslava, *Meth. Biotechnol.*, **18**, 239 (2005).
[13] E. R. A. Almeida and E. Cerda-Olmedo, *Curr. Genetics*, **53**, 129 (2008).
[14] F. J. Murillo, I. L. Calderon, I. Lopez-Diaz and E. Cerda-Olmedo, *Appl. Env. Microbiol.*, **36**, 639 (1978).
[15] B. J. Mehta, L. M. Salgado, E. R. Bejarano and E. Cerda-Olmedo, *Appl. Env. Microbiol.*, **63**, 3657 (1997).
[16] E. Cerda-Olmedo, *FEMS Microbiol. Rev.*, **25**, 503 (2001).
[17] E. Navarro, J. M. Lorca-Pascual, M. D. Quiles-Rosillo, F. E. Nicolas, V. Garre, S. Torres-Martinez and R. M. Ruiz-Vazquez, *Mol. Genet. Genomics*, **266**, 463 (2001).
[18] A. Velayos, M. A. Lopez-Matas, M. J. Ruiz-Hidalgo and A. P. Eslava, *Fungal Genet. Biol.*, **22**, 19 (1997).
[19] EFSA, *The EFSA Journal*, **212**, 1 (2005).
[20] Vitatene Inc., *Application for the approval of lycopene from Blakeslea trispora, under the EC regulation No 258/97 of the European Parliament*, (2003).
[21] J. D. Jones, T. M. Hohn and T. D. Leathers, *Soc. Indust. Microbiol. Annual Meeting*, p. 91 (2004).
[22] T. D. Leathers, J. D. Jones and T. M. Hohn, *US Patent* 6,696,282 (2004).
[23] E. Del Rio, F. G. Acien, M. C. Garcia-Malea, J. Rivas, E. Molina-Grima and M. G. Guerrero, *Biotechnol. Bioeng.*, **100**, 397 (2008).
[24] J. Fabregas, A. Otero, A. Maseda and A. Dominguez, *J. Biotechnol.*, **89**, 65 (2001).
[25] M. A. Borowitzka, J. M. Huisman and A. Osborn, *J. Appl. Phycol.*, **3**, 295 (1991).

[26] M. Kobayashi, T. Kakizono and S. Nagai, *J. Ferm. Bioeng.*, **71**, 335 (1991).

[27] R. Sarada, T. Usha and G. A. Ravishankar, *Process Biochem.*, **37**, 623 (2002).

[28] T. R. Sommer, W. Pott and N. M. Morrissy, *Aquaculture*, **94**, 79 (1991).

[29] K. Ukibe, T. Katsuragi, Y. Tani and H. Takagi, *FEMS Microbiol. Letts.*, **286** 241 (2008).

[30] G. H. An, B. G. Jang and M. H. Cho, *J. Biosci. Bioeng.*, **92**, 121 (2001).

[31] J. M. Cruz and J. C. Parajo, *Food Chem.*, **63**, 479 (1998).

[32] G. T. Hayman, B. M. Mannarelli and T. D. Leathers, *J. Indust. Microbiol. Biotechnol.*, **14**, 389 (1995).

[33] J. D. Fontana, B. Czeczuga, T. M. B. Bonfim, M. B. Chociai, B. H. Oliveira, M. F. Guimaraes and M. Baron, *Bioresource Technol.*, **58**, 121 (1996).

[34] J. Ramirez, M. L. Nunez and R. Valdivia, *J. Indust. Microbiol. Biotechnol.*, **24**, 187 (2000).

[35] E. Longo, C. Sieiro, J. B. Velazquez, P. Calo, J. Cansado and T. G. Villa, *Biotech Forum Europe*, **9**, 565 (1992).

[36] L. B. Flores-Cotera, R. Martin and S. Sanchez, *Appl. Microbiol. Biotechnol.*, **55**, 341 (2001).

[37] Z. Palágyi, L. Ferenczy and C. Vagvölgyi, *World J. Microbiol. Biotechnol.*, **17**, 95 (2001).

[38] J. Ramirez, H. Guttierez and A. Gschaedler, *J. Biotechnol.*, **88**, 259 (2001).

[39] Y. S. Liu, J. Y. Wu and K. P. Ho, *Biochem. Eng. J.*, **27**, 331 (2006).

[40] K. P. Ho, C. Y. Tam and B. Zhou, *Biotechnol. Lett.*, **21**, 175 (1999).

[41] H. Y. Chan and K. P. Ho, *Biotechnol. Lett.*, **21**, 953 (1999).

[42] L. Rubinstein, A. Altamirano, L. D. Santopietro, M. Baigori and L. I. C. D. Figueroa, *Folia Microbiol.*, **43**, 626 (1998).

[43] J. C. Verdoes, G. Sandmann, H. Visser, M. Diaz, M. van Mossel and A. J. J. van Ooyen, *Appl. Env. Microbiol.*, **69**, 3728 (2003).

[44] T. J. Fang and J. M. Wang, *Process Biochem.*, **37**, 1235 (2002).

[45] M. Vazquez, *Food Technol. Biotechnol.*, **39**, 123 (2001).

[46] E. A. Johnson, *Int. Microbiol.*, **6**, 169 (2003).

[47] P. R. David, *US Patent* 7,309,602 (2007).

[48] A. Yokoyama, H. Izumida and W. Miki, *Biosci. Biotech. Biochem.*, **58**, 1842 (1994).

[49] A. Yokoyama and W. Miki, *FEMS Microbiol. Lett.*, **128**, 139 (1995).

[50] A. Yokoyama, K. Adachi and Y. Shizuri, *J. Nat. Prod.*, **58**, 1929 (1995).

[51] A. Tsubokura, H. Yoneda and H. Mizuta, *Int. J. Syst. Bacteriol.*, **49**, 277 (1999).

[52] P. Calo, T. D. Miguel, C. Sieiro, J. B. Velazquez and T. G. Villa, *J. Appl. Bacteriol.*, **79**, 282 (1995).

[53] S. Alcantara and S. Sanchez, *J. Ind. Microbiol. Biotechnol.*, **23**, 697 (1999).

[54] D. Shepherd, J. Dasek, M. Suzanne and C. Carels, *US Patent* 3,951,743 (1976).

[55] P. Bhosale and P. S. Bernstein, *J. Indust. Microbiol. Biotechnol.*, **31**, 565 (2004).

[56] P. Bhosale, A. J. Larson and P. S. Bernstein, *J. Appl. Microbiol.*, **96**, 623 (2004).

[57] P. Bhosale, I. V. Ermakov, M. R. Ermakova, W. Gellermann and P. S. Bernstein, *Biotech. Lett.*, **25**, 1007 (2003).

[58] M. V. Jagannadham, M. K. Chattopadhyay, C. Subbalakshmi, M. Vairamani, K. Narayanan, C. M. Rao and S. Shivaji, *Arch. Microbiol.*, **173**, 418 (2000).

[59] S. Rosa-Putra, A. Hemmerlin, J. Epperson, T. J. Bach, L. H. Guerra and M. Rohmer, *FEMS Microbiol. Lett.*, **204**, 347 (2001).

[60] D. L. Gierhart, *US Patent* 5,308,759 (1994).

[61] A. Berry, D. Janssens, M. Hümbelin, J. P. M. Jore, B. Hoste, I. Cleenwerck, M. Vancanneyt, W. Bretzel, A. F. Mayer, R. Lopez-Ulibarri, B. Shanmugam, J. Swings and L. Pasamontes, *Int. J. Syst. Evol. Microbiol.*, **53**, 231 (2003).

[62] M. Hümbelin, A. Thomas, J. Lin, J. Jore and A. Berry, *Gene*, **297**, 129 (2002).

[63] A. J. Schocher and O. Wiss, *US Patent* 3,891,504, (1975).

[64] A. Berry, W. Bretzel, M. Hümbelin, R. Lopez-Ulibarri, A. F. Mayer and A. A. Yeliseev, *US Patent* 0266518 (2005).
[65] A. Berry, W. Bretzel, M. Hümbelin, R. Lopez-Ulibarri, A. F. Mayer and A. Yeliseev, *World Patent* WO 2002099095 (2002).
[66] G. Raguenes, X. Moppert, L. Richert, J. Ratiskol, C. Payri, B. Costa and J. Guezennec, *Curr. Microbiol.*, **49**, 145 (2004).
[67] D. Asker, T. Beppu and K. Ueda, *Int. J. Syst. Evol. Micriobiol.*, **58**, 601 (2008).
[68] J. Lorquin, F. Molouba and B. L. Dreyfus, *Appl. Environ. Microbiol.*, **63**, 1151 (1997).
[69] L. Hannibal, J. Lorquin, N. A. D'Ortoli, N. Garcia, C. Chaintreuil, C. Masson-Boivin, B. Dreyfus and E. Giraud, *J. Bacteriol.*, **182**, 3850 (2000).
[70] D. Asker and Y. Ohta, *J. Biosci. Bioeng.*, **88**, 617 (1999).
[71] D. Asker and Y. Ohta, *Int. J. Syst. Evol. Microbiol.*, **52**, 729 (2002).
[72] D. Asker and Y. Ohta, *Appl. Microbiol. Biotechnol.*, **58**, 743 (2002).
[73] T. de Miguel, C. Sieiro, M. Poza and T. G. Villa, *Int. Microbiol.*, **3**, 107 (2000).
[74] T. de Miguel, C. Sieiro, M. Poza and T. G. Villa, *J. Agric. Food Chem.*, **49**, 1200 (2001).
[75] P. Veiga-Crespo, L. Blasco, F.R. dos Santos, M. Poza and T. G. Villa, *Int. Microbiol.*, **8**, 55 (2005).
[76] S. L. Wang, D. J. Chen, B. W. Deng and X. Z. Wu, *Yeast*, **25**, 251 (2008).
[77] E. D. Simova, G. I. Frengova and D. M. Beshkova, *J. Indust. Microbiol. Biotechnol.*, **31**, 115 (2004).
[78] P. Buzzini, A. Martini, M. Gaetani, B. Turchetti, U. M. Pagnoni and P. Davoli, *Enzyme Microb. Technol.*, **36**, 687 (2005).
[79] H. Sakaki, T. Nakanishi, K.Y. Satonaka, W. Miki, T. Fujita and S. Komemushi, *J. Biosci. Bioeng.*, **89**, 203 (2000).
[80] J. Tinoi, N. Rakariyatham and R. L. Deming, *Process Biochem.*, **40**, 2551 (2005).
[81] N. Misawa and H. Shimada, *J. Biotechnol.*, **59**, 169 (1998).
[82] M. J. Kang, Y. M. Lee, S. H. Yoon, J. H. Kim, S. W. Ock, K. H. Jung, Y. C. Shin, J. D. Keasling and S. W. Kim, *Biotechnol. Bioeng.*, **91**, 636 (2005).
[83] A. Das, S. H. Yoon, S. H. Lee, J. Y. Kim, D. K. Oh and S. W. Kim, *Appl. Microbiol. Biotechnol.*, **77**, 505 (2007).
[84] W.R. Farmer and J. C. Liao, *Biotechnol. Prog.*, **17**, 57 (2001).
[85] G. Sandmann, M. Albrecht, G. Schnurr, O. Knörzer and P. Böger, *TIBTECH*, **17**, 233 (1999).
[86] Y. Miura, K. Kondo, T. Saito, H. Shimada, P. D. Fraser and N. Misawa, *Appl. Env. Microbiol.*, **64**, 1226, (1998).
[87] S. Yamano, T. Ishii, M. Nakagawa, H. Ikenaga and N. Misawa, *Biosci. Biotechnol. Biochem.*, **58**, 1112 (1994).
[88] C. Wang, M. K. Oh and J. C. Liao, *Biotechnol. Prog.*, **16**, 922 (2000).
[89] D. Umeno and F. H. Arnold, *J. Bacteriol.*, **186**, 1531 (2004).
[90] D. Umeno and F. H. Arnold, *Appl. Environ. Microbiol.*, **69**, 3573 (2003).
[91] G. Sandmann, *ChemBioChem*, **3**, 629 (2002).
[92] C. Schmidt-Dannert, *Curr. Opin. Biotechnol.*, **11**, 255 (2000).
[93] M. Albrecht, S. Takaichi, S. Steiger, Z. Y. Wang and G. Sandmann, *Nature Biotechnol.*, **18**, 843 (2000).
[94] J. M. Jez and J. P. Noel, *Nature Biotechnol.*, **18**, 825 (2000).
[95] H. Ernst, *Pure Appl. Chem.*, **74**, 1369 (2002).

Carotenoids
Volume 5: Nutrition and Health

Chapter 6

Genetic Manipulation of Carotenoid Content and Composition in Crop Plants

Paul D. Fraser and Peter M. Bramley

A. Introduction

Over the past 50 years, modern plant breeding has focused on improved productivity, through increased yield and adaptation to biotic and abiotic stress. In comparison, the enhancement of quality traits such as improved nutritional content and aesthetic colour has been neglected. Now, however, consumers increasingly demand improved food quality and safety and, as a consequence, plant breeding has been forced to address these issues. One example of this is to enhance the levels and types of carotenoids in fruits and vegetables, not only for aesthetic purposes, but also because of the increasing evidence that fruit and vegetables containing high levels of dietary carotenoids are associated with health benefits [1]; such crops are sometimes categorized as 'functional foods' [2].

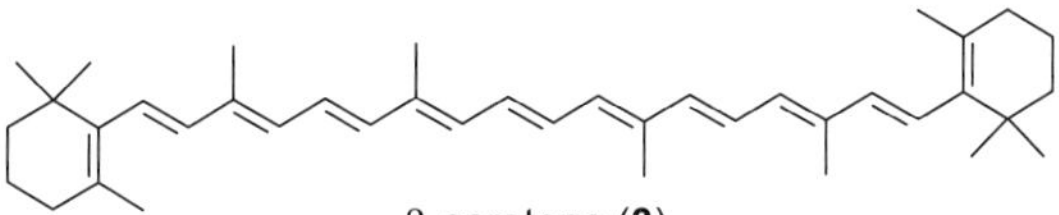

β-carotene (**3**)

The value of carotenoids to human health is supported by a significant body of evidence, as discussed in later chapters in this *Volume*, much of it based on associations between dietary carotenoids and risk of the onset of chronic disease states. β-Carotene (**3**) is the most potent precursor of vitamin A, deficiency of which will cause blindness and eventually death [3].

The xanthophylls zeaxanthin (**119**) and lutein (**133**) have been associated with reduced risk of macular degeneration [4], whilst lycopene (**31**) is associated with the reduction of certain cancers such as prostate cancer [5]. Astaxanthin (**404-406**) has more recently received attention as a carotenoid that may confer preventative effects against cardiovascular disease [6]. The most advantageous effects of carotenoids on health occur when they are eaten in a fruit or vegetable matrix [7], presumably because of synergistic effects with the other health-promoting phytochemicals present in the food. These findings have had a big impact on national health policies of most Western countries, resulting in the recommendation that individuals should consume large quantities of fruits and vegetables ('five a day'), which contain health-promoting phytochemicals, such as carotenoids [8].

Commercially, carotenoids are used in the food, feed, pharmaceutical and cosmetic industries. Although chemical synthesis is the method most often used to produce carotenoids industrially, natural production of carotenoids from plants can offer a more cost-effective and environmentally favourable option.

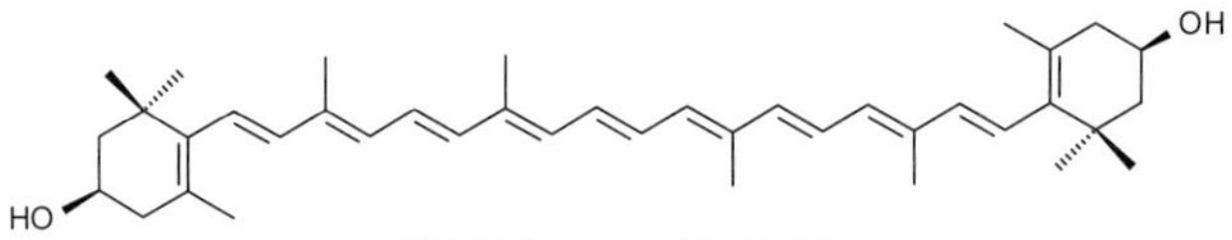

(3*R*,3'*R*)-zeaxanthin (**119**)

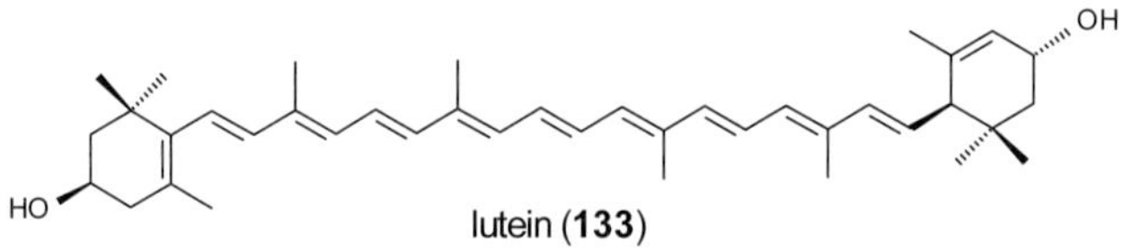

lutein (**133**)

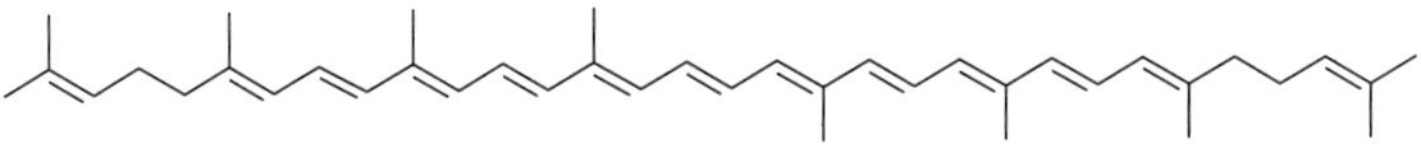

lycopene (**31**)

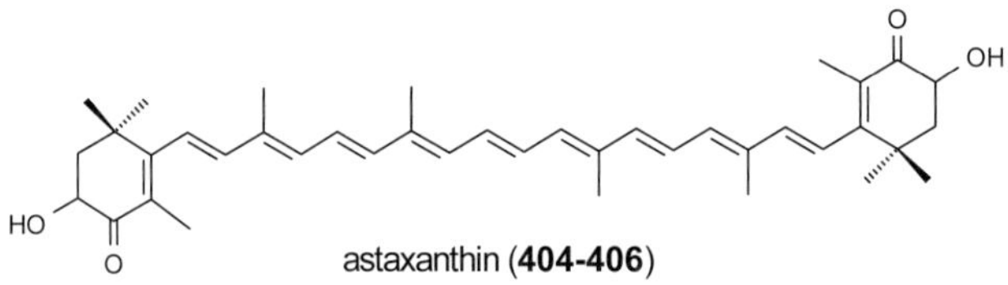

astaxanthin (**404-406**)

B. Strategies for Enhancing Carotenoids in Crop Plants

1. General considerations

The predominant aim of enhancing carotenoids in crop plants is to provide tangible benefits to the quality of human and animal life. In order to achieve this goal there are several prerequisites that should be considered. Addressing these issues at an early conceptual stage will place 'proof of concept' approaches on sound foundations for subsequent scientific developments. The disease state to be addressed through dietary intake needs to be considered, as well as the strength of the experimental and medical evidence supporting the perceived health benefits [8,9]. From a commercial viewpoint, the market needs to be evaluated and the most suitable crop chosen for the countries in which the crop will be used. For example, as the case of the high β-carotene 'Golden Rice' has shown, a local variety of the crop should be used [10]. These factors influence the choice of crop and the target carotenoid(s) within the crop. Synergy with other health-promoting phytochemicals must also be considered. Although not essential, it is advantageous if the crop plant is a staple dietary component, ideally with an established endogenous carotenoid pathway and a known basal carotenoid profile. The generation of genetically modified (GM) crops, especially those possessing traits such as improved nutritional quality, has been restricted by public concerns. The time it may take for these attitudes to change is an important factor that must be considered and has an important bearing on proof of concept, intellectual property and development. Production of carotenoids in non-food crops, followed by bio-fortification of the food chain with supplements, is an alternative means of supplying the consumer with enhanced carotenoid intake [11].

2. Experimental strategies

There are two basic approaches available to generate crop plants with enhanced carotenoid compositions, namely conventional plant breeding, and genetic modification (GM, also termed metabolic or genetic engineering, or genetic manipulation). Over recent years, there have been significant scientific advances in both approaches, due to the development of new technologies. For example, the development of introgression populations and genome sequencing has facilitated efficient molecular marker-assisted breeding [12,13], whilst more efficient transformation vectors and plastid transformation protocols are now widely used. Ideally, the crop plant to be utilized should be amenable to both breeding approaches.

Conventional breeding of tomato has resulted in a wide range of varieties with different carotenoid profiles. These include the high pigment mutants *hp-1* and *hp-2* which have elevated levels of carotenoids, but have weak stems and poor vigour, thus making them unsuitable for commercial exploitation [14]. More recently, a concerted effort to screen the genetic diversity of the tomato has been undertaken, leading to collections of saturated mutant libraries [15] and introgression lines [16] for which metabolic profiles for carotenoid levels can be determined.

The advantage of genetic engineering over conventional plant breeding is the ability to target and transfer gene(s) in a controlled manner and, therefore, in a much shorter time. In addition, the genes can be transferred from unrelated species, including bacteria. There are many reports of the successful elevation of carotenoid levels in crop plants, by use of a variety of genes or cDNAs and promoters. Examples of these are described in section C, and are summarized in Table 1.

Table 1. Examples of genetically modified crops with altered levels of carotenoids.

Inserted gene/cDNA	**Promoter**	**Variety**	**Carotenoid phenotype**	**Ref.**
Rice				
Psy cDNA from daffodil	CaMV 35S	Japonica taipei 309	Phytoene (0.3 μg/gFW) accumulation in endosperm	[17]
Psy cDNA from daffodil	Glutelin	Japonica taipei 309	Phytoene (0.6 μg/gFW) accumulation in endosperm	[17]
crtI (*E. uredovora*) + *Psy* + *Lcy-b* (daffodil)	CaMV 35S Glutelin	Japonica taipei 309	β-Carotene (1.6 μg/gFW) accumulation in endosperm	[18]
crtI + *Psy* + *Lcy-b*	CaMV 35S Glutelin	Indica varieties	β-Carotene (1.6 μg/gFW) accumulation in endosperm	[19]
crtI + *Psy*	Glutelin	Indica	β-Carotene (6.8 μg/gFW) accumulation	[20]
Tomato				
Tomato *Psy-1* antisense	CaMV 35S	Ailsa Craig	100-fold reduction in carotenoids, 3 to 5-fold increase in gibberellins	[21]
E. uredovora, crtI	CaMV 35S	Ailsa Craig	2 to 4-fold increase in β-carotene (20-45 μg/gFW), decreased lycopene	[22]
E. uredovora, crtB	CaMV 35S	Ailsa Craig	2 to 3-fold increases in phytoene, lycopene and β-carotene	[23]
Tomato *Psy-1* sense	PG	Ailsa Craig	Sense suppression, premature lycopene accumulation, dwarf plants (decreased gibberellins and increased ABA)	[24]

Table 1, continued

Inserted gene/cDNA	**Promoter**	**Variety**	**Carotenoid phenotype**	**Ref.**
Tomato				
Yeast *ySAMdc*	E8	VF 36	Increased proportion of β-carotene, 3-fold increase in lycopene	[25]
Tomato *Lcy-b* sense and antisense	Pds	Moneymaker	Up to 7-fold increase in β-carotene	[26]
Tomato *Lcy-b* + Pepper *CrtR-b*	Pds	Moneymaker	Up to 30% increase in lycopene; β-cryptoxanthin: 5 μg/gFW; zeaxanthin 13μg/gFW	[27]
Paracoccus crtW + *crtZ*	CaMV 35S	Ailsa Craig	Ketocarotenoids in leaf	[28]
Tomato *Cry-2*	CaMV 35S	Moneymaker	Hp phenotype	[29]
E. coli Dxps	CaMV 35S and fibrillin	Ailsa Craig	Increase in carotenoids	[30]
Tomato *Det-1*	RNAi + fruit specific promoters	Moneymaker	Increase in carotenoids and flavonoids	[31]
Canola				
E. uredovora, crtB	Napin	Cv 212/86 Quantum	50-fold increase in carotenoids (lutein): 1400 μg/gFW	[32]
Carrot				
E. herbicola, crt genes	CaMV 35S	-	2 to 5-fold increase in carotenoids	[33]
Potato				
Tobacco *Zep*, antisense and sense	GBSS	Freya and Baltica	130-fold increase in zeaxanthin (4 μg/g FW); 5.7-fold increase in total carotenoid (6 μg/gFW)	[34]

Table 1, continued

Inserted gene/cDNA	Promoter	Variety	Carotenoid phenotype	Ref.
Potato				
E. uredovora, crtB	Patatin	Desiree Mayan Gold	7-fold increase in carotenoids 4-fold increase in carotenoids	[35]
E. coli Dxps	Patatin	Desiree	2-fold increase in carotenoids	[36]
Potato *e-Lcy*, antisense	Patatin	Desiree	14-fold increase in β-carotene; 2.5-fold increase in total carotenoids	[37]
Algal *Bkt-1*	Patatin	*S. tuberosum*, *S. phureja*	Accumulation of ketocarotenoids	[38]
Synechocystis crtO + antisense *crtZ*	CaMV 35S	Desiree	Accumulation of ketocarotenoids, *e.g.* astaxanthin	[39]
Erwinia crtB, *crtI* and *crtY*	CaMV 35S or Patatin	Desiree	20-fold increase in β-carotene	[40]
Antisense *Chy-1* and *Chy-2*	Patatin	Desiree	38-fold increase in β-carotene	[41]

All transformations were carried out through *Agrobacterium*-mediated protocols, apart from those in [17], which used microprojectile bombardment. Examples of the transformation of tobacco can be found in [42].
Abbreviations: FW, fresh weight: *Psy*, phytoene synthase; *Pds*, phytoene desaturase; *crtI*, phytoene desaturase; *crtB*, phytoene synthase; *crtO*, carotenoid 4-oxygenase ('ketolase'); *crtW*, β-carotene 4-oxygenase (' ketolase'); *crtY*, lycopene β-cyclase; *crtZ*, β-carotene hydroxylase; β-*Lcy*, lycopene β-cyclase; ε-*Lcy*, lycopene ε-cyclase; *Chy 1 and 2*, β-carotene hydroxylases; *Det-1*, de-etiolated 1; *Zep,* zeaxanthin epoxidase; y*SAM*dc, yeast S-adenosyl methionine decarboxylase; *Dxps*, 1-deoxy-D-xylulose 5-phosphate synthase; *Bkt*-1, β-carotene 4-oxygenase ('ketolase'); CaMV 35S, cauliflower mosaic virus promoter; PG, polygalacturonase; GBSS, granule-bound starch synthase.

3. Optimizing conditions

In order to optimize and control the changes in carotenoid content and composition in crop plants, several prerequisites should be addressed, including the location and activities of enzymes, flux control coefficients, gene expression profiles, carotenoid catabolism, interaction with the biosynthesis of other isoprenoids, regulatory and end-product sequestration mechanisms. A general framework [43] and one addressed more specifically to carotenoid biosynthesis [44] have been described.

a) Choice of crop

In the early studies on genetic modification, the choice of crop was often limited to those that could be transformed efficiently by *Agrobacterium*-based protocols. It is now the case, however, that most crops can be modified effectively in this way, so the choice of crop and species relates to the carotenoids in the wild type, and the species used in the diet in a particular country. In addition, the utility of a plastid transformation system to alter tomato fruit carotenoid content has been demonstrated [45]. As shown in Table 1, the most popular crops used are tomato, rice and potato.

b) Choice of biosynthetic step(s) to target

It is well established that the regulation of carotenogenesis involves the coordinated flux of isoprenoid units into the C_{40} carotenoids and other isoprenoids such as sterols, gibberellins, phytol and terpenoid quinones [46]. An understanding of the complexities of regulation of the pathway is desirable to guide attempts to introduce changes in plant carotenoids by genetic manipulation. However, our understanding of the regulation of the carotenoid pathway is incomplete, so the choice of which step to target cannot be based solely on the current scientific evidence.

Levels of specific carotenoids can also be increased by up-regulation and down-regulation of carotenogenic genes. Qualitative engineering approaches focus primarily on altering the carotenoid composition of a crop. Typically, this is done by utilizing an existing precursor pool in the plant or tissue and redirecting this precursor into the formation of carotenoids. These carotenoids may not be endogenous to the crop undergoing manipulation (*e.g.* astaxanthin in potato [39]). Where no endogenous carotenoids are present, *e.g.* in rice endosperm [18], qualitative and quantitative engineering is required and more than one biosynthesis enzyme must be amplified. To facilitate these approaches, appropriate vectors must be available for multi-gene constructs, for example, vectors that generate self-cleavable poly-proteins [28]. Alternatively, co-transformation or crossing of individual transgenic lines can be used. Carotenoid levels in a crop can also be elevated as a consequence of altering an enzyme or a structural or regulatory protein, in a pathway or biological process which is not directly involved in carotenoid biosynthesis but which nevertheless influences carotenoid formation [31]. In order to achieve down-regulation, antisense and RNAi technology must be feasible in the wild-type crop, as shown recently with tomato [31].

Of the crops that have been manipulated genetically with respect to carotenogenesis (Table 1), probably the most extensively studied is the tomato fruit. Phytoene synthase is significantly up-regulated in ripening tomato fruit [47]. The fruit-specific isoform PSY-1 exhibits the highest flux control coefficient of the enzymes in the carotenoid pathway [23] and has, therefore, often been the target for transformation (Table 1). However, upon introduction of an extra phytoene synthase (*CrtB* from *Erwinia uredovora*), the flux control coefficient for this step decreases, suggesting that control is altered following perturbations of the pathway itself [23]. The expression of the *E. coli* 1-deoxy-D-xylulose 5-phosphate synthase (*Dxps*) in tomato has also been reported. This is thought to be the rate-limiting enzyme in the methylerythritol phosphate (MEP) pathway [48,49], so its up-regulation should increase the formation of the end-product carotenoids. This was indeed the case, albeit with only a modest (1.6-fold) increase in carotenoids in ripe fruit [30].

Phytoene desaturation has also been chosen as a target for genetic engineering. For example, transformation with the *CrtI* gene from *Erwinia* resulted in fruit showing an orange phenotype due to an increase in β-carotene [22].

Transgenic tomatoes have been produced that contain carotenoids not normally present in the fruit. These include ones that contain zeaxanthin (**119**) and β-cryptoxanthin (**55**), through the expression of two cDNAs: the *Arabidopsis β-Lcy* and *Capsicum* β-carotene hydroxylase (*β-Chy*), both with the tomato *Pds* promoter [27]. Tomato has been transformed with two genes from *Paracoccus*, namely the carotene 4,4'-oxygenase (*crtW*) and 3,3'-hydroxylase (*crtZ*), in an attempt to produce ketocarotenoids such as astaxanthin in fruit. Although some ketocarotenoids were found in leaf tissue, none were detected in ripe fruit [28].

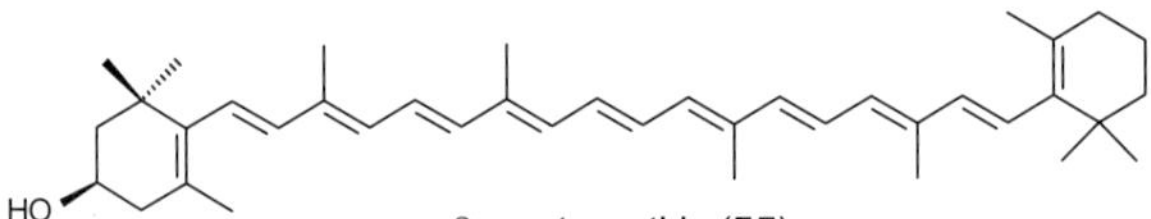

β-cryptoxanthin (**55**)

c) Choice of promoter and gene/cDNA

Most of the carotenoid biosynthesis genes have now been isolated from bacteria, fungi, algae and higher plants, and characterized. To have such a collection of biosynthetic genes, displaying functional similarity but differing homologies at the nucleotide level, is advantageous, because technical problems associated with co-suppression (sense suppression and/or gene silencing) can be alleviated. In addition, it is postulated (or known in some cases), that heterologously expressed enzymes are less susceptible to the regulatory controls, such as allosteric regulation, protein modification or association, that are found with the endogenous system. Only two genes involved in carotenoid sequestration are known (fibrillin and the *Or* genes) and no carotenoid-specific regulatory genes have been isolated from plants.

Various promoters have been used, as outlined in Table 1. These range from constitutive promoters such as CaMV 35S to fruit-ripening specific promoters such as polygalacturonase (PG) and fibrillin. In early studies, the importance of the temporal and spatial expression of the transgene with respect to pleiotropic effects was perhaps overlooked. Experience now suggests that the use of a constitutive promoter usually causes pleiotropic effects that can be detrimental to plant vigour, whilst more specific promoters such as PG, Pds and fibrillin allow metabolic changes to be limited to the fruit itself. An example of this phenomenon was found when the tomato *Psy-1* cDNA was used with the CaMV 35S constitutive promoter, which caused virtually complete absence of fruit carotenoids in some of the transgenic lines [24]. In these cases, the phenotype was very similar to that found with an antisense construct of the same cDNA [21]. Those lines that did not exhibit co-suppression had pleiotropic phenotypes of dwarfism and premature fruit pigmentation, the former being caused by a significant reduction in gibberellins [50]. Co-suppression was successfully avoided by using a synthetic cDNA with low homology to the endogenous gene (<60%) [51], or by using the bacterial homologue of phytoene synthase from *Erwinia uredovora* [23]. Both strategies resulted in increased carotenoid levels in ripe fruit. Probably the most effective promoters with respect to enhancing carotenoid levels in fruit, without detrimental effects, are those involved in very early fruit ripening [31].

d) Targeting of the transgenic protein

When bacterial genes are used for genetic engineering of carotenoids in plants, the transformation vectors must include a plastid transit sequence upstream of the gene of interest. This allows specific targeting of the transgenic protein to the plastid and the subsequent import of the protein in an enzymically active form. Several sequences have been used successfully, including the small subunit of Rubisco (SSU) [22,28], and a modified sequence from *Psy-1* of tomato [24].

C. Examples of the Application of Metabolic Engineering to Carotenoid Formation in Crop Plants

1. Tomato

Developing tomato fruit at first contain chloroplasts and the associated carotenoids but then, as ripening proceeds, chromoplasts develop within the cells and a massively increased (400-fold) accumulation of lycopene occurs. Other acyclic carotenes such as phytoene (**44**) and phytofluene (**42**) also accumulate [47].

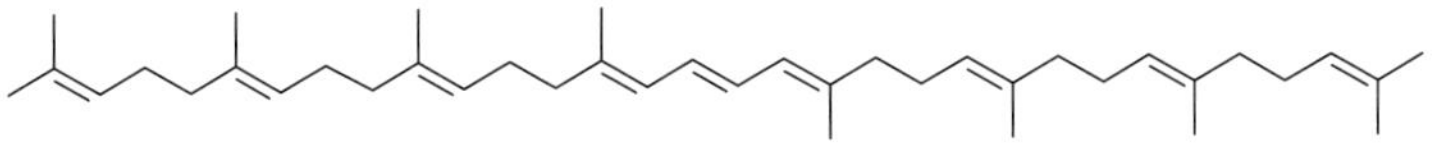

phytoene (**44**)

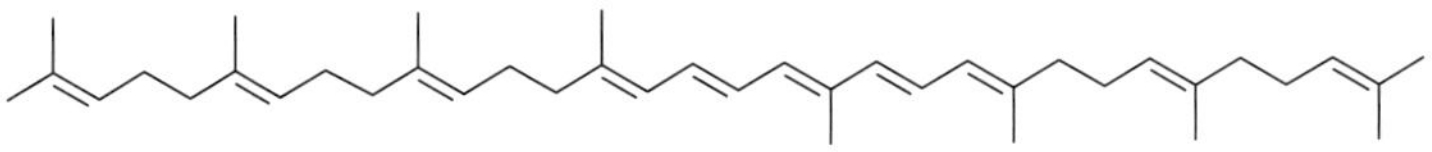

phytofluene (**42**)

Expression studies have revealed that a number of carotenogenic genes are up-regulated during fruit ripening, *e.g.* phytoene synthase-1 (*Psy-1*) [47], carotene isomerase (*CRTISO*) [52], and lycopene β-cyclase [26]. Thus, evidence from gene expression, enzyme activity, flux control values and metabolite levels suggests that phytoene synthase, the first committed step in the formation of carotenoids, exerts the greatest control of flux throughout the pathway [47]. In contrast to the up-regulation of phytoene formation, the down-regulation of lycopene cyclization is also an important factor in facilitating lycopene accumulation in ripe fruit [24].

This knowledge has enabled two strategies to be developed for the quantitative engineering of lycopene content in tomato, namely up-regulation of phytoene synthase and down-regulation of lycopene β-cyclase. Use of an endogenous copy of *Psy-1* and the CaMV 35S promoter resulted in transgenic progeny with detrimental pleiotropic effects [24], especially a dwarf phenotype due to the elevation of carotenoids and abscisic acid (ABA), but reduced gibberellin (GA) levels. In contrast, both ABA and GA levels were reduced in the *Psy-1* antisense fruit [21]. Collectively, these data suggest that the equilibrium of the pool of geranylgeranyl diphosphate (GGDP) is perturbed and more GGDP is channelled into the carotenoid pathway, thus redirecting it from the GA pathway. More recent metabolomic studies have illustrated that effects also extend to intermediary metabolism [53].

Transgenic plants expressing the *E. uredovora* phytoene synthase (*crtB*) showed no pleiotropic effects and ripe fruit contained 2-3 fold increases in carotenoids [23]. These lines prove that amplification of the step in the pathway that has the highest flux control coefficient results in a quantitative increase once co-suppression has been avoided. Characterization of these crtB transgenic plants indicated that the endogenous pathway can compensate for the increased enzyme activity and fluctuations in precursor/product equilibrium by redistributing the balance of control within the pathway. In this case, it appeared, from the accumulation of phytoene, that the subsequent desaturase had become the limiting step [23]. The depletion of prenyl diphosphates and subsequent reduction in GA levels suggests that, in developing fruit, these precursors (or specifically GGDP) are limiting. However, transgenic tomato plants over-expressing the *Erwinia* GGDP synthase (*crtE*) showed no significant increase in end-product carotenoids.

A feed-forward regulatory mechanism associated with *Psy-1* gene expression has been suggested after up-regulation of *Psy-1* with exogenously supplied deoxy-D-xylulose 5-phosphate [49]. On the basis of these findings, transgenic plants over-expressing the *E. coli* deoxy-D-xylulose 5-phosphate synthase (*Dxps*) under constitutive and fruit-ripening enhanced promoters showed moderate increases in end-product carotenoids, but 2-3 fold increases in phytoene levels, suggesting a shift in the equilibrium between precursors and products of the pathway and in the point of control.

The *Erwinia* phytoene desaturase (*crtI*) can convert phytoene into (all-*E*)-lycopene directly. Transgenic tomato plants expressing the *crtI* under constitutive control yielded orange-coloured fruit due to 2 to 4-fold increased β-carotene levels, thus providing 50-100% of the RDA for provitamin A per ripe fruit. Levels of lutein (**133**), zeaxanthin (**119**), neoxanthin

(**234**), antheraxanthin (**231**) and tocopherols were also increased [22] whilst carotenoid intermediates in the pathway to lycopene were all decreased. Gene expression analyses showed a reduction in *Psy-1*, but elevation in the two lycopene β-cyclase genes. Therefore the *crtI* gene product induces subsequent steps in the pathway in a feed-forward manner, but the resulting metabolites appear to be involved in a feedback-inhibition mechanism. Elevations in the β-carotene content of tomato fruit without reduction of lycopene have also been achieved through the expression of lycopene β-cyclase genes by use of either a plastid-based [45] or nuclear-based [26] transformation procedure. Lycopene levels in ripe fruit are also increased (2-fold) by down-regulating *β-Lcy* and *CYC-B* expression through anti-sense technology with the *Pds* and *CYC-B* promoters, respectively [26]. In both cases the carotenoid composition of vegetative tissues was unaffected.

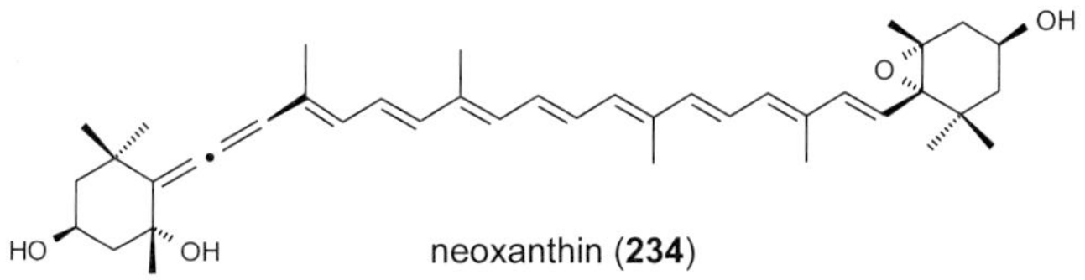

neoxanthin (**234**)

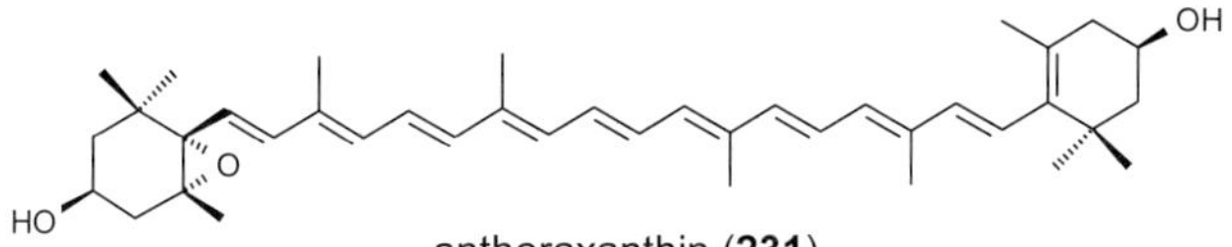

antheraxanthin (**231**)

It is clear from the examples given above and from Table 1 that the manipulation of a specific step or steps in the biosynthetic pathway has been the principal focus of efforts to engineer genetically carotenoid formation in tomato. However, examples of pleiotropic engineering have been reported. For example, high-lycopene transgenic tomato plants resulting from alteration of polyamine levels have been described [25]. The objective of the study was to extend vine longevity, which in turn elevated lycopene levels. More recently, the manipulation of components operating in the light signal transduction pathway and photoreceptors has been reported [29,31]. These studies have shown that the levels of several health-promoting phytochemicals can be elevated. The mode of action of the transgenes in generating such phenotypes is unknown but it appears that increased plastid area during early fruit development is a key factor [29]. This, in turn, suggests that the sequestration and storage within the cell is influential in the accumulation of carotenoids. In order to increase the carotenoid storage potential in plants, cDNAs encoding gene products responsible for plastid division have been expressed in tomato [54]. Despite dramatic effects on plastid morphology to create larger plastids, no increase in carotenoids was observed. The *Capsicum* fibrillin gene product has been shown to facilitate carotenoid sequestration. Expression in tomato, however, resulted only in a moderate elevation of carotenoids [55].

As an alternative to elevating the synthesis of carotenoids, transgenic plants in which the cleavage dioxygenases are down-regulated have been generated [56]. Although the lines showed reduced levels of apocarotenoids, the carotenoid content of the fruit was not altered significantly.

2. Potato

Potato is the most widely consumed vegetable and thus a staple food source for many populations. Potato tubers have a low carotenoid content, with most varieties containing violaxanthin (**259**), antheraxanthin and lutein. Expression of the *Erwinia* phytoene synthase (*crtB*) in a tuber-specific manner led to 6-fold higher carotenoid levels in the flesh of tubers, including a compositional change that resulted in the accumulation of nutritionally significant levels of β-carotene compared to trace levels in the controls [35]. Total carotenoid content has also been increased in potato tubers by down-regulating the endogenous zeaxanthin epoxidase both by sense and antisense suppression [34]. Astaxanthin (**406**) and other ketocarotenoids have been produced in potato tubers through the expression of the β-carotene 4,4′-oxygenase (*Bkt*) from *Haematococcus* [38]. Despite this manipulation of levels of β-ring xanthophylls, which act as ABA precursors, no pleiotropic effects have been reported in potato tubers. Potato tubers heterologously expressing a bacterial *Dxps*, however, showed altered tuber morphology and early tuber sprouting. This phenotype is attributed to increased levels of cytokinins, which are derived biosynthetically from plastid-derived IDP/DMADP [36]. The overall carotenoid content of these potato lines was increased 2-fold, with phytoene being elevated 6-7 fold. Recently potato tubers producing lycopene and very high β-carotene contents have been generated through the silencing of a lycopene ε-cyclase [37] and heterologous expression of a mini-pathway [40].

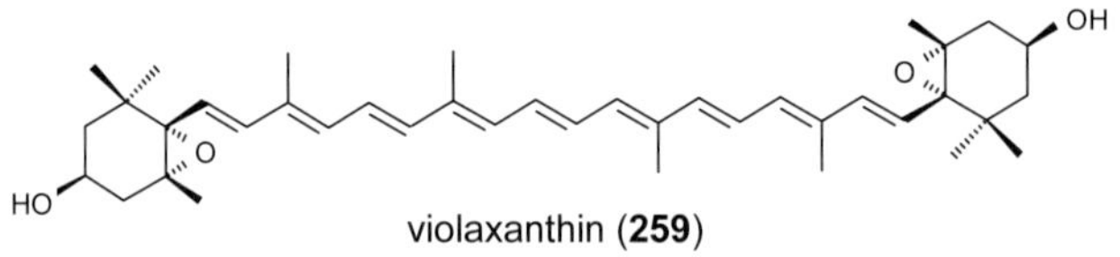

violaxanthin (**259**)

3. Carrot

Carrot roots are a significant source of provitamin A carotenes in the diet. The first report of successful genetic manipulation of carotenoid biosynthesis in a crop plant was reported in carrot, where the *crt* genes from *E. herbicola* were introduced, resulting in 2 to 5-fold increases in root β-carotene content [33]. Apart from this, the study of carotenoid genesis in carrot surprisingly seems to have been neglected.

4. Rice

Vitamin A deficiency is the most common dietary problem affecting children worldwide, and is responsible for 2 million deaths annually (see *Chapter 9*). Rice is the staple foodstuff in many regions, especially Asia, but rice endosperm does not contain carotenoids. The quantitative and qualitative engineering of rice endosperm to produce β-carotene at an appropriate level that could alleviate vitamin A deficiency has been achieved [18] and is now at the development stage [10]. To reach this stage, it was first determined that GGDP was formed and could be utilized [17]. In order to metabolize GGDP to β-carotene, three biosynthetic cDNAs were co-transformed with two vectors. The daffodil (*Narcissus pseudonarcissus*) phytoene synthase and lycopene β-cyclase cDNAs were placed under endosperm-specific control (glutelin promoter) and *Erwinia* phytoene desaturase (*crtI*) under constitutive control. Transformants contained lutein, zeaxanthin, β-carotene and α-carotene (**7**) in their endosperm, with the total carotenoid content being about 1.6 μg/g endosperm. The variety of rice produced was termed 'Golden Rice' [18]. Through the systematic evaluation of different phytoene synthase(s), the carotenoid content of the Golden Rice has now reached 16-26 μg/g endosperm. This variety has been designated 'Golden Rice II' [57] and the levels of β-carotene in the endosperm should provide the RDA of provitamin A in an average rice meal (300 g). The carotenoid phenotypes have also been bred into local cultivars and nutritional and risk assessments are under way.

5. Canola (rape seed)

Canola is not a direct dietary source of carotenoids, but canola vegetable oils are used to prepare many foodstuffs. There have been few basic studies on carotenogenesis in wild-type canola embryos, presumably because of the low carotenoid content. The carotenoid present is typically lutein. The carotenoid content of canola embryos was elevated dramatically (50-fold) by transformation with the *Erwinia* phytoene synthase (*crtB*), expressed in a seed-specific manner [32]. Transgenic canola has also been generated with multiple steps in the biosynthetic pathway amplified. There were qualitative changes in carotenoid composition and additional products were seen following this amplification. Manipulation of phytoene synthase had the greatest influence. The effect of manipulation of additional or multiple gene products did not surpass the effect of the bacterial phytoene synthase alone [58].

D. Conclusions and Perspectives

Over the past 10-15 years, a considerable amount of knowledge has been acquired that facilitates the metabolic engineering of carotenoids in agricultural crops, creating the potential to improve human health through nutritional enhancement. In several crops such as tomato, potato, rice and recently maize [59], the feasibility or 'proof of concept' investigations have been performed successfully. It is now important to carry out safety evaluations, ascertain agronomical properties and assess the nutritional impact of these phenotypes. Transfer of the traits to elite and geographically important varieties can then be carried out. The development of Golden Rice is a very important barometer to the acceptance not only of carotenoid-enhanced crops, but also the feasibility of GM approaches to future developments. The advances in molecular breeding offer an alternative approach if the consumer continues to reject GM crops.

Plant biology is undergoing a period of rapid innovation encompassing interdisciplinary approaches such as the 'omic' technologies which require considerable data handling skills. It is important that such technologies are utilized to evaluate fully metabolic engineering experiments, identify traits in modern breeding programmes and assist nutritional and safety assessments for the enhancement not only of carotenoids, but also of other important nutrients.

References

[1] P. D. Fraser and P. M. Bramley, *Progr. Lipid Res.*, **43**, 228 (2004).

[2] I. Johnson and G. Williamson, *Phytochemical Functional Foods*, Woodhead Publishing, Cambridge (2003).

[3] J. H. Humphrey, K. P. West Jr and A. V. Sommer, *WHO Bulletin*, **70**, 225 (1992).

[4] J. M. Seddon, U. A. Ajami, R. D. Sperato, R. Hiller, N. Blair, T. C. Burton, M. D. Farher, E. S. Gragoudas, J. Haller, D. T. Miller, L. A. Yannazzi and W. Willett, *J. Am. Med. Assoc.*, **272**, 1413 (1994).

[5] E. Giovannucci, *J. Natl. Cancer Inst.*, **91**, 317 (1999).

[6] G. Hussein, H. Goto, S. Oda, U. Sankawa, K. Matsumoto and H. Watanabe, *Biol. Pharm. Bull.*, **29**, 684 (2006).

[7] R. H. Liu, *J. Nutr.*, **134**, 3479C (2004).

[8] T. J. Key, A. Schatzkin, W. C. Willett, N. E. Allen, E. A. Spencer and R. C. Travis, *Public Health Nutr.*, **7**, 187 (2004).

[9] M. A. Grusak and D. DellaPenna, *Annu. Rev. Plant Physiol. Plant Mol. Biol.*, **50**,133 (1999).

[10] I. Potrykus, *Plant Physiol.*, **125**, 1157 (2001).

[11] G. I. Frengova, E. D. Simova and D. M. Beshkova, *Z. Naturforsch. C.*, **61**, 571 (2006).

[12] G. H. Toenniessen, *J. Nutr.*, **132**, 2493S (2002).

[13] Y-S. Liu, A. Gur, G. Ronen, M. Causse, R. Damidaux, M. Buret, J. Hirschberg and D. Zamir, *Plant Biotech. J.*, **1**, 195 (2003).

[14] G. P. Soressi, *Rep. Tomato Genet. Crop*, **25**, 21 (1975).

[15] N. Menda, Y. Semel, D. Peled, Y. Eshed and D. Zamir, *Plant J.*, **38**, 861 (2004).

[16] A. Gur, Y. Semel, A. Cahaner and D. Zamir, *Trends Plant Sci.*, **9**, 107 (2004).

[17] P. K. Burkhardt, P. Beyer, J. Wunn, A. Kloti, G. A. Armstrong, M. Schledz, J. von Lintig and I. Potrykus, *Plant J.*, **11**, 1071 (1997).

[18] X. Ye, S. Al-Babili, A. Klotz, J. Zhang, P. Lucca, P. Beyer and I. Potrykus, *Science*, **287**, 303 (2000).

[19] K. Datta, N. Baiskh, N. Oliva, L. Torriz, E. Abrigo, J. Tan, M. Rai, S. Rehana, S. Al-Babili, P. Beyer, I. Potrykus and S. Datta, *Plant Biotech. J.*, **1**, 81 (2003).

[20] S. Al-Babili, T. T. C. Hoa and P. Schaub, *J. Exp. Bot.*, **57**, 1007 (2006).

[21] C. R. Bird, J. A. Ray, J. D. Fletcher, J. M. Boniwell, A. S. Bird, C. Teulières, I. Blain, P. M. Bramley and W. Schuch, *BioTechnol.*, **9**, 635 (1991).
[22] S. Römer, P. D. Fraser, J. Kiano, C. A. Shipton, N. Misawa, W. Schuch and P. M. Bramley, *Nature Biotech.*, **18**, 666 (2000).
[23] P. D. Fraser, S. Römer, C. A. Shipton, P. B. Mills, J. W. Kiano, N. Misawa, R. G. Drake, W. Schuch and P. M. Bramley, *Proc. Natl. Acad. Sci. USA*, **99**, 1092 (2002).
[24] M. R. Truesdale, *Ph.D. Thesis*, University of London, (1994).
[25] R. A. Mehta, T. Cassol, N. Li, A. K. Handa and A. K. Mattoo, *Nature Biotech.* **20**, 613 (2002).
[26] C. Rosati, R. Aquilani, S. Dharmapuri, P. Pallara, C. Marusic, R. Tavazza, F. Bouvier, B. Camara and G. Giuliano, *Plant J.*, **24**, 413 (2000).
[27] S. Dharmapuri, C. Rosati, P. Pallara, R. Aquilani, F. Bouvier, B. Camara and G. Giuliano, *FEBS Lett.*, **519**, 30 (2002).
[28] L. Ralley, E. M. A. Enfissi, N. Misawa, W. Schuch, P. M. Bramley and P. D. Fraser, *Plant J.*, **39**, 477 (2004).
[29] L. Giliberto, G. Perrotta, P. Pallara, J. L. Weller, P. D. Fraser, P. M. Bramley, A. Fiore, M. Tavazza and G. Giuliano, *Plant Physiol.*, **137**, 199 (2005).
[30] E. M. A. Enfissi, P. D. Fraser, L. M. Lois, A. Boronat, W. Schuch and P. M. Bramley, *Plant Biotech. J.*, **3**, 17 (2005).
[31] G. R. Davuluri, A. van Tuinen, P. D. Fraser, A. Manfredonia, R. Newman, D. Burgess, D. A. Brummell, S. R. King, J. Palys, J. Uhlig, P. M. Bramley, H. M. J. Pennings and C. Bowler, *Nature Biotech.*, **23**, 890 (2005).
[32] C. K. Shewmaker, J. A. Sheehy, M. Daley, S. Colburn and D. Y. Ke, *Plant J.*, **20**, 401 (1999).
[33] R. Hauptmann, W. H. Eschenfeldt, J. English and F. L. Brinkhaus, US Patent 5618988 (1997).
[34] S. Römer, J. Lübeck, F. Kauder, S. Steiger, C. Adomat and G. Sandmann, *Metabol. Eng.*, **4**, 263 (2002).
[35] L. J. M. Ducreux, W. L. Morris, P. E. Hedley, T. Shepherd, H. V. Davies, S. Millam and M. A. Taylor, *J. Exp. Bot.*, **56**, 81 (2005).
[36] W. L. Morris, L. J. M. Ducreux, P. Hedden, S. Millam and M. A. Taylor, *J. Exp. Bot.*, **57**, 3007 (2006).
[37] G. Diretto, R. Tavazza, R. Welsch, D. Pizzichini, F. Mourgues, V. Papacchioli, P. Beyer and G. Giuliano, *BMC Plant Biol.*, **6**, 13 (2006).
[38] W. L. Morris, L. J. M. Ducreux, P. D. Fraser, S. Millam and M. A. Taylor, *Metabol. Eng.*, **8**, 253 (2006).
[39] T. Gerjets and G. Sandmann, *J. Exp. Bot.*, **57**, 3639 (2006).
[40] G. Diretto, S. Al-Babili, R. Tavazza, V. Papacchioli, P. Beyer and G. Giuliano, *PLOS One*, **4**, c350 (2007).
[41] G. Diretto, R. Welsch, R. Tavazza, F. Mourgues, D. Pizzichini, P. Beyer and G. Giuliano, *BMC Plant Biol.*, **7**, 11 (2007).
[42] P. M. Bramley, in *Plant Genetic Engineering* (ed. R. P. Singh and P. K. Jaiwal), Vol. 1, p. 229, Sci. Tech. Pub., LLC, USA (2003).
[43] K. M. Davies, *Mutation Research,* **622**, 122 (2007).
[44] G. Sandmann, S. Römer and P. D. Fraser, *Metabol. Eng.*, **8**, 291 (2006).
[45] D. Wurbs, S. Ruf and R. Bock, *Plant J.* **49**, 276 (2007).
[46] P. M. Bramley, *J. Exp. Bot.*, **53**, 2107 (2002).
[47] P. D. Fraser, M. R. Truesdale, C. R. Bird, W, Schuch and P. M. Bramley, *Plant Physiol.*, **105**, 405 (1994).
[48] L. M. Lois, M. Rodriguez-Concepcion, F. Gallego, N. Campos and A. Boronat, *Plant J.*, **22**, 503 (2000).
[49] M. Rodriguez-Concepcion, I. Ahamada, E. Diez-Juez, S. Sauret-Gueto, L. M. Lois, F. Gallego, L. Carretero-Paulet, N. Campos and A. Boronat, *Plant J.*, **27**, 213 (2001).
[50] R. G. Fray, A. Wallace, P. D. Fraser, D. Valero, P. Hedden, P. M. Bramley and D. Grierson, *Plant J.*, **8**, 693 (1995).
[51] R. G. Drake, C. R. Bird and W. Schuch, US Patent WO97/46690 (1996).

[52] T. Isaacson, G. Ronen, D. Zamir and J. Hirschberg, *Plant Cell*, **14**, 333 (2002).

[53] N. Telef, L Stammitti-Bert, A Mortain-Bertrand, M. Mauciurt, J.-P. Carde, D. Rolin and P. Gallusci, *Plant Mol. Biol.*, **62**, 453 (2006).

[54] P. J. Cookson, J. W. Kiano, C. A. Shipton, P. D. Fraser, S. Römer, W. Schuch, P. M. Bramley and K. A. Pyke, *Planta*, **217**, 896 (2003).

[55] A. J. Simkin, J. Gaffe, J-P Alcaraz, J.-P. Carde, P. M. Bramley, P. D. Fraser and M. Kuntz, *Phytochemistry*, **68**, 1545 (2007).

[56] M. E. Auldridge, D. R. McCarty and H. J. Klee, *Curr. Opin. Plant Biol.*, **9**, 315 (2006).

[57] J. A. Payne, C. A. Shipton, S. Chaggar, R. M. Howells, M. J. Kennedy, G. Verson, S. Y. Wright, E. Hinchliffe, J. L. Adams, A. L. Silverstone and R. Drake, *Nature Biotechnol.*, **23**, 482 (2005).

[58] M. P. Ravanello, D. Ke, J. Alvarez, B. Huang and C. K. Shewmaker, *Metabol. Eng.*, **5**, 255 (2003).

[59] C. Zhu, S. Naqvi, J. Breitenbach, G. Sandmann, P. Christou and T. Capell, *Carotenoid Sci.*, **12**, 61 (2008).

Carotenoids
Volume 5: Nutrition and Health

Chapter 7

Absorption, Transport, Distribution in Tissues and Bioavailability

Kirstie Canene-Adams and John W. Erdman Jr.

A. Introduction

Carotenoids can be detected in human blood and tissues, though usually only at quite low concentration. They are not biosynthesized in the human body but have to be provided in the diet, in food or as supplements. The processes by which the ingested carotenoids are absorbed, transported in the body and deposited in tissues and organs are of fundamental importance in relation to any effect of carotenoids on human health. The term '**bioavailability**' is used to refer to how much of a consumed carotenoid is accessible for utilization in normal physiological functioning, metabolism, or storage and it encompasses absorption, *i.e.* how carotenoids become soluble and incorporated into mixed micelles, transport, *i.e.* how carotenoids are moved into the intestinal enterocytes, packaged into chylomicrons and moved throughout the blood in lipoproteins, and the tissue-specific accumulation in the human body. The conversion of β-carotene and other carotenoids into vitamin A is treated in *Chapter 8*.

For humans, the major source of carotenoids in our diet is fruits and vegetables. The bioavailability of carotenoids from these is affected by many factors, notably the food matrix, location of carotenoids in the individual food sources, effect of food processing, and other dietary factors, as well as human factors such as age and infections. There is also wide variation between individuals. To evaluate and study carotenoid bioavailability is, therefore, not straightforward, but various methods are available. Strengths and weaknesses of these methods are assessed in Section **D** below.

B. Absorption, Transport, and Storage in Tissues

1. Overview

Figure 1 illustrates the complexity of the entry and movement of carotenoids in the human body. Carotenoids are partially released from the food matrix by chewing and the action of stomach acids and digestive enzymes. After being released from the matrix in which they are consumed, the carotenoids must be solubilized into micelles, absorbed, and packaged into chylomicrons before being transported and stored in various tissues. The entry of the ingested carotenoids into the intestinal mucosal cells is defined as **carotenoid uptake**, and movement of the carotenoids from the mucosal cells into the lymphatic blood system as **absorption**. Since carotenoids are fat-soluble, they are absorbed in a similar fashion to other dietary lipids, and inhibitors of cholesterol transport have been shown also to inhibit carotenoid transport [1].

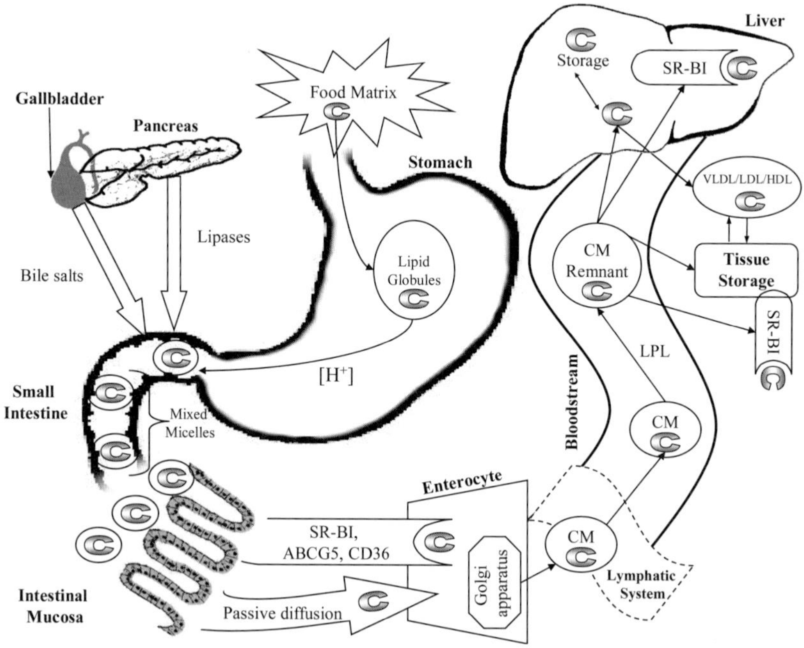

Fig. 1. Diagram tracing the passage of carotenoid molecules (C) from food to blood and tissues. Carotenoids are solubilized in lipid globules, the carotenes in the triacyglycerol-rich core, the xanthophylls near the surface monolayer. Lipids in the small intestine cause the release of bile acids and lipases. Carotenoids are taken up by the enterocytes *via* passive diffusion or *via* scavenger receptor class B type I protein (SR-BI), and enter the lymphatic system in chylomicrons (CM) which, in the bloodstream, are broken down by lipoprotein lipase (LPL). The CM remnants are taken up by the liver. SR-BI in tissues is thought to aid in tissue uptake of carotenoids. Released carotenoids are either stored in the liver or secreted into the circulation in very low density (VLDL) or low density (LDL) lipoproteins. Carotenoids may be taken into tissues directly from CM remnants.

2. Solubilization and incorporation into micelles

After carotenoids are released from the food matrix, they are incorporated into lipid globules in the stomach, where gastric mixing acts to form a lipid emulsion [2]. Only trace amounts (0-1.2%) of carotenoids have been found in the aqueous phase of the stomach [3]. The stomach's role in carotenoid absorption is to initiate the transfer of carotenoids from the food matrix to the lipid portion of the meal. In the lipid phase of a meal the different carotenoid types can differ in their location. Often the carotenes are buried in the triacylglycerol-rich core of the oil drops, whereas the xanthophylls, bearing hydroxy or other functional groups, are more polar, and are, therefore, more likely to reside near the surface monolayer, together with proteins, phospholipids and partially ionized fatty acids. Localization of carotenoids influences the next step of carotenoid digestion, namely transfer into mixed micelles [4], and thus affects carotenoid bioavailability. It has been suggested that the xanthophylls, being located near the surface, may be more easily incorporated into the lipid droplets than the carotenes, which must penetrate into the lipid core [5].

This lipid-carotenoid emulsion then enters the duodenum, where the fat causes the secretion of bile acids from the gall bladder and of lipases from the pancreas (Fig. 1). The surfactant bile acids act to reduce the size of the lipid drops, resulting in the creation of mixed micelles (Fig. 2); in this process, carotenoids are solubilized along with dietary fat.

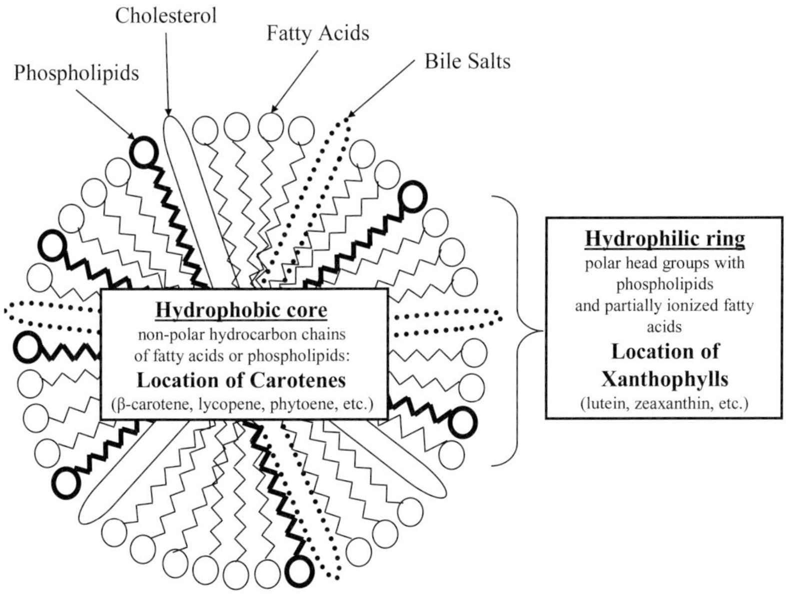

Fig. 2. Carotenoids in mixed micelles. Carotenes are located in the triacyglycerol-rich core, whereas xanthophylls are on the surface monolayer with proteins, phospholipids and partially ionized fatty acids.

The extent to which carotenoids are taken into the micelles depends on the polarity of the carotenoid and the fatty acid composition, chain length and level of saturation of the fat. The effect of lipids on carotenoid bioavailability are described in more detail later in the chapter. Both bile salts and pancreatic enzymes are essential for efficient incorporation of carotenoids into micelles. Micelles provide a vehicle for the lipids, and therefore the lipid-soluble carotenoids, to diffuse across the unstirred water layer.

To determine the effect of the human digestive tract on carotenoid absorption and bioavailability, ten healthy men were fed liquid meals containing sunflower oil (40 g), whey proteins, sucrose, soy lecithin, and either tomato puree, chopped spinach, or pureed carrots as sources of 10 mg lycopene (**31**), 10 mg lutein (**133**), or 10 mg β-carotene (**3**), respectively [3]. The carrot and spinach meals led to an increase in chylomicron β-carotene and lutein, respectively, but the tomato-containing meal resulted in no significant variation in chylomicron lycopene. The response to β-carotene was not the same for all isomers; (13*Z*)-β-carotene was solubilized into micelles to a greater extent than the (all-*E*)-isomer. This study confirmed that approximately 7% of carotenoids are recovered in the micellar phase of the duodenum and that lycopene is less efficiently transferred to micelles than either β-carotene or lutein.

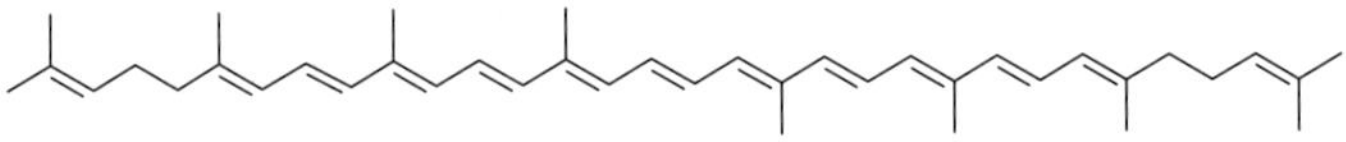

lycopene (**31**)

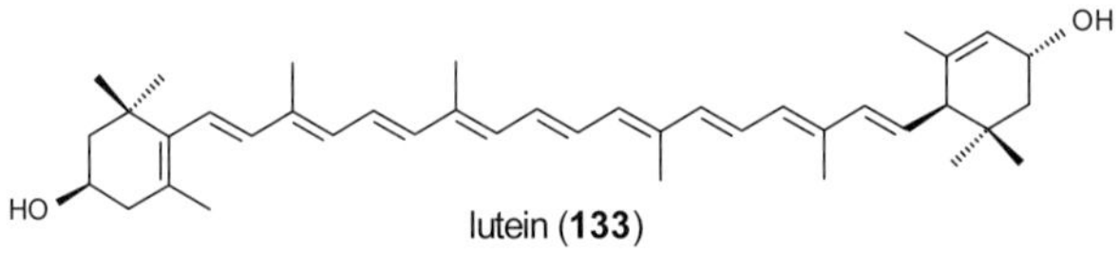

lutein (**133**)

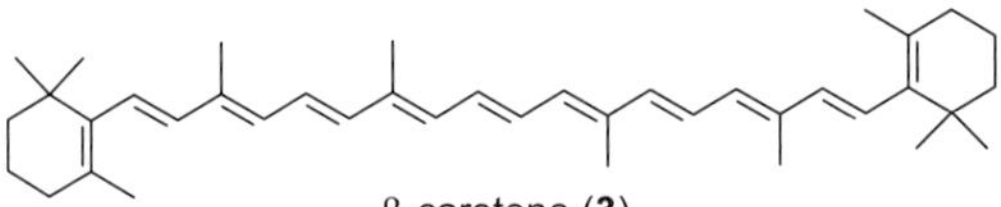

β-carotene (**3**)

Three groups of gerbils were given, by oral gavage, either pure (all-*E*)-β-carotene, (9*Z*)-β-carotene, or (13*Z*)-β-carotene in oil [6]. Geometrical isomerization could occur in the digestive tract because of heat, pH, or gastric microflora, and the presence of isomers other than the one fed was observed in the digestive tract six hours after dosing, yet further isomerization in the small intestine was seen only with the groups administered the Z (*cis*) isomers. The mucosal scrapings of the small intestine showed higher levels of β-carotene in the groups that had been given (all-*E*)-β-carotene than in the groups given either of the *Z* isomers, which seems to indicate a preferential uptake of (all-*E*)-β-carotene.

The *Z* isomers of lycopene are secreted in chylomicrons, suggesting a possible post-enterocyte origin of these isomers. A ferret model of absorption revealed an increase (though not statistically significant) in *Z* isomers of lycopene in the small intestine contents relative to the stomach contents [7]. There was a significant increase, however, in the proportion of *Z* isomers in the small intestine contents and the mucosal lining, due to increased incorporation of the *Z* isomers into bile acid micelles. (all-*E*)-Lycopene has been shown to be isomerized to various *Z* isomers in the acidic environment of the gastric milieu [8]. Furthermore, the acidic environment of the stomach has been shown to create breakdown products which may have biological actions such as anti-proliferation [9].

3. Intestinal absorption

The mixed micelle must then enter the unstirred water layer of the microvillus cell membrane of the enterocyte, where the micelles can diffuse into the membrane and release fat and carotenoids into the cytosol of the cell [2].

Until recently, it was assumed that the process of carotenoid absorption occurs *via* passive diffusion [10]. Recent research, however, indicates the possible involvement of an active process for uptake of carotenoids *via* Scavenger Receptor class B type I protein (SR-BI), which is partially responsible for the transport of lipids and cholesterol from lipoproteins to tissues and from tissues to lipoproteins [11,12]. The SR-BI contains a large extracellular domain and is anchored to each side of the plasma membrane by transmembrane domains adjacent to short cytoplasmic N-terminal and C-terminal domains [13]. The main role of SR-BI is not yet understood, but evidence indicates that it plays a role in selective uptake of cholesterol esters. SR-BI is found in human adrenals, ovaries, placenta, kidneys, prostate, and liver [14], and throughout the small intestine. Another protein, cluster determinant 36 (CD36), a surface membrane glycoprotein involved in the uptake of long-chain fatty acids and oxidized low-density lipoproteins, is found mainly in the duodenum and jejunum [15]. Because SR-BI and CD36 have similar functions in lipid mobilization, it has been suggested that CD36 might also play a role in movement of carotenoids into cells.

It is not known how SR-BI binds lipoproteins to the cell surface to transport cholesterol, lipids or carotenoids into cells. A cholesterol-transport inhibitor, ezetimibe, was shown also to inhibit carotenoid transport in the Caco-2 cell model either by interfering physically with transporters such as SR-BI, Niemann-Pick type C1-Like Protein 1 (NPC1L1) (a recently discovered protein thought to play a role in the absorption of cholesterol), and the ATP-binding cassette transporter subfamily, or by downregulating the expression of these proteins [1]. This inhibitory effect was smaller for carotenoids of greater polarity. When SR-BI transgenic mice, in which SR-BI was over-expressed in the intestine, were fed a diet enriched with 0.25 g/kg (all-*E*)-lycopene for one month, their plasma lycopene was almost ten times higher than in controls, signifying that SR-BI aids in lycopene transport across the intestinal walls [16]. Inhibition of SR-BI *via* a blocking antibody only partially affected lycopene uptake *via* Caco-2 cells, but neither an NPC1L1-blocking antibody nor ezetimibe had any effect on lycopene transport. Also, lycopene did not increase the mRNA levels for SR-BI or NPC1L1 in the mouse intestines or in Caco-2 cells, indicating that there must also be another mechanism of lycopene transport such as passive diffusion, or CD36. In Caco-2 TC-7 cells, lutein was shown to be transported *via* SR-BI, as indicated by saturation and impairment of uptake at low temperature (4°C) [17]. Many biological processes are slower with decreased temperature, and this inhibition of lutein transport could be ascribed to the impairment of one or more receptors/transporters at this low temperature. Because some lutein was still absorbed at 4°C, however, these results also suggest that a fraction of the lutein may be absorbed by passive diffusion. The absorption rate from the basolateral side to the apical side was much slower than the reverse, indicating that there is indeed a transporter or a receptor for a transporter, which may be SR-BI, on the apical membrane aiding in carotenoid transport. The *ninaD* mutant of *Drosophila* has a nonsense mutation in a gene encoding for SR-BI, and lacks carotenoids because of the inability to absorb them [18]. Studies both *in vivo* and *in vitro* showed that two brush border membrane class B scavenger receptors, SR-BI and CD36, facilitate the absorption of β-carotene. Studies with the anti-SR-BI antibody pAB150 or with SR-BI –/– mice, showed inhibition of β-carotene uptake [19]. In none of these studies, however, was there complete inhibition of carotenoid transport, suggesting that more than one mechanism is important and passive diffusion or additional transporters may be involved.

The ATP-Binding Cassette G5 (ABCG5), found in the liver and intestines, uses ATP to drive molecules such as lipids across cell membranes. Persons with the human genetic variant of the ABCG5, the C/C genotype, had higher plasma response to lutein from eggs than did those with the C/G genotype [20]. Consequently, the ABCG5 could be another factor which has some role in the transport of carotenoids into the enterocytes, and eventually into the liver.

A variety of factors can enhance the pace of diffusion. These include acidification of the luminal contents, addition of fatty acids, and decreasing the thickness of the unstirred water layer [10,21]. An acidic environment increases the concentration of hydrogen ions and these suppress the negative surface charge of both the micelle and luminal absorptive cell membrane, thus allowing increased diffusion of micelles and diffused lipids [8]. At high doses of carotenoids, it is thought that two concentration gradients determine the rate of absorption, namely (i) movement of carotenoids from the micelle to the brush border membrane and (ii) removal of carotenoids from this membrane to intracellular locations [22]. Carotenoids which are taken up by the enterocytes, but not incorporated into chylomicrons, are sloughed off during enterocyte turnover and enter the lumen of the gastrointestinal tract [23]. Both the maturity of the enterocyte and the morphology of the mucosa influence the rate of enterocyte turnover and the release of carotenoids back into the lumen. In a study with rats, pre-feeding a diet rich in carotenoids, in this case from tomato powder, was found to decrease the absorption of a subsequent single oral dose of lycopene, and consequently the further accumulation of that carotenoid in tissues [24].

4. Transport in blood

a) Incorporation into chylomicrons

In the Golgi apparatus of the enterocytes, carotenoids and lipids are formed into chylomicrons, large lipoprotein particles which, after being released into the lymphatic system, deliver exogenous lipids and lipid-soluble components from the intestines to other organs in the body. How carotenoids are translocated into the Golgi apparatus is not yet known; intracellular binding proteins may be involved [2]. It is only newly absorbed carotenoids that are packaged in chylomicrons for circulation in the lymphatic system. Once in the lymphatic system, the carotenoid-containing chylomicrons are delivered to the circulation *via* the thoracic duct.

b) Other lipoproteins

Chylomicrons in the bloodstream are degraded by lipoprotein lipase, leaving chylomicron remnants which are quickly taken up by the liver. In the fed state, the liver will store or secrete the carotenoids in very low density lipoproteins (VLDL) and low density lipoproteins (LDL). In the fasted state, plasma carotenes are found in LDL, whereas the more polar carotenoids (xanthophylls) are located mainly in LDL and high density lipoproteins (HDL), and a small proportion in VLDL. These triacylglycerol-rich lipoproteins serve as carotenoid transporters in the blood, but they do not carry carotenoids to equal extents. LDL transport accounts for about 55%, HDL for 31%, and VLDL for 14% of total blood carotenoids. The relative proportions of the main carotenoids in the different lipoproteins is given in Table 1 [25]. As with micelles, the carotenes are found in the hydrophobic core of these lipoproteins and the more polar carotenoids closer to the surface. Some carotenoids may be released from the postprandial circulating triacylglycerol-rich lipoproteins and taken up directly by extrahepatic tissues [24] (Fig. 1). Specific factors that regulate tissue uptake, recycling of carotenoids back to the liver, and excretion are not understood.

Table 1. Distribution (as percentage of the total) of total and individual carotenoids in human blood lipoprotein fractions [25]).

Lipoprotein	Total carotenoid	Lutein (**133**)/ Zeaxanthin (**119**)	β-Crypto-xanthin (**55**)	Lycopene (**31**)	α-Carotene (**7**)	β-Carotene (**3**)
VLDL	14	16	19	10	16	11
LDL	55	31	42	73	58	67
HDL	31	53	39	17	26	22

It would be expected that, as serum cholesterol and triacylglycerol levels increase, the ability to transport carotenoids would be increased. Interestingly, however, an inverse relationship was found between serum triacylglycerol concentration and lycopene levels [26]. This is suggested to be because of the food intake patterns; people who have high fat intake and hence high blood lipid levels consistently have a low intake of carotenoid-containing foods such as fruits and vegetables. Another study, however, with participants who were known to be responders or non-responders to dietary cholesterol, showed that the plasma response to cholesterol could predict the plasma response to carotenoids [27]. Those who were hyper-responders to consumed egg cholesterol also had higher baseline plasma levels of lutein, zeaxanthin (**119**), α-carotene (**7**), and β-carotene, as well as a greater plasma response to oral provision of carotenoids. Interestingly, some had a hyper-response to lutein but not to β-carotene, as was seen in the feeding study with yellow carrots described in Section **C**.5.c [28].

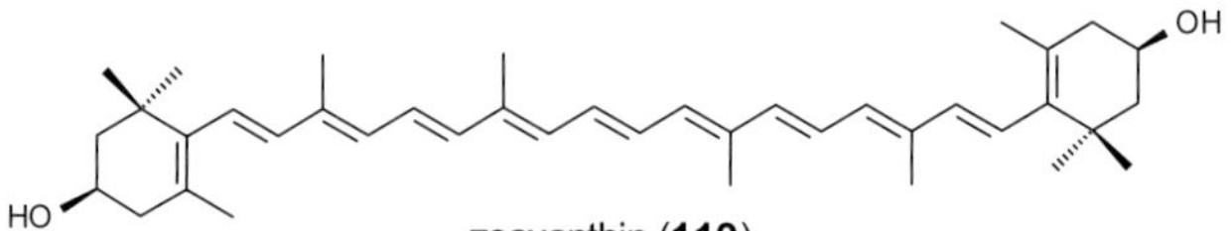

zeaxanthin (**119**)

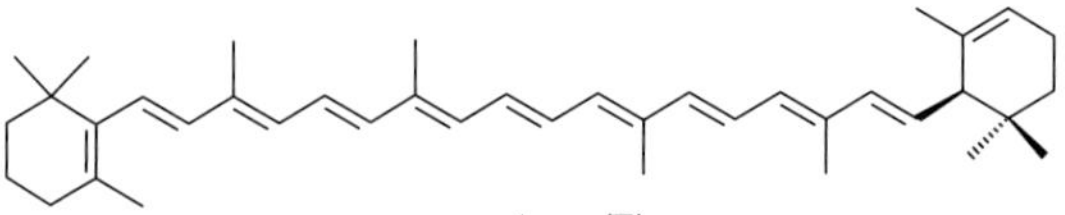

α-carotene (**7**)

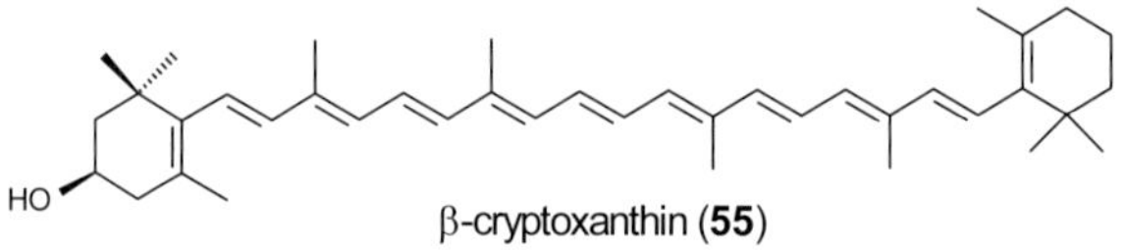

β-cryptoxanthin (**55**)

5. Accumulation and distribution in tissues

a) General features

All the carotenoids that are found in human serum also accumulate in other organs and tissues, but in substantially different concentrations. The liver, adrenals, and reproductive tissues generally have ten times higher carotenoid concentrations than other tissues, including adipose tissue. The five most prominent carotenoids found in US citizens are β-carotene, α-carotene, lycopene, lutein, and β-cryptoxanthin (**55**) [29]. It is thought that the differential tissue uptake of carotenoids such as lycopene and β-carotene is dependent on the amount of LDL receptors and SR-BI [18]. The isomeric pattern of a particular carotenoid in tissues does not necessarily reflect the pattern of isomers in the food source. For example, tomatoes contain >95% (all-*E*)-lycopene yet *Z* isomers account for >50% of blood lycopene and >75% of tissue lycopene. This indicates that there is selective uptake by tissues, not only of different carotenoids but also of geometrical isomers of a particular carotenoid. This selective uptake is aided by the interaction of lipoproteins with receptors, *e.g.* SR-BI, and the degradation of lipoproteins by extra-hepatic enzymes *e.g.* lipoprotein lipase. The primary accumulation sites for the largest amounts of carotenoids are the liver and adipose tissue.

b) Blood

After a meal, carotenoids appear first in the chylomicron fraction of the blood, but the peak blood concentration of carotenoids occurs 24-48 hours after consumption. This reflects the time needed for transport of carotenoid-containing chylomicrons to the liver and then the secretion of carotenoids as components of lipoproteins. Table 2 shows typical levels of carotenoids found in human plasma. Plasma makes up about 55% of the blood volume and contains mostly water and proteins including fibrinogen, globulins, and human serum albumin. Serum refers to blood plasma from which the clotting factors have been removed. There is an agreement between carotenoid concentrations measured in the serum and Li-heparin plasma, thus values across studies can be compared [30]. This *Chapter* will refer to serum or plasma levels as the blood levels of carotenoids.

When people are fed carotenoid-depleted diets, reduction of levels of the blood carotenoids follows first-order kinetics with reported half-lives of 76 days for lutein, 45 days for α-carotene, 39 days for β-cryptoxanthin, 38 days for zeaxanthin, 37 days for β-carotene, and 26 days for lycopene [31]. Another study showed that, when a lycopene-free diet is consumed, it takes only two weeks for blood to show a 50% reduction in lycopene concentration, with (all-*E*)-lycopene being cleared more rapidly than the *Z* isomers [32]. In addition, when subjects consumed foods rich in lycopene, blood lycopene levels reached a plateau after two weeks of daily consumption [32].

Blood carotenoids fluctuate significantly during the menstrual cycle so, when studies are performed in which blood carotenoid concentrations are an end point, the menstrual phase of the female participants should be taken into consideration. The phases of the cycle are as follows: menses for 1-2 days, early follicular for the next 4-6 days, the late follicular phase when there is a surge in luteinizing hormone about 11 days later, and then the mid-luteal phase which occurs approximately 8 days after the surge in luteinizing hormone. Not every carotenoid responds in the same way with each phase of the menstrual cycle. For example, α-carotene levels in the LDL fraction of the blood were lower in the early follicular phase than in the late follicular or luteal phases, whereas β-carotene, lutein, and zeaxanthin blood levels were highest in the late follicular phase [33,34] and plasma lycopene and phytofluene concentrations reached their highest levels at the mid-luteal phase. It is claimed, however, that there is not a change in carotenoids as a consequence of the menstrual cycle when adjustment is made for cholesterol levels [40].

Table 2. Concentrations (μmol/L) of carotenoids in human plasma.

Subject	Carotenoid	Concentration	Reference
Male	Lycopene	0.47	[26]
Female	Lycopene	0.43	[26]
Both sexes	Lutein	0.22 - 0.43	[30,35]
Both sexes	Zeaxanthin	0.03 - 0.12	[30,35]
Both sexes	α-Cryptoxanthin	0.09	[30]
Both sexes	β-Cryptoxanthin	0.21-0.37	[30,36]
Both sexes	α-Carotene	0.08 - 0.22	[30,35,36]
Both sexes	β-Carotene	0.35-0.69	[30,36]
Both sexes	Lycopene	0.43-0.66	[30,36]
Pregnant	α-Carotene	0.02	[37]
Pregnant	β-Carotene	0.59	[37]
Pregnant	Lutein	1.61	[37]
Pregnant	Zeaxanthin	0.17	[37]
2 days post-partum	Lutein	0.35	[38]
2 days post-partum	Zeaxanthin	0.05	[38]
2 days post-partum	β-Cryptoxanthin	0.41	[38]
2 days post-partum	α-Carotene	0.19	[38]
2 days post-partum	β-Carotene	0.65	[38]
2 days post-partum	Lycopene	0.52	[38]
19 days post-partum	Lutein	0.27	[38]
19 days post-partum	Zeaxanthin	0.04	[38]
19 days post-partum	β-Cryptoxanthin	0.31	[38]
19 days post-partum	α-Carotene	0.20	[38]
19 days post-partum	β-Carotene	0.73	[38]
19 days post-partum	Lycopene	0.49	[38]
Infant : breast fed	Lycopene	0.04	[39]
Infant : breast fed	α-Carotene	0.01	[39]
Infant : breast fed	β-Carotene	0.06	[39]
Infant : formula fed	Lycopene	undetectable	[39]
Infant : formula fed	α-Carotene	undetectable	[39]
Infant : formula fed	β-Carotene	0.03	[39]

c) Liver

The liver is a major accumulation site for carotenoids, in part because of its large size. It has been thought that there could be one or more carotenoid-binding proteins in the liver which concentrate various carotenoids and allow the storage or assembly of lipoproteins in the liver tissue. SR-BI mRNA, for example, is found in very high levels in the human liver [14], so SR-BI could act as a selective transporter to move carotenoids from circulating lipoproteins to the liver. Additionally, a carotenoid-protein complex from the livers of rats fed β-carotene has been partially characterized [41]. The subcellular distribution indicated that most of the complex was found in the mitochondrial and lysosomal fractions, indicating that this carotenoid-protein complex is part of the membrane fraction of the liver cell. Twenty-four hours after [^{14}C]-lycopene was administered to rats, 80% of the radioactivity in hepatic tissue was present in lycopene (all-*E* and *Z* isomers) and 20% in polar metabolites. The total radioactivity decreased in the liver after 24 hours but no decrease in the polar metabolites was seen [42]. Further investigation by HPLC-MS showed that those polar metabolites included apo-8′-lycopenal (8′-apo-ψ-caroten-8′-al, **491**) and apo-12′-lycopenal (12′-apo-ψ-caroten-12′-al, *1*) [43]. Androgen depletion *via* castration as well as a 20% dietary restriction have been shown to increase hepatic lycopene accumulation two-fold, indicating an effect of hormones on carotenoid metabolism and storage [44,45].

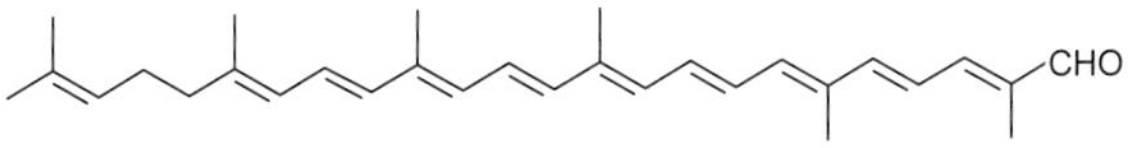

apo-8'-lycopenal (**491**)

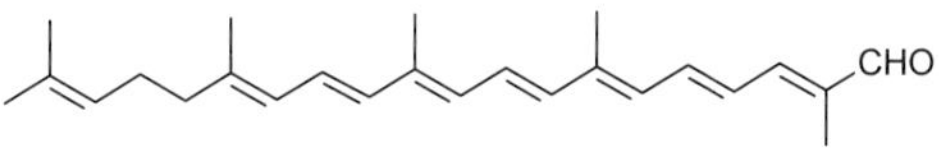

apo-12'-lycopenal (*1*)

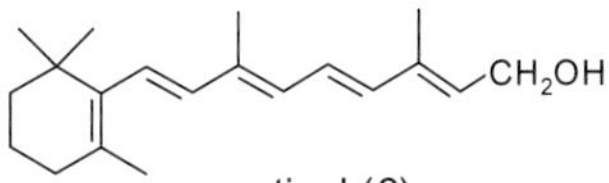

retinol (*2*)

d) Adipose tissue

The large volume of adipose tissue in the human body is a major accumulation site for carotenoids and its carotenoid concentration is considered a marker for long-term intake levels. In a Costa Rican population, concentrations of α-carotene, β-carotene, β-cryptoxanthin, and lycopene in adipose tissues were inversely associated with risk of myocardial infarction [46].

e) Eyes

Only a few carotenoids, notably (3*R*,3′*R*,6′*R*)-lutein (**133**) and (3*R*,3′*S*, *meso*)-zeaxanthin (**120**), and also some lycopene (**31**), and their metabolites, can be found in the human eyes [47] (*Chapter 15*). In particular, these carotenoids are found in the area of the highest visual acuity, the macula lutea of the central retina [48]. Because of the specific uptake of lutein and zeaxanthin into the macula there was great speculation that there were binding protein(s) present in the eye, and some have now been discovered [49]. The Pi isoform of glutathione S-transferase (GSTP1) has been shown to be the xanthophyll-binding protein in the human macula; it binds (3*R*,3′*S*)-zeaxanthin more strongly than lutein [50].

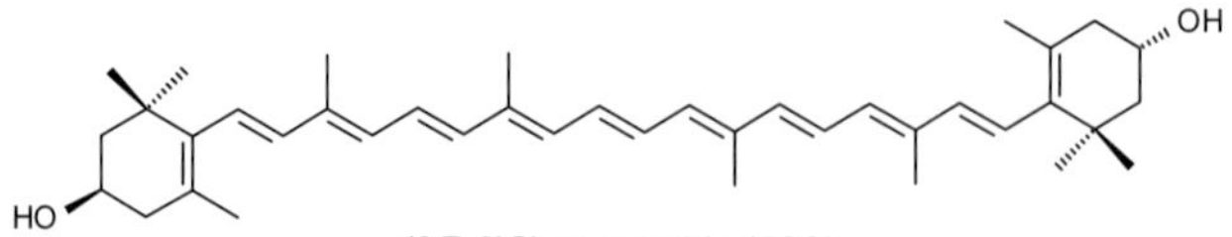

(3*R*,3'*S*)-zeaxanthin (**120**)

Membrane-bound RPE65 is expressed in the retinal pigment epithelium (RPE) and acts as a binding protein for (all-*E*)-retinyl esters, which are involved in the visual cycle [51]. Membrane-bound RPE65 has been found to be a fairly specific retinoid-binding protein directed at long chain esters of (all-*E*)-retinol (*2*) [52]. Retinoids and carotenoids were identified in the bovine ciliary epithelium including retinyl esters, (all-*E*)-retinol, and β-carotene, but not (11*Z*)-retinoids. The absence of (11*Z*)-retinoids suggests that the function of retinoid-processing proteins in the ciliary epithelium differs from that in the retina.

f) Breast milk and colostrum

Thirty-four carotenoids, including thirteen *Z* isomers and eight metabolites, have been identified in breast milk and the serum of lactating women; the concentrations of some carotenoids are given in Table 3 [53].

Table 3. Concentrations of carotenoids (μmol/L) in human milk and colostrum, and in infant formula milk.

Form of infant feed	Carotenoid	Concentration	Reference
Colostrum	Lycopene	0.23-0.51	[38,39]
Colostrum	α-Carotene	0.11-0.17	[38,39]
Colostrum	β-Carotene	0.42-0.47	[38,39]
Colostrum	Lutein	0.16	[38]
Colostrum	Zeaxanthin	0.03	[38]
Colostrum	β-Cryptoxanthin	0.24	[38]
Breast Milk (~1 mo postpartum)	α-Carotene	0.02-0.03	[38,39]
Breast Milk (~1 mo postpartum)	β-Carotene	0.08-0.11	[38,39]
Breast Milk (~1 mo postpartum)	Lutein	0.09	[38]
Breast Milk (~1 mo postpartum)	Zeaxanthin	0.02	[38]
Breast Milk (~1 mo postpartum)	β-Cryptoxanthin	0.06	[38]
Breast Milk (~1 mo postpartum)	Lycopene	0.06	[38]
Infant Formula	β-Cryptoxanthin	0.01-0.02*	[39]
Infant Formula	Lycopene	Not detectable*	[39]
Infant Formula	α-Carotene	Not detectable*	[39]
Infant Formula	β-Carotene	0.07-0.36*	[39]

* Of eight formula samples tested, only four contained β-carotene and three contained β-cryptoxanthin. Lycopene and α-carotene were not detected in any of the formula samples.

The provitamin A carotenoids are an important source of vitamin A for the infant, yet the carotenoid content in breast milk is also thought to provide protection against respiratory and gastrointestinal diseases and to improve overall health of the infant. The colostrum, the initial *post-partum* breast milk, has a distinct yellow colour due to an approximately five times higher carotenoid content than in later milk [39]. It is thought that one of the reasons why the colostrum is so rich in carotenoids is the mobilization of lipids stored in the breast, but the mechanisms behind the transfer of carotenoids from the lactating breast tissue to the breast milk are not understood.

The concentration of carotenoids in breast milk decreases to the normal milk levels after approximately a month (Table 4). The concentrations of α-carotene, lycopene, and β-cryptoxanthin decreased over the first month of lactation to levels around 5-10% of that of plasma [54]. On the other hand, concentrations of lutein in milk remained at the same levels (around 30% of the plasma lutein). Lutein made up 25% of total carotenoids at day four *post partum*, but 50% of total carotenoids by day 32.

Table 4. Concentrations of carotenoids in human milk (nmol/g milk fat) [37].

Human Breast Milk	Carotenoid	Concentration
~1 month *postpartum*	α-Carotene	0.03
~1 month *postpartum*	β-Carotene	0.82
~1 month *postpartum*	Lutein	7.41
~1 month *postpartum*	Zeaxanthin	1.04
3 months *postpartum*	α-Carotene	0.02
3 months *postpartum*	β-Carotene	0.89
3 months *postpartum*	Lutein	9.77
3 months *postpartum*	Zeaxanthin	1.23

In some studies, supplementation with β-carotene has not been shown to increase the β-carotene levels in the milk, indicating that either the concentration of this carotenoid is tightly regulated, or it is already at saturation levels in human breast milk [54]. Yet supplementation with a source of high bioavailability, red palm oil, which is rich in both α-carotene and β-carotene, did increase the α-carotene, β-carotene, and retinol concentrations of breast milk without altering the concentrations of lutein and zeaxanthin [37,55]. Due to their stable concentrations, it was suggested that lutein and zeaxanthin are actively secreted into breast milk. Also, there are proportionally more carotenes in the women's plasma than in breast milk, but a higher lutein and zeaxanthin concentration in breast milk than in plasma [37].

The carotenoid content of breast milk is lower in the first than in subsequent lactations. Multiparous mothers had a mean total carotenoid content of 2.18 ± 1.94 μg/mL colostrum, compared to 1.14 ± 1.32 μg/mL for first-time mothers [56]. Many kinds of infant formula are not supplemented with carotenoids at all, or supplemented at a similar level to that in breast milk [39]. The plasma carotenoids levels of α-carotene, β-carotene, and lycopene are lower in formula-fed infants than in infants who are breast fed (Table 3) [39].

g) Breast

It is known that lactation can lower a woman's risk of breast cancer, and it was suggested that one mechanism to explain this could be the mobilization of carotenoids to the breast when lactating [57]. Breast nipple aspirate fluid can be obtained from breast tissue and its analysis is useful for increasing understanding of the impact of diet on breast carotenoids. In the Nutrition and Breast Health Study, breast nipple aspirate fluid was collected from pre-menopausal, non-pregnant women; carotenoid levels were significantly higher in women who lactated 6 months or longer than in those who had lactated for a shorter time or who had never lactated [58]. Another study to measure carotenoid availability in nipple aspirate fluid found that total carotenoids ranged from 0.4 to 4.0 µg/mL, with a mean level of 1.9 ± 1.2 [59]. From this, the hypothesis was proposed that lactation may be protective against breast cancer by enhancing the delivery of chemopreventive substances, which may include carotenoids, from the blood to breast tissue cells.

h) Male reproductive tissues

i) Prostate. An inverse association between tomato intake and prostate cancer risk has been established [60,61]. Whilst whole tomato products are more effective than lycopene alone in reducing risk of prostate cancer [62-64], lycopene concentrations are used as a general biomarker for tomato intake levels. The dorsolateral lobe of the rat prostate shows considerable homology with the human prostate site for cancer, so there have been extensive studies with rodent models. Three hours after [^{14}C]-lycopene was provided to rats, 69% of the radioactivity in the dorsolateral prostate was recovered as polar products of lycopene, and after one week this increased to 82%, suggesting extensive metabolism of lycopene in the prostate [24,42]. There was also preferential uptake of phytofluene (**42**) from tomato powder into the prostate, though to a lesser extent than lycopene [65].

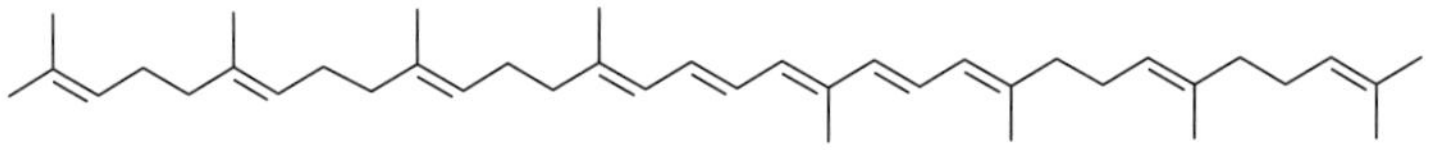

phytofluene (**42**)

Human prostates from the Health Professionals Follow-Up Study were analysed for carotenoid levels [66]. Lycopene and (all-*E*)-β-carotene were the predominant carotenoids present, with concentrations as shown in Table 5. (9*Z*)-β-Carotene, α-carotene, lutein, α-cryptoxanthin, zeaxanthin, and β-cryptoxanthin were also detectable in prostate tissue.

Table 5. Concentrations of carotenoids in human tissues (nmol/g)

Tissues	Carotenoid	Concentration	Reference
Skin: Back	Total carotenoids	0.23	[72]
Skin: Forehead	Total carotenoids	0.26-0.6	[72,79]
Skin: Inner arm	Total carotenoids	0.1-0.21	[72,79]
Skin: Palm of hand	Total carotenoids	0.32-0.71	[72,79]
Skin: Back of hand	Total carotenoids	0.29-0.35	[72,79]
Skin	Lycopene	0.48	[80]
Adipose	Lycopene	0.34-0.70	[46,80]
Adipose	α-Carotene	0.37	[46]
Adipose	β-Carotene	0.98	[46]
Adipose	β-Cryptoxanthin	0.33	[46]
Adipose	Lutein + Zeaxanthin	1.10	[46]
Prostate	Lycopene	0.6	[66]
Prostate	(all-*E*)-β-Carotene	0.5	[66]
Prostate	α-Cryptoxanthin[1]	0.2	[66]
Prostate	Zeaxanthin	0.2	[66]
Prostate	β-Cryptoxanthin	0.1	[66]
Prostate	(9*Z*)-β-Carotene	0.4	[66]
Prostate	α-Carotene	0.4	[66]
Prostate	Lutein	0.3	[66]
Prostate cancer	Lycopene	0.9	[66]
Prostate cancer	(all-*E*)-β-Carotene	0.60	[66]
Prostate cancer	α-Cryptoxanthin	0.3	[66]
Prostate cancer	Zeaxanthin	0.3	[66]
Prostate cancer	β-Cryptoxanthin	0.2	[66]

[1]Reported as α-cryptoxanthin. Distinction was not made between β,ε-caroten-3-ol (**60**) and β,ε-caroten-3′-ol (**62**).

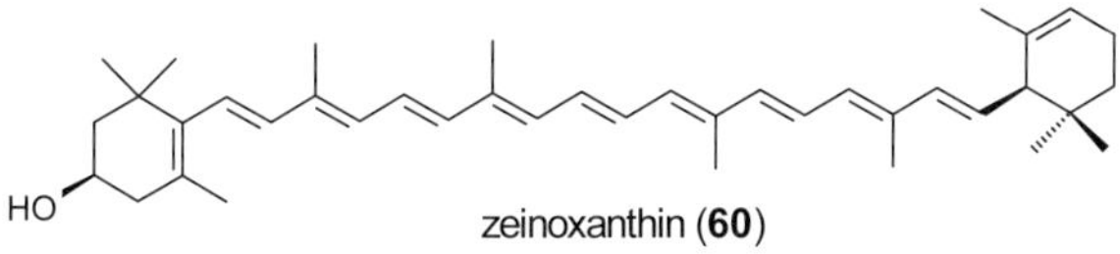

zeinoxanthin (**60**)

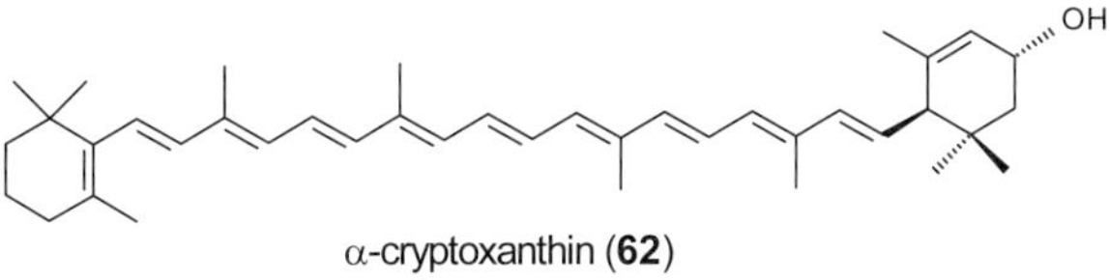

α-cryptoxanthin (**62**)

As described previously, mRNA for SR-BI, a protein of *ca.* 80 kDa, can be found in numerous human tissues. Interestingly, a smaller, *ca.* 50 kDa, protein is found in the human prostate but it is not yet known if this is a cross-reactive protein or a degradation product of SR-BI [14]. If there is an active carotenoid transporter in the human prostate, it could explain why such high levels of carotenoids are found in this tissue.

ii) Testes. The young, normal, human testes express SR-BI in the spermatids, the cytoplasm of the Sertoli cells, the site of the blood-testes barrier and spermatogenesis, and the surface of the Leydig cells which synthesize testosterone [67]. The young testes do not express CD36, but aging and pathological Sertoli and Leydig cells of the testes do [67]. The presence of these two lipid transporters in the testes suggests a mechanism to bring free cholesterol to the tissue for testosterone synthesis and spermatogenesis. Male rats provided with either phytofluene, lycopene, or whole tomato powder showed *ca.* 40-50% lower serum testosterone concentration than control-fed rats [68]. It may be that SR-BI facilitates the entry of carotenoids, as well as free cholesterol, into the testes, so less cholesterol substrate was available for testosterone synthesis.

iii) Semen. Lycopene is also found in human seminal fluid; it is incorporated into the lipid-rich prostasomes and secreted from the prostate into semen, where it is proposed to act as an antioxidant protecting spermatozoa from damage [69]. These prostasomes are essential to male fertility by modulating and regulating the microenvironment of spermatozoa *via* the exchange of lipids. Immuno-infertile men have shown decreased levels of β-carotene, retinol, and lycopene compared to fertile men [70]. Oral lycopene provision is reported to improve sperm concentration, motility, and morphology, and it was suggested that reducing reactive oxygen species (ROS) is one mechanism by which lycopene may improve male fertility [71].

i) Skin

Some carotenoids have been shown to prevent skin damage by offering protection against UV radiation [72,73] (*Chapter 16*). In a placebo-controlled study, twelve volunteers received either β-carotene (24 mg/day from an algal source), or 24 mg/day of a carotenoid mixture consisting of β-carotene, lutein and lycopene (8 mg/day each), or placebo, for 12 weeks [74]. The intake of carotenoids increased total carotenoids in the skin, and erythaema intensity after UV light exposure was less in both groups which received carotenoids. Consumption of tomato paste for 10 weeks has also been shown to reduce levels of skin burning [75].

j) Adrenals

Carotenoid concentrations are higher in the adrenal glands than in other tissues analysed [6,45,76]. Rats fed a tomato powder diet had higher levels of phytoene (**44**) and phytofluene (**42**) than of lycopene in the adrenal tissue [65,77].

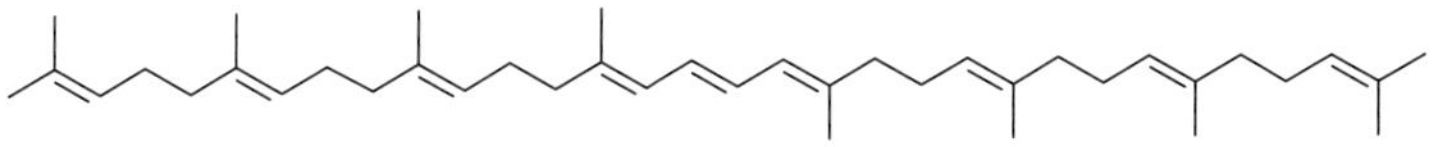

phytoene (**44**)

The uptake of cholesterol by LDL receptors from lipoproteins occurs when core cholesterol is taken into cells without the uptake of the lipoprotein particle itself. Steroidogenic tissues, such as the adrenal glands and gonads, use this pathway to obtain cholesterol to produce steroid hormones. The selective uptake of cholesterol into the adrenals is due to the high concentration of SR-BI [78], which allows for the transfer of cholesterol from lipoproteins into cells. The parallel uptake of carotenoids and cholesterol into the adrenals could explain the high concentration found there.

C. Bioavailability

1. Introduction

The term 'Bioavailability' is used to indicate how much of a consumed nutrient or dietary constituent is accessible for utilization in normal physiological functioning, metabolism, or storage. The bioavailability of carotenoids has been defined as the proportion of an ingested carotenoid that is taken up by the intestinal enterocytes and transported in the bloodstream [81]. Often, studies of the bioavailability of carotenoids from a given food source report the relative bioavailability compared to a source of known high bioavailability, *e.g.* comparison of β-carotene in carrots with that dissolved in oil. Carotenoids still embedded in their food matrix cannot be absorbed efficiently. It is critical, therefore, to understand the roles that the food matrix, food processing, other dietary or host factors, and digestion within the gastrointestinal tract play in carotenoid absorption and bioavailability (reviewed in [82]). The mnemonic acronym 'SLAMENGHI' was developed as a convenient way to list the major contributors which affect carotenoid bioavailability. SLAMENGHI stands for **S**pecies of carotenoid, **L**inkages at the molecular level, **A**mount of carotenoid, **M**atrix, **E**ffectors, **N**utrient status, **G**enetics, **H**ost-related factors, and **I**nteractions among these variables [83].

To begin the process of absorption, ingested carotenoids, after release from the food matrix, are taken up by the intestinal mucosal cells, and are transported from the mucosal cells into the lymphatic blood system. Then they must be solubilized into micelles, absorbed, and packaged into chylomicrons, before being transported and stored in various tissues. This section classifies the features of carotenoid bioavailability into three main groups, the food matrix, effects of food processing, and other dietary and host factors. It should be noted, however, that these factors interrelate to affect overall carotenoid bioavailability. A range of carotenoids will be examined, both carotenes and xanthophylls.

2. Effect of food matrix

a) Carotenoids in fruits and vegetables

The major source of carotenoids in the human diet is fruits and vegetables (*Chapter 3*). Carrots, squash, and dark green leafy vegetables are common sources of β-carotene, carrots of α-carotene, tomatoes and watermelon of lycopene, kale, peas, spinach, and broccoli of lutein, and red peppers, oranges, and papayas of β-cryptoxanthin and zeaxanthin [29]. The food matrix is often the major factor that determines bioavailability, and the release of carotenoids from this matrix is considered to be the first step needed to facilitate absorption. The more the food matrix is disrupted the greater the possibility of carotenoid absorption.

Food products are not the only source of carotenoids; some people also consume carotenoid supplements (*Chapter 4*). Many studies have shown that the bioavailability of carotenoids from commercial supplements is greater than that from food sources. There are some exceptions, however, including reports of similar bioavailability of synthetic β-carotene and that from natural red palm oil [84], presumably because of the high solubility of β-carotene in the oil. In contrast, other studies have shown a greater bioavailability of lutein from lutein-enriched eggs and processed spinach than from supplements [85].

b) Location of carotenoids

To understand why carotenoids from different plant sources have different degrees of bioavailability, we must consider that, in plants, carotenoids are located in various organelles and complexes in chloroplasts and chromoplasts [86]. For instance, in green leafy vegetables, β-carotene, lutein and zeaxanthin are located in the chloroplasts, together with chlorophyll, in pigment-protein complexes [87] (*Volume 4, Chapter 14*). In other sources, carotenes often have a lower bioavailability than the xanthophylls [88,89]. This may, in part, be explained by the location and physical state of these carotenoids in the plant tissue. The carotenes in carrot roots are in crystalline form and encased in membranous sheets of large proteins [90]. Raman spectroscopy showed that the carotenoids in carrot root were not uniformly distributed, but the highest concentrations were found in the phloem and xylem parenchyma [91]. Lycopene is known to be present as crystals in mature red tomatoes [92,93].

To evaluate the degree to which the plant tissue affects carotenoid bioavailability, three types of green vegetables, *i.e.* broccoli (flowers), green peas (seeds), and spinach (leaves), were fed to seventy-two volunteers at a level of 300 g daily for four days [94]. A greater increase in plasma β-carotene was seen after feeding green peas and broccoli than after feeding spinach, despite the fact that the spinach had a 10-fold higher amount of β-carotene. To investigate how the location of carotenoids affects bioavailability, tomato skin was added to tomato sauce and fed to healthy male volunteers [95]. The results indicated that β-carotene had much higher bioavailability than lycopene, and that the carotenoids from tomato skin were just as effective as those in the tomato flesh. Thus, the food matrix and intracellular location and physical state of carotenoids can play a significant part in carotenoid bioavailability.

3. Effect of food processing

Numerous studies have been performed to establish how the degree and form of food processing breaks down the food matrix and alters carotenoid bioavailability. Food processing includes techniques of heat treatment and mechanical homogenization. It is generally accepted that mild processing procedures break cell walls, release carotenoids from intracellular organelles, disrupt the carotenoid-protein complexes, and decrease particle size, so allowing greater uptake of carotenoids. Food processing can also inactivate oxidizing enzymes that can cause degradation of carotenoids. Excess thermal processing, however, can result in isomerization and oxidation of carotenoids (see *Chapter 3*).

It has been estimated that homogenization or heating can increase carotenoid bioavailability by as much as six times [86]. The effect of processing on carotenoid bioavailability from tomatoes has been evaluated extensively, and it was found that the bioavailability of lycopene from tomato paste was greater than that of lycopene from raw tomatoes [96]. To understand further the effect of processing, tomatoes were subjected to boiling, addition of vegetable oils, chopping and agitation, to determine the susceptibility of carotenoids to undergo isomerization [97]. In this study, oil and chopping did not promote thermal isomerization of (all-*E*)-lycopene or other acyclic and monocyclic carotenoids, whereas (all-*E*)-β-carotene and (all-*E*)-lutein underwent some isomerization to the *Z* forms. In other studies, however, heating tomato products did increase formation of (*Z*)-lycopene [98]. An additional study confirmed that the processing of tomatoes affects the bioavailability of lycopene [32]. A 65% greater plasma response was found from tomato sauce than from tomato juice, although the sauce contained only 20% more lycopene. When subjects were provided with tomato soup or juice, similar blood concentrations of lycopene were seen despite there being 42% less lycopene in the soup. From the carotenoid response in triacylglycerol-rich lipoproteins after a single consumption of tomatoes *versus* the change in fasting plasma carotenoid concentrations after 4 days in healthy humans, it was concluded that the bioavailability of lycopene from tomatoes was improved by mechanical homogenization and/or heat treatment [99].

Four spinach products, namely whole leaf spinach, minced spinach, and enzymically liquefied spinach with or without added fibre, were tested to determine the effect of food processing on bioavailability of β-carotene and lutein [100]. The relative β-carotene bioavailability, when compared to that of a β-carotene supplement, was 5.1, 6.4, 9.5, and 9.3% for the whole leaf, minced, liquefied, and liquefied with fibre, respectively. Bioavailability for lutein was 45, 52, 55, and 54% for the whole leaf, minced, liquefied, and liquefied with fibre, respectively. Thus, disruption of the food matrix by food processing was found to increase the bioavailability of β-carotene, but not of lutein. A similar trend was seen with carrots; more β-carotene was absorbed from cooked, pureed carrots than from raw carrots [101]. In healthy women, feeding heat-processed and pureed carrots and spinach caused serum β-carotene to be three times higher than when the same dietary level of β-carotene was consumed in the raw food sources [102]. In a population of women at risk for breast cancer, serum concentrations of lutein and α-carotene, but not of β-carotene, β-cryptoxanthin, or lycopene, were higher in women consuming vegetable juice, rather than cooked or raw vegetables [103].

It can be concluded that mild heating methods, such as steaming, increase carotenoid bioavailability from fruits and vegetables, but extreme heating, such as rapid boiling or frying, causes carotenoids to undergo isomerization and oxidation, and food processing can improve carotenoid bioavailability by disrupting the structural matrix and reducing particle sizes.

4. Structure and isomeric form of the carotenoid

Differences in structure of carotenoids can affect their bioavailability. For instance, HPLC analysis of lymph in ferrets has indicated a greater absorption of lutein and zeaxanthin than of lycopene and β-carotene [7]. This could be because the xanthophylls are more polar, and so may be incorporated preferentially on the outer portion of micelles, or because they are more easily taken up by the enterocytes [104].

The bioavailability of free xanthophylls and their acyl esters may be different, but conflicting results have been obtained. The carboxyl ester lipase (also known as cholesterol esterase or bile-salt-stimulated lipase), which is found in exocrine pancreatic secretions, hydrolyses esters of hydroxycarotenoids, including lutein, capsanthin (**335**), zeaxanthin, and β-cryptoxanthin [105]. Such hydrolysis of zeaxanthin esters in the small intestine has been reported to increase bioavailability by increasing the amount of free zeaxanthin in micelles, thus allowing greater uptake by the intestinal epithelial cells [105]. Other studies, however, have shown no difference between the bioavailability of free lutein and of lutein esters, presumably because there is sufficient esterase activity in human intestines to hydrolyse the esters efficiently [85,106].

The *E/Z* (*cis/trans*) isomeric form can also affect the bioavailability, as discussed below for β-carotene and lycopene.

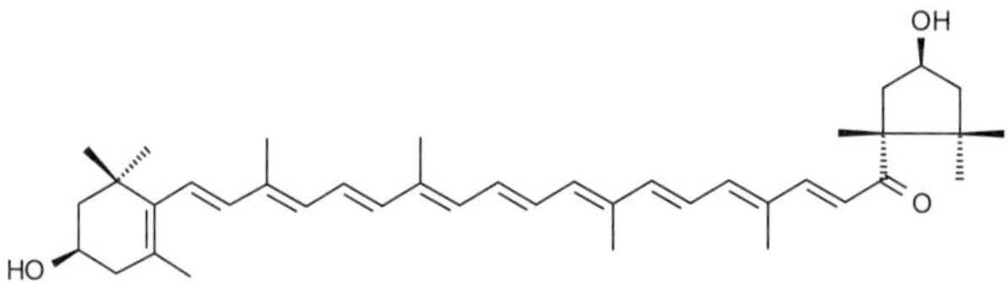

capsanthin (**335**)

a) β-Carotene

In fruits and vegetables, β-carotene is predominantly found in the (all-*E*) form, with small amounts of *Z* isomers present [107]. Heat processing increases the amount of *Z* isomers, especially (9*Z*)-β-carotene and (13*Z*)-β-carotene [108,109]. Typically, the β-carotene isomer profile in blood is similar to that in fruit and vegetables, with a high proportion of (all-*E*)-β-carotene and smaller amounts of (13*Z*)-β-carotene, which accounts for only about 7% of total plasma β-carotene. (9*Z*)-β-Carotene does not accumulate to any significant degree in plasma of humans [110] or animals, even when (9*Z*)-β-carotene is fed [6]. Tissues, however, show different β-carotene isomeric patterns, with significant levels of *Z* isomers present. These isomer patterns could be due to a variety of factors, including differential bioavailability, absorption, transport, and uptake. Additionally, isomerization of β-carotene in the gastrointestinal tract or within tissues could explain the different ratios of β-carotene isomers present in tissues compared to food sources and blood.

Studies with gerbils, based on the gain in liver β-carotene and vitamin A, revealed a bioavailability of 38% for an oral dose of (all-*E*)-β-carotene compared with 27% and 32%, respectively, for (9*Z*)-β-carotene and (13*Z*)-β-carotene, relative to retinol [6]. A model of digestion *in vitro* showed that these *Z* isomers of β-carotene were incorporated into micelles 2-3 times more efficiently than (all-*E*)-β-carotene [111].

b) Lycopene

In tomatoes, (all-*E*)-lycopene is by far the predominant isomer but, in organ tissues, there are approximately equivalent amounts of (all-*E*) and *Z* isomers. Whereas (all-*E*)-β-carotene seems to be the isomer with greatest bioavailability, for lycopene it is the *Z* isomers [8]. It is thought that *Z* isomers of lycopene, having >75% greater solubility in mixed micelles, are better absorbed than the all-*E* isomer [7]. The *Z* isomers of lycopene are less prone to aggregate or crystallize than the all-*E* form, and thus are more likely to be incorporated into a bile acid micelle [112]. The isomerization of lycopene into *Z* isomers is thought to increase lycopene bioavailability, and the stomach acid may play a major role in this isomerization [8]. In ferrets, after a single oral dose of lycopene (92% all-*E*), the lymph contained >75% of lycopene as *Z* isomers, whilst storage tissues only contained 50% of lycopene in the *Z* forms, indicating preferential uptake of (*Z*)-lycopene and considerable isomerization between the isomers [7].

5. Effects of other dietary factors

Other factors in the diet can also have an impact on carotenoid bioavailability. Some of these, such as dietary fat, can increase bioavailability. Others, such as high levels of some kinds of dietary fibre, have a detrimental effect and decrease carotenoid bioavailability. Furthermore, there are large differences between individuals. This section describes both positive and negative dietary influences on the bioavailability of carotenoids.

a) Dietary fat

It is well known that fat must be consumed to optimize the absorption of carotenoids. Lipids in the small intestine stimulate the release of bile salts from the gall bladder and enlarge the bile salt micelle, thus increasing solubilization of carotenoids. It has been assumed that the fat must be consumed in the same meal as the carotenoids, but some fat from a previous meal may remain in the intestines and aid the absorption of carotenoids consumed later [113].

Optimal carotenoid absorption has been shown to occur with as little as 3-5 g of fat per meal [114]. Carotenoid absorption was better when carotenoid-containing foods were consumed with full-fat salad dressings than with reduced-fat salad dressings [115]. Also, astaxanthin (**404-406**), a characteristic carotenoid of seafood, has a higher bioavailability when provided in a lipid formulation [116]. Gerbils consuming 30% of their total energy as fat showed enhanced post-absorptive conversion of β-carotene into vitamin A compared to those consuming only 10% of their diet as fat [117].

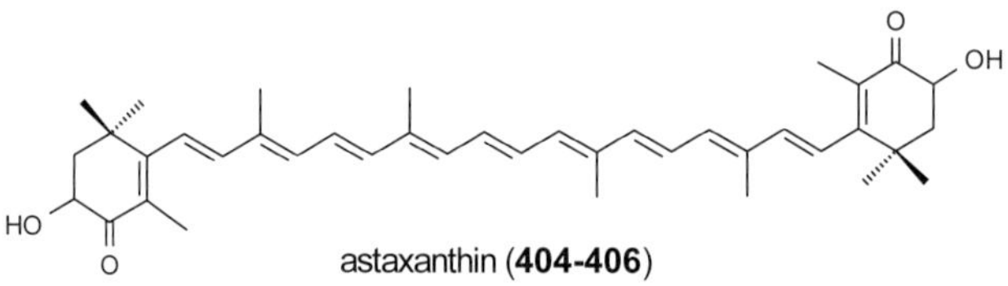

astaxanthin (**404-406**)

The type of fat may influence the rate and effectiveness of carotenoid absorption. Avocado oil was found to increase carotenoid absorption significantly when added to salsa or salads compared to the control which was fat-free salad dressing; this was attributed to the type of lipid present in the avocado fruit, which predominantly has monounsaturated (18:1) fatty acyl chains [118]. In studies performed *in vitro*, when lipids containing the monounsaturated acyl chains of oleic acid were present, the rate of β-carotene uptake was increased, whereas when polyunsaturated lipids containing linoleic and linolenic acid chains were present, β-carotene exhibited decreased rates of absorption [10].

The importance of fat in increasing carotenoid bioavailability suggests that a β-carotene supplement in an oil matrix would have greater bioavailability than water-miscible β-carotene beadlets, but the opposite was seen in a human trial [119].

b) Inhibitors in the diet

i) Sucrose-polyesters ('Fake-fats'). Olestra™ is a sucrose-polyacyl ester used in savoury snacks as a fat substitute. It is neither hydrolysed by gastrointestinal enzymes nor absorbed, but it interferes with the absorption of fat-soluble compounds such as vitamins and carotenoids. Specifically, Olestra™ has been shown to inhibit bioavailability of carotenoids when consumed in the same meal as carotenoid-containing foods, but there is not such a significant inhibition by Olestra™ eaten in snack foods at a different time. Olestra™ has been shown to inhibit the absorption of molecules with octanol-water partition coefficients greater than 7.5; this includes the phytosterols and carotenoids [120]. It was calculated that there would be a 6-10% reduction in β-carotene bioavailability as a result of consuming olestra-containing snack foods [120]. In adults, Olestra™ consumption was associated with statistically significant reductions in serum α-carotene (14.1%), β-carotene (10.1%), and lycopene (11.7%) [121]. The decrease in the bioavailability of carotenoids seen with Olestra™ is comparable to that caused by other dietary inhibitors or by not having fat in the diet. While supporters of Olestra™ have said that these inhibitions of carotenoid bioavailability are not relevant to human health, the long term impact of Olestra™ consumption and the consequent decreased absorption of carotenoids and other fat-soluble micronutrients is not known.

ii) Statins and plant sterols. Statins are commonly prescribed to reduce serum cholesterol levels; they block cholesterol biosynthesis by inhibiting the enzyme hydroxymethylglutaryl-coenzyme A reductase (HMG-CoA reductase). Statins are also known to reduce serum carotenoid levels, though not by the same mechanism. To evaluate the long-term effects, atorvastatin or simvastatin was given to patients for 52 weeks [122]. After twelve weeks of statin therapy, serum β-carotene levels were significantly reduced, but after 52 weeks returned to baseline levels. Thus, it appears that the negative impact of statins on carotenoid bioavailability is only temporary. The drug Zetia®, which reduces cholesterol transport, has also been shown to reduce carotenoid transport in Caco-2 cells [1], but the long-term effects in humans have not been evaluated.

Plant sterols reduce the absorption of cholesterol in the gut by competing for incorporation into mixed micelles or by decreasing hydrolysis of cholesterol esters in the small intestine. However, they also affect carotenoid bioavailability. Plant sterols and their esters were tested at a level of 2.2 g/day for one week to determine their effects on β-carotene bioavailability [123]. Both reduced bioavailability of β-carotene by approximately 50% but the reduction was greater with the esters. Other researchers have found a reduction, albeit much smaller, in serum carotenoids as a result of the consumption of plant sterols [124,125].

iii) Dietary fibres. Dietary fibres, such as pectin, gels, cellulose, and bran, have been shown to decrease the bioavailability of carotenoids, by entrapping carotenoids and intermingling with bile acids, resulting in decreased absorption and increased faecal excretion of fats and fat-soluble substances [82]. The effect on carotenoid bioavailability depends on the fibre's particle size, gel formation, and capacity for binding water and bile acids. Soluble fibre enhances the viscosity of gastric contents and slows gastric emptying, which disrupts lipid and carotenoid absorption because bile salts and cholesterol are trapped in the gel phase instead of forming micelles [126,127].

Pectin, guar gum, alginate, cellulose, and wheat bran were tested to see how these commonly-consumed dietary fibres affect the bioavailability of β-carotene, lycopene, and lutein [128]. All the fibres inhibited lycopene and lutein absorption but only the water-soluble fibres pectin, guar, and alginate, were found to decrease absorption of β-carotene, whereas in gerbils, citrus pectin, but not oat gum fibre, decreased β-carotene bioavailability [117].

Not only purified fibre, but also fruit and vegetables as sources of this fibre, cause reductions in carotenoid bioavailability. Dietary fibre, lignin, and resistant proteins found in green leafy vegetables inhibited the release of β-carotene and lutein [129] and citrus pectin reduced plasma β-carotene responses [126]. The presence of fibre in fruits and vegetables could, in part, explain why the bioavailability of carotenoids from fruits and vegetables is lower than that of a purified supplement.

c) Interactions between carotenoids

When several carotenoids are provided together they, theoretically, could have an inhibitory effect on the absorption, metabolism, and transport of each other so that serum responses of particular carotenoids may be diminished. However, there is some controversy about the biological significance of interactions between carotenoids, especially ones from food sources. It is plausible that interaction among the carotenoids could be due to competition for uptake into enterocytes or incorporation into chylomicrons [130]. In the intestinal mucosa, carotenoids that cannot serve as a substrate could either hamper or boost the activity of the carotenoid cleavage enzymes, circulating carotenoids could be exchanged among plasma lipoproteins, or tissue uptake and release of one carotenoid could be enhanced or inhibited by another [82,86]. Another suggestion is that carotenoids could spare each other by acting as antioxidants in the intestinal tract, so that the uptake of those spared carotenoids is increased, and the bioavailability of other carotenoids consumed in the diet is thus improved [86].

Twenty women were fed either (i) 96 g/day tomato puree (containing 15 mg lycopene and 1.5 mg β-carotene), (ii) 92 g/day cooked chopped spinach (containing 12 mg lutein and 8 mg β-carotene), (iii) 96 g/day tomato puree plus 92 g/day chopped spinach, (iv) 96 g/day tomato puree plus a lutein pill (12 mg lutein), or (v) 92 g/day chopped spinach plus a lycopene pill (15 mg lycopene), during three-week periods separated by three-week washout periods [130]. In the short term there was postprandial competitive inhibition among carotenoids for incorporation into chylomicrons but, after three weeks, the overall carotenoid levels in plasma were not altered.

Four women and five men participated in a randomized, blinded, 3 x 3 crossover intervention with seven-day administration of yellow carrots (providing 1.7 mg lutein per day), white carrots as a negative control (no lutein), or a supplement of lutein (1.7 mg/day) in oil as a positive control [28]. Yellow carrots were found to be an excellent whole food source of lutein, and the blood concentrations of other carotenoids, including β-carotene, were maintained, which was not the case with lutein supplementation [28]. This novel food source of lutein provides a good alternative to supplements because it does not have a negative effect on the levels of β-carotene and other circulating carotenoids.

6. Human factors

Individuals vary considerably in their innate ability to absorb carotenoids. Some are considered 'low-responders' or 'non-responders.' A number of other factors can also affect how an individual responds to dietary carotenoid intakes, *e.g.* age, parasite infections.

a) 'Non-responders'

Some individuals show small or no significant increase in blood β-carotene after the ingestion of a single high dose of β-carotene or after consuming a β-carotene-rich diet for several weeks. These individuals are known as 'non-responders.' The lack of a response could be due to impaired uptake of β-carotene by the intestinal mucosa, excessive conversion of β-carotene into vitamin A, poor incorporation of β-carotene into chylomicrons, or an error in lipoprotein metabolism. Some aspects of this have been proven experimentally [131].

Seventy-nine healthy, young, male subjects were enlisted to consume a meal containing 120 mg β-carotene, following which the concentrations of serum β-carotene, retinyl palmitate, and triacylglycerols were measured [131]. There was large variability in incorporation of β-carotene into chylomicrons but, in this study population, there were no true non-responders to pharmacological doses of β-carotene. It was concluded that the ability to respond to β-carotene is an intrinsic characteristic of each individual, which can be explained by genetic variations in β-carotene absorption, chylomicron metabolism, current β-carotene status, and the activity of intestinal β-carotene 15,15′-oxygenase. The subject is still controversial. Some researchers accept that there are non-responders and low responders to β-carotene [132,133], whereas others do not believe this to be the case. This area of research needs more investigation, and the mapping of the human genome may shed light on genetic variations which contribute to individual differences in carotenoid bioavailability.

b) Age

Young (20-35 years) and older (60-75 years) healthy adults were fed vegetable sources of carotenoids, and chylomicron carotenoid levels were measured nine hours after feeding [134]. Age had no effect on the amount of (all-*E*)-β-carotene, (*Z*)-β-carotene, α-carotene, or lutein present in the chylomicrons, but there was a significant ($P < 0.04$) 40% decrease of the chylomicron triacylglycerol-adjusted lycopene response in the older subjects. It was suggested that absorption of lycopene is influenced more than that of some other carotenoids by the changes which occur in the digestive tract over a lifetime [134]. The incidence of atrophic gastritis increases with age [135]. This results in reduced stomach acidity and, in turn, carotenoid absorption, which requires a low pH for greater efficiency, is diminished, as described below [21,82]. In male rats, increases in hydrogen ion concentration increased the rates of β-carotene absorption by the proximal and distal small intestines [10], indicating that the bioavailability of other carotenoids should also be affected by pH.

c) Parasitic infections

In developing countries, infection with parasites is common and often results in substantial reductions in carotenoid bioavailability [88,114]. Parasites may compete with the host for the carotenoids, or disrupt the normal gastrointestinal function, thereby reducing carotenoid bioavailability. Meals containing β-carotene and fat were fed to children 3-6 years of age for three weeks; some children received de-worming treatments before the feeding period, others did not. The greatest increase in serum retinol occurred with β-carotene, added fat and de-worming treatment. When meals containing additional β-carotene were given, added fat resulted in a further enhancement of retinol concentrations in the serum, but only if infection was low [114]. In children affected with *Ascaris*, an intestinal parasitic worm (helminth), transit time from mouth to caecum was shorter when the intensity of infection was greater [136]. A cross-sectional survey, in which intestinal infection and plasma levels of vitamin A and carotenoids were measured simultaneously, showed a significant inverse relationship.

Low plasma carotenoid concentrations in malaria sufferers were found to be influenced strongly by reduced levels of carrier molecules such as retinol-binding protein (RBP) and by plasma cholesterol, suggesting possible mechanisms by which parasitic infection affects carotenoid uptake [137]. For example, a drastic decrease in serum lycopene was seen in men and women infected with *Schistosoma mansoni* compared to non-infected controls [138]. The liver is a major accumulation site for lycopene and also the main site for *S. mansoni* infection and hepatitis. It was suggested that lycopene is utilized to protect against reactive oxygen and nitrogen species produced as a result of the infection; this could result in lower circulating levels of lycopene. After treatment for *Plasmodium falciparum* malaria, children's plasma levels of provitamin A carotenoids, non-provitamin A carotenoids, and retinol increased [139]. Provitamin A carotenoids can also reduce the negative health effects of many disease conditions in third world countries by alleviating disease symptoms and incidence, and improving immunity by improving the child's growth, reducing diarrhoea frequency and severity, and altering cytokines [140-142].

D. Methods for Evaluating Carotenoid Bioavailability

Various methods including *in vitro* digestion models, human intestinal cell lines, and the postprandial chylomicron response in humans have been used to investigate how carotenoids are absorbed and what factors improve or hinder absorption [82]. Food frequency questionnaires and 24-hour dietary recalls have bias and error so, to validate carotenoid bioavailability properly, additional markers should be used, such as plasma response [143].

1. Oral-faecal balance

As the method of choice about 50 years ago, the oral-faecal balance compared the amount of carotenoid consumed orally to the amount excreted in faeces. Whilst this method is straightforward to perform, there are major limitations. For example, it does not take into account either carotenoid degradation in the upper and lower gastrointestinal tract or the excretion of endogenous carotenoids. This method is also cumbersome, imprecise, time consuming, expensive to perform, and does not take into account the influence of the gut microorganisms on carotenoid bioavailability, thus giving rise to considerable variation which greatly restricts the value of oral-faecal studies.

A more recent variation is to analyse human exfoliated colonic epithelial cells extracted from faecal matter [144]. To obtain exfoliated colonic epithelial cells, approximately 500 mg stool is added to transport medium, filtered through a 40 μm filter, centrifuged at 200 x g for 10 minutes, then the recovered cells are washed and collected by centrifugation at high speed (1000 x g) [144]. Colonic epithelial cells isolated while subjects were on a low β-carotene diet showed a decrease in β-carotene, but a single dose of β-carotene resulted in an increase in colonic epithelial cell β-carotene, and this correlated with plasma levels [144]. This strong relationship between β-carotene in the diet, plasma, and the colonic epithelial cells suggests that this method is a useful and non-invasive way to assess bioavailability of carotenoids. Further validation using more subjects with a wider range of characteristics such as age and gender is still needed, however.

2. Blood response

Changes in the serum or plasma concentrations of carotenoids resulting from a test meal or diet are often used to estimate carotenoid bioavailability; the relative bioavailability can be compared between various carotenoids, food sources, dietary preparations, or different test subjects. This method is similar to that of the pharmacokinetic measure used for drugs, and often uses the area under the curve (AUC) of carotenoids in plasma over time [145]. An advantage of this technique is that it is relatively easy to acquire and analyse blood samples from test subjects. Yet, there are limitations to this method for the assessment of carotenoid bioavailability. For example, there could be large variability between individuals, the serum is a transient tissue and levels of carotenoids give no indication of the flux in intestinal absorption, degradation, tissue uptake, and release from tissue stores. In addition, the provitamin A carotenoids can be metabolized to retinyl esters during absorption, and therefore would not be accounted for. As mentioned previously, there can be greater variability with female participants as their blood carotenoid levels are altered during the menses cycle.

3. Triacylglycerol-rich fraction response

Analysis of the triacylglycerol rich fraction in blood a few hours after a meal gives an excellent evaluation of the bioavailability of carotenoids from that meal. Carotenoid concentrations can be measured in chylomicrons and/or VLDLs to estimate bioavailability plus conversion into retinyl esters. After the serum is collected from whole blood, the serum with the highest triacylglycerol concentration (most often that taken 3 hours after the test meal) is used to separate the lipoproteins by ultracentrifugation into three groups: (i) triacylglycerol-rich lipoproteins, containing chylomicrons, chylomicron remnants, and VLDL, (ii) LDL, and (iii) HDL [146]. Estimating carotenoid bioavailability in this way accounts for conversion into retinyl esters and makes it possible to distinguish between recently absorbed carotenoids (in lipoprotein group i) and those from endogenous stores (in the other lipoprotein groups), information that is not given by simple serum analysis. On the other hand, this method does not discriminate between carotenoids in VLDLs that originate from the liver stores, and those in chylomicrons from the intestine. A benefit of this model, however, is that the whole food source of carotenoids can be evaluated, including the effect of the food matrix (Section **C**.2) [3].

4. Digestion methods *in vitro*

Digestion studies *in vitro* can be used to gain better understanding of how carotenoids are released from the food matrix, and for screening the effects of various food processing techniques. For example, the uptake of carotenoids from a meal into micelles, the contributions of the gastric and intestinal phases in the transfer of carotenoids from food, and the effects of acids, enzymes, and bile salts on the release of carotenoids from the food matrix, can be studied. The digestion system *in vitro* provides an alternative to animal and human research subjects, gives a rapid estimation of carotenoid bioavailability from foods, and allows for the careful control of experiments to investigate the impact of food processing and dietary factors on carotenoid bioavailability [5]. There are some negative aspects to measurements of bioavailability *in vitro*, such as limited solubility of carotenoids in cell culture conditions, and test conditions that could create artefacts and might not mimic the conditions within human gastrointestinal tracts [145].

The contribution of cell culture models to our understanding of carotenoid absorption by the gastrointestinal tract has been reviewed [29]. A commonly applied method to determine the bioavailability of carotenoids *in vitro* is the use of Caco-2 cells, which are differentiated cultures derived from human colonic carcinoma, and are used as a model for the human intestinal epithelium. Monolayers of Caco-2 cells can be used to assess the uptake of carotenoids in micelles, the intestinal metabolism, and the transfer across the basolateral membrane of enterocytes [5]. The relationship between micellar composition and carotenoid uptake was tested for fifteen different carotenoids in the Caco-2 cell line. The phospholipid composition of the micelle greatly affected the carotenoid uptake, in a way that was particularly dependent on the lipophilicity of the carotenoid [147]. For example, the uptake of carotenoids was reduced in micelles containing phosphatidylcholine; this effect was greater for lutein than for β-carotene, a finding attributed to strong associations with long-chain acyl moieties on phosphatidylcholine in the micelle [147]. In contrast, lysophosphatidylcholine enhanced carotenoid uptake by Caco-2 cells, not because of the reduced size of lysophosphatidylcholine-containing micelles, but possibly by stimulating intracellular processing of lipids into chylomicrons and their secretion [147].

Overall, a linear relationship was found between the lipophilicity of a carotenoid and its uptake by Caco-2 cells, signifying a simple diffusion mechanism. The cell line, Caco-2 TC-7, is preferred for such studies because it shows a greater degree of homogeneity to the human colon than does the Caco-2 line and it possesses β-carotene 15,15′-oxygenase activity [17,148].

5. Stable isotopes

Stable isotopic labelling for studying carotenoid bioavailability is excellent for the assessment of dietary status over a long period [149]. Similar to the oral-faecal balance method, stable isotope methods determine the net amount of a carotenoid absorbed [145]. With stable isotopes, however, it is possible to distinguish the dosed carotenoids from endogenous stores, determine the extent of intestinal vitamin A conversion, estimate the absolute absorption and post-absorption metabolism, and use doses which will not affect endogenous carotenoid pools (see *Chapter 8*). Furthermore, isotopic tracer techniques are highly sensitive and can be used to investigate the bioavailability, bioconversion, and bioefficacy of carotenoids from foods or from supplements of pure carotenoids, in human subjects [150,151].

6. Raman spectroscopy

Raman spectroscopy can be applied to measure carotenoids *in situ* in skin and in the retina [152,153] (see *Chapter 10*). Macular pigment density of lutein and zeaxanthin can be measured non-invasively and correlations can be made with carotenoid intakes, as a functional indicator of the bioavailability of these two carotenoids. A portable Raman device is available for measuring carotenoids in the stratum corneum layer of the palm of the hand [153]. It is claimed that this allows correlations to be made between tissue carotenoid levels and the risk of degenerative diseases related to oxidative stress, such as cancers and macular degeneration. Before this method can gain wide acceptance, rigorous validation is needed (see *Chapter 10*).

E. The Future

There is still much to be learned about carotenoid bioavailability. Much of the work to date has concentrated on β-carotene or, to some extent, lycopene and lutein. These studies need to be extended to other carotenoids. Populations other than healthy volunteers with good vitamin A status should also be investigated [149]. Much more research is required to understand the process by which carotenoids enter the intestinal mucosa; is it by passive diffusion, or with the assistance of transporter proteins such as SR-BI or ABCG5? Are other dietary structural or genetic factors important? With the mapping of the human genome, greater understanding of genetic polymorphisms that alter carotenoid bioavailability can be expected, and this may provide an explanation for the existence of carotenoid 'non-responders' and other alterations in carotenoid metabolism.

References

[1] A. During, H. D. Dawson and E. H. Harrison, *J. Nutr.*, **135**, 2305 (2005).
[2] D. M. Deming and J. W. Erdman Jr., *Pure Appl. Chem.*, **71**, 2213 (1999).
[3] V. Tyssandier, E. Reboul, J. F. Dumas, C. Bouteloup-Demange, M. Armand, J. Marcand, M. Sallas and P. Borel, *Am. J. Physiol. Gastrointest. Liver Physiol.*, **284**, G913 (2003).
[4] P. Borel, P. Grolier, M. Armand, A. Partier, H. Lafont, D. Lairon and V. Azais-Braesco, *J. Lipid Res.*, **3**, 250 (1996).
[5] D. A. Garrett, M. L. Failla and R. J. Sarama, *J. Agric. Food Chem.*, **47**, 4301 (1999).
[6] D. M. Deming, S. R. Teixeira and J. W. Erdman Jr., *J. Nutr.*, **132**, 2700 (2002).
[7] A. C. Boileau, N. R. Merchen, K. Wasson, C. A. Atkinson and J. W. Erdman Jr., *J. Nutr.*, **129**, 1176 (1999).
[8] R. Re, P. D. Fraser, M. Long, P. M. Bramley and C. Rice-Evans, *Biochem. Biophys. Res. Commun.*, **281**, 576 (2001).
[9] A. Asai, M. Terasaki and A. Nagao, *J. Nutr.*, **134**, 2237 (2004).
[10] D. Hollander and P. E. Ruble Jr., *Am. J. Physiol.*, **235**, E686 (1978).
[11] B. Jian, M. de la Llera-Moya, Y. Ji, N. Wang, M. C. Phillips, J. B. Swaney, A. R. Tall and G. H. Rothblat, *J. Biol. Chem.*, **273**, 5599 (1998).
[12] P. G. Yancey, M. de la Llera-Moya, S. Swarnakar, P. Monzo, S. M. Klein, M. A. Connelly, W. J. Johnson, D. L. Williams and G. H. Rothblat, *J. Biol. Chem.*, **275**, 36596 (2000).
[13] E. Reaven, Y. Cortez, S. Leers-Sucheta, A. Nomoto and S. Azhar, *J. Lipid Res.*, **45**, 513 (2004).
[14] G. Cao, C. K. Garcia, K. L. Wyne, R. A. Schultz, K. L. Parker and H. H. Hobbs, *J. Biol. Chem.*, **272**, 33068 (1997).
[15] M. V. Lobo, L. Huerta, N. Ruiz-Velasco, E. Teixeiro, P. de la Cueva, A. Celdran,A. Martin-Hidalgo, M. A. Vega and R. Bragado, *J. Histochem. Cytochem.*, **49**, 1253 (2001).
[16] M. Moussa, J. F. Landrier, E. Reboul, O. Ghiringhelli, C. Comera, X. Collet, K. Frohlich, V. Bohm and P. Borel, *J. Nutr.*, **138**, 1432 (2008).
[17] E. Reboul, L. Abou, C. Mikail, O. Ghiringhelli, M. Andre, H. Portugal, D. Jourdheuil-Rahmani, M. J. Amiot, D. Lairon and P. Borel, *Biochem. J.*, **387**, 455 (2005).
[18] C. Kiefer, E. Sumser, M. F. Wernet and J. Von Lintig, *Proc. Natl. Acad. Sci. USA*, **99**, 10581 (2002).

[19] A. van Bennekum , M. Werder, S. T. Thuahnai, C. H. Han, P. Duong, D. L. Williams, P. Wettstein, G. Schulthess, M. C. Phillips and H. Hauser, *Biochemistry*, **44**, 4517 (2005).
[20] K. L. Herron, M. M. McGrane, D. Waters, I. E. Lofgren, R. M. Clark, J. M. Ordovas and M. L. Fernandez, *J. Nutr.*, **136**, 1161 (2006).
[21] G. Tang, C. Serfaty-Lacrosniere, M. E. Camilo and R. M. Russell, *Am. J. Clin. Nutr.*, **64**, 622 (1996).
[22] R. S. Parker, *FASEB J.*, **10**, 542 (1996).
[23] T. W. M. Boileau, A. C. Moore and J. W. Erdman Jr., in *Antioxidant Status, Diet, Nutrition and Health* (ed. A. M. Papas), p. 133, CRC Press, New York (1999).
[24] S. Zaripheh and J. W. Erdman Jr., *J. Nutr.*, **135**, 2212 (2005).
[25] B. A. Clevidence and J. G. Bieri, *Meth. Enzymol.*, **214**, 33 (1993).
[26] V. Ganji and M. R. Kafai, *J. Nutr.*, **135**, 567 (2005).
[27] R. M. Clark, K. L. Herron, D. Waters and M. L. Fernandez, *J. Nutr.*, **136**, 601 (2006).
[28] K. L. Molldrem, J. Li, P. W. Simon and S. A. Tanumihardjo, *Am. J. Clin. Nutr.*, **80**, 131 (2004).
[29] A. During and E. H. Harrison, *Arch. Biochem. Biophys.*, **430**, 77 (2004).
[30] B. Olmedilla-Alonso, F. Granado-Lorencio and I. Blanco-Navarro, *Clin. Biochem.*, **38**, 444 (2005).
[31] B. J. Burri, T. R. Neidlinger and A. J. Clifford, *J. Nutr.*, **131**, 2096 (2001).
[32] C. M. Allen, S. J. Schwartz, N. E. Craft, E. L. Giovannucci, V. L. De Groff and S. K. Clinton, *Nutr. Cancer*, **47**, 48 (2003).
[33] M. R. Forman, E. J. Johnson, E. Lanza, B. I. Graubard, G. R. Beecher and R. Muesing, *Am. J. Clin. Nutr.*, **67**, 81 (1998).
[34] M. R. Forman, G. R. Beecher, R. Muesing, E. Lanza, B. Olson, W. S. Campbell, P. McAdam, E. Raymond, J. D. Schulman and B. Graubard, *Am. J. Clin. Nutr.*, **64**, 559 (1996).
[35] K. J. Yeum, S. L. Booth, J. A. Sadowski, C. Liu, G. Tang, N. I. Krinsky and R. M. Russell, *Am. J. Clin. Nutr.*, **64**, 594 (1996).
[36] *Dietary Reference Intakes for Vitamin C, Vitamin E, Selenium, and Carotenoids*, The National Academies Press (2000).
[37] G. Lietz, G. Mulokozi, J. C. Henry and A. M. Tomkins, *J. Nutr.*, **136**, 1821 (2006).
[38] F. J. Schweigert, K. Bathe, F. Chen, U. Buscher and J. W. Dudenhausen, *Eur. J. Nutr.*, **43**, 39 (2004).
[39] O. Sommerburg, K. Meissner, M. Nelle, H. Lenhartz and M. Leichsenring, *Eur. J. Pediatr.*, **159**, 86 (2000).
[40] C. L. Rock, M. A. Demitrack, E. N. Rosenwald and M. B. Brown, *Cancer Epidemiol. Biomarkers Prev.*, **4**, 283 (1995).
[41] M. R. Lakshman and C. Okoh, *Meth. Enzymol.*, **214**, 74 (1993).
[42] S. Zaripheh, T. W. Boileau, M. A. Lila and J. W. Erdman Jr., *J. Nutr.*, **133**, 4189 (2003).
[43] M. Gajic, S. Zaripheh, F. Sun and J. W. Erdman Jr., *J. Nutr.*, **136**, 1552 (2006).
[44] T. W. Boileau, S. K. Clinton, S. Zaripheh, M. H. Monaco, S. M. Donovan and J. W. Erdman Jr., *J. Nutr.*, **131**, 1746 (2001).
[45] T. W. Boileau, S. K. Clinton and J. W. Erdman Jr., *J. Nutr.*, **130**, 1613 (2000).
[46] E. K. Kabagambe, J. Furtado, A. Baylin and H. Campos, *J. Nutr.*, **135**, 1763 (2005).
[47] F. Khachik, L. Carvalho, P. S. Bernstein, G. J. Muir, D. Y. Zhao and N. B. Katz, *Exp. Biol. Med.*, **227**, 845 (2002).
[48] J. T. Landrum and R. A. Bone, *Arch. Biochem. Biophys.*, **385**, 28 (2001).
[49] F. Gonzalez-Fernandez, *J. Endocrinol.*, **175**, 75 (2002).
[50] P. Bhosale, A. J. Larson, J. M. Frederick, K. Southwick, C. D. Thulin and P. S. Bernstein, *J. Biol. Chem.*, **279**, 49447 (2004).
[51] D. R. Gollapalli, P. Maiti and R. R. Rando, *Biochemistry*, **42**, 11824 (2003).
[52] P. Maiti, D. Gollapalli and R. R. Rando, *Biochemistry*, **44**, 14463 (2005).

[53] F. Khachik, C. J. Spangler, J. C. Smith Jr., L. M. Canfield, A. Steck and H. Pfander, *Anal. Chem.*, **69**, 1873 (1997).

[54] C. P. Gossage, M. Deyhim, S. Yamini, L. W. Douglass and P. B. Moser-Veillon, *Am. J. Clin. Nutr.*, **76**, 193 (2002).

[55] G. Lietz, C. J. Henry, G. Mulokozi, J. K. Mugyabuso, A. Ballart, G. D. Ndossi, W. Lorri and A. Tomkins, *Am. J. Clin. Nutr.*, **74**, 501 (2001).

[56] S. Patton, L. M. Canfield, G. E. Huston, A. M. Ferris and R. G. Jensen, *Lipids*, **25**, 159 (1990).

[57] C. A. Clarke, D. M. Purdie and S. L. Glaser, *BMC Cancer*, **6**, 170 (2006).

[58] Z. Djuric, D. W. Visscher, L. K. Heilbrun, G. Chen, M. Atkins and C. Y. Covington, *Breast J.*, **11**, 92 (2005).

[59] C. M.-G. A. Covington, D. Lawson, I. Eto and C. Grubbs, *Adv. Exp. Med. Biol.*, **501**, 143 (2001).

[60] J. K. Campbell, K. Canene-Adams, B. L. Lindshield, T. W. Boileau, S. K. Clinton and J. W. Erdman Jr., *J. Nutr.*, **134**, 3486S (2004).

[61] E. Giovannucci, *J. Nutr.*, **135**. 2030S (2005).

[62] T. W. Boileau, Z. Liao, S. Kim, S. Lemeshow, J. W. Erdman Jr. and S. K. Clinton, *J. Natl. Cancer Inst.*, **95**, 1578 (2003).

[63] S. K. Clinton, *J. Nutr.*, **135**, 2057S (2005).

[64] K. Canene-Adams, B. L. Lindshield, S. Wang, E. H. Jeffery, S. K. Clinton and J. W. Erdman Jr., *Cancer Res.*, **67**, 836 (2007).

[65] J. K. Campbell, N. J. Engelmann, M. A. Lila, and J. W. Erdman Jr., *Nutrition Res.,* **27**, 794 (2007).

[66] S. K. Clinton, C. Emenhiser, S. J. Schwartz, D. G. Bostwick, A. W. Williams, B. J. Moore and J. W. Erdman Jr., *Cancer Epidemiol. Biomarkers Prev.*, **5**, 823 (1996).

[67] M. I. Arenas, M. V. Lobo, E. Caso, L. Huerta, R. Paniagua and M. A. Martin-Hidalgo, *Human Pathol.*, **35**, 34 (2004).

[68] J. K. Campbell, C. K. Stroud, M. T. Nakamura, M. A. Lila and J. W. Erdman Jr., *J. Nutr.*, **136**, 2813 (2006).

[69] A. Goyal, G. H. Delves, M. Chopra, B. A. Lwaleed and A. J. Cooper, *Int. J. Androl.*, **29**, 528 (2006).

[70] P. Palan and R. Naz, *Arch. Androl.*, **36**, 139 (1996).

[71] N. P. Gupta and R. Kumar, *Int. Urol. Nephrol.*, **34**, 369 (2002).

[72] S. Alaluf, U. Heinrich, W. Stahl, H. Tronnier and S. Wiseman, *J. Nutr.*, **132**, 399 (2002).

[73] W. Stahl, U. Heinrich, O. Aust, H. Tronnier and H. Sies, *Photochem. Photobiol. Sci.*, **5**, 238 (2006).

[74] U. Heinrich, C. Gartner, M. Wiebusch, O. Eichler, H. Sies, H. Tronnier and W. Stahl, *J. Nutr.*, **133**, 98 (2003).

[75] W. Stahl, U. Heinrich, S. Wiseman, O. Eichler, H. Sies and H. Tronnier, *J. Nutr.*, **131**, 1449 (2001).

[76] S. Zaripheh, T. W. Boileau, M. A. Lila and J. W. Erdman Jr., *J. Nutr.*, **133**, 4189 (2003).

[77] J. K. Campbell, S. Zaripheh, M. A. Lila and J. W. Erdman Jr., *FASEB J.*, **19**, A472 (2005).

[78] M. A. Connelly and D. L. Williams, *Trends Endocrinol. Metabolism*, **14**, 467 (2003).

[79] W. Stahl, U. Heinrich, H. Jungmann, J. von Laar, M. Schietzel, H. Sies and H. Tronnier, *J. Nutr.*, **128**, 903 (1998).

[80] Y. Walfisch, S. Walfisch, R. Agbaria, J. Levy and Y. Sharoni, *Br. J. Nutr.*, **90**, 759 (2003).

[81] S. J. Schwartz, *Sight and Life Newsletter*, **3/2004**, 8 (2004).

[82] K. J. Yeum and R. M. Russell, *Annu. Rev. Nutr.*, **22**, 483 (2002).

[83] J. J. Castenmiller and C. E. West, *Annu. Rev. Nutr.*, **18**, 19 (1998).

[84] K. H. van het Hof, C. Gartner, A. Wiersma, L. B. Tijburg and J. A. Weststrate, *J. Agric. Food Chem.*, **47**, 1582 (1999).

[85] H. Y. Chung, H. M. Rasmussen and E. J. Johnson, *J. Nutr.*, **134**, 1887 (2004).

[86] K. H. van Het Hof, C. E. West, J. A. Weststrate and J. G. Hautvast, *J. Nutr.*, **130**, 503 (2000).

[87] C. A. Tracewell, J. S. Vrettos, J. A. Bautista, H. A. Frank and G. W. Brudvig, *Arch. Biochem. Biophys.*, **385**, 61 (2001).
[88] S. de Pee, C. E. West, D. Permaesih, S. Martuti, Muhilal and J. G. Hautvast, *Am. J. Clin. Nutr.*, **68**, 1058 (1998).
[89] C. E. West, A. Eilander and M. van Lieshout, *J. Nutr.*, **132**, 2920S (2002).
[90] A. Frey-Wyssling and F. Schwegler, *J. Ultrastruct. Res.*, **13**, 543 (1965).
[91] M. Baranska, H. Schulz, R. Baranski,T. Nothnagel and L. P. Christensen, *J. Agric. Food Chem.*, **53**, 6565 (2005).
[92] W. M. Harris and A. R. Spurr, *Am. J. Bot.*, **56**, 369 (1969).
[93] W. P. Mohr, *Ann. Bot.*, **44**, 427 (1979).
[94] K. H. van het Hof, L. B. Tijburg, K. Pietrzik and J. A. Weststrate, *Br. J. Nutr.*, **82**, 203 (1999).
[95] E. Reboul, P. Borel, C. Mikail, L. Abou, M. Charbonnier, C. Caris-Veyrat, P. Goupy, H. Portugal, D. Lairon and M. J. Amiot, *J. Nutr.*, **135**, 790 (2005).
[96] C. Gartner, W. Stahl and H. Sies, *Am. J. Clin. Nutr.*, **66**, 116 (1997).
[97] M. Nguyen, D. Francis and S. J. Schwartz, *J. Sci. Food Agric.*, **81**, 910 (2001).
[98] A. Agarwal, H. Shen, S. Agarwal and A. V. Rao, *J. Med. Food*, **4**, 9 (2001).
[99] K. H. van het Hof, B. C. de Boer, L. B. Tijburg, B. R. Lucius, I. Zijp, C. E. West, J. G. Hautvast and J. A. Weststrate, *J. Nutr.*, **130**, 1189 (2000).
[100] J. J. Castenmiller, C. E. West, J. P. Linssen, K. H. van het Hof and A. G. Voragen, *J. Nutr.*, **129**, 349 (1999).
[101] O. Livny, R. Reifen,I. Levy, Z. Madar, R. Faulks, S. Southon and B. Schwartz, *Eur. J. Nutr.*, **42**, 338 (2003).
[102] C. L. Rock, J. L. Lovalvo, C. Emenhiser, M. T. Ruffin, S. W. Flatt and S. J. Schwartz, *J. Nutr.*, **128**, 913 (1998).
[103] A. J. McEligot, C. L. Rock, T. G, Shanks, S. W. Flatt, V. Newman, S. Faerber and J. P. Pierce, *Cancer Epidemiol. Biomarkers Prev.*, **8**, 227 (1999).
[104] J. W. Erdman Jr., *Am. J. Clin. Nutr.*, **70**, 179 (1999).
[105] C. Chitchumroonchokchai and M. L. Failla, *J. Nutr.*, **136**, 588 (2006).
[106] P. E. Bowen, S. M. Herbst-Espinosa, E. A. Hussain and M. Stacewicz-Sapuntzakis, *J. Nutr.*, **132**, 3668 (2002).
[107] H. Schulz, M. Baranska and R. Baranski, *Biopolymers*, **77**, 212 (2005).
[108] W. J. Lessin, G. Catigani and S. Schwartz, *J. Agric. Food Chem.*, **45**, 3728 (1997).
[109] D. F. S. S. Minhthy Nguyen, *J. Sci. Food Agric.*, **81**, 910 (2001).
[110] R. M. Faulks, D. J. Hart, P. D. Wilson, K. J. Scott and S. Southon, *Clin. Sci.*, **93**, 585 (1997).
[111] M. G. Ferruzzi, J. L. Lumpkin, S. J. Schwartz and M. Failla, *J. Agric. Food Chem.*, **54**, 2780 (2006).
[112] G. Britton, *FASEB J.*, **9**, 1551 (1995).
[113] J. D. Ribaya-Mercado, *Nutr. Rev.*, **60**, 104 (2002).
[114] F. Jalal, M. C. Nesheim, Z. Agus, D. Sanjur and J. P. Habicht, *Am. J. Clin. Nutr.*, **68**, 623 (1998).
[115] M. J. Brown, M. G. Ferruzzi, M. L. Nguyen, D. A. Cooper, A. L. Eldridge, S. J. Schwartz and W. S. White, *Am. J. Clin. Nutr.*, **80**, 396 (2004).
[116] J. Mercke Odeberg, A. Lignell, A. Pettersson and P. Hoglund, *Eur. J. Pharm. Sci.*, **19**, 299 (2003).
[117] D. M. Deming, A. C. Boileau, C. M. Lee and J. W. Erdman Jr., *J. Nutr.*, **130**, 2789 (2000).
[118] N. Z. Unlu, T. Bohn, S. K. Clinton and S. J. Schwartz, *J. Nutr.*, **135**, 431 (2005).
[119] C. Fuller, D. N. Butterfoss and M. L. Failla, *Nutr. Res.*, **21**, 1209 (2001).
[120] D. A. Cooper, D. R. Webb and J. C. Peters, *J. Nutr.*, **127**, 1699S (1997).
[121] M. L. Neuhouser, C. L. Rock, A. R. Kristal, R. E. Patterson, D. Neumark-Sztainer, L. J. Cheskin and M. D. Thornquist, *Am. J. Clin. Nutr.*, **83**, 624 (2006).

[122] T. Vasankari, M. Ahotupa, J. Viikari, I. Nuotio, T. Strandberg, H. Vanhanen, H. Gylling, T. Miettinen and J. Tikkanen, *Ann. Med.*, **36**, 618 (2004).
[123] M. Richelle, M. Enslen, C. Hager, M. Groux, I. Tavazzi, J. P. Godin, A. Berger, S. Metairon, S. Quaile, C. Piguet-Welsch, L. Sagalowicz, H. Green and L. B. Fay, *Am. J. Clin. Nutr.*, **80**, 171 (2004).
[124] H. A. Colgan, S. Floyd, E. J. Noone, M. J. Gibney and H. M. Roche, *J. Human Nutr. Dietetics*, **17**, 561 (2004).
[125] Å. L. F. N. Amundsen, N. van der Put and L. Ose, *Eur. J. Clin. Nutr.*, **58**, 1612 (2004).
[126] C. L. Rock and M. E. Swendseid, *Am. J. Clin. Nutr.*, **55**, 96 (1992).
[127] M. Eastwood and D. Kritchevsky, *Annu. Rev. Nutr.*, **25**, 1 (2005).
[128] J. Riedl, J. Linseisen, J. Hoffmann and G. Wolfram, *J. Nutr.*, **129**, 2170 (1999).
[129] J. Serrano, I. Goni and F. Saura-Calixto, *J. Agric. Food Chem.*, **53**, 2936 (2005).
[130] V. Tyssandier, N. Cardinault, C. Caris-Veyrat, M. J. Amiot, P. Grolier, C. Bouteloup, V. Azais-Braesco and P. Borel, *Am. J. Clin. Nutr.*, **75**, 526 (2002).
[131] P. Borel, P. Grolier, N. Mekki, Y. Boirie, Y. Rochette, B. Le Roy, M. C. Alexandre-Gouabau, D. Lairon and V. Azais-Braesco, *J. Lipid Res.*, **39**, 2250 (1998).
[132] S. J. Hickenbottom, J. R. Follett, Y. Lin, S. R. Dueker, B. J. Burri, T. R. Neidlinger and A. J. Clifford, *Am. J. Clin. Nutr.*, **75**, 900 (2002).
[133] T. van Vliet, W. H. Schreurs and H. van den Berg, *Am. J. Clin. Nutr.*, **62**, 110 (1995).
[134] N. Cardinault, V. Tyssandier, P. Grolier, B. M. Winklhofer-Roob, J. Ribalta, C. Bouteloup-Demange, E. Rock and P. Borel, *Eur. J. Nutr.*, **42**, 315 (2003).
[135] R. M. Russell, *J. Nutr.*, **131**, 1359S (2001).
[136] D. L. Taren, M. C. Nesheim, D. W. Crompton, C. V. Holland, I. Barbeau, G. Rivera, D. Sanjur, J. Tiffany and K. Tucker, *Parasitol.*, **95**, 603 (1987).
[137] B. S. Das, D. I. Thurnham and D. B. Das, *Am. J. Clin. Nutr.*, **64**, 94 (1996).
[138] C. Eboumbou, J.-P. Steghens, O. M. S. Abdallahi, A. Mirghani, P. Gallian, A. van Kappel, A. Qurashi, B. Gharib and M. De Reggi, *Acta Tropica*, **94**, 99 (2005).
[139] V. Nussenblatt, G. Mukasa, A. Metzger, G. Ndeezi, W. Eisinger and R. D. Semba, *J. Health Popul. Nutr.*, **20**, 205 (2002).
[140] V. Persson, F. Ahmed, M. Gebre-Medhin and T. Greiner, *Eur. J. Clin. Nutr.*, **55**, 1 (2001).
[141] E. Villamor, R. Mbise, D. Spiegelman, E. Hertzmark, M. Fataki, K. E. Peterson, G. Ndossi and W. W. Fawzi, *Pediatrics*, **109**, e6 (2002).
[142] K. Z. Long, T. Estrada-Garcia, J. L. Rosado, J. I. Santos, M. Haas, M. Firestone, J. Bhagwat, C. Young, H. L. DuPont, E. Hertzmark and N. N. Nanthakumar, *J. Nutr.*, **136**, 1365 (2006).
[143] L. Natarajan, S. W. Flatt, X. Sun, A. C. Gamst, J. M. Major, C. L. Rock, W. Al-Delaimy, C. A. Thomson, V. A. Newman and J. P. Pierce, for the Women's Healthy Eating and Living Study Group, *Am. J. Epidemiol.*, **163**, 770 (2006).
[144] T. Gireesh, P. P. Nair and P. R. Sudhakaran, *Br. J. Nutr.*, **92**, 241 (2004).
[145] R. P. Heaney, *J. Nutr.*, **131**, 1344S (2001).
[146] C. Dubois, M. Armand, V. Azais-Braesco, H. Portugal, A. M. Pauli, P. M. Bernard, C. Latge, H. Lafont, P. Borel and D. Lairon, *Am. J. Clin. Nutr.*, **60**, 374 (1994).
[147] T. Sugawara, M. Kushiro, H. Zhang, E. Nara, H. Ono and A. Nagao, *J. Nutr.*, **131**, 2921 (2001).
[148] A. During, G. Albaugh and J. C. Smith Jr., *Biochem. Biophys. Res. Commun.*, **249**, 467 (1998).
[149] B. J. Burri and A. J. Clifford, *Arch. Biochem. Biophys.*, **430**, 110 (2004).
[150] M. van Lieshout, C. E. West and R. B. van Breemen, *Am. J. Clin. Nutr.*, **77**, 12 (2003).
[151] A. Lienau, T. Glaser, G. Tang, G. G. Dolnikowski, M. A. Grusak and K. Albert, *J. Nutr. Biochem.*, **14**, 663 (2003).
[152] B. R. Hammond and B. R. Wooten, *J. Biomed. Opt.*, **10**, 054002 (2005).
[153] I. V. Ermakov, M. Sharifzadeh, M. Ermakova and W. Gellermann, *J. Biomed. Opt.*, **10**, 064028 (2005).

Carotenoids
Volume 5: Nutrition and Health

Chapter 8

Carotenoids as Provitamin A

Guangwen Tang and Robert M. Russell

A. Introduction

In 1930, Moore discovered that β-carotene (**3**) could be converted *in vivo* into vitamin A [1]. Since then, the vitamin A values of β-carotene and other provitamin A carotenoids, particularly α-carotene (**7**) and β-cryptoxanthin (**55**), have been investigated by various techniques.

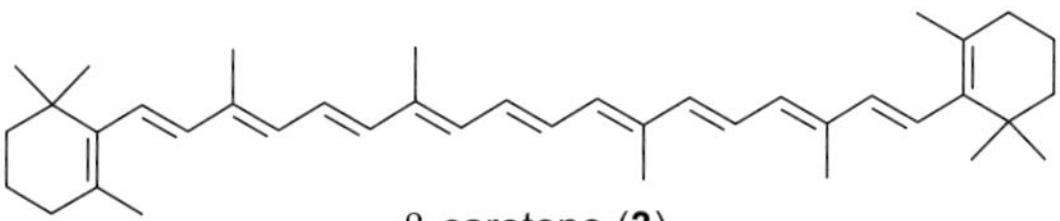

β-carotene (**3**)

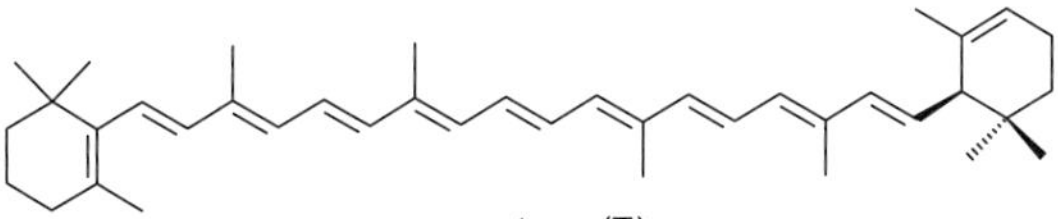

α-carotene (**7**)

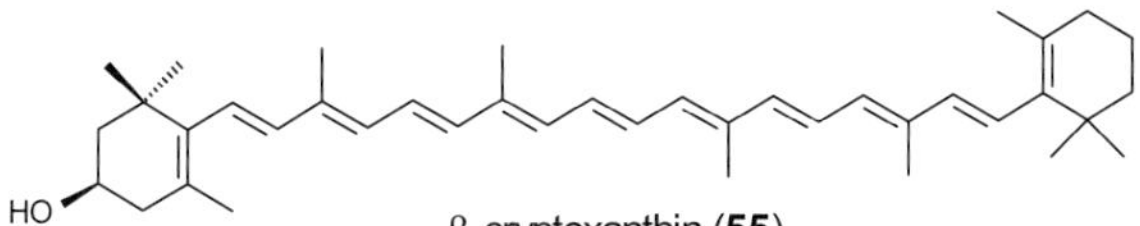

β-cryptoxanthin (**55**)

As discussed in *Chapter 9*, vitamin A nutrition is of worldwide interest; deficiency of the vitamin remains a problem in developing countries, affecting 75 to 140 million children [2]. Deficiency of vitamin A (VAD) can result in visual malfunction such as night blindness and xerophthalmia [3], and can impair immune function [4], resulting in an increased incidence and/or severity of respiratory infections, gastrointestinal infections [5], and measles [6]. Vitamin A levels in HIV-positive children are lower than those in HIV-negative children [7].

Humans obtain vitamin A from their diets. In developing countries, provitamin A carotenoids in vegetables and fruits may provide more than 70% of daily vitamin A intake [8]. In contrast, in Western societies, where sources of the pre-formed vitamin A, *i.e.* eggs, meat, fish, and dairy products, are consumed extensively, provitamin A carotenoids derived from plants may provide less than 30% of daily vitamin A intake [9]. In extensive programmes to reduce or prevent clinical vitamin A deficiency in developing countries, doses of chemically synthesized vitamin A have been given periodically to populations at risk, and this has been demonstrated to be an efficient and safe strategy [10-14]. However, supplementation programmes rely on periodic mass distribution, which is difficult to sustain because of high distribution costs. Food-based interventions to increase the availability of foods rich in provitamin A have been suggested as a realistic and sustainable alternative to tackle vitamin A deficiency globally [15], but the efficacy of carotenoid-rich foods in the prevention of vitamin A deficiency has been questioned in some recent studies [16,17].

In Western populations, interest in studying the vitamin A value of dietary carotenoids has been aroused after epidemiological data have shown that diets rich in carotenoid-containing foods are associated with reduced risk of certain types of chronic diseases such as cancer [18], cardiovascular disease [19], age-related macular degeneration [20,21] and cataract [22,23] (see *Chapters 13-15*). Any disease-preventing activity of β-carotene and other provitamin A carotenoids could be ascribed either to their conversion into retinoids or to activity as intact molecules. The results of several human intervention studies, however, indicate that high-dose supplementation with β-carotene, either alone [24] or with vitamin E [25] or with vitamin A [26], does not decrease the risk of cancer or cardiovascular disease, and might even be harmful to smokers or former asbestos workers. Thus, it may be that β-carotene and other carotenoids may be health-promoting when taken at physiological levels in foods, but may have adverse properties when given in high doses and under highly oxidative conditions. The issue of the efficiency of conversion of provitamin A carotenoids into vitamin A and other retinoids is therefore of interest in both developing and developed countries.

As is well known, after an oral dose of β-carotene, both intact β-carotene and its metabolite, retinol (*1*), can be found in the circulation. In humans, conversion of β-carotene into vitamin A takes place in the intestine and in other tissues. The ratio of the amount of β-carotene given in an oral dose to the amount of vitamin A derived from this β-carotene dose is defined as the β-carotene to vitamin A conversion factor.

B. Conversion into Vitamin A *in vitro*

As illustrated in Fig. 1 and described in detail in *Volume 4, Chapter 16*, two pathways have been proposed for the conversion of β-carotene into vitamin A in mammals. The central cleavage pathway [27,28] leads to the formation of two molecules of vitamin A aldehyde (retinal, *2*) and hence vitamin A itself, retinol (*1*) from one β-carotene molecule by cleavage of the C(15,15') double bond, whereas the excentric pathway leads to the formation of a single molecule of retinal (and thus retinol) by a stepwise oxidation of β-carotene beginning at another of the double bonds of the polyene chain [29,30].

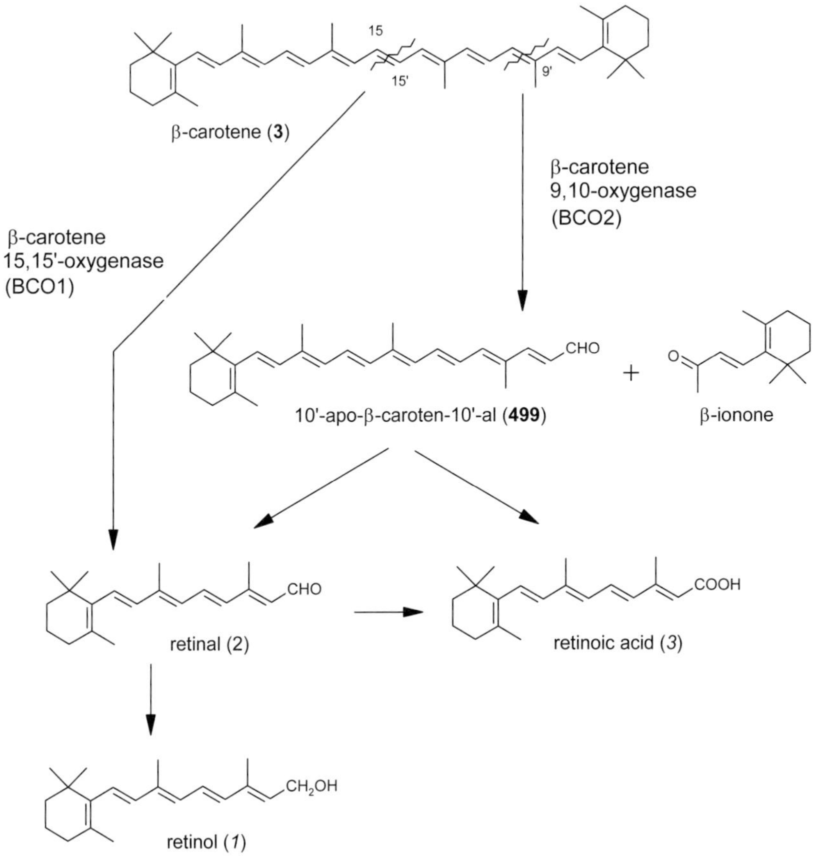

Fig. 1. The formation of vitamin A (retinol, *1*), retinal (*2*) and retinoic acid (*3*) by central or excentric cleavage of β-carotene (**3**) by β-carotene 15,15′-oxygenase (BCO1) and β-carotene 9,10-oxygenase (BCO2), respectively.

A β-carotene 9,10-oxygenase has been identified. The enzymic conversion of β-carotene into retinoic acid (*3*), retinal (*2*), 12′-apo-β-caroten-12′-al (**507**), 10′-apo-β-caroten-10′-al (**499**), and 8′-apo-β-caroten-8′-al (**482**) by mammalian tissues *in vitro* has been demonstrated [31]. In addition, the appearance of the metabolites 13-apo-β-caroten-13-one (C_{18}-ketone, *4*) and 14′-apo-β-caroten-14′-al (**513**), formed in significant amounts during the incubation of mammalian tissues with β-carotene, has been reported [32]. A recent study confirmed that both central and excentric cleavage of β-carotene take place in the post-mitochondrial fraction of rat intestinal cells, but the relative activity of the two pathways depends on the presence or absence of an antioxidant such as α-tocopherol [33].

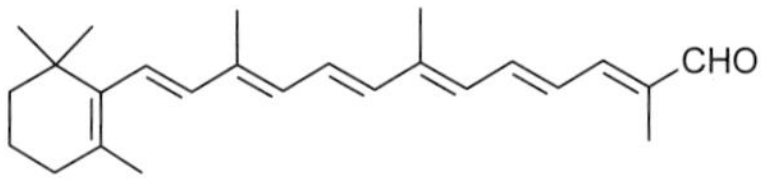

12′-apo-β-caroten-12′-al (**507**)

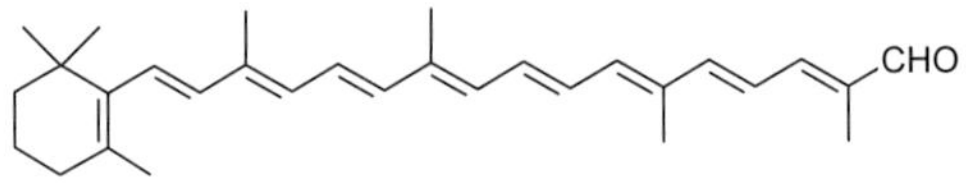

8′-apo-β-caroten-8′-al (**482**)

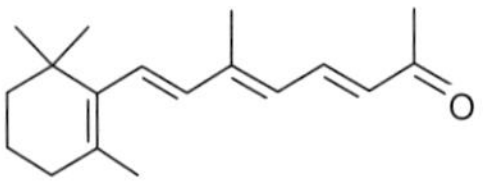

'C_{18}-ketone' (*4*)

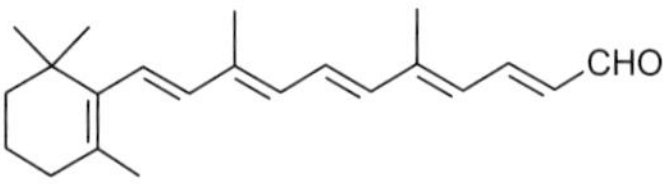

14′-apo-β-caroten-14′-al (**513**)

In 2000, the enzyme β-carotene 15,15′-oxgenase that cleaves β-carotene to retinal was identified in chicken intestinal mucosa and subsequently sequenced and expressed in two different cell lines [34]. In addition, the existence of different types of cleavage enzymes of β-carotene in mouse [35] and human [36] was reported. Very recently, both central (β-carotene 15,15′-oxygenase, BCO1) and excentric (β-carotene 9,10-oxygenase, BCO2) cleavage enzymes have been reported in small intestine, liver, skin, eye, and other tissues [36]. The existence of at least two different β-carotene oxygenases makes estimation of the vitamin A value of β-carotene complex. The genes and enzymes, their regulation and the reaction mechanisms are discussed in *Volume 4, Chapter 16*.

C. The Conversion of Provitamin A Carotenoids into Vitamin A *in vivo*: Methods to Determine Conversion Factors

In relation to the value of β-carotene and other carotenoids as dietary precursors of vitamin A, a key and controversial question concerns the efficiency of the enzymic conversion of the carotenoids into vitamin A *in vivo*. The many food tables that list the precise carotenoid content of fruit and vegetables (see *Chapter 3*) tell only part of the story; the efficiency with which the body can obtain vitamin A from these sources is another vital factor. The absorption, transport and other factors that influence the bioavailability of carotenoids are described in *Chapter 7*. Many different numerical values (conversion factors) have been reported for the formation of vitamin A from β-carotene and other provitamin A carotenoids, either obtained from the diet or provided as supplements, and several different methods have been used to determine these conversion factors. The most useful of these methods, and the results obtained by their use, are described and evaluated below.

1. Measuring radioactivity recovered in lymph and blood after feeding radio-isotopically labelled β-carotene

A few studies have been carried out to investigate the conversion rate of radioactive β-carotene in humans. Two early studies [37,38] reported the absorption and conversion of β-carotene in adult subjects. An oral dose of labelled β-carotene was given and thoracic duct lymph was collected. In one study [37], the total radioactivity recovered in the lymph of two adult subjects given a labelled β-carotene dose was 8.7% and 16.8%. Of this, 22-30% of the absorbed radioactivity was recovered in β-carotene, and 61-71% in retinyl esters. In another study [38], the mean total radioactivity recovered in the lymph of four adult patients after taking a labelled β-carotene dose was 23.1% (range 8.7-52.3%). In this case, 1.7-27.9% of the absorbed radioactivity was recovered in β-carotene, and 68.2-87.9% in retinyl esters (one outlier was omitted). From these results, it is reasonable to speculate that the absorption of pure β-carotene in humans is in the range of 10-20%, and that about 70% of the absorbed radioactivity from labelled β-carotene is recovered in retinyl esters.

In recent years, the development of very sensitive accelerator mass spectrometry (AMS) has made it possible to use minute doses of [^{14}C]-β-carotene to study the presence of metabolites of [^{14}C]-β-carotene in human plasma, urine and faeces samples [39]. The absorption of the β-carotene was estimated at 43%, and 62% of this absorbed β-carotene was converted into vitamin A. Vitamin A values of 0.53, 0.62, and 0.54 mol from 1 mol of β-carotene were calculated, though a number of assumptions were made in the calculation, *e.g.* that 77% of absorbed β-carotene is cleaved through excentric cleavage [39,40].

2. Measuring the repletion doses of β-carotene and vitamin A needed to reverse vitamin A deficiency in vitamin A depleted adults

A depletion study [41] was conducted on sixteen healthy subjects between the ages of 19 and 34 years (seven additional subjects served as positive controls). After twelve months of depletion, only three of the subjects were vitamin A deficient; both a blood concentration below 0.35 μmol/L (10 μg/dL) and deterioration in dark adaptation were used to define 'unmistakably deficient' subjects. Of the three subjects with these 'unmistakable' signs of vitamin A deficiency, two were given β-carotene and one was given pre-formed vitamin A. Daily doses of 1,500 μg of β-carotene or 390 μg of retinol for 3 weeks to 6 months were sufficient to reverse vitamin A deficiency in these subjects. Therefore, from this human study, the β-carotene:vitamin A equivalence was determined to be 3.8:1 by weight. In 1974, another extensive and well controlled vitamin A depletion-repletion study in human subjects was reported [42]. Eight healthy male subjects between 31 and 43 years of age were depleted in vitamin A within 359-771 days. Depletion was defined by a plasma retinol level below 0.3 μmol/L (10 μg/dL) and clinical signs of vitamin A deficiency (dark adaptation impairment, abnormal electroretinogram, or follicular hyperkeratosis). Five subjects were then given vitamin A and three subjects given β-carotene. Daily doses of 600 μg retinol or 1200 μg of β-carotene were required to cure vitamin A deficiency. In this study, the β-carotene to vitamin A equivalence was, therefore, 2:1 by weight. In these studies, all subjects had been made deficient in vitamin A, so it cannot be determined whether a 3.8 μg or 2 μg equivalence of β-carotene to 1 μg of retinol is applicable in vitamin-A-sufficient individuals.

The results of earlier studies in 1939 and 1940 [43,44] are in question because of the lack of standardization of the experimental approaches and endpoints.

On the basis of these investigations with synthetic β-carotene in humans, and the lack of any precise data on the bioavailability or bioconversion of carotenoids from foods, the availability of β-carotene from the diet has been taken as one-third of the provitamin A carotenoids ingested, with a maximum conversion of absorbed β-carotene of 50% on a weight basis [9]. Since other provitamin A carotenoids (α-carotene, β-cryptoxanthin, *etc.*) can provide one molecule of vitamin A, they are expected to exhibit approximately half the vitamin A activity of β-carotene [45]. Therefore, the retinol equivalence of carotenoids in food has generally been assumed and accepted as being: 6 μg of (all-*E*)-β-carotene, or 12 μg of other provitamin A carotenoids are equivalent to 1 μg of retinol (1 retinol equivalent, RE) [9,46,47]. By using these assumptions, the NHANES (National Health and Nutrition Examination Survey) of 1970-1980 in the U.S. showed that the median adult dietary intake of vitamin A was 624 RE, with *ca.* 25% coming from carotenoids and *ca.* 75% coming from preformed vitamin A sources, as calculated from food composition tables [9] and the conversion factor of 6:1 for β-carotene to retinol conversion.

3. Measuring changes of serum vitamin A levels after feeding synthetic β-carotene or food rich in provitamin A carotenoids

There are several reasons why the vitamin A activities of provitamin A carotenoids provided in food had not been studied quantitatively in humans until recently. It was found that plasma β-carotene concentration could not be altered by eating a meal containing up to 6 mg of β-carotene in a food matrix [48,49]. Therefore, doses of unlabelled β-carotene of 6 mg or less could not be used to study β-carotene absorption or conversion, because of the insensitivity of the blood response. Past studies reported that supplementation with 12-180 mg of β-carotene is required to investigate the blood or chylomicron β-carotene response in humans [48-50]. The conversion of β-carotene into vitamin A cannot be estimated accurately in well-nourished humans by assessing changes in serum retinol after supplementation with unlabelled β-carotene, because newly-formed retinol cannot be distinguished from retinol derived from body reserves; it is well known that blood retinol concentrations are homeostatically controlled in a well-nourished individual. Nevertheless, many investigations with populations who normally have low vitamin A intake have reported blood retinol responses to acute or chronic β-carotene supplements [16,17,51]. Changes in serum retinol levels were seen [52] in vitamin A deficient (~ 0.7 μmol/L) anaemic schoolchildren aged 7-11 years, who were fed one of four supplements: (i) 556 RE/day from retinol-rich foods, n = 48; (ii) 509 RE/day from fruits, n = 49; (iii) 684 RE/day from vegetables, n = 45; or (iv) 44 RE/day from low-retinol and low-carotene foods, n = 46. The supplements were fed six days per week for 9 weeks, and the changes in serum retinol were then assessed to determine a relative conversion efficiency of β-carotene from vegetables or fruits compared with that from food rich in preformed vitamin A (egg, chicken liver, fortified margarine, and fortified chocolate milk). Those consuming fruit (diet ii) or vegetables (diet iii) showed increases of 0.12 μmol/L and 0.07 μmol/L, respectively, in serum retinol whereas the group consuming foods rich in preformed vitamin A (diet i) showed an increase of 0.23 μmol/L. The relative mean conversion factor of vegetable β-carotene into retinol was calculated, by weight, as 26:1 and that of β-carotene from orange-coloured fruit as 12:1. Use of a similar approach [53] showed that, for breast-feeding women, the conversion factors of β-carotene into retinol were, by weight, 12:1 for fruit and 28:1 for green leafy vegetables.

4. Measuring changes in body stores of vitamin A after feeding dietary provitamin A carotenoids (paired DRD test)

As shown in the previous Section, for populations with marginal to normal vitamin A status, the changes of serum retinol may not be a sensitive indicator of vitamin A status. Instead, isotope dilution techniques can be used to measure changes of total body stores of vitamin A. A deuterated retinol dilution (DRD) method was used in a study of children with marginal to normal vitamin A status, who participated in a food-based intervention with either green-yellow vegetables or light-coloured vegetables with low carotene content [54]. The serum carotenoid concentrations of children fed green-yellow vegetables increased, whilst the serum concentration of vitamin A did not change. In contrast, the isotope dilution tests carried out before and after the vegetable intervention showed that the body stores of vitamin A were stable in the group fed green-yellow vegetables, but decreased in the group fed light-coloured vegetables. Over a 10-week period, a loss of 7 mg vitamin A from body stores was seen in the children fed light-coloured vegetables containing little β-carotene, but 275 mg β-carotene from green-yellow vegetables prevented this loss. From this paired DRD test, it was calculated that 27 μg β-carotene from vegetables was equivalent to 1 μg retinol. This conversion factor is similar to that reported in other studies for carotenoids from vegetables [54].

The paired DRD technique has also been used [55] to measure change in the vitamin A pool size after 60-day supplementation with 750 RE/day as either retinyl palmitate, β-carotene, sweet potato, or Indian spinach, compared with a control containing no retinol or carotene. Vitamin A equivalency factors of 6:1 for β-carotene in oil, 10:1 for β-carotene in Indian spinach, and 13:1 for β-carotene in sweet potato were determined.

A recent study used mixed-vegetable intervention and the paired DRD test to measure the changes in vitamin A pool size [56]. The results showed that the conversion factors were better than 12:1 for β-carotene and 24:1 for other provitamin A carotenoids.

5. Measuring intestinal absorption by analysis of postprandial chylomicron fractions after feeding synthetic β-carotene or food rich in provitamin A carotenoids

In another approach, postprandial chylomicron (PPC) response curves of β-carotene and retinyl esters in blood were measured following a single dose of β-carotene supplement in oil or from vegetables [57-59]. In these studies, triacylglycerol-rich lipoproteins (TRL) with density less than 1.006 g/mL were separated and analysed to evaluate the absorption efficiency of β-carotene (intact and, after central cleavage, as retinyl palmitate). The TRL fraction of blood lipoprotein contains both VLDL (very low density lipoproteins) and chylomicrons. However, the postprandial TRL fraction contains mainly chylomicron particles. The efficiency of absorption of β-carotene by each subject was calculated by measuring the areas under the curve (AUC, nmol.h/L) of β-carotene and retinyl ester concentrations in postprandial TRL fractions collected hourly. These curves were compared with hypothetical AUC after an intravenous dose of the same amount of β-carotene, assuming that the β-carotene disappearance follows a first-order elimination from blood with a chylomicron remnant half-life of 11.5 min [58]. Total absorption of β-carotene was measured as the sum of the AUC of β-carotene and retinyl palmitate, with the assumption that 1 molecule of β-carotene is converted into 1 molecule of retinyl palmitate.

On the basis of a postprandial chylomicron (PPC) study in ten young men aged 20-24 years [57], the mean absorption of 15 mg β-carotene (as 10% water-soluble beadlets) was reported as 17% (2.6 mg), and the conversion of absorbed β-carotene into retinyl palmitate as 52-83% (1.6 mg). A similar approach [59] in six men and six women aged 20-25 years gave a value of 8% (3.2 mg) for the mean absorption of β-carotene from a capsule of palm oil extract containing 40 mg β-carotene, while the conversion of absorbed β-carotene into retinyl palmitate was 40% (1.3 mg). These studies showed relatively similar β-carotene AUC responses, but up to a two-fold discrepancy in the reported AUC values for retinyl esters formed from the β-carotene dose, possibly due to variable recovery of the TRL fraction and the dynamic nature of chylomicron secretion and clearance. When a similar approach was used to evaluate the utilization of β-carotene from vegetables [59], little or no β-carotene response was observed in the TRL fraction after equivalent doses of 15 mg carotenoids from cooked carrots, tomato paste, or spinach were given. Thus, the suitability of the PPC method for studying the conversion of a normal dietary level of β-carotene from food is uncertain.

To compensate for the variability of TRL recovery, deuterium-labelled vitamin A has been used [60] as an extrinsic standard. A subject was given raw carrots containing 9.8 μmol (5 mg) β-carotene and 5.2 μmol (2.8 mg) α-carotene together with 7 μmol (2 mg) $[^{2}H_{4}]$-retinyl acetate, and the concentrations of β-carotene, α-carotene, and labelled and unlabelled retinyl esters in the TRL were measured at various time points up to 7 hours. With the assumption that absorption of labelled retinyl acetate was about 80% of the dose, it was calculated that 0.8 μmol of the carrot β-carotene was absorbed intact and that 1.5 μmol of unlabelled retinyl esters were formed from the carrot dose. The mass equivalency of carrot β-carotene to vitamin A was, therefore, 13:1 (without considering the contribution from 5.2 μmol of α-carotene to vitamin A). If the contribution of α-carotene is considered, the ratio is higher (16:1), assuming that α-carotene has half the activity of β-carotene.

6. Measuring blood response kinetics after feeding β-carotene labelled with stable isotopes

a) Single dose

For studying the absorption and conversion of β-carotene in humans, a sensitive method involving administration of β-carotene labelled with either $[^{13}C]$-β-carotene or $[^{2}H]$-β-carotene and analysis by MS has been used [60-64]. In one study, 1 mg of per-labelled $[^{13}C_{40}]$-β-carotene was given to a middle-aged male subject. The isotope ratios were determined by gas chromatography-combustion-isotope ratio mass spectrometry (GC-C-IRMS) [61]. On a molar basis, 64% of the absorbed $[^{13}C]$-β-carotene in the circulation was recovered in retinyl esters, 21% in retinol, and 14% in intact β-carotene.

In another study [62], 73 μmol (*ca.* 40 mg) of [$^2H_8$]-β-carotene was given to a male subject, and plasma samples were drawn over a 24-day period. The isotope ratio of [$^2H_8$]-β-carotene:unlabelled β-carotene in plasma was determined. A strong correlation was reported between the ratios of [$^2H_8$]-β-carotene:unlabelled β-carotene in the plasma determined by either lengthy HPLC or MS/MS methods. The HPLC method, however, was able to detect as little as 1.87 pmol of [$^2H_8$]-β-carotene, whereas the detection limit for the MS/MS method was 100 pmol. Compartmental analysis [63] of these data showed that 22% of the β-carotene dose was absorbed, 17.8% as intact β-carotene and 4.2% as retinol. That is, 1 μg dietary β-carotene was equivalent to 0.054 μg retinol. When 37 μmol [$^2H_6$]-β-carotene (*ca.* 20 mg) and 30 μmol [$^2H_6$]-retinyl acetate (10 mg) were fed to eleven healthy, non-smoking, female subjects aged 19 to 39 years, only six of the volunteers showed a measurable response ($\geq$ 0.01 μmol.h/L for [$^2H_3$]-retinol and/or [$^2H_6$]-β-carotene) to the labelled β-carotene dose [64]. The mean absorption of intact [$^2H_6$]-β-carotene was 6.1% for the six normal responders and <0.01% for the five non-responders. The mean absorption of [$^2H_6$]-β-carotene as [$^2H_3$]-retinol was not reported, but the data indicate that *ca.* 10% of the total absorbed [$^2H_6$]-β-carotene was converted into [$^2H_3$]-retinol. The lower absorption value found in this study was attributed to the use of doses that were neither 'solubilized nor emulsified' [64].

b) Multiple doses

To analyse [^{13}C]-labelled β-carotene, an LC/ESI-MS (liquid chromatography/electrospray ionization-mass spectrometry) method was developed [65] with a detection limit for β-carotene between 1 and 2 pmol. However, the response of ESI-MS *versus* β-carotene concentration was not linear. Later, an LC/APCI-MS (liquid chromatography/atmospheric pressure chemical ionization-mass spectrometry) method was developed [66] and used to study the metabolism of [$^{13}C_{10}$]-β-carotene in children 8-11 years of age who had been given multiple doses of [$^{13}C_{10}$]-β-carotene, mostly as *Z* isomers, (80 μg/day) and retinyl palmitate (80 μg/day) to reach an enrichment plateau in the circulation (plateau isotope enrichment technique). The results showed that 2.4 μg β-carotene (mostly as (*Z*)-β-carotene) in oil could be converted into 1 μg retinol [66]. In another study that used HPLC/APCI-MS to determine the enrichment of intact β-carotene from a deuterium-labelled β-carotene dose, the detection limit of β-carotene was 50 pg [67]. Thus, physiological doses of β-carotene could be used to study the absorption of labelled β-carotene in humans.

c) Use of labelled retinyl acetate as a reference

An isotope reference method to determine the retinol equivalence of β-carotene in humans was also developed [68,69]. By using a known amount of [$^2H_8$]-retinol as a reference and comparing its blood response to the amount of [$^2H_4$]-retinol formed *in vivo* from [$^2H_8$]-β-carotene, the vitamin A value of the vitamin A precursor or a food can be determined. This 'isotope reference method' can be used to define in humans the vitamin A activity of various vitamin A precursors, *e.g.* synthetic β-carotene or provitamin A carotenoids in vegetables, fruits or algae, as shown in Fig. 2.

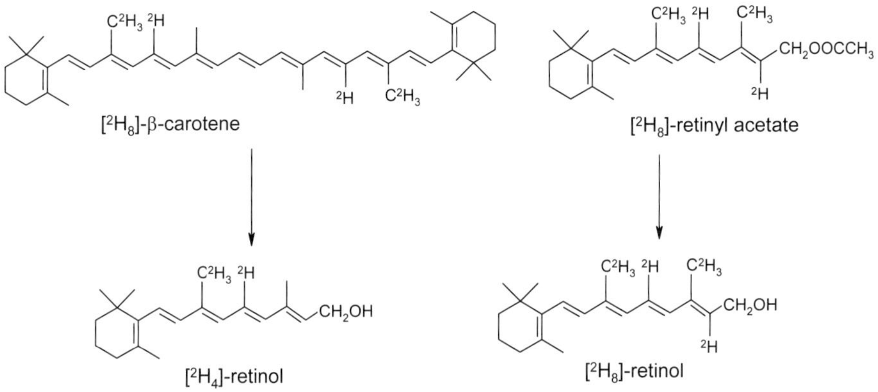

Fig. 2. Scheme to illustrate the origin of [$^2H_4$]-retinol and [$^2H_8$]-retinol detected in serum after feeding [$^2H_8$]-β-carotene and [$^2H_8$]-retinyl acetate.

In one study, two dosage levels (a pharmacological dose, 126.0 mg [$^2H_8$]-β-carotene, and a physiological dose, 6.0 mg [$^2H_8$]-β-carotene) were used 2.5 years apart in an adult female volunteer to study dose effects on the conversion of β-carotene into vitamin A [68]. Blood samples were collected over 21 days. β-Carotene and retinol were extracted from serum and isolated by HPLC. The retinol fraction was converted into a trimethylsilyl ether derivative [69], which was analysed by GC/ECNCI-MS (gas chromatography/electron capture negative chemical ionization-mass spectrometry). The [$^2H_4$]-retinol response in the circulation reached a peak 24 hours after the [$^2H_8$]-β-carotene dose was given, with a higher percent enrichment after the physiological dose than after the pharmacological dose. From this, it was calculated that 6 mg of [$^2H_8$]-β-carotene (11.2 μmol) was equivalent to 1.6 mg of retinol (*i.e.* 3.8 mg of β-carotene was equivalent to 1 mg of retinol), whereas 126 mg of [$^2H_8$]-β-carotene (235 μmol) was equivalent to 2.3 mg of retinol (*i.e.* 55 mg β-carotene was equivalent to 1 mg retinol). These results demonstrate the feasibility of using a stable isotope reference method to study the retinol equivalence of provitamin A carotenes, and show that there is an inverse dose-dependent efficiency of bioconversion of β-carotene into retinol.

The bioavailability of 6 mg (11.2 μmol) synthetic [$^2H_8$]-β-carotene was studied in 22 adult subjects (10 men and 12 women). To avoid possible absorption competition between [$^2H_8$]-β-carotene and [$^2H_8$]-retinyl acetate, the two tracers were given separately [70]. On day 1, the subjects were given 6 mg of [$^2H_8$]-β-carotene in corn oil with a high-fat liquid beverage (25% total energy was from fat). Serum samples were collected at 0, 3, 5, 7, 9, 11, and 13 hrs after the [$^2H_8$]-β-carotene dose. On days 2 and 3, fasting serum samples were collected. Then, on day 4, volunteers were given 3.0 mg (8.9 μmol) of [$^2H_8$]-retinyl acetate (equivalent to 2.6 mg retinol) in corn oil with the same high-fat liquid beverage as was used on day 1. Serum samples were collected from 0 to 13 hrs (as on day 1) after the dose of [$^2H_8$]-retinyl acetate was given. From days 5 to 10, daily fasting serum samples were collected. After day 10, subjects were free living and their fasting serum samples were collected weekly for 8 weeks. Serum samples were analysed by HPLC and GC/ECNCI-MS. A representative serum response of the [$^2H_4$]-retinol from the [$^2H_8$]-β-carotene doses and the [$^2H_8$]-retinol from the reference [$^2H_8$]-retinyl acetate is presented in Fig. 3. The AUCs of [$^2H_4$]-retinol and [$^2H_8$]-retinol percent enrichment response for all subjects were obtained and a conversion factor for each subject was calculated. The values ranged between 2.4:1 and 20.2:1, with an average of 9.1:1 (by weight). In a similar study conducted in a healthy Chinese population, the same conversion factor of 9.1:1 was observed, with a range from 3.8:1 to 22.8:1 [71].

In a similar study of a subject given 30 μmol of [$^2H_6$]-β-carotene (16.2 mg) and a reference dose of [$^2H_6$]-retinyl acetate (10.2 mg) in olive oil (11 g), 15.9 μg of β-carotene was found to be equivalent to 1 μg of retinol [72).

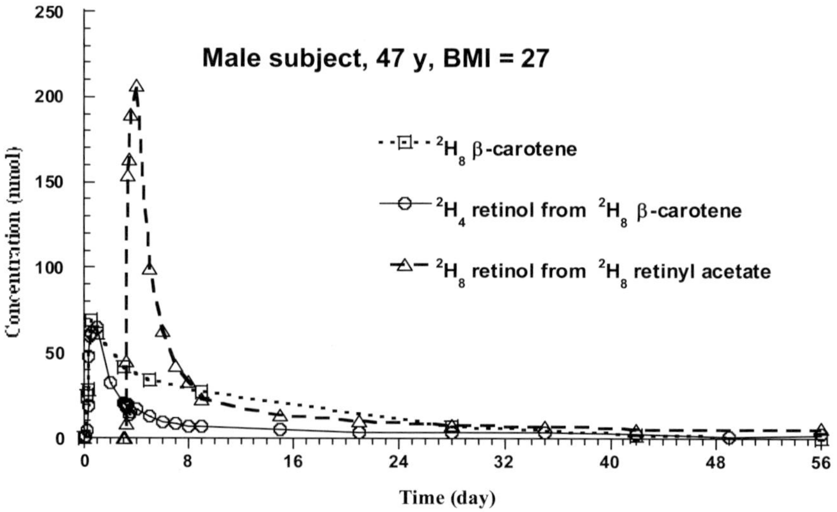

Fig. 3. Graph to illustrate the changes in concentration of [$^2H_8$]-β-carotene, [$^2H_4$]-retinol and [$^2H_8$]-retinol in serum of a male subject, age 47 years, after feeding [$^2H_8$]-β-carotene on day 0 and [$^2H_8$]-retinyl acetate on day 4. BMI: body mass index.

7. Feeding intrinsically labelled dietary provitamin A carotenoids in food

It has been common practice to assess the vitamin A value of a food from the amounts of preformed vitamin A and provitamin A carotenoids contained in that food. As discussed in Section **D** and in *Chapter 7*, major factors that affect the bioavailability of food carotenoids and the bioconversion of food carotenoids into vitamin A in humans are the food matrix, food preparation, and the fat content of a meal. Absorption of carotenoids and vitamin A from various food matrices has not been well studied because, until recently, isotopically labelled foods that can be fed to humans were not available. Therefore, in order to achieve an accurate assessment of carotenoid bioabsorption and a subsequent vitamin A value from a food source, food material is required in which the carotenoids have been endogenously or intrinsically labelled with a low abundance stable isotope. This allows presentation of the carotenoids in their normal cellular compartments, and the isotopic label makes it possible to identify those serum carotenoids (or derived retinol), which come from the specific food in question.

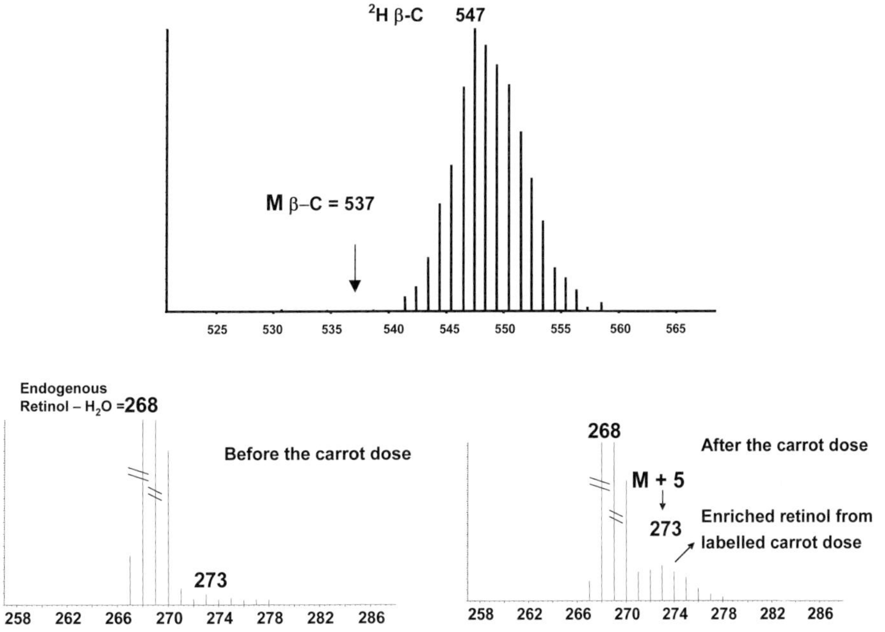

Fig. 4. Top: molecular ion region of the mass spectrum of β-carotene isolated from carrots grown in water enriched with ^2H_2O (25 atom %). A range of pseudo-molecular ions $[M+1]^{\cdot+}$ is seen, the most abundant being that at *m/z* 547, due to $[^2H_{10}]$-β-carotene. Bottom left: detail of the mass spectrum of unlabelled retinol, showing the $[M\text{-}H_2O]$ ion at *m/z* 268. Bottom right: detail of the mass spectrum of retinol formed after consumption of $[^2H]$-enriched carrots containing $[^2H]$-labelled β-carotene, of which the main species is the $[^2H_{10}]$ isotopomer.

Plant carotenoids can be intrinsically labelled either with ^{13}C from $^{13}CO_2$, or with ^{2}H from $^{2}H_2O$. To achieve high enrichments of the carotenoid pool, the plants must be maintained on a constant supply of the isotope throughout their entire growth period. Labelling with $^{13}CO_2$ requires a closed atmospheric system that can be regulated for humidity, temperature, CO_2 and O_2 concentrations. For $^{2}H_2O$ labelling, on the other hand, plants can easily be grown hydroponically [73] on a nutrient solution with a fixed ^{2}H atom percentage. No special facilities for the growth system are required but, by enclosing hydroponically labelled plants in a closed chamber in which the atmospheric water vapour is also enriched with $^{2}H_2O$, improved labelling is achieved, and costs can be reduced by recovering transpired water *via* a condensing system. Water with 25% atom excess of ^{2}H generates a range of isotopomers of carotenoids, with peak enrichment in the $^{2}H_{10}$ species. Figure 4 (top) demonstrates the isotope profile of β-carotene from carrot grown hydroponically with 25 atom % $^{2}H_2O$ and analysed by LC/APCI-MS. The highest abundance peak is at *m/z* 547 [(M + 1) + 10]. In Fig. 4 (lower), the GC/ECNCI-MS analysis confirms that the labelled retinol formed from the labelled carrot dose has the most abundant enrichment peak at *m/z* 273 [(M + 1) + 5].

Spinach and carrots were harvested 32 and 60 days, respectively, after initiating the hydroponic growth in the $^{2}H_2O$-enriched medium. The spinach leaves (or carrots) were steamed in thin layers for 10 minutes. The cooked vegetables were immersed in cold water (1 litre water per 200 g vegetable) for 2 minutes, and then drained, pureed, sealed in a plastic container, and stored at -70°C before being used for the analysis of contents and for human consumption experiments.

Seven men (average age 56 years) each took the spinach and carrot in separate meals 3 months apart [74], to avoid possible interference between the doses, which were given in a random order. A fasting blood sample (10 ml) was drawn on day 0. Then, a liquid formula breakfast was given (25% energy from fat). In the middle of this meal, the subject took an oral dose of either spinach (300 g, thawed), or carrot (100 g, thawed). On day 7, the volunteer repeated the procedures described for day 0 of the study, except that he received as a reference dose a 3.0 mg [$^{2}H_8$]-retinyl acetate capsule together with a liquid formula meal. No vitamin supplements or large amounts of either β-carotene or vitamin A in the diet were permitted during this period. The process was repeated on day 90 with the other vegetable. The serum samples were analysed by GC/ECNCI-MS to determine the isotopic enrichment of retinol formed from the labelled vegetables. The enrichment of each isotopomer was counted in the calculation. The 300 g labelled spinach and 100 g labelled carrots each contained *ca.*11 mg (all-*E*)-β-carotene, and it was assumed that α-carotene and (*Z*)-β-carotene, which were also present, have half the activity of (all-*E*)-β-carotene. The retinol equivalences were determined to be 21 μg spinach β-carotene or 15 μg carrot β-carotene to 1 μg retinol.

With a similar approach, ten men (average age 48 years) each took 5 g dried *Spirulina* powder, containing 4.3 mg β-carotene [75]. When compared to a reference dose of 2.0 mg [$^{13}C_{10}$]-retinyl acetate in oil (capsule), 4.5 mg *Spirulina* β-carotene provided 1 mg retinol.

Another recent report demonstrated the absorption of β-carotene from intrinsically labelled kale and the formation of labelled retinol formed from the labelled kale β-carotene [76], but no conversion factor was estimated.

8. Conversion factors of β-carotene into retinol in humans: Summary

A summary of the major human studies to determine conversion factors for β-carotene, either synthetic or as a plant food constituent, into retinol is presented in Table 1. These data show that the conversion efficiency of vegetable β-carotene is very variable and poorer than previously thought.

Table 1. Summary of the results of studies to determine the conversion factor for β-carotene (β-C) in oil or in food sources. (n = number of subjects)

Food matrix	Method	Dose	Conv. factor	Ref
β-C in oil capsule n = 3	Depletion/repletion; adults	Repletion daily with 390 μg vitamin A (n = 1) or 1500 μg β-C (n = 2)	3.8 : 1	[41]
β-C in oil capsule n = 5	Depletion/repletion; adults	Repletion daily with 600 μg vitamin A (n = 2) or 1200 μg β-C (n = 3)	2 : 1	[42]
β-C and vitamin A in oil capsules n = 35	Enrichment plateau in school children (age 8-11) with normal or marginal vitamin A status	Twice daily, 80 μg [$^{13}C_{10}$]-β-C and 80 μg [$^{13}C_{10}$]-vitamin A in oil capsules for 21 days	1.5 or 2.4 : 1	[66]
β-C and vitamin A in oil capsule n = 1	Comparing AUC response to the β-C dose and the vitamin A reference dose in adults	β-C in oil, 6 mg β-C in oil, 126 mg	3.8 : 1 55 : 1	[68]
β-C and vitamin A in oil capsule n = 22	Comparing AUC response to the β-C dose and the vitamin A reference dose in adults	β-C in oil, 6 mg	9 : 1	[70,71]
β-C and vitamin A in oil capsule n = 1	Comparing AUC response to the β-C dose and the vitamin A reference dose in an adult	16.8 mg [$^{2}H_{6}$]-β-C in oil capsule and 10.2 mg [$^{2}H_{6}$]-vitamin A	15.9 : 1	[72]
Fruits, n = 49 Vegetables, n = 45 Retinol-rich foods, n = 48	Changes of serum retinol conc. in vitamin A-deficient (~0.7 μmole/L) anaemic school children (age 7-11)	Fruits: 509 RE/day Vegetables: 684 RE/day Vitamin A-rich foods: 556 RE/day	12 : 1 26 : 1	[52]

Table 1, continued.

Food matrix	**Method**	**Dose**	**Conv. factor**	**Ref.**
Green/yellow vegetables, n = 10 Light coloured vegetables, n = 8	Total body stores of vitamin A before and after the vegetable intervention in school children (age 5.3 -6.5) with normal or marginal vitamin A status	Green/yellow vegetables (206 mg calculated *E*-β-C) to prevent the decrease of 7.7 mg in liver stores	27 : 1	[54]
Sweet potato, Indian spinach, β-carotene capsule, or retinyl palmitate, (all, n = 14)	Mean changes of total body stores of vitamin A before and after a 60-day intervention in adult men compared with the mean changes in the retinyl palmitate group	Sweet potato, 750 μg RE Indian spinach, 750 μg RE β-carotene capsule, 750 μg RE retinyl palmitate, 750 μg RE	13 : 1 10 : 1 6 : 1	[55]
[^{2}H]-Labelled spinach, and vitamin A in oil capsule, n = 14	Comparing AUC responses to the spinach and the vitamin A reference dose in adults	Calculated 11 mg *E*-β-C from 300 g pureed, cooked spinach, and 3 mg [$^2H_8$]-vitamin A	21 : 1	[74]
[^{2}H]-Labelled carrot, and vitamin A in oil capsule, n = 7	Comparing AUC responses to the carrot and the vitamin A reference dose in adults	Calculated 11 mg *E*-β-C from 100 g pureed and cooked carrot, and 3 mg [$^2H_8$]-vit A	15 : 1	[74]
Fruit, n = 69 Leafy vegetables , n = 70 Retinol-rich foods, n = 70 Control, n = 68	Changes of serum retinol concentration in lactating women after taking fruit, vegetables or preformed vitamin A	Fruit, 4.8 mg *E*-β-C Vegetables, 5.6 mg *E*-β-C Retinol-rich diet, 610 μg retinol Control, 0.6 mg β-C and 1 μg retinol	12 : 1 28 : 1	[53]
Spirulina powder, n = 10	Comparing AUC responses to the *Spirulina* and the vit A reference dose in adults	4.3 mg *Spirulina E*-β-C	4.5 : 1	[75]

These findings illustrate that the vitamin A value of individual plant foods in humans is in need of further investigation. The β-carotene to vitamin A conversion factor is used as a guideline for dietary recommendations to aid in the fight to combat vitamin A deficiency worldwide, but there is wide variation between conversion factors reported in different studies and between individuals in a particular study. A value of at least 12:1 seems a more realistic guideline than the long-accepted 6:1.

D. Factors that Affect the Bioabsorption and Conversion *in vivo*

1. Vitamin A status

The efficacy of carotenoids as provitamin A is affected by vitamin A status. The activity of the intestinal β-carotene cleavage enzyme in vitamin A-sufficient rats is only half that in vitamin A-deficient rats [77]. Another study showed that the carotene cleavage is affected by the vitamin A concentration of the rats' diet [78]. Similar indications come from human studies *in vivo*. For example [79], after intervention with 40 g amaranth, children aged 2-6 years with initial serum retinol <25 μg/dL increased their serum retinol by 12.6 μg/dL, whilst those with initial serum retinol >25 μg/dL increased their serum retinol only by 6.2 μg/dL. In another study [51], children aged 7-12 years, with an average serum retinol concentration of 34 μg/dL, considered adequate, were given a β-carotene supplement (6 mg/day) or carrots containing 6 mg β-carotene per day. Neither intervention resulted in a change in the serum retinol concentration. These observations were further confirmed by a recent report [80] that, when provided with provitamin A carotenoids, children with inadequate vitamin A status (<25 μg/dL) showed the greatest increase in serum vitamin A concentration, whilst children with serum retinol >25 μg/dL showed very little or no response. As mentioned earlier, however, the change in serum retinol concentration before and after an intervention is not a good indicator, because the vitamin A formed from the supplement may contribute to increased body (liver) stores of vitamin A, but not to the serum retinol concentration of subjects with normal vitamin A status.

2. Food matrix

There are striking differences in the bioavailability of carotenoids and vitamin A from various food matrices [48,49]. The efficiency of absorption and uptake of β-carotene is discussed in *Chapter 7*. β-Carotene in spinach is present in protein complexes [81] located in chloroplasts. β-Carotene in carrots is largely in the form of carotene crystals in chromoplasts [81]. Different conversion factors have been observed for β-carotene from spinach and carrots. Several studies have shown that the carotene:retinol equivalency from fruits and vegetables is in the range of 12-27 μg of carotene to 1 μg of retinol [47,52]. These studies have shown that the food matrix affects the bioavailability of vitamin A and that carotenoids in fruit have better bioavailability than those in vegetables [14]. In the transgenic 'Golden Rice', β-carotene is in the yellow-coloured endosperm [82]. Rice endosperm contains starch and protein, and cooked rice is easy to digest. Thus, the efficiency of absorption and bioconversion of β-carotene from Golden Rice is predicted to be greater than that of β-carotene from spinach and carrot.

3. Food preparation

Food preparation practices have some effect on the bioavailability of carotenoids [83]. In a relevant study, subjects received, over a 3-week period, either a control diet (10 subjects), the control diet supplemented with β-carotene, or one of four spinach products (12 subjects per group): namely, (i) whole leaf spinach with an almost intact food matrix; (ii) minced spinach with the matrix partially disrupted; (iii) enzymically liquefied spinach in which the matrix was further disrupted, and (iv) liquefied spinach to which dietary fibre (10 g/kg wet weight) was added. Consumption of spinach in any of these forms significantly increased serum concentrations of (all-*E*)-β-carotene, (*Z*)-β-carotene and, consequently, total β-carotene and retinol. Serum total β-carotene responses, however, *i.e.* changes in serum concentrations of β-carotene from the start to the end of the intervention period, differed significantly between the groups fed whole leaf and liquefied spinach, and between the groups fed minced and liquefied spinach. Addition of dietary fibre to the liquefied spinach had no effect on serum carotenoid responses. The relative bioavailability of β-carotene from the spinach preparations compared with that of β-carotene from the carotenoid supplement was 5.1% for whole leaf spinach, 6.4% for minced spinach, 9.5% for liquefied spinach, and 9.3% for liquefied spinach plus added dietary fibre. Therefore, enzymic disruption of the matrix (cell wall structure) enhanced the bioavailability of β-carotene from whole leaf and minced spinach.

Another study reported that processing carrots as puree or by boiling and mashing can improve the bioavailability of carotenes and the vitamin A value [84].

4. Other carotenoids

It has been reported [85] that plasma β-carotene response is reduced in the presence of lutein (**133**), but no information was given on whether the conversion of β-carotene to retinol was also affected. Use of the postprandial chylomicron method to evaluate the effect of other carotenoids on the absorption and cleavage of β-carotene demonstrated that lutein, but not lycopene (**31**), led to a reduction in β-carotene absorption, though neither of these carotenoids affected the formation of retinyl palmitate [59].

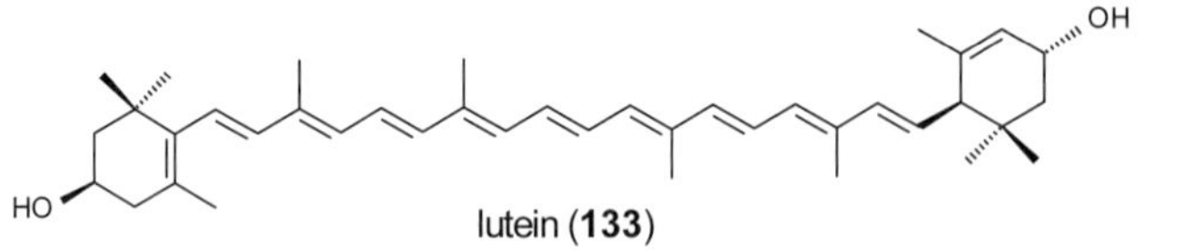

lutein (**133**)

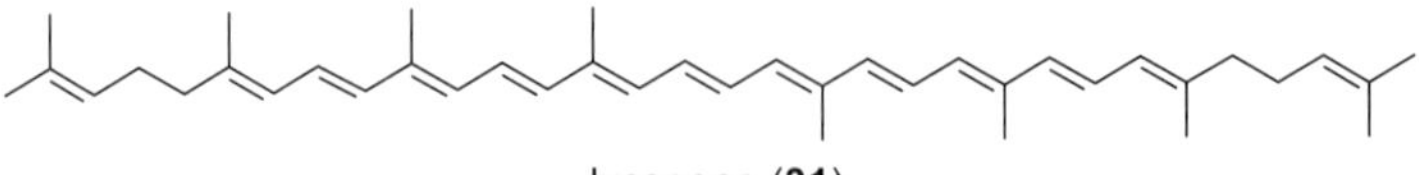

lycopene (**31**)

5. Protein malnutrition

The β-carotene 15,15′-oxygenase and 9,10-oxygenase enzymes have been found in intestine, liver, eye, and other tissues. Populations with protein malnutrition may, therefore, be deficient in these enzymes and will thus have diminished capability to convert β-carotene into vitamin A. In support of this, it has been reported that the activity of β-carotene cleavage enzymes in protein-deficient rats was significantly lower than in protein-adequate rats [86].

6. Intraluminal infections

It is common that populations at heightened risk of vitamin A deficiency are also likely to have a high prevalence of parasitic infestation and to rely on a high intake of plant food as provitamin A source. Data on whether parasitic infection affects vitamin A nutrition are somewhat conflicting [86]. The extent to which ascaris/hookworm infections affect the absorption of vitamin A and/or bioconversion of dietary provitamin A carotenoids to vitamin A remains to be determined.

7. Fat and fibre

The effects of fat content of a meal on the bioavailability of β-carotene have been investigated [87] (see *Chapter 7*). It has generally been accepted that a higher fat content in the diet facilitates the formation of intestinal micelles that are needed for absorption of vitamin A and carotene. A recent study [88] assessed the accumulation of β-carotene and vitamin A, derived from the β-carotene doses, in liver, kidney, and adrenal tissue of Mongolian gerbils that were given a β-carotene-deficient diet for 1 week, followed by one of eight isocaloric, semi-purified diets supplemented with carrot powder (1 μg β-carotene, 0.5 μg α-carotene/kJ diet) for 2 weeks (12 animals per group). Increasing dietary fat from 10% to 30% of total energy resulted in higher vitamin A tissue levels and lower β-carotene stores in the liver, suggesting that consumption of high-fat diets enhances conversion of β-carotene into vitamin A. Consumption of citrus pectin resulted in lower hepatic vitamin A stores and higher hepatic β-carotene stores compared with all other groups, suggesting lower conversion of β-carotene into vitamin A. In contrast, consumption of oat gum resulted in higher vitamin A and lower β-carotene stores in the liver, compared with values seen for gerbils fed citrus pectin. Further, the level of dietary fat consumed with soluble fibre had no interactive effects on hepatic vitamin A, β-carotene or α-carotene stores. These results demonstrate that absorption of β-carotene is affected independently by dietary fat level and type of soluble fibre, and suggest that these dietary components independently modulate the conversion of β-carotene into vitamin A.

A recent study [56] investigated how consumption of dietary fat at 7, 15, or 29 g/day with mixed vegetables containing 4.2 mg provitamin A per day affects total vitamin A pool size and the serum concentration of carotenoids. No difference was observed between groups taking various levels of dietary fat. Therefore, the requirement of dietary fat for optimal absorption of carotenoids appeared to be minimal.

E. Conversion in Tissues other than Intestine

Liver, fat, lung and kidney are capable of converting β-carotene into retinoids [31]. In addition to incubation studies with human and animal tissues, mathematical modelling has shown that, in order to fit a physiological compartmental model, the intestine and liver must be equally important in the conversion of β-carotene [63]. In the study in which 6 mg labelled β-carotene in corn oil was supplied to humans [70], the post-absorption conversion of β-carotene (the conversion of β-carotene after the intestinal absorption) *in vivo* was 7.8, 13.6, 16.4, and 19.0% at days 6, 14, 21, and 53, respectively.

F. Vitamin A Value of α-Carotene and (*cis*)-β-Carotenes

Other provitamin A carotenoids can also be converted into vitamin A *in vivo*. α-Retinol (*5*) was detected in livers of Mongolian gerbils fed α-carotene, and twice the amount of α-carotene than of β-carotene was needed to maintain vitamin A status in those gerbils [89]. In another study on gerbils, it was reported that the relative vitamin A values of (9*Z*)-β-carotene and (13*Z*)-β-carotene were 38% and 62%, respectively, of that of (all-*E*)-β-carotene [90]. The differences in the vitamin A value may be related to the intestinal absorption efficiency for the various isomers of β-carotene [91] or due to the different efficiencies of the isomers as substrates for the cleavage enzymes.

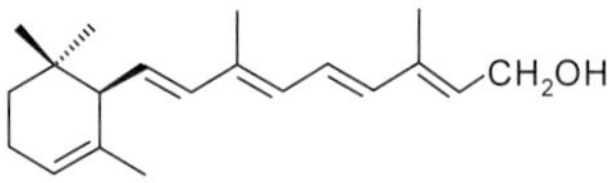

α-retinol (*5*)

G. Formation of Retinoic Acid from β-Carotene

Retinoic acid (*3*) plays an important role in the prevention and therapy of cancers, through its control of gene expression [92] (*Chapter 18*). β-Carotene can be converted into retinoic acid *via* an excentric cleavage pathway in ferret intestine [93,94] and in human intestinal mucosa [95]. The concentration of (all-*E*)-retinoic acid in the serum of rabbits fed β-carotene was found to be higher than in those fed no β-carotene [96]. Direct formation of retinoic acid from β-carotene has been reported in hepatic stellate cells [97] and in rat intestine, kidney, liver, lung and testes [98]. A recent report, however, stated that the conversion of β-carotene into retinoic acid remains to be demonstrated in humans [99].

The pathway of formation of retinoic acid from β-carotene (*via* retinal or not), and the factors which affect the formation, warrant further investigation (see also *Chapter 18*).

H. Conclusion

Provitamin A carotenoids (mainly β-carotene) can provide vitamin A nutrition for humans. β-Carotene is converted enzymically into vitamin A in various tissues, and the small intestine is the prominent site for the conversion. The post-absorption conversion of absorbed β-carotene into vitamin A by tissues other than intestine is also likely and needs to be studied carefully.

The present reported values for β-carotene to vitamin A conversion show wide variation from 2 μg β-carotene:1 μg retinol for synthetic pure β-carotene in oil to 27 μg β-carotene:1 μg retinol for β-carotene from vegetables. Factors that affect β-carotene conversion to vitamin A include host nutrition status (vitamin A and protein nutrition), dietary fat and fibre content (macronutrient), food matrix (*e.g.* vegetables, fruits), and host intestinal health (parasitic infection and other infections). In an effort to increase the production of popular foods with better bioconversion factors, scientists are working to produce β-carotene-enriched staple foods through natural breeding and/or bioengineering techniques. Examples of such foods are Golden Rice, high β-carotene yellow maize, and high β-carotene ground nuts. These new food products will need rigorous scientific evaluation of their ability to provide vitamin A for combating vitamin A deficiency worldwide.

In human studies, the vitamin A value of pure β-carotene or of β-carotene in food can be determined quantitatively by using stable isotope techniques to study intrinsically labelled compounds and plants in conjunction with the paired DRD tests. It is not practical, however, to determine actual conversion factors for every population, individual or diet, in widely differing conditions. From the data that have been obtained in the various studies, it seems reasonable to think that a guideline conversion factor of at least 12:1 should ensure adequate provision of vitamin A from provitamin A carotenoids in food.

References

[1] T. Moore, *Biochem. J.*, **24**, 696 (1930).

[2] United Nations Administrative Committee on Coordination Sub-Committee on Nutrition (ACC/SCN), *The 4th Report on the World Nutrition Situation - Nutrition Throughout the Life Cycle, 2000.* (http://www.unsystem.org/SCN/archives/rwns04/begin.htm#Contents)

[3] A. Sommer, *Nutritional Blindness: Xerophthalmia and Keratomalacia*, Oxford University Press, New York (1982).

[4] A. Ross and C. Stephensen, *FASEB J.*, **10**, 979 (1996).
[5] N. Usha, A. Sankaranarayanan, B. Walia and N. Ganguly, *J. Pediatr. Gastroenterol. Nutr.*, **13**, 168 (1991).
[6] A. Sommer and K. P. West Jr., *Vitamin A Deficiency. Health, Survival, and Vision*. Oxford University Press, New York (1996).
[7] J. Jason, L. K. Archibald, O. C. Nwanyanwu, A. L. Sowell, L. Anne, I. Buchanan, J. Larned, M. Bell, P. N. Kazembe, H. Dobbie and W. R. Jarvis, *Clin. Diagnostic Lab. Immunol.*, **9**, 616 (2002).
[8] K. Ge, F. Zhai and H. Ye, *Acta Nutr. Sinica*, **17**, 123 (1995).
[9] J. Olson, *Am. J. Clin. Nutr.*, **45**, 704 (1987).
[10] A. Sommer, I. Tarwotjo, E. Djunaedi, K. P. West Jr, A. A. Loeden, R. Tilden and L. Mele, *Lancet*, **1**, 1169 (1986).
[11] L. Rahmathullah, B. Underwood, R. Thulasiraj, R. C. Milton, K. Ramaswamy, R. Rahmathullah and G. Babu, *New Eng. J. Med.*, **323**, 929 (1990).
[12] K. P. West Jr., R. Pokhrel, J. Katz, S. C. LeClerq, S. K.Khatry, S. R. Shrestha, E. K. Pradhan, J. M. Tielsch, M. R. Pandey and A. Sommer, *Lancet*, **338**, 67 (1991).
[13] K. P. West Jr., J. Katz , S. K. Khatry, S. C. LeClerq, E. K. Pradhan, S. R. Shrestha, P. B. Connor, S. M. Dai, P. Christian, R. P. Pokhrel and A. Sommer, on behalf of the NNPS-2 Study Group, *Br. Med. J.*, **318**, 570 (1999).
[14] E. Villamow and W. W. Fawzi, *J. Infect. Disease*, **182 (Suppl. 1)**, S122 (2000).
[15] B. A. Underwood and P. Arthur, *FASEB J.*, **10**, 1040 (1996).
[16] S. de Pee, C. E. West, Muhilal, D. Karyadi and J. G. A. J. Hautvast, *Lancet*, **346**, 75 (1995).
[17] J. Bulux, J. D. Serrano, A. Giuliano, R. Perez, C. Y. Lopez, C. Rivera, N. W. Solomons and L. M. Canfield, *Am. J. Clin. Nutr.*, **59**, 1369 (1994).
[18] G. van Poppel and R. Goldbohm, *Am. J. Clin. Nutr.*, **62**, 1393S (1995).
[19] J. M. Gaziano and C. H. Hennekens, *Ann. NY Acad. Sci.*, **691**, 148 (1993).
[20] J. Goldberg, G. Flowerdew, E. Smith, J. Brody and M. Tso, *Am. J. Epidemiol.*, **128**, 700 (1988).
[21] J. M. Seddon, U. A. Ajani, R. D. Sperduto, R. Hiller, N. Blair, T. C. Burton, M. D. Farber, E. S. Gragoudas, J. Haller and D. T. Miller, *JAMA*, **272**, 1413 (1994).
[22] S. E. Hankinson, M. J. Stampfer, J. M. Seddon, G. A. Colditz, B. Rosner, F. E. Speizer and W. C. Willett, *Br. Med. J.*, **305**, 335 (1992).
[23] P. F. Jacques, L. T. Chylack, R. B. McGandy and S. C. Hartz, *Arch. Ophthalmol.*, **106**, 337 (1988).
[24] C. H. Hennekens, U. E. Buring and R. Peto, *New Engl. J. Med.*, **330**, 1080 (1994).
[25] The α-Tocopherol, β-Carotene Cancer Prevention Study Group, *New Engl. J. Med.*, **330**, 1029 (1994).
[26] G. S. Omenn, G. Goodman, M. D. Thongquist, J. Balmes, M. R. Cullen, A. Glass, J. P. Keogh, F. L. Meyskens, B. Valanis, J. H. Williams, S. Barnhart and S. Hammar, *New Engl. J. Med.*, **334**, 1150 (1996).
[27] J. A. Olson and O. Hayaishi, *Proc. Natl. Acad. Sci. USA*, **54**, 1364 (1965).
[28] D. S. Goodman, H. S. Huang and T. Shiratori, *J. Biol. Chem.*, **241**, 1929 (1966).
[29] J. Glover and E. R. Redfearn, *Biochem. J.*, **58**, 15 (1954).
[30] J. Ganguly and P. S. Sastry, *Wld. Rev. Nutr. Diet.*, **45**, 198 (1985).
[31] X. D. Wang, G. Tang, J. G. Fox, N. I. Krinsky and R. M. Russell, *Arch. Biochem. Biophys.*, **285**, 8 (1991).
[32] G. Tang, X. D. Wang, R. M. Russell and N. I. Krinsky, *Biochemistry*, **30**, 9829 (1991).
[33] K.-J. Yeum, A. L. A. Ferreira, D. Smith, N. I. Krinsky and R. M. Russell, *Free Radic. Biol. Med.*, **29**, 105 (2000).
[34] J. von Lintig and K. Vogt, *J. Biol. Chem.*, **275**, 11915 (2000).
[35] C. Kiefer, S. Hessel, J. M. Lampert, K. Vogt, M. O. Lederer, D. E. Breithaupt and J. von Lintig, *J. Biol. Chem.*, **276**, 14110 (2001).
[36] A. Lindqvist, Y.-K. He and S. Anderson, *J. Histochem. Cytochem.*, **53**, 1403 (2005).

[37] D. S. Goodman, R. Blomstrand, B. Werner, H. S. Huang and T. Shiratori, *J. Clin. Invest.*, **45**,1615 (1966).

[38] R. Blomstrand and B. Werner, *Scand. J. Clin. Lab. Invest.*, **19**, 339 (1967).

[39] R. S. Dueker, Y. Lin, A. B. Buchholz, P. D. Schneider, M. W. Lamé, H. J. Segall, J. S. Vogel and A. J. Clifford, *J. Lipid Res.*, **41**, 1790 (2000).

[40] S. L. Lemke, S. R. Dueker, J. R. Follett, Y. Lin, C. Carkeet, B. A. Buchholz, J. S. Vogel and A. J. Clifford, *J. Lipid Res.*, **44**, 1591 (2003).

[41] E. M. Hume and H. A. Krebs, *Medical Research Council Special Report Series, No. 264*, His Majesty's Stationery Office, London (1949).

[42] H. E. Sauberlich, R. E. Hodges, D. L. Wallace, H. Kolder, J. E. Canham, J. Hood, N. Racia and L. K. Lowry, *Vit. Horm.*, **32**, 251 (1974).

[43] L. E. Booher, E. C. Calliston and E. M. Hewston, *J. Nutr.*, **17**, 317 (1939).

[44] K. H. Wagner, *Ztschrf. Physiol. Chem.*, **264**, 153 (1940).

[45] J. C. Bauernfeind, *J. Agr. Food Chem.*, **20**, 456 (1972).

[46] FAO/WHO, Report of a Joint Food and Agriculture Organization/World Health Organization Experts Committee. *FAO Nutrition Meeting Report Series No. 41. WHO Technical Report Series No. 362.* World Health Organization, Geneva (1967).

[47] NRC (National Research Council), *Recommended Dietary Allowances, 9th Revised Edition, Report of the Committee on Dietary Allowances*, Food and Nutrition Board, Division of Biological Sciences, Assembly of Life Sciences, National Academy of Sciences, Washington, D.C. (1980).

[48] E. D. Brown, M. S. Micozzi, N. E. Craft, J. G. Bieri, G. Beecher, B. K. Edwards, A. Rose, P. R. Taylor and J. C. Smith Jr., *Am. J. Clin. Nutr.*, **49**, 1258 (1989).

[49] E. D. Brown, A. Rose, N. Craft, K. E. Seidel and J. C. Smith, *Clin. Chem.*, **35**, 310 (1989).

[50] E. J. Johnson and R. M. Russell, *Am. J. Clin. Nutr.*, **56**, 128 (1992).

[51] J. Bulux, J. Q. de Serrano, R. Perez, C. Rivera and N. W. Solomons, *Am. J. Clin. Nutr.*, **49**, 173 (1998).

[52] S. de Pee, C. E. West, D. Permaesih, S. Martuti, K. Muhilal and G. Hautvast, *Am. J. Clin. Nutr.*, **68**, 1058 (1998).

[53] N. C. Khan, C. E. West, S.de Pee. D. Bosch, H. D. Pluong, P. J. M. Hulshof, H. H. Khoi, H. Verhoef and J. G. A. J. Hautvast, *Am. J. Clin. Nutr.*, **85**, 1112 (2007).

[54] G. Tang, X. Gu, Q. Xu, X. Zhao, J. Qin, C. Fjeld, G. G. Dolnikowski, R. M. Russell and S. Yin, *Am. J. Clin. Nutr.*, **70**, 1069 (1999).

[55] M. J. Haskell, K. M. Jamil, F. Hassan, J. M. Peerson, M. I. Hossain, G. J. Fuchs and K. H. Brown, *Am. J. Clin. Nutr.*, **80**, 705 (2004).

[56] J. D. Ribaya-Mercado, C. C. Maramag, L. W. Tengco, G. G. Dolnikowski, J. B. Blumberg and F. S. Solon, *Am. J. Clin. Nutr.*, **85**, 1041 (2007).

[57] T. van Vliet, W. H. P. Schreurs and H. van den Berg, *Am. J. Clin. Nutr.*, **62**, 110 (1995).

[58] M. E. O'Neill and D. I. Thurnham, *Br. J. Nutr.*, **79**, 149 (1998).

[59] H. van den Berg and T. van Vliet, *Am. J. Clin. Nutr.*, **68**, 82 (1998).

[60] R. S. Parker, J. E. Swanson, C.-S. You, A. J. Edwards and T. Huang, *Proc. Nutr. Soc.*, **58**, 1 (1999).

[61] R. S. Parker, J. E. Swanson, B. Marmor, K. J. Goodman, A. B. Spielman, J. T. Brenna, S. M. Viereck and W. K. Canfield, in *Carotenoids in Human Health*, (ed. L. M.. Canfield, N. I. Krinsky and J. A Olson), p. 86, New York Academy of Sciences, New York, NY (1993).

[62] S. R. Dueker, A. D. Jones, G. M. Smith and A. J. Clifford, *Anal. Chem.*, **66**, 4177 (1994).

[63] J. Novotny, S. R. Dueker, L. Zech and A. J. Clifford, *J. Lipid Res.*, **36**, 1825 (1995).

[64] Y. Lin, S. R. Dueker, B. J. Burri, T. R. Reidlinger and A. J. Clifford, *Am. J. Clin. Nutr.*, **71**, 1545 (2000).

[65] R. B. van Breemen, *Anal. Chem.*, **67**, 2004 (1995).

[66] M. van Lieshout, C. E. West, D. Permaesih, Y. Wang, X. Xu, R. B. van Breemen, A. F. L. Creemers, M. A. Verhoeven and J. Lugtenburg, *Am. J. Clin. Nutr.*, **73**, 949 (2001).

[67] G. Tang, B. Andrien, G. Dolnikowski and R. M. Russell, *Meth. Enzymol.*, **282**, 140 (1997).
[68] G. Tang, J. Qin, G. G. Dolnikowski and R. M. Russell, *Eur. J. Nutr.*, **39**, 7 (2000).
[69] G. Tang, J. Qin and G. G. Dolnikowski, *J. Nutr. Biochem.*, **9**, 408 (1998).
[70] G. Tang, J. Qin, G. G. Dolnikowski and R. M. Russell, *Am. J. Clin. Nutr.*, **78**, 259 (2003).
[71] Z. Wang, X. Zhao, S. Yin, R. M. Russell and G. Tang, *Br. J. Nutr.*, **91**, 121 (2004).
[72] J. Sabrina, L. Hickenbottom, L. Shawna, S. R. Dueker, Y. Lin, J. R. Follett, C. Carkeet, B. A. Buchholz, J. S. Vogel and A. J. Clifford, *Eur. J. Nutr.*, **41**, 141 (2002).
[73] M. Grusak, *J. Nutr. Biochem.*, **8**, 164 (1997).
[74] G. Tang, J. Qin, G. G. Dolnikowski, R. M. Russell and M. G. Grusak, *Am. J. Clin. Nutr.*, **82**, 821 (2005).
[75] J. Wang, Y. Wang, Z. Wang, L. Li, J. Qin, W. Lai, Y. Fu, P. M. Suter, R. M. Russell, M. A. Grusak, G. Tang, and S. Yin. *Am. J. Clin. Nutr.*, **87**, 1730 (2008).
[76] J. A. Novotny, A. C. Kurilich, S. J. Britz and B. A. Clevidence, *Lipid Res.*, **46**,1896 (2005).
[77] L. Villard and C. Bates, *Br. J. Nutr.*, **56**, 115 (1986).
[78] T. van Vliet, M. F. van Vlissingen, F. van Schaik and H. van den Berg, *J. Nutr.*, **126**, 499 (1996).
[79] V. R. Lala and V. Reddy, *Am. J. Clin. Nutr.*, **23**, 110 (1970).
[80] F. McEvoy and W. Lynn, *J. Biol. Chem.*, **248**, 4568 (1973).
[81] T. W. Goodwin, *The Biochemistry of the Carotenoids. Vol. 1. Plants*. Chapman and Hall, London and New York (1980).
[82] X. Ye, S. Al-Babili, A. Kloti, J. Zhang, P. Lucca, P. Beyer and I. Potrykus, *Science*, **287**, 303 (2000).
[83] J. M. J. Castenmiller, C. E. West, J. B. H. Linssen, K. H. van het Hof and A. G. J. Voragen, *J. Nutr.*, **129**, 349 (1999).
[84] A. J. Edwards, C. H. Nguyen, C.-S.You, J. E. Swanson, C. Emenhiser and R. S. Parker, *J. Nutr.*, **132**, 159 (2002).
[85] D. Kostic, W. S. White and J. A. Olson, *Am. J. Clin. Nutr.*, **62**, 604 (1995).
[86] S. G. Parvin and B. Sivakuma, *J. Nutr.*, **130**, 573, (2000).
[87] F. Jalal, M. C. Nesheim, Z. Agus, D. Sanjur and J. P. Habicht, *Am. J. Clin. Nut.*, **68**, 623 (1998).
[88] D. M. Deming, A. C. Boileau, C. M. Lee and J. W. Erdman Jr., *J. Nutr.*, **130**, 2789 (2000).
[89] S. A. Tanumihardjo and J. A. Howe, *J. Nutr.*, **135**, 2622 (2005).
[90] D. M. Deming, D. H. Baker and J. W. Erdman Jr., *J. Nutr.*, **132**, 2709 (2002).
[91] D. M. Deming, S. R. Teixeira and J. W. Erdman Jr., *J. Nutr.*, **132**, 2700 (2002).
[92] P. Chambon, *FASEB J.*, **10**, 940 (1996).
[93] X. D. Wang, R. M. Russell, R. P. Marini, G. Tang, G. G. Dolnikowski, J. G. Fox, and N. I. Krinsky, *Biochim. Biophys. Acta*, **1167**, 159 (1993).
[94] X. Hebuterne, X. D. Wang, D. E. H. Smith, G. Tang and R. M. Russell, *J. Lipid Res.*, **37**, 482 (1996).
[95] X.-D Wang, N. I. Krinsky, G. Tang and R. M. Russell, *Arch. Biochem. Biophys.*, **293**, 298 (1992).
[96] Y. Folman, R. M. Russell, G. Tang and G. Wolf, *Br. J. Nutr.*, **62**, 195 (1989).
[97] R. B. Martucci, A. L. Ziulkoski, V. A. Fortuna, R. M. Guaragna, F. C. R. Guma, L. C. Trugo and R. Borojevic, *J. Cell. Biochem.*, **92**, 414 (2004).
[98] J. L. Napoli and K. R. Race, *J. Biol. Chem.*, **263**, 17372 (1988).
[99] C. C. Ho, F. F. de Moura, S.-H. Kim and A. J. Clifford, *Am. J. Clin. Nutr.*, **85**, 770 (2007).

Carotenoids
Volume 5: Nutrition and Health

Chapter 9

Vitamin A and Vitamin A Deficiency

George Britton

A. Introduction

Vitamin A is an essential factor for development, growth, health and survival. Vitamin A (retinol, *1*) and its chemically and metabolically related forms retinal (*2*) and retinoic acid (*3*) play essential roles in such diverse processes as vision and cell regulation.

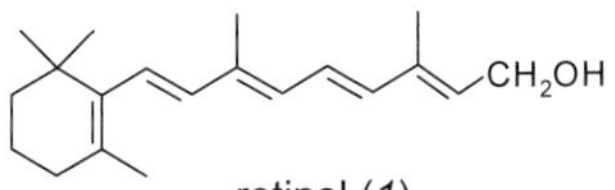

retinol (*1*)

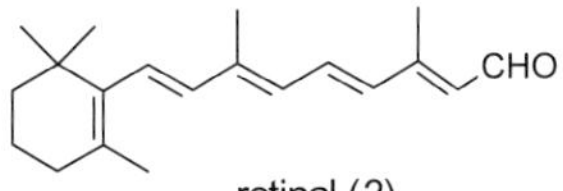

retinal (*2*)

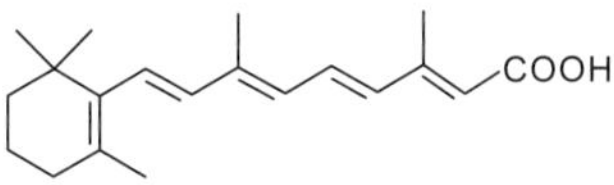

retinoic acid (*3*)

The role of retinal as the chromophore of the visual pigments such as rhodopsin has been investigated intensively and is summarized in *Volume 4, Chapter 15*. It is well understood, therefore, why prolonged deficiency of vitamin A can lead to reversible night blindness that may be followed by irreversible loss of sight. It is also well understood that vitamin A is a factor in maintaining immunocompetence and that retinoic acid is an essential hormone-like factor in regulation of gene expression in relation to growth and development. Not surprisingly then, vitamin A deficiency (VAD) can have very serious consequences.

Although much of this book deals with those aspects of nutrition and health that are of most concern in richer countries where food is plentiful, we must not forget that ensuring adequate supplies of vitamin A or provitamin A in poor countries to prevent the scourge of vitamin A deficiency remains the most important life-or-death aspect of carotenoids and health. To lose sight of this fact is unforgivable.

Any attempt to cover the subject of vitamin A and vitamin A deficiency comprehensively would take up at least a full volume, not just a short chapter. The subject of vitamin A deficiency has a long history and an extensive literature, including a number of important books [1-4], and reports from International Agencies [5,6]. These should be consulted for details of symptoms and clinical lesions, and for results of surveys and trials in various parts of the world. Sight and Life, in addition to initiating and supporting programmes, provides authoritative and very readable reviews and research reports on all aspects of VAD as well as specific topical reports and developments from all over the world in *Sight and Life Newsletter* (now *Magazine*).

This *Chapter* is not a new analysis of the subject and its primary literature; this has been digested and evaluated by leading authorities who are much more experienced and expert. Rather, it will summarize the main features in a way that seems appropriate for a chapter in a book on carotenoids. It should also be considered together with the detailed treatment of the conversion of the provitamin carotenoids into vitamin A, and evaluation of conversion factors, presented in *Chapter 8*. The topic is of global importance and is a matter of life or death for millions, especially young children. Vitamin A deficiency and the battle to overcome it perhaps has the most important consequences of any involvement of carotenoids in human health.

B. Vitamin A

1. Basic biochemistry

The biochemistry of vitamin A is discussed in detail in reviews (*e.g.* [7,8]). Only a brief summary of the main features is given here.

Vitamin A, retinol (*1*), may be obtained from the diet either as 'pre-formed' vitamin A itself, or as the provitamin, β-carotene (**3**) or other carotenoids containing an unsubstituted β end group, especially α-carotene (**7**) and β-cryptoxanthin (**55**). Nutritional aspects of the conversion of β-carotene into vitamin A are described in *Chapter 8*, including a detailed evaluation of conversion factors and methods for determining them, and of the factors that regulate or influence the efficiency of the conversion. The biochemistry and molecular biology of the central (BCO1) and excentric (BCO2) cleavage enzymes are described in *Volume 4, Chapter 16*. Information in *Chapter 18* complements and extends the outline of the roles of vitamin A, of retinal (*2*) in vision, and of retinoic acid (*3*) as a hormone-like regulator of gene expression in growth and development that was given in *Volume 4, Chapter 15*.

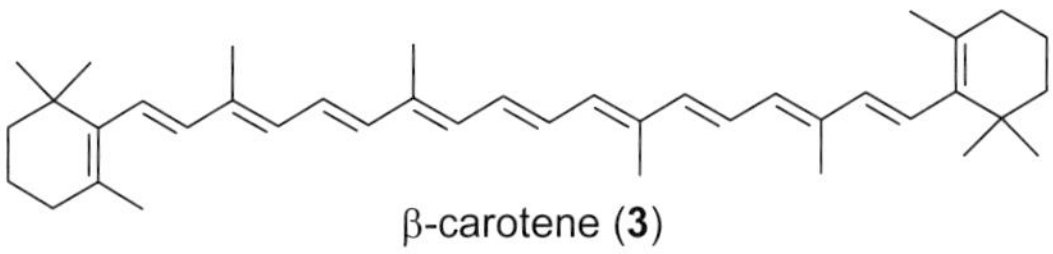

β-carotene (**3**)

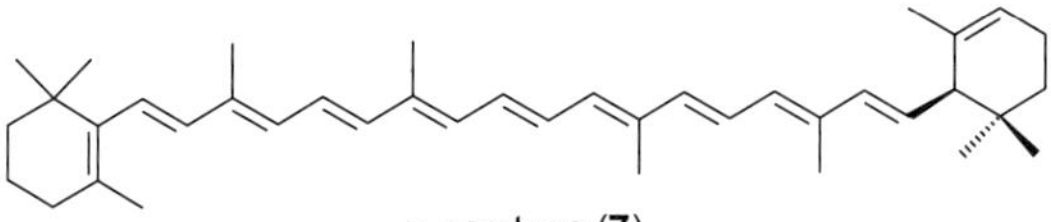

α-carotene (**7**)

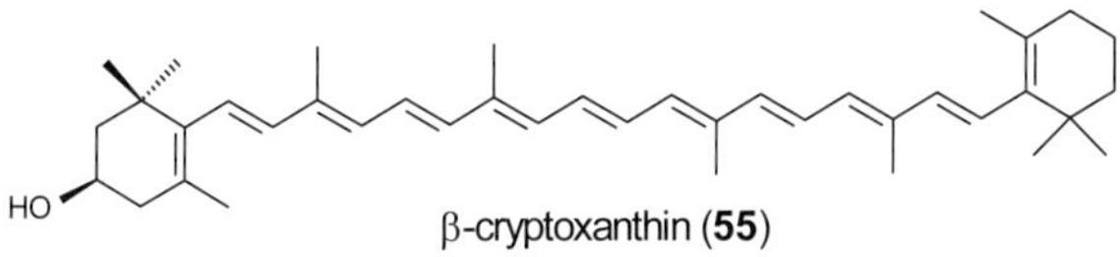

β-cryptoxanthin (**55**)

The enzymic conversion of the provitamin carotenoids into vitamin A is strictly controlled. A major regulatory factor is blood retinol concentration (*Chapter 8*) so that retinol is only formed from the provitamin when it is needed to maintain an adequate concentration and cannot build up to toxic levels.

In the intestine, retinol, either formed from the provitamin or obtained direct from the diet, is absorbed along with other lipids and transported in chylomicrons to the liver where it is stored as esters, mainly the palmitate [9,10]. Some 10-20% is stored in specialized lipid globules in hepatocytes, but the bulk is stored, also in lipid globules, in stellate cells. Some other tissues have some limited capacity to store retinyl palmitate, also in stellate cells [7-10]. When vitamin A intake is low, the efficiency of recycling is high, so that losses from the body are minimized [11].

When vitamin A is required by tissues, the stored retinyl palmitate is hydrolysed and the free retinol is delivered to the tissues by the specific transporter retinol-binding protein, RBP [12-14]. The 21.2 kDa protein, apo-RBP, is made and stored in the liver. Holo-RBP, with the retinol ligand bound, forms a 1:1 complex with another protein, transthyretin, and this complex delivers the retinol to the cells [15]. The blood concentration of RBP is strictly controlled at a constant plasma concentration around 2 μmol/L in a well-nourished adult, so that the amount of retinol delivered to the tissues is limited, as a safeguard against toxicity. Other, related proteins, cellular retinol-binding proteins (cRBPs) and cellular retinoic acid-binding proteins (cRABPs) are responsible for the controlled transport of these ligands within cells [9]. In extreme protein malnutrition, the body may not have sufficient amounts of these proteins, so tissues may be deficient in vitamin A even if the dietary supply of vitamin A is adequate.

When large amounts of retinol are ingested, *e.g.* in high-dose supplements, high concentrations of retinyl palmitate are incorporated in chylomicrons and some may bypass the controlled RBP transport system and be delivered to tissues, along with other lipids. The high doses lead to a marked increase in retinyl ester concentration in plasma, though the controlled RBP concentration is maintained.

2. Vitamin A status and requirements

It is customary to use serum retinol concentration, usually given as µg/dL or µmol/L, where 1 µmol = 286 µg, to define vitamin A status [4]. The usual correlation is that a retinol concentration above 20 µg/dL (0.7 µmol/L) is considered as 'normal', 10-20 µg/dL (0.35-0.7 µmol/L) low and <10 µg/dL (<0.35 µmol/L) deficient. Values around 20 µg/dL are often considered as 'marginal'. These are not universal absolute values; deficiency symptoms may be seen in some individuals with >20 µg/dL, but not in some other individuals with low values. In a population with low to marginal average concentration, cases of sub-clinical manifestations (Section **C**.3) are likely to occur and, with 'moderate deficiency', *ca.* 10 µg/dL, cases are almost certain and signs of xerophthalmia (Section **C**.1) are likely to emerge.

To express both preformed vitamin A and provitamin A concentrations in food *etc.*, several terms are used. The retinol equivalent (RE) was introduced and is defined as equivalent to 1 µg retinol or 6 µg β-carotene or 12 µg of other provitamin A carotenoids, based on the then accepted conversion factor [16]. This was superseded by the RAE (retinol activity equivalent) to allow for the fact that the conversion factor for β-carotene from food was much poorer than previously thought [17]. 1 RAE (1 µg retinol) is equivalent to 12 µg (all-*E*)-β-carotene or 24 µg other carotenoids. Also used, especially for expressing the size of supplements, is the International Unit (IU) which is defined as equivalent to 0.3 µg retinol or 0.6 µg β-carotene.

Because of variability due to many factors, it is not possible to give a definite universal value, most evaluations indicate that a daily intake of around 300-375 RE of retinol is necessary to maintain adequate liver stores and is considered safe for infants [8].

3. Hypervitaminosis A: toxicity

It is well known that, in excess, vitamin A is extremely toxic. Intake of a large single dose (>0.7 mmol, 200 mg, 660 000 IU, by adults, half this dose by children) may cause some rapid, acute effects including nausea, vomiting, headache, muscular incoordination, and blurred vision [18,19]. Some infants are affected by a dose of only 0.1 mmol. The symptoms are usually transient, lasting only about a day. Extremely large doses (*ca.* 500 mg) rapidly cause drowsiness, skin exfoliation and itching. Repeated intake of large doses or recurrent intake of smaller doses, *i.e.* 0.13 µmol, 3.75 mg, 12 500 IU (>10 x RDA), leads to chronic hypervitaminosis, and severe effects such as defective bone structure and osteoporosis [20], and liver damage. Fatal consequences are likely.

Excess vitamin A also causes serious teratogenic effects [21,22]. It is likely that a single extremely large dose or a week of high daily doses of 30-90 mg in early pregnancy will lead to foetal malformation and birth defects. Healthy women who routinely eat a diet containing adequate amounts of vegetables and fruit do not require supplements of vitamin A during pregnancy. When dietary intake is low and supplements are advisable because of low vitamin A status, the daily intake of preformed retinol from all sources should not exceed 10 μmol (3 mg, 10 000 IU).

C. Consequences of Vitamin A Deficiency

Historically, vitamin A deficiency has been associated with blindness, especially in young, pre-school children. Indeed, night blindness, the inability to see in dim light, was for long used as the first indicator of vitamin A deficiency. It could easily be explained; if vitamin A is deficient, it is not possible to make up for losses during the visual cycle of the retinal-opsin visual pigment rhodopsin in retinal rod cells (*Volume 4, Chapter 15*). This condition could be reversed rapidly by providing vitamin A supplements. If the vitamin A deficiency is more severe and prolonged, it leads to structural changes in the eye, as described in the various stages of xerophthalmia (see Section **C.**1), and to permanent, irreversible blindness.

Serious and debilitating as they are, these effects are not in themselves life threatening. It is now recognized that vitamin A deficiency has other profound consequences, leading to increased morbidity (susceptibility to serious life-threatening diseases) and mortality (death from these diseases).

A history of studies of xerophthalmia and its control [23] provides a stimulating introduction to the topic, and highlights lessons that have been learned and lessons still not learned.

1. Xerophthalmia

The term 'xerophthalmia' literally means 'dry eyes', due to all causes, not only vitamin A deficiency. During the past 30 years or so, the definition has been standardized [24] so that 'xerophthalmia' now includes all ocular signs and symptoms of vitamin A deficiency, from night blindness to keratomalacia (successive softening, ulceration and necrosis of the cornea). A manual provides a guide to recognition of the signs and lesions characteristic of the various stages of xerophthalmia [25]. A series of stages of increasing severity have been classified [6], as listed in Table 1, and related to vitamin A status, defined by serum vitamin A levels [4]. Note that this is a continuum, and there is some overlap between the vitamin A levels associated with the various stages. The Table shows the mean ranges determined in most surveys, but some values are outside these ranges [4]. A study in Indonesia found a significant incidence of mild xerophthalmia in pre-school children with serum retinol levels above 20 μg/dL [24].

Table 1. Stages of xerophthalmia in order of increasing severity (based on data in [4]).

Stage	**Serum retinol (μg/dL)**
Normal	>20
Night blindness (XN)	10-20
Conjunctival xerosis (X1A)	10-20
Bitot's spots (X1B)	10-20
Corneal xerosis (X2)	5-10
Corneal ulceration/keratomalacia, <33% (X3A)	5-10
Corneal ulceration/keratomalacia, >33% (X3B)	5-10
Corneal scar (XS)	5-10
Xerophthalmic fundus (XF)	5-10

The first four stages (XN, X1A, X1B and X2) are usually reversible by provision of vitamin A. The later effects are much more severe and the damage cannot be reversed.

2. Keratinization

a) Eye tissues

In the more advanced stages of xerophthalmia, the epithelial surfaces of the conjunctiva and cornea undergo keratinizing metaplasia and become dry, hardened and scaly as abnormal keratin synthesis occurs. This leads to irreversible damage to these and other eye tissues and blindness becomes permanent.

b) Other epithelial tissues

One of the functions of vitamin A is to maintain the condition of the skin and various mucus-secreting epithelial tissues. The keratinizing metaplasia associated with vitamin A deficiency extends not only to destruction of the cornea and other eye tissues but to other mucus-secreting soft epithelial tissues, notably the respiratory and genito-urinary tracts, which become keratinized. The terminal differentiation of skin keratinocytes is markedly affected and larger, harder keratins are produced [26].

3. Subclinical, systemic effects

In addition to xerophthalmia, which is a relatively late manifestation of the slow depletion of vitamin A stores [27], vitamin A deficiency leads to anaemia, growth retardation, and increased incidence and severity of morbid infections, thereby resulting in reduced childhood survival, the most severe consequence [1-4]. These effects may appear before any sign of xerophthalmia is detected. Increasing vitamin A status is now considered to be one of the most cost-effective measures for reducing childhood mortality, which is currently about 14 million *per annum*.

There is overwhelming evidence that vitamin A status influences the incidence or severity of a variety of infections, particularly diarrhoea, measles, urinary tract infections and some forms of respiratory diseases. The keratinizing effect of vitamin A deficiency on epithelial tissues and linings may impair the natural barriers against infection. But the rapid response to treatment of existing infections indicates stimulation by vitamin A of defences against established infections, *i.e.* an immune response.

Whereas the appearance of xerophthalmia is readily detected, in its absence vitamin A deficiency is more difficult to diagnose. The association between severe xerophthalmia and increased mortality has been recognized for a long time. Xerophthalmia is not the direct cause of this mortality, however: both are consequences of vitamin A deficiency. The relationship between VAD and infection is complicated by the fact that not only does VAD increase susceptibility to infection but frequently infection leads to reduction in vitamin A status and lowered serum retinol concentration. Also vitamin A deficiency is difficult to dissociate from general malnutrition (protein - energy malnutrition, PEM).

Even when vitamin A deficiency is only marginal, and no visible signs of xerophthalmia are detected, other effects of the deficiency may be serious. Providing sufficient vitamin A to raise the serum retinol concentration from 18-20 μg/dL to 30 μg/dL can reduce mortality rates by as much as 50% [4]. The most serious consequences of the deficiency are described briefly below.

a) Measles

In poor countries, measles is a severe and life-threatening disease [4]. The incidence and severity of measles and its associated pneumonia and diarrhoea in young children are strikingly related to vitamin A status and are increased by VAD. Measles infection has a deleterious effect on serum vitamin A level, leading to severe vitamin A deficiency. Treatment with vitamin A promotes recovery and reduces mortality by 50%. It affects the severity of the disease rather than the incidence of measles.

b) Diarrhoea/dysentry

Because of often appalling living conditions, diarrhoea is a common affliction in poor communities, especially among children, and has a high death rate [4]. The severity of this serious condition is closely related to vitamin A status, and the severe diarrhoea exacerbates the depletion of vitamin A supplies. Vitamin A supplementation reduces diarrhoea-specific mortality. Improvement of the vitamin A status of deficient populations protects pre-school age children from severe, dehydrating, life-threatening diarrhoea, but may have little impact on the frequency of 'trivial' (*i.e.* not life-threatening) diarrhoeal episodes.

c) Respiratory infections

Here the situation is not so clear but results are consistent with a relationship between vitamin A deficiency and risk of respiratory infection. Also, results following supplementation suggest a potential reduction in severity [28]. There are some discrepancies that at first sight are not easy to reconcile, such as between the increased susceptibility of vitamin A deficient children to severe respiratory disease and the failure of vitamin A supplementation to reduce respiratory-related deaths, except in cases of measles.

d) HIV and AIDS

Vitamin A levels are depressed in cases with HIV infection. Mortality among AIDS patients is higher when serum retinol levels are low [29]. Mortality of infants born to HIV-infected mothers is >90% if the maternal serum retinol level is below 20 μg/dL [30].

e) Other infections

Urinary tract infections are commonly reported in cases of vitamin A deficiency and they usually respond to treatment with vitamin A [31,32]. There is also an increased risk of middle-ear infections [33].

f) Immune response

A basic introduction to the human immune response system is included in *Chapter 17*. Many studies have shown that vitamin A improves immune competence in experimental animals. A role for vitamin A in stimulating the immune system has been known for many years. Measles and VAD both impair the immune response. In humans, vitamin A deficiency leads to a reduction in various immune parameters, particularly natural killer (NK) cells. This reduction is reversed by treatment with vitamin A, and especially with retinoic acid. Specific effects are reported on stimulating maintenance of lymphoid organs and cells. In cell-mediated immunity, the effect seems to be on the production of NK cells *etc*., rather than functional impairment.

Effects on the humoural immune system involve dysregulation of signalling processes, not the efficiency of antibody production. The relationship between immunocompetence and vitamin A status has been reviewed [34].

D. Scale of Vitamin A Deficiency

1. Global distribution

Vitamin A deficiency is a global, international public health problem. In 1996 it was known to occur, at different levels of severity, in 73 countries [4,35]. The highest risk is associated with tropical and sub-tropical regions, and VAD is particularly serious in Africa, South and South-East Asia, and parts of Central America. Clinical symptoms are especially prevalent in parts of Saharan/sub-Saharan Africa, the Indian subcontinent and the Philippines. Around 200 million children are estimated to be at risk of sub-clinical vitamin A deficiency, and about 125 million actually deficient, with a death rate of 1-2.5 million each year [4,35]. About 5-10 million develop xerophthalmia and about half a million go blind each year. Within a region, the deficiency typically occurs in clusters, in villages, districts or provinces where environmental conditions and living practices are similar. If at least one child in a village or homestead is known to have xerophthalmia, the risk of others in the same village or homestead developing vitamin A deficiency is higher. There can, however, be distinct differences between villages and districts that are neighbours but have different climatic conditions or different cultural practices. Knowledge of such clustering is a great help in designing and implementing VAD prevention programmes.

2. Contributing factors

a) Age

Although children of all ages and even some adults are at risk, vitamin A deficiency is especially severe and most prevalent among children of pre-school age, <6 years old. The prevalence is usually not so great in the first 6-12 months because of supplies from the mother's milk, but there may be a rapid rise on weaning, especially on to a simple rice-based diet containing little or no vitamin A or provitamin carotenoids [36].

The prevalence of mild xerophthalmia increases with age through the pre-school years [37]. Moderate to severe deficiency also increases, associated with chronic dietary inadequacy.

The prevalence of sub-clinical vitamin A deficiency, estimated by serum levels, can also be expected to increase with age during early childhood.

b) Socioeconomic status

Not surprisingly, it is people from the lowest socioeconomic strata who are most vulnerable to vitamin A deficiency. These are people with few possessions, poor housing and sanitation, a low level of education, and a subsistence-level life, with inadequate food supply. Such people are at 1.5 to 3 times higher risk than more fortunate members of their community [4].

c) Seasonality

The incidence and severity of vitamin A deficiency may vary with season and climatic conditions. Many foods are seasonal, so food availablity and quality obviously depend on climatic factors. At the peak season, there will be better supplies of provitamin A carotenoids. But children often experience a 'growth spurt' associated with increased caloric availability immediately after a rice harvest [38], when there may be no concomitant increase in dietary carotenoids, so the need for extra vitamin A to support growth may not be met.

Water contamination, parasitic infestations, flies *etc.* also lead to seasonal peaks in infectious diseases that are influenced by or can exacerbate VAD.

E. Strategies to Combat VAD

The underlying cause of vitamin A deficiency is a diet that lacks sufficient amounts of preformed vitamin A or sustained levels of provitamin A carotenoids. It is obvious that if the intake of vitamin A, either pre-formed or as the provitamin, is below the minimum requirement, VAD will result, with the likely consequences discussed above. There is thus an urgent need to boost vitamin A status in individuals and populations at risk. A long-term sustainable strategy that would ensure an adequate supply of vitamin A or the provitamin from the normal diet so that VAD does not occur in the first place would be ideal. However, in cases of acute VAD or risk of acute VAD, a different strategy is needed to give a rapid boost to vitamin A levels and status by administration of large-dose supplements. A third strategy, fortification of food with added vitamin A or provitamin A, combines elements of the other two.

For detailed assessment of these approaches and description of some programmes that have been evaluated, see [4,35,39]. Here, just an outline of the main features will be given, especially in relation to the application of intact provitamin A carotenoids.

1. Supplements

a) Vitamin A

Individuals, aged 12 months or more, suffering from the consequences of VAD, such as measles, diarrhoea or other infections, as well as ones showing signs of xerophthalmia, are typically treated by immediate administration of a high dose (200 000 IU, 60 mg) of vitamin A, usually as retinyl palmitate. Smaller doses, usually 100 000 or 25 000 IU, are given to infants aged 6-12 months or less than 6 months, respectively. Vitamin A status, as serum retinol concentration, rises rapidly, liver stores are replenished, and dramatic improvements in health are often seen. In the absence of any other measures to increase the provision of vitamin A or the provitamin, by dietary improvement or fortification, the supplementation is typically repeated every 3-6 months, to maintain the improvements. Repeated supplementation at these intervals is assumed to be safe; in any case the benefits outweigh any risk of toxicity. When a population is identified as having marginal/low vitamin A status and VAD is diagnosed in some individuals, a public health programme of supplementation is recommended.

b) Provitamin carotenoids

Supplementation with the provitamin A, β-carotene, would remove the risk of vitamin A toxicity [40]. The conversion is controlled and there is no risk of vitamin A building up to toxic levels, but the carotene would need to be given in a form with high bioavailability (see *Chapter 8*).

Bioavailability studies with stable isotopic labelling have shown that β-carotene in oil is absorbed efficiently and the conversion efficiency is high (as good as 2.6:1) [41], so that β-carotene in this form can be considered as almost a full equivalent of vitamin A on a weight basis. Some trials have been undertaken. In a comparative study in Orissa State in India, periodic dosing with red palm oil had the same effect on vitamin A status as did the administration of a high dose (200 000 IU) supplement of retinyl palmitate [42]. Although vitamin A supplements may be supplied to young infants either direct or through breast milk after supplementation of the mother, direct supplementation of the infant with red palm oil would not be satisfactory because the high requirement for secretion of lipases and bile salts needed to deal with the large volume of oil would not be met by a digestive capacity suited primarily to human milk with its specialized fat content and composition. The indirect approach has been shown to be satisfactory, however. Studies in Honduras and Tanzania have demonstrated an improvement in the vitamin A status of both mother and infant following supplementation of the mother with β-carotene [43,44].

2. Fortification

There have been various programmes to increase dietary vitamin A intake by fortification, *i.e.* adding vitamin A to commonly consumed food ingredients. The most extensive trials have been undertaken with sugar, in Central America [45], or monosodium glutamate, in Indonesia [46] and the Philippines [47]. Fortification of other food vehicles, such as cereals, condiments and dairy products is under consideration. It is difficult to reach the poorest, highest-risk communities in remote areas, who live a long distance from the markets and cannot afford the fortified products unless the extra cost is subsidized.

The food is usually fortified with vitamin A, but high doses again would lead to risk of vitamin A toxicity. A programme of continued fortification with low doses is difficult to sustain. In principle, fortification with provitamin A carotene should be effective and safe but the strong colour of the carotene may impair consumer acceptance.

3. Dietary improvement

The ideal long-term sustainable strategy would be to ensure that everyone obtained sufficient vitamin A or provitamin A from the normal food components of the diet, especially provitamin carotene from vegetables and fruit.

a) Home gardens

Programmes have been initiated to encourage people to grow more vegetables in home, community or school gardens, to grow varieties with a higher vitamin A nutritional content and to improve cultivation conditions, within the constraints of the local climate and environment. A large-scale horticulture initiative in Bangladesh has led to an improvement in vitamin A status in a number of communities [48]. Dark green leafy vegetables contain sufficient β-carotene to meet the needs of virtually any population, but the bioavailability is not good, and the products are not readily accepted, especially by the most vulnerable group, young children. Fruits, *e.g.* mangoes, are good sources, but may be too expensive for the poorest families who are most at risk. There are some good carotene-rich local sources, *e.g.* 'buriti' and other rich local sources in South America [49,50], 'Karat' bananas in Micronesia [51], the Palmyra palm fruit in Bangladesh [52], and 'gac' fruit (*Momordica cochinchinensis*) in Vietnam and neighbouring countries [53]. The greater use of these should be encouraged and their introduction into other locations perhaps considered. Also, the wider use of the carotene-rich orange-fleshed sweet potato, instead of white varieties would be beneficial [54].

b) 'Biofortification'

Another approach is the use of plant breeding and genetic modification techniques to improve the nutrient quality, including provitamin A content, of foods that are used as dietary staples by many people. This strategy has been termed 'biofortification' [35]. A good example is the development of a GM strain of carotene-producing 'Golden rice' [55], though, again, the colour is a disadvantage; many populations associate quality with a pure white rice. There are also concerns about bioavailability, which has not been established. Other possible targets include potatoes, cassava, bananas and various cereals [56]. With this approach it is necessary to find the optimum balance between many factors, such as nutritient content, cultivation requirements, bioavailability from the product as it is consumed, consumer acceptance and economic advantages. The wider public concern about the safety and environmental impact of GM crops and practices must also be taken into account.

As mentioned before, some plant oils, especially red palm oil but also oil of 'gac' fruit contain a high concentration of carotene in a form that is absorbed efficiently. Use of these in cooking or as dressings could boost vitamin A status substantially, although the orange-red colour that they impart would not be appreciated in some food.

c) Post-harvest treatment

The importance of good treatment of fruit and vegetables post harvest to conserve provitamin A content should not be overlooked. Losses during transport, storage, cooking and processing can be high (see *Chapter 3*). To minimize destructive effects, prolonged heating should be avoided, as should exposure to strong light and air during drying, storage and transport [57]. The greatest risk of destruction of vital provitamin A carotene comes from the traditional and widespread practice of drying in air and from transport in the open in the heat of the day, in conditions of high ambient temperature and intense sunlight. Storage conditions are often not good; there is no refrigeration and ambient temperatures are high. Cooking facilities may be limited and may be determined by long-established tradition. Harsh but popular cooking conditions such as deep-frying, prolonged boiling and baking can cause particularly severe losses.

The destruction that can be caused by cooking must be balanced against improvement in bioavailabilty due to disrupting, weakening or softening the structural matrix of the food, which is a major determining factor in bioavailability

4. Strategy overall

There is a long standing argument about whether VAD should be treated by vitamin A supplementation or by a food-based programme to increase provitamin A consumption, and different factions tend to promote one at the exclusion of the other. But why should there be this argument? Surely, when several ways are available to tackle the problem, it is realistic and logical to use all of these as appropriate for particular circumstances. All have merits and benefits. All may have limitations and disadvantages.

It seems logical and sensible to make use of all available strategies.

In principle, dietary improvement to increase the availability of vitamin A and of provitamin carotene in a normal diet, from vegetables, fruit and staples, including 'biofortified' strains, augmented if necessary by fortified products, would be an ideal solution. Augmentation could be with vitamin A but fortification with provitamin carotenoids should be given greater consideration. Increased consumption of animal products such as milk, eggs, fat and butter, would provide more preformed vitamin A and carotene, but these products are not readily accessible to many poor families and communities. Increased production of primary sources of carotene – vegetables, fruit, staples – in home, community or school gardens is achievable.

When VAD is acute and rapid action is needed because patients are suffering life-threatening infections, the administration of high-dose supplements of vitamin A gives a rapid boost to vitamin A status and can have a dramatic effect on alleviating symptoms. Single high-dose supplementation is also used to boost vitamin A status in populations at serious risk of VAD, identified by the incidence of mild xerophthalmia, infectious disease and/or low serum retinol concentrations. To be effective in the longer term, this strategy requires the subsequent administration of follow-up doses, to maintain vitamin A sufficiency and liver stores. This raises the obvious concern about the toxicity of large doses of vitamin A. It may also be difficult to reach remote communities and to ensure adherence to the supplementation programme. This strategy is one of intervention therapy, not of sustainable dietary improvement.

F. Underlying Causes

In simple terms we know that vitamin A deficiency results when the intake of vitamin A or the provitamin is insufficient, so vitamin A status needs to be improved. But what is the underlying reason for the low vitamin A intake and status? Why should a particular individual, family, community or population be vitamin A deficient when their neighbours are not? Are they not aware of the problem and its treatment? Why are they not obtaining sufficient vitamin A? Are supplies of vitamin A-sufficient food adequate and affordable? Are they just not making full use of available supplies? Is this because of personal preference or is it determined by custom or tradition? Is lifestyle a factor? If the reasons are known, the scientific basis exists for treatment. A good illustration of this comes from a study of two tribes in Orissa, India. These tribes were living under similar conditions and eating a generally similar diet. A high incidence of xerophthalmia was seen in the children of one tribe, but not in the other. It was found that, in the tribe in which xerophthalmia was common, the infants were weaned early onto a carotenoid-free rice-based diet whereas in the other tribe breast-feeding was continued for much longer [58].

G. Conclusions

1. Place for carotenoid research

Many important challenges remain for carotenoid science. The expertise and experimental tools exist to identify carotene-rich food sources and to enhance crop plants by breeding and GM programmes. Much effort is being directed to optimizing methods to determine bioefficacy for particular sources under natural conditions; stable isotope methods to assess carotenoid uptake and conversion are proving very useful (see *Chapter 8*).

Associated with this is the challenge to identify sources and forms with high bioavailability for use in supplements and for fortification. Red palm oil is an excellent example of a natural material for this, but other carotene-rich oils merit exploration. A wide variety of formulations have been designed for various commercial applications of purified synthetic or natural carotenoids, solubilized or dispersed in various media, as colourants or as an easily assimilated form in animal feed for agriculture and aquaculture. The knowledge and technology exist so, with a similar research effort, carotenoid products and formulations could surely be devised for effective use in supplements and for fortification.

As discussed in *Chapter 3*, the great precision usually reported in food composition tables can be misleading. Analytical results recorded are for a particular sample grown in a particular place under particular, often optimized conditions. The values given may, therefore, bear little resemblance to the real values in actual food that is being eaten in the household in a community at risk. Proper guidance on this is needed and it would be so useful to develop a simple inexpensive method that could be used to determine rapidly the carotene content in such real samples, even in the most remote places.

We know much about carotenoids but there are still serious gaps in knowledge about the human subjects, particularly in regard to the great variability between individuals, not just between different ethnic groups and populations in different parts of the world, but between individuals in the same community. Some differences are due to environmental factors and cultural traditions but many answers may lie in the unseen genetic factors. With the mapping of the human genome, new technologies of molecular biology and molecular genetics hold the key to solving these mysteries. An important example is understanding the basis of 'responders' and 'non-responders' [59]. Identifying the genetic and other factors that determine how efficiently an individual absorbs and stores carotenoids and converts them into vitamin A would open the door to real progress in defining the needs of individuals and populations. Recent work has revealed that genetic variations (single nucleotide polymorphisms, SNPs) can have a profound influence on the efficiency of the β-carotene-cleaving enzymes [60].

2. Political, educational, cultural

There are areas where more knowledge and understanding are needed, but generally the science base is solid. In many ways, the main battle is not scientific but cultural or economic. At a local level, it can be extremely difficult to overcome or change eating practices that are rooted deep in culture, tradition or religion. Developing education programmes is particularly important to inform about the problem of vitamin A deficiency and its consequences and to encourage acceptance of intervention measures and the adoption of good nutritional practices. Economic reality means that the most vulnerable families may not be able to afford the kinds of food that would ensure them adequate supplies of vitamin A.

It has taken the dedicated efforts of many scientists and others battling against all kinds of difficulties to implement programmes, inform local populations, influence political thought and convince funding agencies of the urgency of action. Without these individuals and the international action of various agencies and bodies such as WHO, UNICEF, Helen Keller International, USAID, Sight and Life and Harvest Plus, and the effectiveness of IVACG and other meetings as a forum for communication, dissemination of knowledge and planning of the implementation of international intervention programmes, the great progress that has been made could not have been made. Emphasis now is likely to be on sustainable measures and action is likely to be shaped by the growing realization that provision of adequate vitamin A is part of the need for a wider integrated programme to ensure adequate availability of all micronutrients. In richer countries there is much interest in 'functional foods' that provide sufficient amounts of substances that are associated with various health benefits, especially reduction in risk of serious diseases. With a food-based approach to providing adequate supplies of provitamin A carotenoids, these poorer people would also be in a position to benefit from health-promoting effects of other carotenoids and other micronutrients that the food, especially fruit and vegetables, provides.

References

[1] D. S. McLaren, *Malnutrition and the Eye*, Academic Press, New York (1963).

[2] A. Sommer, *Nutritional Blindness: Xerophthalmia and Keratomalacia*, Oxford University Press, New York (1982).

[3] J. C. Bauernfeind (ed.), *Vitamin A Deficiency and its Control*, Academic Press, Orlando (1986).

[4] A. Sommer and K. P. West Jr., *Vitamin A Deficiency: Health, Survival, and Vision*, Oxford University Press, New York and Oxford (1996).

[5] WHO/USAID, *Vitamin A Deficiency and Xerophthalmia, WHO Tech. Report Ser.*, 590, WHO, Geneva (1976).

[6] Joint WHO/UNICEF/USAID/Helen Keller International IVACG Meeting Report, *Control of Vitamin A Deficiency and Xerophthalmia, WHO Tech. Report Ser., 672*, WHO, Geneva (1982).

[7] W. S. Blaner and J. A. Olson, in *The Retinoids: Biology, Chemistry and Medicine, 2nd. Edn.* (ed. M. B. Sporn, A. B. Roberts and D. S. Goodman), p. 229, Raven Press, New York (1994).

[8] J. A. Olson, in *Vitamin A Deficiency: Health, Survival, and Vision* (ed. A. Sommer and K. P. West Jr.), p. 221, Oxford University Press, New York (1996).
[9] D. E. Ong, M. E. Newcomer and F. Chytil, in *The Retinoids: Biology, Chemistry and Medicine, 2nd. Edn.* (ed. M. B. Sporn, A. B. Roberts and D. S. Goodman), p. 283, Raven Press, New York (1994).
[10] N. Wongsiriroj and W. S. Blaner, *Sight and Life Newsletter 2/2004*, 2 (2004).
[11] R. Blomhoff, M. H. Green and K. R. Norum, *Ann. Rev. Nutr.*, **12**, 37 (1992).
[12] W. S. Blaner, L. Quadro, M. Gottesman and M. V. Gamble, *Sight and Life Newsletter 3/2001*, 7 (2001).
[13] E. H. Harrison, *Ann. Rev. Nutr.*, **25**, 87 (2005).
[14] N. Wongsiriroj and W. S. Blaner, *Sight and Life Magazine 3/2007*, 32 (2007).
[15] D. S. Goodman, in *The Retinoids, Volume 2* (ed. M. B. Sporn, A. B. Roberts and D. S. Goodman), p. 42, Academic Press, New York (1984).
[16] J. C. Bauernfeind, in *The Safe Use of Vitamin A*, p. 44, IVACG Nutr. Foundn., Washington DC (1980).
[17] J. N. Hathcock, D. G. Hattan, M. Y. Jenkins, I. T. McDonald, P. R. Sundaresan and V. L. Wilkenin, *Am. J. Clin. Nutr.*, **52**, 183 (1990).
[18] K. J. Rothman, L. L. Moore, M. R. Singer, U. S. Nguyen, S. Mannino and A. Milunsky, *New Engl. J. Med.*, **333**, 1369 (1995).
[19] J. Adams, *Neurotoxicol. Teratol.*, **15**, 193 (1993).
[20] H. Melhus, K. Michaelsson, A. Kindmark, R. Bergström, K. Holmberg, H. Mallmin, A. Wolk and S. Ljunghall, *Ann. Intern. Med.*, **12**, 770 (1998).
[21] FAO/WHO, *Requirements of Vitamin A, Thiamine, Riboflavin and Niacin, FAO Food and Nutrition Ser., 8*, FAO, Rome (1967).
[22] Food and Nutrition Board Standing Committee of the Scientific Committee of Dietary Reference Intakes of Vitamin A, Vitamin K, Arsenic, Boron, Chromium, Copper, Iodine, Iron, Manganese, Molybdenum, Nickel, Silicon, Vanadium and Zinc, National Academy of Sciences Institute of Medicine, Washington DC (2001).
[23] D. S. McLaren, *The Control of Xerophthalmia: A Century of Contributions and Lessons*, Sight and Life, Basel (2004).
[24] A. Sommer, G. Hussaini, Muhilal, I. Tarwotjo, D. Susanto and J. S. Saroso, *Am. J. Clin. Nutr.*, **33**, 887 (1980).
[25] D. S. McLaren and M. Frigg, *Sight and Life Manual on Vitamin A Deficiency Disorders (VADD), 2nd. Edn.*, Sight and Life, Basel (2001).
[26] E. Fuchs, *Biochem. Soc. Trans.*, **19**, 1112 (1991).
[27] H. N. Green and E. Mellanby, *Br. Med. J.*, **20**, 691 (1928).
[28] R. Biswas, A. B. Biswas, B. Manna, S. K. Bhattacharya, R. Dey and S. Sarkar, *Eur. J. Epidemiol.*, **10**, 57 (1994).
[29] R. D. Semba, N. M. H. Graham, W. T. Caiaffa, J. B. Margolick, L. Clement and D. Vlahov, *Arch. Intern. Med.*, **153**, 2149 (1993).
[30] R. D. Semba, P. G. Miotti, J. D. Chiphangwi, A. J. Saah, J. K. Canner, G. A. Dallabetta and D. R. Hoover, *Lancet*, **343**, 1593 (1994).
[31] C. E. Bloch, *Am. J. Dis. Child.*, **27**, 139 (1924).
[32] K. H. Brown, A. Gaffar and S. M. Alamgir, *J. Pediatr.*, **95**, 651 (1979).
[33] M. Lloyd-Puryear, J. H. Humphrey, K. P. West, K. Aniol, F. Mahoney, J. Mahoney and D. G. Keenum, *Nutr. Res.*, **9**, 1007 (1989).
[34] A. C. Ross, in *Vitamin A Deficiency: Health, Survival, and Vision* (ed. A. Sommer and K. P. West Jr.), p. 251, Oxford University Press, New York and Oxford (1996).
[35] B. A. Underwood, *Sight and Life Newsletter2/2006*, 10 (2006).
[36] B. A. Underwood, *Vitamin A Deficiency in Infancy. Report of the XV IVACG Meeting, Arusha, Tanzania*, The Nutrition Foundation, Washington DC (1993).

[37] Muhilal, J. Tarwotjo, B. Kodyat, S. Herman, D. Permaesih, D. Karyadi, S. Wilbur and J. M. Tielsch, *Eur. J. Clin. Nutr.*, **48**, 708 (1994).
[38] K. H. Brown, R. E. Black and S. Becker, *Am. J. Clin. Nutr.*, **36**, 303 (1982).
[39] N. W. Solomons, *Sight and Life Newsletter 3/2002*, 87 (2002).
[40] N. W. Solomons, *Nutr. Rev.*, **56**, 309 (1998).
[41] M. van Lieshout, C. E. West, D. Permaesih, Y. Wang, X. Xu, R. B. van Breemen, A. F. L. Creemers, M. A. Verhoeven and J. Lugtenburg, *Am. J. Clin. Nutr.*, **73**, 949 (2001).
[42] S. Mohapatra and R. Manorama, *Asia Pacific J. Clin. Nutr.*, **66**, 246 (1997).
[43] L. M. Canfield, R. G. Kaminsky, D. L. Taren, E. Shaw and J. K. Sander, *Eur. J. Nutr.*, **40**, 30 (2001).
[44] G. Lietz, C. J. K. Henry, G. Mulokozi, J. Mugyabuso, A. Ballart, G. Ndossi, W. Lorri and A. Tomkins, *Food Nutr. Bull.*, **21**, 215 (2000).
[45] G. Arroyave, in *Vitamin A Deficiency and its Control* (ed. J. C. Bauernfeind), p. 405, Academic Press, Orlando (1986).
[46] Muhilal, D. Permaesih, Y. R. Idjradinata, Muherdiyantiningsih and D. Karyadi, *Am. J. Clin. Nutr.*, **48**, 1271 (1988).
[47] F. S. Solon, M. C. Latham, R. Guirriec, R. Florentino, D. F. Williamson and J. Aguilar, *Food Technol.*, **39**, 71 (1985).
[48] M. W. Bloem, N. Huq, J. Gorstein, S. Burger, T. Khan, N. Islam, S. Baker and F. Davidson, *Eur. J. Clin. Nutr.*, **50 (Suppl.)**, 62 (1996).
[49] H. T. Godoy and D. B. Rodriguez-Amaya, *Arq. Biol. Tecnol.*, **38**, 109 (1995).
[50] D. B. Rodriguez-Amaya, *Sight and Life Newsletter 4/2002*, 3 (2002).
[51] L. Englberger, *Sight and Life Newsletter 2/2002*, 28 (2002).
[52] A. A. Shamim, M. G. Mawla, Z. Islam and M. S. Rahman, *Sight and Life Newsletter 2/2003*, 21 (2003).
[53] L. T. Vuong, *Food Nutr. Bull.*, **21**, 173 (2000).
[54] J. W. Low, M. Arimond, N. Osman, B. Cunguara, F. Zano and D. Tschirley, *J. Nutr.*, **137**, 1329 (2007).
[55] S. Al-Babili, X. Ye, P. Lucca, I. Potrykus and P. Beyer, *Nature Biotechnol.*, **18**, 750 (2000).
[56] I. Potrykus, *Plant Physiol.*, **125**, 1157 (2001).
[57] D. B. Rodriguez-Amaya, *Carotenoids and Food Preparation: The Retention of Provitamin A Carotenoids in Prepared, Processed and Stored Foods*, OMNI, Arlington (1997).
[58] D. S. McLaren, *J. Trop. Pediatr.*, **2**, 135 (1956).
[59] Z. Wang, S. Yin, X. Zhao, R. M. Russell and G. Tang, *Br. J. Nutr.*, **91**, 121 (2004).
[60] F. Tourniaire, W. Leung, C. Méplan, A.-M. Minihane, S. Hessel, J. Von Lintig, J. Flint, H. Gilbert, J. Hesketh and G. Lietz, *Carotenoid Sci.*, **12**, 57 (2008).

Carotenoids
Volume 5: Nutrition and Health

Chapter 10

Epidemiology and Intervention Trials

Susan T. Mayne, Margaret E. Wright and Brenda Cartmel

A. Introduction to Epidemiology

Determining the health effects of carotenoids in humans is a challenging yet high priority area of research. Epidemiology is the study of the distribution and determinants of disease in human populations. Epidemiologists who study carotenoids are thus interested in determining if carotenoid intake or carotenoid status is associated with risk of various disease endpoints. As carotenoids are known to have antioxidant functions in plants, and evidence suggests that oxidative stress could be involved in the aetiology of chronic diseases such as cancer, heart disease, cataract and macular degeneration, much of the epidemiological research on carotenoids has emphasized links with risk of these and other chronic diseases, as summarized in *Chapters 13-15*.

Epidemiological studies all have in common the fact that they examine associations between exposure to some factor, in this case carotenoid intake/status, and the disease outcomes of interest. As will be detailed below, exposure assessment can be undertaken by collecting dietary data (asking subjects to recall their intake of carotenoid-containing foods and supplements) and/or more objective measurements of carotenoid status, including those obtained in the laboratory. Whilst much of the earlier epidemiological research on carotenoids and health used the traditional questionnaire-based approach, current research is relying increasingly on laboratory measurements to determine exposure objectively (biochemical and molecular epidemiology).

B. Types of Epidemiological Studies

Some general types of epidemiological study designs are summarized in Fig. 1. Epidemiological studies include both observational studies and experimental/intervention trials. The distinction between these two types of study design is important. In observational studies, there is no attempt by the researcher to modify the exposure status of the study subjects with regard to carotenoids or any other factor. In contrast, intervention trials are essentially an experimental design where the exposure status of the study subjects to the factor of interest is manipulated. For carotenoids, this includes both carotenoid supplementation trials, and also trials where subjects are asked to increase consumption of carotenoid-rich foods. Observational and intervention research both provide valuable information and are critical to our understanding of the health effects of carotenoids. Both designs also have important limitations that are detailed within each study design discussion.

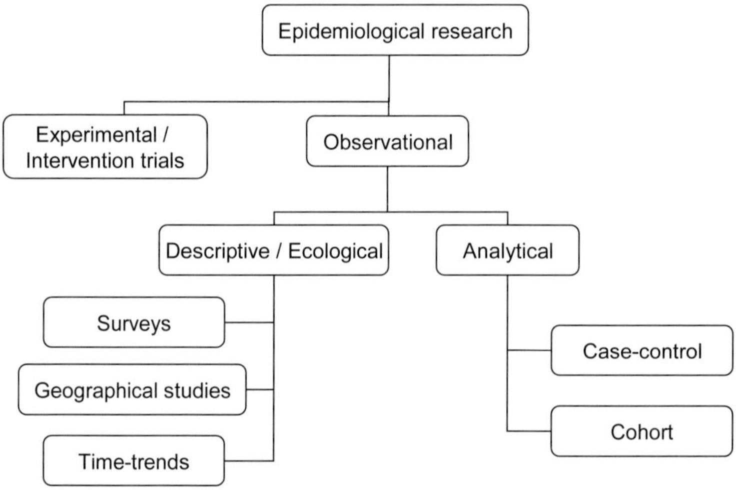

Fig. 1. Summary of epidemiological study designs described in the text.

1. Observational study designs

a) Descriptive epidemiology

The aim of descriptive epidemiology is to describe patterns of exposure and/or disease in a population. In carotenoid research, descriptive epidemiology methods are used to describe carotenoid intake patterns in various populations (by age and sex), or to describe typical blood or tissue levels of carotenoids in various populations. Many studies from different parts of the world have set out to ascertain plasma carotenoid concentrations, both for total carotenoids and for individual carotenoids. In the United States, the best source of data for the descriptive epidemiology of carotenoids comes from a national nutrition survey known as NHANES (National Health and Nutrition Examination Survey).

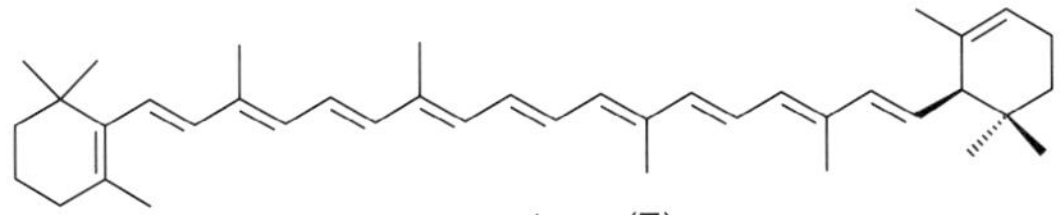

α-carotene (**7**)

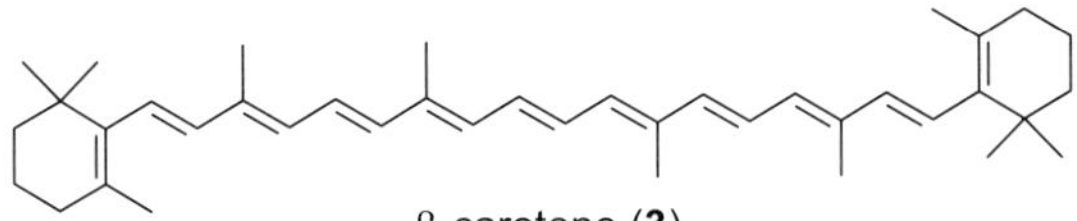

β-carotene (**3**)

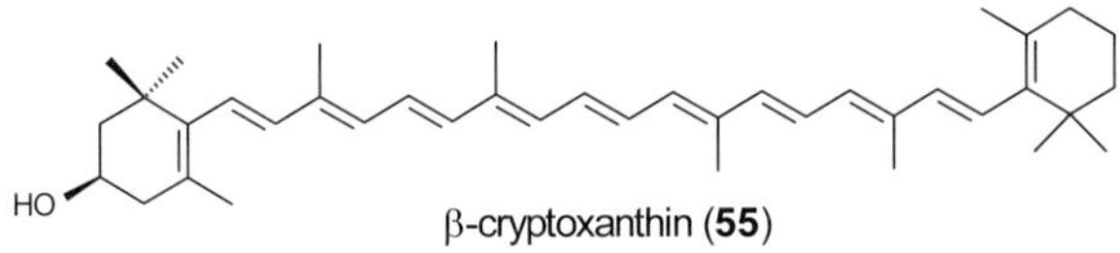

β-cryptoxanthin (**55**)

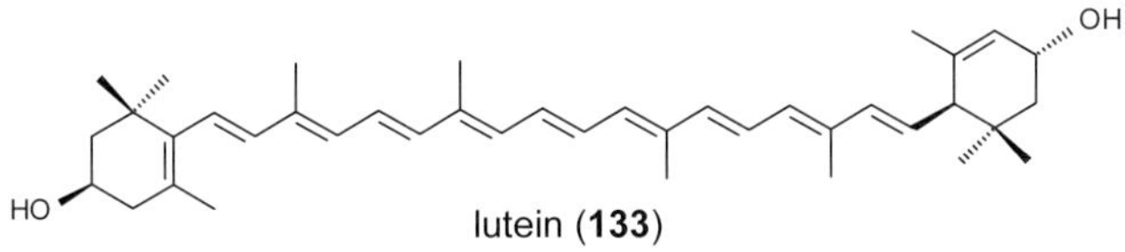

lutein (**133**)

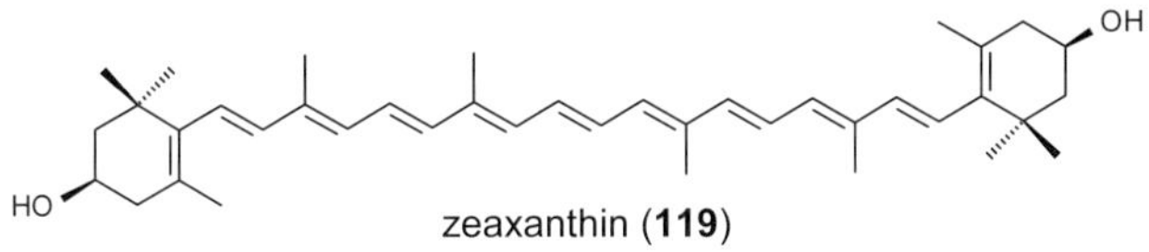

zeaxanthin (**119**)

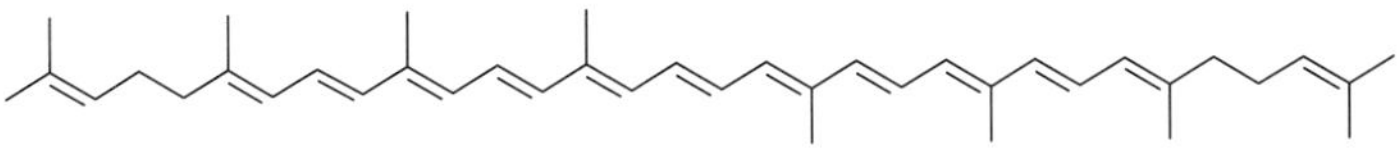

lycopene (**31**)

There have been several waves of NHANES surveys; NHANES III included both dietary data and biochemical measurements of various plasma carotenoids for a probability sample, selected to create a population sample from which inferences can be made to the overall U.S.

population. The carotenoid intake data include estimated intakes of α-carotene (**7**), β-carotene (**3**), β-cryptoxanthin (**55**), lutein (**133**) + zeaxanthin (**119**), and lycopene (**31**), and are reported for various age-specific and sex-specific groups [1].

The dietary intake estimates are based on dietary data obtained from nearly 30,000 Americans, and are therefore robust estimates of intake. Median intakes (50th percentile) of carotenoids from NHANES III are summarized in Table 1. The median serum data for these carotenoids are summarized in Table 2.

Table 1. Usual intake of carotenoids (μg/day) from food. The data are taken from the NHANES III survey (1988-1994), showing medians (50th percentile) and selected other percentiles.

Carotenoid		**Percentile**	
	10th	50th	90th
β-Carotene (**3**)	774	1,665	3,580
α-Carotene (**7**)	2	36	1,184
Lutein (**133**) + Zeaxanthin (**119**)	714	1,466	3,021
β-Cryptoxanthin (**55**)	24	88	319
Lycopene (**31**)	3,580	8,031	16,833

Data are based on all individuals excluding pregnant and lactating women (n=28,575) and are taken from reference [1].

Table 2. Serum concentrations of carotenoids (μg/dL) of persons aged 4 years and older. The data are taken from the NHANES III survey (1988-1994), showing medians (50th percentile) and selected other percentiles.

Carotenoid		**Percentile**	
	10th	50th	90th
β-Carotene (**3**)	6.4	14.7	35.1
α-Carotene (**7**)	1.3	3.4	9.2
Lutein (**133**) + Zeaxanthin (**119**)	11.1	18.9	33.0
β-Cryptoxanthin (**55**)	4.0	8.0	16.4
Lycopene (**31**)	11.9	22.4	36.1

Data are taken from reference [2].

These data, compiled by age-specific and sex-specific groupings, are based on a sample size in excess of 20,000 Americans, with all samples analysed in one laboratory [2]. Thus, the NHANES III data are a valuable source of information on typical carotenoid status in a well-nourished population. Carotenoid levels in blood reflect dietary intake, so data for the U.S. may not be an appropriate comparison for countries with different carotenoid intake patterns, but are included here as a reference point.

Descriptive studies have also been done to establish tissue levels of carotenoids [2]; these studies tend to be based upon convenience samples (samples selected for relative ease of access) with a relatively small sample size (usually fewer than 100 subjects).

Descriptive data on a population's typical intake of carotenoids are sometimes used as a basis for ecological studies, in which the intake of carotenoids across populations might be compared with disease patterns across those same populations. Other types of ecological studies include (i) time trends studies, in which trends in carotenoid intake within a population over time might be compared with trends in disease incidence within the same population over time, and (ii) geographical studies, where, for example, carotenoid intake in different parts of a country or region is compared with disease incidence patterns across that country or region. There are many differences other than nutrient intake across populations, so ecological studies are only appropriate for generating new hypotheses, not for suggesting causality.

b) Analytical epidemiology

Analytical epidemiology studies include both case-control and cohort studies. These are the two study designs used most commonly to identify health effects of carotenoids.

i) Case-control studies. In case-control studies, cases with a particular disease are identified and interviewed, as is a comparison group of subjects who do not have the disease of interest. The control group is generally selected to reflect the age and gender distribution of the case subjects. For carotenoid research, the cases are asked to report on their usual consumption of carotenoid-rich foods in some stated period of time **before** the onset of their disease, and the controls for a similar period in the past. Ideally, case-control studies are population-based, meaning that both the cases and the controls are sampled from a defined study population. In contrast to population-based case-control studies, hospital-based case-control studies recruit both cases and controls from one or more hospitals. It is a requirement that controls do not have the disease under study, but they may be afflicted with one or more conditions that led to a hospital admission. One of the limitations with the hospital-based approach for studying carotenoids is that inadequate intake of these nutrients could be related to risk of numerous chronic diseases (not just the one being studied) so that selecting an appropriate control group can be difficult. Thus, population-based case-control studies of carotenoids and disease are considered more informative than hospital-based case-control studies.

Case-control studies are an efficient study design, but the presence of disease in cases might affect the reported carotenoid-containing food intake, as well as affecting circulating carotenoid concentrations, thereby precluding biochemical epidemiological studies of carotenoids. For example, patients with gastrointestinal diseases may have altered their diet in the months preceding diagnosis, because of the disease symptoms; it may be difficult for these cases to recall accurately their normal diets before the onset of disease. This is an important limitation to case-control studies. As case-control studies are less expensive and more efficient than cohort studies, most of the earlier literature on health effects of carotenoids was derived from case-control studies. More recently, however, data are becoming widely available from numerous large cohort studies of diet and health, conducted around the world.

ii) Cohort studies. The basic cohort design involves recruiting a large population, obtaining dietary and other data on that population, and then following the population forward in time, generally for many years, for the development of future disease. Some cohort studies obtain dietary data only at baseline (when the cohort is constructed), whilst others collect updated dietary intake data at some points during follow-up. There are many well-known cohort studies in the area of nutrition and health; a few of the many that have contributed to the literature on carotenoids and health are the U.S. Nurses Health Study, the U.S. Health Professionals' Follow-Up Study, the Alpha-Tocopherol, Beta-Carotene Cancer Prevention (ATBC) study from Finland, the Women's Health Initiative cohort from the U.S., and the European Prospective Investigation into Cancer and Nutrition (EPIC) study.

In contrast to the case-control approach, in the cohort study design, dietary data are obtained from apparently healthy study participants **before** the development of disease is detected. This provides important temporal information, because the nutrient intake pattern preceded the development of disease rather than being a consequence of the disease. Thus, cohort studies of carotenoids and health are generally considered less biased than case-control studies. In practice, however, it is difficult to study rarer diseases by cohort designs; even the largest cohorts may have too few cases of a particular disease occurring during the follow-up to allow for robust epidemiological research. For these reasons, epidemiologists continue to conduct both case-control and cohort studies to identify health effects of carotenoids, recognizing the strengths and limitations of each approach.

2. Intervention trials

Intervention trials are essentially an experimental design in which the carotenoid exposure status of the study subjects is manipulated, and the resulting effect on some endpoint is evaluated. For carotenoids, this includes both carotenoid supplementation trials with carotenoids alone or in combination with other nutrients, and also trials in which subjects are asked to increase consumption of carotenoid-rich foods, either total fruits and vegetables, or specific sub-groups of fruits and vegetables, such as tomato products.

a) Supplementation trials

β-Carotene was the first carotenoid to be widely available for supplementation purposes; consequently, most of the completed large-scale carotenoid supplementation trials have tested β-carotene. More recently, supplements of lutein and zeaxanthin are being used in intervention trials aimed at reducing progression of eye diseases (see *Chapter 15*), and lycopene supplements are being used in intervention trials in relation to prostate cancer (biomarker trials to date are summarized in *Chapter 13*). The conduct of these trials is relatively straightforward; subjects are assigned randomly to receive the carotenoid supplement or not. Placebo pills that look identical to the carotenoid supplement provide a comparison group (randomized, placebo-controlled, blinded trial). The doses of carotenoids studied in most human trials are under 50 mg/day and can be formulated into one capsule to be taken daily. Compliance has generally been quite good. In one study that used daily supplementation with 50 mg β-carotene over several years [3], excellent compliance was found, as assessed by returned blister packs, *e.g.* 81% took >90% of the pills and 94% took >75% during the first year of intervention; compliance remained high with 87% taking >90% of their pills during year 4.

Whilst it is straightforward to conduct carotenoid supplementation trials, interpreting the results can be more complex, because the doses studied are often supra-physiological (*e.g.* 50 mg β-carotene per day versus typical median dietary intakes of <2 mg/day, Table 1). Also, resulting plasma concentrations are often vastly in excess of those achieved through normal dietary intake, reflecting both the higher dose and higher bioavailability of carotenoids from supplements. So, results obtained from intervention trials with dietary supplements are necessarily limited to the dose studied. Also, whilst most chronic diseases take decades to develop, most of the carotenoid supplementation trials have a duration of less than one decade, with the exception of the Physicians' Health Study, which studied 12 years of supplementation with β-carotene [4]. So, a lack of effect on a chronic disease endpoint may simply reflect a relatively short intervention duration compared with the period of time involved in development and progression of the chronic disease.

b) Food-based interventions

The other approach for conducting carotenoid intervention trials is *via* food-based interventions. Some trials have randomized subjects to a diet high in fruit and vegetables, or even a specifically high-carotenoid diet, and then followed the study subjects forward in time for either disease development or modulation of some biomarker of interest. Intervention trials are usually only initiated if promising data from observational epidemiology (along with supportive evidence from animal/mechanistic studies) suggests an advantage to high carotenoid intake. Because nearly all of the observational epidemiology research on carotenoids reflects health effects of carotenoids from foods, the food-based design has the advantage of being a more direct test of results obtained in observational studies. However, adherence is a substantial barrier to these interventions. Considering increases in plasma carotenoid as a biomarker of adherence, it is evident that, in some trials, *e.g.* in an ongoing trial involving breast cancer survivors [5], carotenoid intake and status increased substantially with food-based interventions. Other trials, however, have had much less success in producing significant alterations in plasma carotenoids, although some of them have included fruit and vegetable interventions as part of an overall dietary intervention [6]. The characteristics of the population studied (gender, smoking status, overall health, other behaviour such as alcoholic beverage consumption) as well as the design and intensity of the intervention arc all likcly to influence adherence to dietary recommendations for increased consumption.

A modification of the food-based approach for carotenoid intervention trials involves a single food source, rather than an overall dietary change, to increase consumption. Examples of this are studies in which interventions based on tomato sauce are used to increase lycopene intake from foods [7], or interventions based on red palm oil to increase intake of carotenes [8,9]. In this design, the researchers typically provide the intervention food to the study subjects, facilitating adherence to the intervention.

3. Exposure assessment in epidemiological studies

a) Dietary assessment

Both observational studies and intervention trials of carotenoids include intake assessment through the diet. The most common method for assessing dietary intake of carotenoids from foods is the food frequency questionnaire, which asks subjects to characterize their usual frequency of consumption of various food items in the diet, including carotenoid-containing foods, primarily fruits and vegetables but sometimes also mixed dishes that contain carotenoids.

i) Food frequency questionnaires. Most food frequency questionnaires assess the overall diet, not just carotenoid-containing foods in the diet. It is important to include mixed dishes in carotenoid intake assessment; for example, in a recent study in a U.S. population, spaghetti/lasagna/other pasta were the top food sources of dietary lycopene [10].

While there are many 'off-the-shelf' food frequency questionnaires in use by nutritional epidemiologists, it must be recognized that investigators working on carotenoids may need to modify existing questionnaires in order better to capture data on the carotenoids of interest. For example, many food frequency questionnaires do not differentiate between different types of lettuces commonly consumed in salads. However, the lutein and zeaxanthin content of darker green 'lettuces' (kale, collard greens, spinach) is substantially higher than that of other green lettuces (butterhead, romaine, iceberg, other green) [11] so modification of the questionnaire may be necessary to distinguish foods that are similar but have different carotenoid content.

Obtaining data on the frequency of consumption is only the first step in intake assessment; the frequency data must then be converted into estimated daily carotenoid intakes by linking the questionnaire data to a food composition database. In the U.S., a carotenoid composition database has been developed [12] that includes data on α-carotene, β-carotene, β-cryptoxanthin, lutein + zeaxanthin, and lycopene in approximately 4,000 food items. This database is publicly available [13] and is updated as new information becomes available.

It is well known that all dietary questionnaires, including food frequency questionnaires, have some measurement error associated with them. Some researchers have even challenged the usefulness of the food frequency questionnaire, given its inherent measurement error [14]. Other dietary assessment methods such as 24-hour recalls and food diaries are also available. Fortunately for carotenoid researchers, blood carotenoid concentrations provide a reference biomarker against which different dietary questionnaires can be assessed for validity. In one recent study, serum carotenoid concentrations were used to examine the validity of fruit and vegetable intake estimated by 14-day weighed records (where subjects are asked to weigh and record all foods consumed over a 14-day period), a 27-item questionnaire and a 180-item questionnaire [15]. The correlation coefficients between serum carotenoids and fruit and vegetable intake were slightly higher for the 14-day weighed records than for the two questionnaires, but no difference was observed between the 180-item and the 27-item questionnaires. Validity coefficients are similar to correlation coefficients but instead use one measurement (in this case plasma carotenoids) as a criterion to evaluate the validity of another measurement (in this case dietary intake). The highest validity coefficients (VC) were observed for vegetable intake (estimated from weighed records, the 180-item questionnaire, and the 27-item questionnaire) when serum α-carotene was used as the criterion biomarker, with VCs of 0.77, 0.58, and 0.51, respectively. These results, along with data from many other studies, suggest that measurement of fruit and vegetable intake, and therefore carotenoid intake, by self-report has acceptable validity within the population studied, especially when combined with biomarkers of carotenoid status.

ii) Dietary supplement questionnaires. Carotenoids can also be consumed as dietary supplements, so dietary intake assessment for epidemiological research often involves a detailed dietary supplement questionnaire, focusing on carotenoid-containing supplements. In the U.S., many multivitamins include β-carotene (as provitamin A), and many now also include lutein. Carotenoids are often also a component of antioxidant-type combination supplements, and some, *e.g.* β-carotene, can be purchased as single nutrient supplements. These sources of carotenoid intake need to be considered in studies of dietary carotenoids and health. There are some particular challenges, however, to doing this properly. In some supplements, the actual amount of vitamin A as β-carotene is not always indicated, so some assumptions may have to be made about carotenoid content. Also, compared to foods, supplements are often consumed erratically, with periods of use and non-use, and frequent switching of supplement brands. This presents some challenges to the accurate estimation of 'usual' carotenoid intake values.

iii) Combined intake assessment. Once intake estimates from foods and supplements have been obtained, it is not clear how that information should be used for exposure assessment. For many nutrients, it is entirely appropriate to combine nutrients from foods with those from supplements in order to arrive at a total intake level of that nutrient. For carotenoids, this is not appropriate, because the bioavailability of carotenoids from supplements is dramatically better than that from most food sources (*Chapter 7*). For example, the absorption of β-carotene in supplements in a form solubilized with emulsifiers and protected by antioxidants can be 70% or more [2], whereas less than 5% bioavailability has been reported for carotenes from raw foods such as carrots [2]. So, it may be logical to keep intake of carotenoids from supplements separate from that from foods when making intake assessments. It must be noted, however, that some foods contain both naturally occurring carotenoids and carotenoids added to supplement the food source either as a source of vitamin A or for food colouration. In this case, the food label does not differentiate between the two, making it difficult, in practice, to derive an estimate of carotenoids from the foods themselves, while excluding those added during food fortification.

iv) Pooling and correlation of data. As mentioned earlier, cohort studies are being used increasingly to identify diet/disease relationships. A challenge in conducting large cohort studies of dietary nutrients like carotenoids is that different dietary questionnaires may need to be used. For example, in the European Prospective Investigation into Cancer and Nutrition Study (EPIC), dietary data are being collected in many countries across Europe, some of which have quite distinct dietary patterns [16]. It is difficult to be sure that intakes in one region (*e.g.* Southern Europe) are assessed similarly to intakes in another region (*e.g.* Northern Europe), due to different dietary patterns in these regions. This type of measurement error is called non-differential measurement error, and means that it may be more difficult to discover a true association between carotenoid intake and disease risk.

A related concern involves the pooling of data from several different case-control and/or cohort studies of carotenoids and health, into a larger study (pooled analyses or meta-analyses). Most of the dietary questionnaires are considered to have some validity for assessing relative levels of intake within a population (*i.e.* classifying who is a high-consumer and who is a low-consumer), but these same questionnaires have limitations in terms of assessing intakes quantitatively; portion sizes are difficult to estimate, more extensive food lists tend to produce over-reporting, *etc.* However, the pooled analyses of carotenoids or carotenoid-containing foods in relation to health need some common measurement of intake (*e.g.* grams of vegetables consumed per day) to compare and combine studies. It is not appropriate simply to categorize into ‘high’ or ‘low’ from a particular population, because a low intake in one population (such as lowest quartile of lycopene in the U.S.) may actually be a similar intake to the highest quartile in another population (*e.g.* lycopene in China). Thus, pooled analyses require a level of quantitative measurement that does not exist in most dietary questionnaires today. For these reasons, pooled analyses based on dietary intake data must be interpreted with great caution.

Despite these challenges in measuring dietary intake for studies of carotenoids and health, dietary measurements do correlate, albeit not highly, with measurements of blood carotenoids by HPLC. Correlation coefficients between dietary carotenoid intake and carotenoid concentrations in blood tend to be modest (approximately 0.2-0.4 in most studies). These coefficients, however, are better than those obtained for other nutritional factors, such as energy, where intake estimates correlate poorly, if at all, with objective biomarkers of intake [17].

b) Biomarker assessment

i) Analysis of blood samples. Given the inherent difficulties in assessing quantitatively carotenoid intakes for human studies, biomarkers are an attractive alternative for determining carotenoid status. To date, blood carotenoid concentration has been the most commonly used biomarker. Carotenoids in plasma or serum can readily be analysed by HPLC, but at significant cost. For large epidemiological studies and clinical intervention trials, involving tens of thousands of subjects, this cost may be prohibitive. Also, the use of blood samples requires study subjects to agree to submit to venipuncture, which may reduce rates of participation and possibly introduce participation bias. The blood sample has to be protected from light and processed relatively quickly to separate the plasma/serum, which then has to be stored frozen to await analysis, adding to the cost and complexity. Furthermore, carotenoid concentrations in blood fluctuate in response to recent dietary intake. Thus, plasma carotenoid concentrations have the advantage of being an objective biomarker of intake, but there are some practical and economical limitations to their use for epidemiological studies.

For investigators who choose to measure plasma carotenoids for large epidemiological studies, laboratory quality control becomes very important, because it may take months, if not years, to complete all the biochemical analyses for large studies, and avoiding drift over time in the laboratory assay is essential. Most biochemical epidemiological studies that measure plasma carotenoid concentration are cohort studies or intervention trials, although some case-control studies will use this approach, more often for diseases that are not likely to affect systemic nutrient levels. If samples from case-control studies are measured, it is imperative to include in each batch samples from both cases and controls, in the same ratio of cases to controls as in the overall study, in order to avoid potential artifacts. Sometimes, in cohort studies, blood samples are collected from all participants at baseline, then subjects are monitored over time to determine who develops the disease of interest. Only the samples from those cases who developed the disease and a sub-sample of the remaining cohort who remained free of disease are then retrieved and analysed. This modified cohort design is called a nested case-control study, as the case-control study is nested within a larger cohort study. As with traditional case-control studies, samples from both cases and controls should be included in each batch of laboratory analyses.

There are some formal quality control programmes in place for laboratories that determine carotenoids for epidemiological studies and other purposes. In the U.S. a government agency, the National Institute of Standards and Technology (NIST), has coordinated a micronutrient quality assurance programme for participating laboratories. Blinded samples are sent to the laboratories, and results are fed back to the NIST programme to assess both the accuracy (in comparison to other laboratories, are the values correct?) and the reproducibility (if a sample is sent at one time point and then again several months later, how closely do the laboratory results agree?). Such quality control programmes have greatly improved the quality of laboratory data obtained on the most commonly occurring carotenoids in human blood.

ii) Analysis of tissue samples. Whilst blood is most commonly used to assess carotenoid status of humans, other tissues can be used. Adipose tissue is thought to be a more stable depot of carotenoids than blood, reflecting the strongly lipophilic nature of carotenoids, and a few epidemiological studies have utilized adipose tissue to assess carotenoid exposure status [18,19]. This approach, however, requires biopsies, more extensive sample preparation, *e.g.* saponification to remove excess lipids, and HPLC analysis. Thus, for large population studies, carotenoid concentrations in adipose tissue are more difficult and more expensive to use than blood carotenoid concentrations as a marker of systemic carotenoid concentrations.

Other tissues that have been used to monitor carotenoid exposure status in humans, by HPLC, include exfoliated oral mucosa, and tissue-specific biopsies (*e.g.* lung biopsies), where researchers measure carotenoids in a target tissue of interest [2]. These approaches are not typically used in epidemiological research.

iii) Non-invasive methods. A newer research approach to assessing carotenoid status of humans involves non-invasive assessment by spectroscopic methods. Resonance Raman (RR) spectroscopy has recently been developed for the non-invasive measurement of carotenoids in the macula (see *Chapter 15*) and also in the skin [20] (see *Chapter 16*). The obvious advantage of this approach is that it is non-invasive, as no biopsies or venipuncture are required. Also, the measurement is very quick, with results obtained almost instantaneously. A limitation to this approach, however, is that, with the exception of lycopene, it is not possible to separate out the contributions of individual carotenoids to the Raman signal.

Before RR spectroscopy measurements of dermal carotenoids can be used as a suitable biomarker in human studies, data on intra-subject and inter-subject variability, and validity are critically needed. A recent study [21] assessed the reproducibility and validity of RR spectroscopy measurements of dermal carotenoids in 75 healthy humans. Exciting light of 488 nm was used to estimate total carotenoids, and light of 514 nm to estimate lycopene separately. Measurements were taken from three sites, the palm, inner arm and outer arm, at baseline and after 1 week, 2 weeks, 1 month, 3 months and 6 months, to maximize seasonal variation. Reproducibilty was assessed by intra-class correlation coefficients (ICCs). For total carotenoids, ICCs across the three body sites for each time point ranged from 0.85 to 0.89, and the ICCs across time were 0.97 (for palm), 0.95 (inner arm) and 0.93 (outer arm).

In a second part of this study, 30 healthy subjects were examined. Dietary carotenoid intake, HPLC analyses of blood carotenoids and RR spectroscopy measurements of dermal carotenoid status (back of hip) were determined. Dermal biopsies (3 mm) were performed and the dermal carotenoids were analysed by HPLC. Total back-of-hip dermal carotenoids assessed by RR spectroscopy were highly and significantly correlated with total dermal carotenoids determined by HPLC of dermal biopsy samples. Correlation with blood carotenoid content determined by HPLC was also good. Similarly lycopene assessed by RR spectroscopy with exciting light of 514 nm was highly and significantly correlated with lycopene assessed by HPLC of dermal biopsies. These studies show that the RR spectroscopy method is reproducible and valid for use as a suitable biomarker for human studies.

Other non-invasive approaches are possible; recently an optical method based on light reflection spectroscopy has been proposed as a method to assess carotenoid levels in skin [22].

The development and validation of biomarkers of carotenoid status that can be used for epidemiological research is a very important priority, because dietary data are known to have significant errors, and can be biased. Due to social desirability biases, subjects may report that they are consuming more carotenoid-containing foods than they truly are. This makes it difficult to interpret studies based solely on dietary measurements of carotenoid intake. Having non-invasive measurements of carotenoid status as objective indicators, to support or refute self-reported dietary data, is important to furthering our understanding of carotenoid and health/disease associations.

c) Assessment of multiple antioxidant nutrients: Antioxidant indices

Interactions between antioxidants are important in biological systems [23]. Examination of multiple antioxidants simultaneously may, therefore, capture antioxidant and disease associations more effectively than other approaches that focus on single nutrients. A dietary antioxidant index has been constructed that summarizes the combined intake of individual carotenoids, flavonoids, tocopherols (vitamin E), vitamin C, and selenium [24]. The index was created by use of principal components analysis, a sophisticated statistical approach that reduces a large number of highly correlated variables (nutrients in this case; correlated because several nutrients such as carotenoids and flavonoids and vitamin C share similar food sources) to a smaller set of components that capture as much of the variability in the data as possible. The index was evaluated in terms of its ability to predict lung cancer risk in a cohort of Finnish male smokers. Risks of lung cancer were lower among men with higher antioxidant index scores. Of note was the finding that the composite index predicted risk similarly to total fruit and vegetable intake, but better than alternative nutrient measurements, including direct summation of intakes of groups of related nutrients, such as carotenoids.

C. Interpretation of Diet-Disease Associations Relevant to Carotenoids

1. Interpreting results of observational studies with carotenoid-containing foods

As most of the carotenoids consumed by typical human populations come from foods, an issue of great importance is to what extent observed effects are due to the carotenoids in the foods, or to the food sources themselves. For example, carrots are the leading food source for α-carotene in the U.S. diet so, in studies that examine α-carotene as a possible protective factor for chronic disease risk, it is difficult to isolate effects of α-carotene from effects of carrots. This is true even for studies that use plasma analysis; plasma α-carotene is a biomarker of carrot consumption. As another example, lycopene is consumed in the diet primarily from tomatoes and tomato products. While lycopene is found in some other foods, *e.g.* watermelon, pink grapefruit, the frequency of consumption of these foods is such that, in many populations, they contribute only modestly to lycopene intake at a population level. Thus, dietary lycopene and plasma lycopene are generally markers of tomato product intake, so it is difficult to know whether associations are driven by lycopene or by tomato products.

Much of the older literature on carotenoids and health failed to recognize this distinction carefully, so that effects were often attributed to specific carotenoids, without appreciation that results could be attributable to other components found in those same carotenoid-rich foods, or from the combination of nutrients found naturally in carotenoid-rich foods, *e.g.* one carotenoid interacting with other carotenoids or other phytochemicals. Today's research should recognize this and be more careful in the interpretation of associations with carotenoids when these are derived from carotenoid-rich foods.

A method that aims to separate the effects of carotenoids from those due to their primary plant food sources has been suggested [25]. In this study, it was found initially that higher intakes of β-carotene, β-cryptoxanthin, lutein + zeaxanthin, and total carotenoids were each associated with lower risks of lung cancer in women residing in rural America. After including total vegetable intake, which is the strongest predictor of lung cancer risk among all fruit and vegetable groupings, in the statistical models, however, the attributed protective effects of carotenoids disappeared. Importantly, vegetable intake remained significantly inversely associated with lung cancer risk in these same models. The authors concluded that vegetable consumption was more strongly associated with a lower risk of lung cancer than intake of any individual carotenoid or total carotenoids, which was concordant with two other studies that also used statistical testing to separate formally the effects of carotenoids from those of plant foods [26,27]. Future epidemiological studies of carotenoids could attempt this approach to understand better if protective effects are more likely to be due to carotenoids *per se*, or reflect the food sources rich in those same carotenoids, as subsequent intervention strategies, *e.g.* provision of nutrients *versus* foods, may differ.

2. Interpreting results of intervention trials with carotenoid-containing foods

Observational studies often find that people who consume more of the carotenoid-rich foods (fruits and vegetables) are at lower risk of various chronic diseases than are people who eat less of these same foods. It is impossible, however, to know whether or not associations are causal in these observational studies. People who eat more fruits and vegetables are less likely to smoke [28] and to be obese, and are more likely to engage in health-promoting behaviour such as physical activity. These 'confounding' factors make it difficult to assert causality from observational research. Researchers attempt to control statistically for confounding, but there remains the possibility that associations are not due to dietary intake specifically, but rather to correlated behaviour, *e.g.* smoking or not.

In order to overcome this limitation, intervention trials can be used, wherein study participants are randomly assigned to a dietary intervention or not. In this randomized trial design, the researchers strive to achieve balance in the intervention and control arm with regard to important confounders such as smoking. Ideally, the only variable being manipulated in these designs is the dietary pattern or specific dietary factor of interest. This is true in principle but, because diets are complex, it turns out that dietary manipulations tend to affect multiple nutrients simultaneously. For example, interventions aimed at increasing the consumption of carotenoid-containing foods in a population are likely to alter not only carotenoid status, but also intake of many other plant-based nutrients (folate, fibre, vitamin C, *etc.*), several of which are under investigation for their own health-promoting properties. Whilst plasma carotenoids are typically used as the biomarker of adherence to trials aimed at increasing intake of fruits and vegetables, it is obvious from the above that concentrations of many other nutrients and phytochemicals are also being modified. If such a dietary intervention is shown to affect rates of chronic disease in comparison to a usual diet group, then it remains inappropriate to conclude that it is carotenoids *per se* that are having disease-fighting properties. For these reasons, supplementation trials are a stronger design for truly evaluating relationships between carotenoids and chronic disease.

3. Interpreting results of carotenoid supplementation trials

Randomized trials of carotenoid supplements have been done with the goal of clearly identifying causal relationships between carotenoids and disease. As noted earlier, β-carotene is by far the most widely studied carotenoid in supplementation trials. Despite the rigour of this experimental design, results must also be interpreted cautiously. This is because, typically, only one dose level of carotenoid can be evaluated within a trial, and results obtained with this may not predict what may happen at a different dose level. As an example, two lung cancer prevention trials that used high-dose supplements of β-carotene (at least 20 mg β-carotene/day) unexpectedly indicated adverse effects on lung cancer risk [29,30]. In the setting of lung cancer prevention, other trials with either lower doses [31,32] or preparations of high-dose β-carotene with lower bioavailability [4] have not revealed this adverse effect. In addition to dose, lifestyle characteristics of the population under study may affect disease prevention efficacy. Thus a combination antioxidant supplement including β-carotene may well have a beneficial effect against cancer in a poorly nourished population from rural China [31], but not in a better nourished French population [32]. Tobacco use [33,34] and alcohol consumption [33] are other factors that may substantially modify the efficacy of carotenoids in disease prevention [35]. These considerations suggest that single trials are inadequate to test associations between carotenoids and disease, and that multiple trials that use different doses in differing populations are needed to allow better understanding and prediction of health effects of carotenoids in diverse populations.

The practicality of conducting multiple trials of carotenoid supplements in diverse populations is questionable, however, due to limited resources and concerns about additional adverse effects. For example, whilst β-carotene has been shown to interact with concurrent tobacco exposure to produce adverse effects, it is possible that adverse interactions with tobacco could extend to other carotenoids such as lycopene, lutein and zeaxanthin. Just as β-carotene was found to exacerbate harmful effects in the lungs of heavy smokers (who are under significant oxidative stress), the same could hold true in the retinas of patients with early macular degeneration (wherein the macula is under significant oxidative stress) with higher-dose supplementation of the carotenoids lutein and zeaxanthin. For these reasons, carotenoid supplementation trials must proceed with great caution and with appropriate Data and Safety Monitoring Committees to monitor the progress of the trial overall.

As any intervention has possible harms and benefits, study designs that maximize the information obtained but expose relatively fewer study subjects to potential harm are an attractive option. For example, considering cancer prevention trials, an attractive study design to gain evidence of efficacy is to limit the trial to persons who have previously had a cancer of interest, and then aim to prevent second cancers, for which they are often at higher risk [36]. This design has been used in carotenoid supplementation trials, for example to evaluate efficacy in the prevention of second cancers of the head and neck [3], or skin [37]. Likewise, trials of carotenoid supplementation for efficacy in prevention of age-related macular degeneration have enrolled patients who already have early stages of the disease, with the goal of slowing disease progression [38]. The benefit of this type of design is that efficacy (at least in disease progression) can be evaluated while exposing fewer study subjects to unknown potential harms associated with higher-dose supplementation.

While carotenoid supplementation trials continue, the limited but disappointing results to date suggest a cautious approach to conducting such trials in the first place, and a careful interpretation, as results (both beneficial and harmful) obtained in one population may not predict the experience in diverse populations.

4. Interpreting results of trials with intermediate endpoints

Trials that are designed to study modulation of a chronic disease endpoint such as cancer incidence or development of macular degeneration are necessarily very large trials, with typically thousands of participants enrolled. As noted above, sample size requirements can be reduced to some extent by choosing populations at very high risk, such as patients with prior cancers. An alternative design is to conduct trials that use intermediate endpoints as a 'signal' of possible preventive or therapeutic efficacy. For example, lycopene has not yet been evaluated in a trial aimed at prostate cancer prevention, but some preliminary evidence of efficacy comes from trials that use various biomarkers of risk for prostate cancer prevention

(*Chapter 13*). These types of biomarker trials are a logical step to take before launching larger disease-prevention trials. However, it is critical that biomarker trials are not used as the final arbiter of efficacy, as it is still not clear that modulation of a biomarker of interest, *e.g.* decreased prostate-specific antigen levels following carotenoid supplementation, is predictive of a decreased risk of prostate cancer.

In some cases, intermediate endpoints may be shown to predict preventive efficacy for carotenoids, but may be based on too few subjects to reveal the full risk-benefit ratio associated with carotenoid supplementation. As an example, supplementation with β-carotene was shown to regress oral precancerous lesions in several trials of patients with such lesions [39], and a subsequent efficacy trial also observed a 31% (non-significant) decreased risk of oral/pharynx/larynx cancers in β-carotene-supplemented subjects [3]. However, the trials aimed at oral precancerous lesions failed to identify the increase in lung cancer risk that was identified in the efficacy trial that had a larger sample size. So, biomarker trials are a useful, but incomplete, approach to evaluating preventive efficacy for carotenoids and other agents.

D. Future Directions

The study of health effects of carotenoids in humans has proven to be a difficult area of research. It is challenging to separate effects of carotenoids from those of the plant food sources in which the carotenoids are concentrated. There are some possible approaches for epidemiologists to take to make progress in this area, to help to understand, from observational studies, whether observed risk reductions are likely to be specific to carotenoids, or to carotenoid-rich foods. Intervention trials, including biomarker trials, trials in high-risk populations, and experimental animal studies, provide additional information about whether observed risk-reducing effects of carotenoids are real, are specific to carotenoids, and are biologically plausible. Thus, it is all of the study designs together that contribute to the totality of evidence concerning health effects of carotenoids in humans. Despite all our best efforts, we must realize that clinical intervention trials are only undertaken when there is a state of equipoise; that is, sufficient evidence to warrant further evaluation of health effects of carotenoids, balanced against sufficient scepticism or possible concern about the use of carotenoids as an interventional approach. So, sometimes we will be right and the intervention will be found to be beneficial, and sometimes we will be proven wrong and the intervention will have no effect or even be proven harmful.

In order to increase the likelihood that carotenoid health effects are successfully identified, the following approach to research development is suggested.

(i) Perform careful observational epidemiological research with measurements of both dietary intake and objective biomarkers of carotenoid status in diverse populations to distinguish carotenoid effects from those of fruit and vegetables.

(ii) Complement these with animal studies, using appropriate animal models of disease, of food extracts *versus* single carotenoids, as has been done, for example, with lycopene, tomato powder, and prostate cancer [40], with the goal of clarifying efficacy and mechanisms of action.

(iii) If evidence continues to support carotenoid-specific effects, embark on human intervention trials cautiously, with intermediate endpoint trials in relevant populations of interest (considering smoking, alcohol drinking, and baseline nutritional status of the population), using more than one dose if possible to examine dose-dependency of effects.

(iv) Embark on secondary prevention trials/therapeutic trials in populations at risk, to identify efficacy and possible adverse effects.

(v) Conduct primary prevention trials in more general populations as the final step in the research process.

Some of these steps could be concurrent (especially steps i-iii) in order to keep research moving forward on several fronts simultaneously, but the key to this more cautious and necessarily more time-consuming approach to elucidate health effects of carotenoids depends on a clearer understanding of carotenoid actions before large, primary prevention trials in human populations are launched.

References

[1] National Academy of Sciences, Institute of Medicine, Food and Nutrition Board, Panel on Micronutrients, *Dietary Reference Intakes for Vitamin A, Vitamin K, Arsenic, Boron, Chromium, Copper, Iodine, Iron, Manganese, Molybdenum, Nickel, Silicon, Vanadium, and Zinc*, National Academy Press, Washington, D.C. (2001).

[2] National Academy of Sciences, Institute of Medicine, Food and Nutrition Board, Panel on Dietary Antioxidants and Related Compounds, *Dietary Reference Intakes for Vitamin C, Vitamin E, Selenium, and Carotenoids*, National Academy Press, Washington, D.C. (2000).

[3] S. T. Mayne, B. Cartmel, M. Baum, G. Shor-Posner, B. G. Fallon, K. Briskin, J. Bean, T. Zheng, D. Cooper, C. Friedman and W. J. Goodwin Jr., *Cancer Res*., **61**, 1457 (2001).

[4] C. H. Hennekens, J. E. Buring, J. E. Manson, M. Stampfer, B. Rosner, N. R. Cook, C. Belanger, F. LaMotte, J. M. Gaziano, P. M. Ridker, W. Willett and R. Peto, *N. Engl. J. Med*., **334**, 1145 (1996).

[5] C. L. Rock, S. W. Flatt, F. A. Wright, S. Faerber, V. Newman, S. Kealey and J. P. Pierce, *Cancer Epidemiol. Biomarkers Prev.*, **6**, 617 (1997).

[6] E. Lanza, A. Schatzkin, C. Daston, D. Corle, L. Freedman, R. Ballard-Barbash, B. Caan, P. Lance, J. Marshall, F. Iber, M. Shike, J. Weissfeld, M. Slattery, E. Paskett, D. Mateski and P. Albert, *Am. J. Clin. Nutr*., **74**, 387 (2001).

[7] L. Chen, M. Stacewicz-Sapuntzakis, C. Duncan, R. Sharifi, L. Ghosh, R. van Breemen, D. Ashton and P. E. Bowen, *J. Natl. Cancer Inst*., **93**, 1872 (2001).

[8] C. S. You, R. S. Parker and J. E. Swanson, *Asia Pac. J. Clin. Nutr*., **11 Suppl**, S438 (2002).

[9] N. M. Zagre, F. Delpeuch, P. Traissac and H. Delisle, *Public Health Nutr*., **6**, 733 (2003).

[10] S. T. Mayne, B. Cartmel, F. Silva, C. S. Kim, B. G. Fallon, K. Briskin, T. Zheng, M. Baum, G. Shor-Posner and W. J. Goodwin Jr., *J. Nutr*., **129**, 849 (1999).

[11] http://www.nal.usda.gov/fnic/foodcomp/Data/SR18/nutrlist/sr18w338.pdf

[12] A. R. Mangels, J. M. Holden, G. R. Beecher, M. R. Forman and E. Lanza, *J. Am. Diet. Assoc.*, **93**, 284 (1993).

[13] http://www.ars.usda.gov/Services/docs.htm?docid=9673

[14] A. R. Kristal, U. Peters and J. D. Potter, *Cancer Epidemiol. Biomarkers Prev.*, **14**, 2826 (2005).

[15] L. F. Andersen, M. B. Veierod, L. Johansson, A. Sakhi, K. Solvoll and C. A. Drevon, *Br. J. Nutr.*, **93**, 519 (2005).

[16] E. Riboli, *J. Nutr.*, **131**, 170S (2001).

[17] A. F. Subar, V. Kipnis, R. P. Troiano, D. Midthune, D. A. Schoeller, S. Bingham, C. O. Sharbaugh, J. Trabulsi, S. Runswick, R. Ballard-Barbash, J. Sunshine and A. Schatzkin, *Am. J. Epidemiol.*, **158**, 1 (2003).

[18] A. F. Kardinaal, P. Van't Veer, H. A. Brants, H. van den Berg, J. van Schoonhoven and R. J. Hermus, *Am. J. Epidemiol.*, **141**, 440 (1995).

[19] K. J. Yeum, S. H. Ahn, S. A. Rupp de Paiva, Y. C. Lee-Kim, N. I. Krinsky and R. M. Russell, *J. Nutr.*, **128**, 1920 (1998).

[20] T. R. Hata, T. A. Scholz, I. V. Ermakov, R. W. McClane, F. Khachik, W. Gellermann and L. K. Pershing, *J. Invest. Dermatol.*, **115**, 441 (2000).

[21] S. T. Mayne, B. Cartmel, S. Scarmo, H. Lin, D. J. Lefell, I. Ermakov, P. Bhosale, P. S. Bernstein and W. Gellermann, *Abstr. 15th Int. Symp. Carotenoids, Okinawa, 2008, Carotenoid Sci.*, **12**, 54 (2008).

[22] W.. Stahl, U. Heinrich, H. Jungmann, J. von Laar, M. Schietzel, H. Sies and H. Tronnier, *J. Nutr.*, **128**, 903 (1998).

[23] M. A. Eastwood, *QJM: An International Journal of Medicine*, **92**, 527 (1999).

[24] M. E. Wright, S. T. Mayne, R. Z. Stolzenberg-Solomon, Z. Li, P. Pietinen, P. R. Taylor, J. Virtamo and D. Albanes, *Am. J. Epidemiol.*, **160**, 68 (2004).

[25] M. E. Wright, S. T. Mayne, C. A. Swanson, R. Sinha and M. C. Alavanja, *Cancer Causes Control*, **14**, 85 (2003).

[26] P. Knekt, R. Jarvinen, L. Teppo, A. Aromaa and R. Seppanen, *J. Natl. Cancer Inst.*, **91**, 182 (1999).

[27] L. Le Marchand, J. H. Hankin, L. N. Kolonel, G. R. Beecher, L. R. Wilkens and L. P. Zhao, *Cancer Epidemiol. Biomarkers Prev.*, **2**, 183 (1993).

[28] J. Dallongeville, N. Marecaux, J. C. Fruchart and P. Amouyel, *J. Nutr.*, **128**, 1450 (1998).

[29] G. S. Omenn, G. E. Goodman, M. D. Thornquist, J. Balmes, M. R. Cullen, A. Glass, J. P. Keogh, F. L. Meyskens, B. Valanis, J. H. Williams, S. Barnhart and S. Hammar, *N. Engl. J. Med.*, **334**, 1150 (1996).

[30] The Alpha-Tocopherol, Beta-Carotene Cancer Prevention Study Group, *N. Engl. J. Med.*, **330**, 1029 (1994).

[31] W. J. Blot, J. Y. Li, P. R. Taylor, W. Guo, S. Dawsey, G. Q. Wang, C. S. Yang, S. F. Zheng, M. Gail, G. Y. Li, Y. Yu, B. Q. Liu, J. Tangrea, Y. H. Sun, F. Liu, J. F. Fraumeni Jr., Y. H. Zhang and B. Li, *J. Natl. Cancer Inst.*, **85**, 1483 (1993).

[32] S. Hercberg, P. Galan, P. Preziosi, S. Bertrais, L. Mennen, D. Malvy, A. M. Roussel, A. Favier and S. Briancon, *Arch. Intern. Med.*, **164**, 2335 (2004).

[33] J. A. Baron, B. F. Cole, L. Mott, R. Haile, M. Grau, T. R. Church, G. J. Beck and E. R. Greenberg, *J. Natl. Cancer Inst.*, **95**, 717 (2003).

[34] S. T. Mayne and S. M. Lippman, *J. Natl. Cancer Inst.*, **97**, 1319 (2005).

[35] D. Albanes and M. Wright, in *Carotenoids in Health and Disease* (ed. N. Krinsky, S. Mayne and H. Sies), p. 531, Marcel Dekker, New York (2004).

[36] S. Mayne and B. Cartmel, *Cancer Epidemiol. Biomarkers Prev.*, **15**, 2033 (2006).

[37] E. R. Greenberg, J. A. Baron, T. A. Stukel, M. M. Stevens, J. S. Mandel, S. K. Spencer, P. M. Elias, N. Lowe, D. W. Nierenberg, G. Bayrd, J. C. Vance, D. H. Freeman Jr., W. E. Clendenning, T. Kwan and the Skin Cancer Prevention Study Group, *N. Engl. J. Med.*, **323**, 789 (1990).

[38] AREDS Report number 8, *Arch. Ophthalmol.*, **119**, 1417 (2001).

[39] S. Mayne, B. Cartmel and D. Morse, in *Head and Neck Cancer: Emerging Perspectives* (ed. J. F. Ensley, J. S. Gutkind, J. R. Jacobs and S. M. Lippman), p. 261, Academic Press, New York (2003).

[40] T. W. Boileau, Z. Liao, S. Kim, S. Lemeshow, J. W. Erdman Jr. and S. K. Clinton, *J. Natl. Cancer Inst.*, **95**, 1578 (2003).

Carotenoids
Volume 5: Nutrition and Health

Chapter 11

Modulation of Intracellular Signalling Pathways by Carotenoids

Paola Palozza, Simona Serini, Maria Ameruso and Sara Verdecchia

A. Introduction

Cancer and cardiovascular disease are the most common causes of death in developed countries. A high dietary intake of fruits and vegetables has been associated with a decreased risk of developing such chronic diseases [1,2]. It has been suggested that carotenoids may play a key role in these beneficial effects of fruit and vegetable consumption. Some observational epidemiological studies (*Chapters 10, 13* and *14*) have indicated that carotenoids may act as protective agents against some lung cancers and a variety of other chronic diseases [3]. These epidemiological data have been supported by several studies performed *in vivo* and *in vitro* [4-6] in which inhibitory effects of carotenoids on tumour growth and cardiovascular diseases have been observed. Some intervention trials, though, in which β-carotene (**3**) was administered as a supplement to individuals at high risk, such as smokers and asbestos workers, have shown no preventive effects or indeed have indicated enhanced incidence of lung cancer [7,8]. In others, however, supplementation with β-carotene was reported to exert beneficial effects [9].

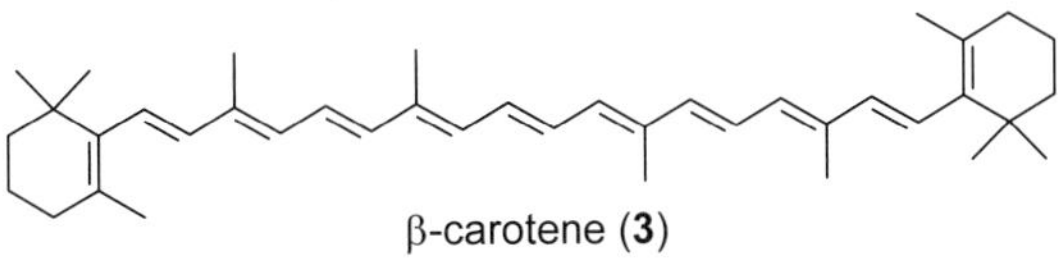

β-carotene (**3**)

The controversial results from these human intervention trials have been discussed widely, along with evidence from other experimental approaches, especially studies of effects of carotenoids on cellular and molecular processes in cultured normal and cancer cells. A feature of cancer is that the normal process of cell division and differentiation is impaired and instead of differentiating into normal functional cells, uncontrolled unfunctional cells proliferate. Differentiation is normally controlled by fundamental processes, namely the cell cycle and apoptosis (programmed cell death), which are regulated by molecular signals. It is well known that the β-carotene metabolite retinoic acid (*1*) is an essential regulatory factor in growth and development. There is now also increasing evidence from studies with cell cultures that intact carotenoids or other metabolites/breakdown products can influence and modulate essential regulatory signalling processes, including the cell cycle and apoptosis [10].

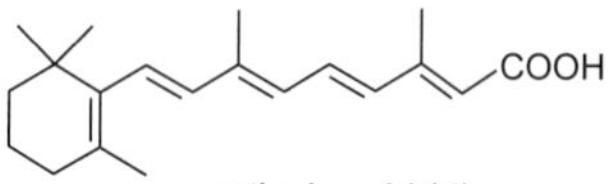

retinoic acid (*1*)

This evidence, especially in relation to the ability of β-carotene and other carotenoids to modulate the expression of proteins and transcription systems involved in cancer cell proliferation, inflammation and atherosclerosis, is discussed in Section **C** of this *Chapter*. This *Chapter* should be considered together with *Chapter 18*, which addresses the question of whether the effects seen are due to metabolites or breakdown products rather than to the intact carotenoids themselves, and with Chapter 17, which describes the specialized effects of carotenoids on the immune response system.

To appreciate fully the possible significance of effects of carotenoids on intercellular communication and signalling and on the cell cycle and apoptosis, the non-specialist reader is encouraged to consult the treatment of these topics in any modern biology or biochemistry textbook. Some of the main features are outlined in the following section (**B**).

B. Intercellular Communication and Signalling

Cell-to-cell communication is essential for all multicellular organisms for growth and development, differentiation and specialization of cells and tissues, maintenance of cell and organ function, regulation of metabolism, and response to external/environmental signals. The billions of cells in any individual human must communicate to coordinate their activities. Any disruption to this communication is likely to have serious consequences.

1. Cell signalling pathways and mechanisms

Chemical signals, *e.g.* hormones and growth factors, are paramount in regulating many cellular processes. The chemical signal must be recognized and acted upon by the cell. This requires receptor proteins with specific binding sites to recognize the signal, and signal transduction mechanisms to translate the signal into an effect. It is common for receptors to be located on the outside of the cell membrane. When the ligand binds, this activates a membrane-bound signal transduction mechanism consisting of several interacting proteins, leading to the release of a second messenger substance which conveys the signal within the cell. In other cases, the signal compound itself can cross the membrane, enter the cell and bind to a receptor in the cytosol. The receptor-ligand complex goes to the nucleus where it binds to a specific response element on a gene, thereby activating the gene, leading to the synthesis of the protein gene product. The regulation is very complex and is mediated by many factors, generally proteins, acting as, for example, transcription factors, that regulate the transcription of the gene DNA region into RNA which directs the synthesis of the required protein. They thus control which genes are active in a cell at a particular time.

2. Gap junction communication

Distinct from this is the direct communication between adjacent cells *via* gap junctions. Gap junctions are composed of structures called connexons in which six proteins (connexins) are arranged in a circle to create a channel through the plasma membranes of two adjacent cells and the small space between them. The channel is too small to permit passage of large molecules such as proteins, but it does allow small subtances, including small signalling molecules, to move from cell to cell.

3. The cell cycle and apoptosis

Organs and tissues are maintained and repaired by cell division (mitosis) to produce new, genetically identical, functional cells. Cells that are no longer functional or are in some way defective are destroyed by a built-in process of programmed cell death, known as apoptosis.

a) The cell cycle

The replication of cells is not a simple one-step process; it involves a series of events known as the cell cycle (Fig. 1). Following cell mitosis, the cell enters the first gap or growth phase, G_1, a period of cell growth and active functioning. Long-lived cells may pause in the cycle and remain and function for some time in a resting phase, G_0. At some point, an appropriate signal may cause the cell to leave the G_0/G_1 phase and progress towards mitosis. The G_1 phase is followed by the S phase in which DNA is synthesized in preparation for mitosis. A second gap or growth phase, G_2, then precedes the M (mitosis) phase in which the cells divide.

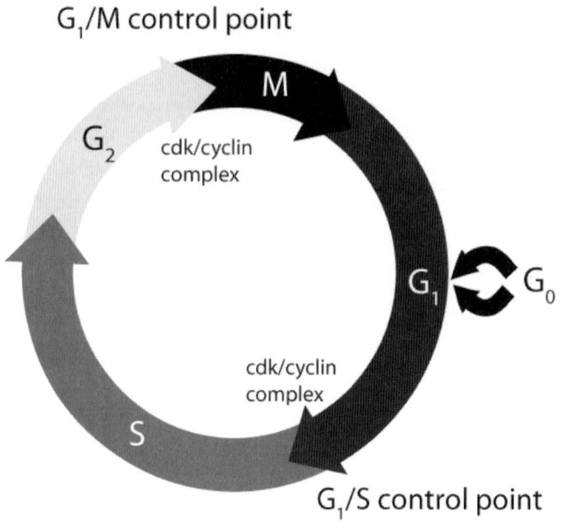

Fig. 1. Simple diagram to illustrate the eukaryotic cell cycle. M designates the mitosis (cell division) phase, G_1 and G_2 the first and second gap or growth phases, and S the synthesis phase in which DNA is replicated. Progression of the cycle is mainly controlled by two checkpoints at which binding of cyclin is required to activate cyclin-dependent kinases (cdk). The activation is induced by growth factors, promoted by some oncogenes, inhibited by tumour suppressors such as Rb and p53, and influenced by nutritional status of the cell.

The cycle must be controlled rigorously to prevent undesired replication of defective cells or proliferation of transformed, non-functional cells as tumours. The progression of the cycle is controlled by irreversible checkpoints. Progress through the checkpoints is determined by molecular signals. If the correct signals are received, then the cycle passes the checkpoint. If the correct signals have not been received, the cycle is halted. The primary checkpoint ($G_0/G_1/S$) determines whether the cell progresses from G_0 into G_1 or G_1 into the S phase which marks the beginning of replication. There are further checkpoints at G_2/M and in the late M phase, but $G_0/G_1/S$ is the most important. The changes require the phosphorylation of proteins, catalysed by cyclin-dependent kinase (cdk) enzymes which are activated by binding regulatory proteins (cyclins). External signals, especially growth factors, received by receptors on the cell, lead to the activation of the cdks so that the cycle can proceed. Growth factors are proteins that are released by certain cells and stimulate other cells to divide. This checkpoint is also influenced by the size and nutritional state of the cell. The activation is dependent on other proteins, especially p53, which can detect damaged or mutated DNA and block division of the defective cells at this checkpoint by preventing binding of cyclins to cdk.

After successful repair of the DNA the blocking action of p53 is removed and the cycle is allowed to proceed. Other proteins (tumour-suppressors) such as Rb and p21 are also involved in this mechanism, which prevents uncontrolled replication (proliferation) that could lead to cancer. Mutation or defects in the control systems have serious consequences; defective p53 is a common feature of many cancers. Also some oncogenes, *e.g. ras*, promote passage through this checkpoint. Any disturbance of the balance, *e.g.* by decreasing the effectiveness of tumour suppressors or increasing the activity of response to growth factors or other signals reduces the efficiency of control of the cell cycle, leading to cell proliferation.

b) Apoptosis

There is a built-in mechanism whereby cells that are no longer functional or are recognized as defective are destroyed and removed from the system. This suicidal programmed cell death, apoptosis, consists of a controlled chain of events which leads to destruction of the cell and recycling of cellular components. Particularly important is the induction of a cascade of caspase enzymes which degrade cellular proteins. Apoptosis is controlled by a number of protein signalling molecules. Some, such as bax and p53, promote apoptosis. Others, such as bcl-2, block it. If the production or activity of pro-apototic signals is impaired, or if anti-apoptotic signals are overproduced or overactive, apoptosis is checked, cells survive and proliferate in a non-functional form, leading to tumour formation.

4. Reactive oxygen species as second messengers

Reactive oxygen species (ROS) have been reported to play a major physiological role in several aspects of intracellular signalling and regulation [11]. It has been demonstrated clearly that ROS interfere with the expression of a number of genes and signal transduction pathways [12]. These ROS influence the redox status of the cell, *i.e.* the balance between oxidative and reductive conditions and processes, and may, according to their concentration, cause either a positive response (cell proliferation) or a negative response (growth arrest or cell death). High concentrations of ROS cause cell death or even tissue necrosis, whereas ROS can promote cell proliferation only at low or transient concentrations of radicals. Low concentrations of superoxide radical and hydrogen peroxide in fact stimulate proliferation and enhance survival in a variety of cell types. The ROS can thus play a very important physiological role as secondary messengers [13]. Other examples of this include regulation of the cytosolic calcium concentration (which itself regulates the above-mentioned biological activities), regulation of protein phosphorylation, and activation of transcription factors such as the nuclear transcription factors NF-κB and the AP-1 family [14].

5. Carotenoids as redox agents

There is evidence to suggest that carotenoids can act as modulators of intracellular redox status. Their ability to function as antioxidants has been known for many years. The conjugated double-bond structure is primarily responsible for the ability of β-carotene to quench singlet oxygen physically without degradation, and for the chemical reactivity of β-carotene with free radicals such as the peroxyl, hydroxyl, and superoxide radicals. Carotenoids have been shown to be able to prevent or decrease oxidative damage to DNA, lipid and protein [4-6].

Carotenoids may also act as pro-oxidants, and increase the total radical yield in a system [15]. The key factors that determine the switch of carotenoids from antioxidant to pro-oxidant are the oxygen partial pressure (pO_2) and the carotenoid concentration [15-18]. At higher pO_2 a carotenoid radical can react with molecular oxygen to generate a carotenoid-peroxyl radical [19] which can act as a pro-oxidant by promoting oxidation of unsaturated lipids. So, although work, mostly with β-carotene, has shown that carotenoids can exhibit antioxidant behaviour at low oxygen partial pressures, usually below 150 Torr, they may lose antioxidant properties, or even become pro-oxidants, at high pressures of oxygen; at high carotenoid concentrations there is also a propensity for pro-oxidant behaviour.

These properties, discussed in detail in *Chapter 12*, are crucial to the ability of carotenoids to influence intracellular redox status and those molecular processes that are regulated by it.

C. Effects of Carotenoids on Cell Signalling and Communication

1. Modulation of cell cycle

Carotenoids are able to control progression of the cell cycle (see Section **B**.3), but there are only a few studies showing that this control can be exerted through a direct modulation of cell cycle-related proteins.

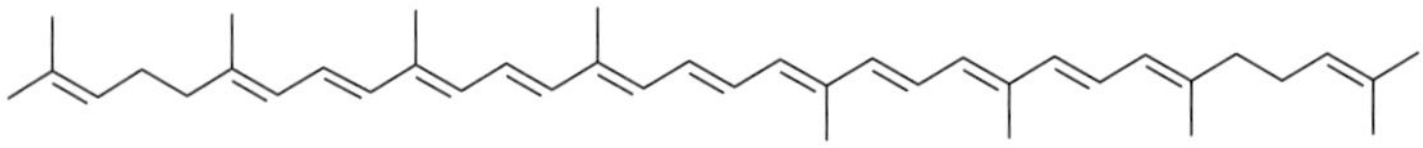

lycopene (**31**)

Growth-inhibitory effects of lycopene (**31**) in both MCF-7 mammary and endometrial cancer cells have been reported [20] to occur through the down-regulation of cyclins D1 and D3. This effect was associated with a reduction in the activity of the cyclin-dependent kinases cdk4 and cdk2 and with the hypophosphorylation of the regulatory protein Rb. Moreover, the down-regulation of cyclin D was accompanied by a retention of protein p27 in the cyclin E-cdk2 complex, resulting in a further inhibition of cdk2 kinase activity [20]. Lycopene also caused inhibition of cell growth through a mechanism involving down-regulation of cyclin D1 but not of cyclin E, at the protein level, and induced an arrest in the cell cycle so that the cells remained in the G_0/G_1 phase. This G_0/G_1 arrest was also observed in lycopene-treated HL-60 cells [21]. On the other hand, in human colon adenocarcinoma cells, β-carotene induced a cell-cycle delay, at the G_2/M checkpoint, by decreasing the expression of cyclin A [22].

It has also been reported that excentric cleavage products of β-carotene can inhibit the growth of oestrogen-receptor positive and negative breast cancer cells, through the down-regulation of cell cycle regulatory proteins such as E2F1 and Rb, and through the inhibition of AP-1 transcriptional activity [23]. In a recent study, LNCaP and PC3 prostate cancer cells treated with lycopene-based agents have been reported to undergo mitotic arrest [24]. It is likely that the reported antiproliferative effects of lycopene were achieved through a block in G_1/S transition mediated by decreased levels of cyclins D1 and E and the kinase cdk4, and suppressed Rb phosphorylation [24]. It was reported recently that, when tomato that had been subjected to a digestion procedure *in vitro* was added to cultured colon (HT-29 and HCT-116) cancer cells, cell cycle progression was arrested at the G_0/G_1 phase [25]. This effect was accompanied by a dose-dependent decrease in the expression of cyclin D1. Tomato digestate contains a complex mixture of compounds besides lycopene, including a large variety of micronutrients and microconstituents such as polyphenols and other non pro-vitamin A carotenoids, so the effects cannot definitely be attributed to lycopene. The observation may, however, support the notion that lycopene could be an important molecule in the regulation of intracellular levels of cyclin D.

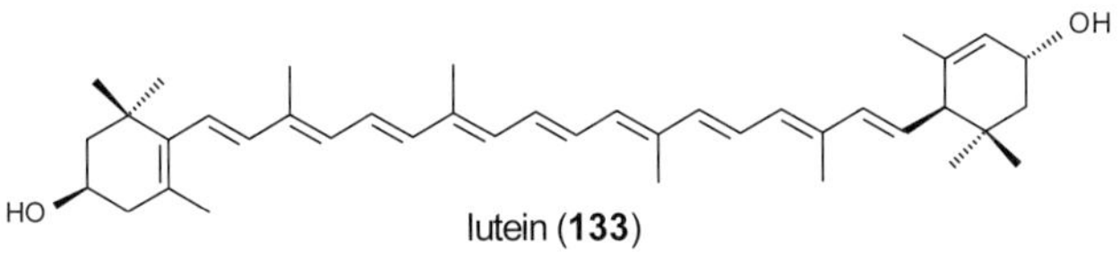

lutein (**133**)

In a recent study of the antiproliferative effect of β-carotene, lycopene and lutein (**133**), the carotenoids suppressed cell growth of human KB cells to different extents by acting as inhibitors of the expression of proliferating cell nuclear antigen (PCNA) and cyclin D1 [26].

2. Modulation of apoptosis

It has been demonstrated that carotenoids are able to induce apoptosis in several cultured cell lines [27], but the mechanisms for this are still under investigation. One possibility is that carotenoids may change the expression of apoptosis-related proteins, including the Bcl-2 family proteins and the caspase proteins. Such effects have been observed both *in vitro* and *in vivo*. Microarray analysis has shown that β-carotene can increase the expression of the pro-apoptotic protein Bax in U-937 cells and HUVEC cells [28,29]. This result was also confirmed by real-time Q-PCR analysis, and supported by flow cytometry apoptosis tests

[28,29]. Inhibition of mouse mammary tumour growth by dietary lutein was attributed to induction of increased expression of Bax and decreased expression of the anti-apoptotic Bcl-2 [30]. It has also been shown that lycopene can induce apoptosis of PC-3 cells, by down-regulating the expression of cyclin D1 and Bcl-2 and up-regulating that of Bax [31]. Experiments performed with auto-oxidative cleavage products of lycopene also demonstrated apoptotic effects [32]. (*E,E,E*)-4-Methyl-8-oxo-nona-2,4,6-trienal (MON, *2*), derived by oxidative cleavage of the C(5,6) and C(13,14) double bonds of lycopene, was shown to induce an enhancement of caspase-8 and caspase-9 activities and a down-regulation of Bcl-2 and Bcl-XL in HL-60 cells [32].

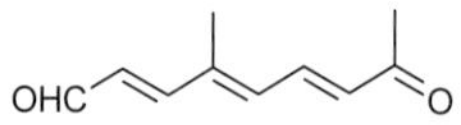

MON (*2*)

3. Modulation of the cell cycle and apoptosis *via* redox-sensitive proteins

It has been suggested that beneficial or harmful effects of carotenoids in relation to cancer as well as other chronic diseases, may occur by modulation of the expression of redox-sensitive regulatory proteins such as p53 and p21WAF1 [33,34], and that the determining factor may be a pro-oxidant rather than a protective antioxidant role. In particular, β-carotene was able to induce a remarkable increase in ROS production in HL-60 leukaemia cells, accompanied by an enhanced expression of p21WAF1 and by a concomitant arrest of cell cycle progression at the G_0/G_1 phase [35]. Moreover, treatment of various cultured cells, including RAT-1 immortalized fibroblasts, Mv1Lu lung, MCF-7 mammary, Hep-2 larynx and LS-174 colon cancer cells, with a combination of β-carotene and cigarette smoke condensate (TAR) induced an increase in the levels of 8-hydroxydeoxyguanosine, which is a well known marker of oxidative DNA damage, and is associated with mutagenesis and carcinogenesis [36]. In these cells, DNA damage was also accompanied by an increased proportion of proliferating cells, due to a de-regulation of p53 expression which, in turn, affected the levels of p21WAF1 and cyclin D1 [36].

It has been demonstrated in different studies that carotenoids may modulate the expression of apoptosis-related proteins by a redox mechanism. Free-radical species, such as singlet oxygen [37] and nitric oxide [38], have been reported to activate caspase-8, an important protein-degrading enzyme involved in the apoptotic cascade. In agreement with this, β-carotene was able to induce caspase-3 activity in several cancer cell lines, mainly by interacting with a signal complex that is located on the cell membrane, and induces caspase-8 activation [39], but also, within the cytoplasm, through a non-receptor signalling pathway, which induces caspase-9 activation, followed by the release of the truncated form of the protein Bid. The latter was then translocated to the mitochondria where it acted as a potent inducer of apoptosis, *via* release of cytochrome c and activation of caspase-9 [39].

Mitochondria are now well established as being critical for processing and integrating pro-apoptotic signals. Diverse apoptotic stimuli can cause mitochondrial dysfunction, leading to pro-oxidative changes in redox homeostasis. The involvement of mitochondria in pro-apoptotic effects of β-carotene has been demonstrated clearly; the carotenoid was able to induce the release of cytochrome c from mitochondria and to alter mitochondrial membrane potential (Δψm) in human leukaemia, colon adenocarcinoma and melanoma cell lines [39]. Carotenoids have been found also to affect mitochondrial functions through an alteration of mitochondrial transmembrane potential, as recently observed in LNCaP human prostate cancer cells treated with lycopene [40]. Treatment with the polar xanthophyll neoxanthin (**234**) has been reported to induce apoptosis in colon cancer cells by a mechanism which involves accumulation of the neoxanthin into the mitochondria and a consequent loss of mitochondrial transmembrane potential and release of cytochrome c and apoptosis-inducing factor (AIF) [41]. Neoxanthin has never been detected in human blood or tissues, however, and is unlikely to be present in cells *in vivo*.

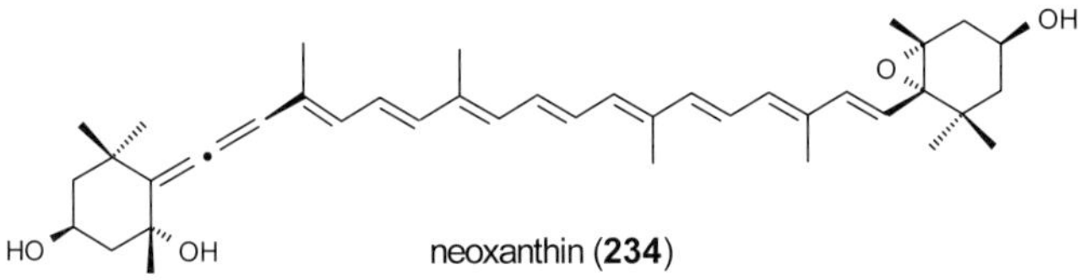

neoxanthin (**234**)

It was reported previously that β-carotene was able to decrease the expression of the anti-apoptotic protein Bcl-2 and that the decrease in levels of this protein was accompanied by an increase in ROS production and by induction of apoptosis. This is particularly interesting in the light of the data that support a role for Bcl-2 in an antioxidant pathway, because this protein prevents programmed cell death by decreasing the formation of ROS and lipid peroxidation products [42].

It has been reported that carotenoids are able to modulate the expression of the heat shock proteins hsp70 and hsp90, which are nuclear binding proteins involved in both oxidative stress and apoptosis. They are produced in response to stress and act to provide defence against stress. In particular, in cervical dysplasia-derived cells, both (9*Z*)-β-carotene and (all-*E*)-β-carotene have been shown to induce an intracellular accumulation of hsp70, accompanied by morphological changes indicative of apoptosis [43]. On the other hand, lycopene was able to decrease the expression of hsp90 in RAT-1 fibroblasts treated with TAR. The decrease of this protein was accompanied by induction of apoptosis through changes in Bad, a member of the Bcl-2 family of proteins [44].

The cyclo-oxygenase enzyme Cox-2 is the rate-limiting enzyme in prostaglandin production from arachidonic acid, and ROS are generated as a side product of this reaction. It has been suggested that Cox-2 may function as an anti-apoptotic protein and that it is modulated by oxidative stress [45]. β-Carotene was able to down-regulate the expression of Cox-2 in colon cancer cells and this effect was accompanied by induction of apoptosis [46]. Concomitant with a dose-dependent decrease in the expression of Cox-2, a dose-dependent decrease in ROS production was observed in cells treated with β-carotene. Since the production of ROS by the peroxidase function of Cox-2 may be necessary for cell proliferation, its inhibition by β-carotene, directly or through cyclo-oxygenase inhibition, may represent a potential mechanism to explain the growth-inhibitory effects of β-carotene in this cell model. These findings suggest, therefore, that two distinct redox-sensitive mechanisms may be implicated in the pro-apoptotic effects of the carotenoid in colon cancer cells. The first, involving an increase in ROS production, occurs at high β-carotene concentration; the second, involving the modulation of Cox-2 expression, occurs at low carotenoid concentration.

4. Modulation of growth factors

The signal for cells to divide is transmitted by growth factors that are delivered in the bloodstream and recognize and bind to receptors on the cell surface. Carotenoids may modulate the expression of growth factors and growth-factor receptors. Lycopene has been reported to decrease the expression of the insulin-like growth factor IGF-1 in lungs of ferrets exposed to cigarette smoke [47]. Moreover, lycopene induced pro-apoptotic effects, through a decreased phosphorylation of Bad. In the same model, lycopene also increased the levels of insulin-like growth factor binding protein-3 (IGFBP-3), which is reported to act as a potent inhibitor of both AKT and mitogen-activated protein kinase (MAPK) signalling pathways [47]. In contrast, it has been suggested [48] that β-carotene may prevent cervical carcinogenesis through an induction of apoptosis mediated by the down-regulation of epidermal growth factor (EGF) receptor in pre-malignant cervical dysplastic cells. A sustained expression of the EGF receptor has been suggested to play a key role in the development of carcinogenesis [49]. It is noteworthy that the non-provitamin A carotenoid astaxanthin (**404-6**) has been reported to be as active as β-carotene in down-regulating EGF binding, suggesting that such a mechanism is independent of the conversion to retinoids [48].

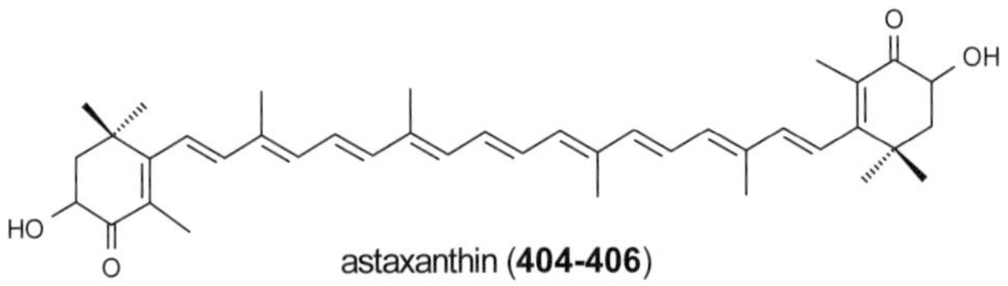

astaxanthin (**404-406**)

Elevated serum concentrations of IGF-1 are associated with an increased risk for cancer, including breast, prostate, colorectal and lung cancers [50,51]. Moreover, IGF-1 up-regulation has been directly implicated in the progression of prostate cancer [52]. It has been reported that lycopene caused a strong reduction in the IGF-1-stimulated growth of MCF-7 breast cancer cells and that this inhibition was associated with an arrest in the G_1-S phase of cell cycle progression [53]. An up-regulation of IGF-binding protein-3 by lycopene was also demonstrated in ferret lungs [47], and supplementation with lycopene decreased the expression of IGF-1 in the MatLyLu Dunning prostate cancer model [54]. In agreement with this, some clinical data show that changes of systemic IGF-1 levels may occur in response to tomato consumption [55]. Consumption of cooked tomatoes was inversely associated with IGF-1 plasma levels [56]. Recently, it has been observed that lycopene decreased IGF-1 expression in normal prostate tissue of young rats [57].

On the other hand, osteoblastic MC3T3-E1 cells treated with β-cryptoxanthin (**55**) showed an increased expression of IGF-1 and the transforming growth factor (TGF)-β1, suggesting that, potentially, the carotenoid may be helpful in the prevention of osteoporosis [58].

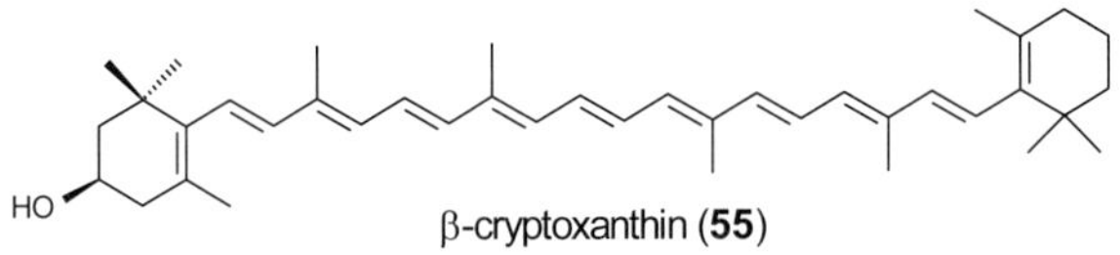

β-cryptoxanthin (**55**)

It has been suggested by some studies *in vivo* that β-carotene and canthaxanthin (**380**) may increase vascular growth and levels of TGFα [59]. In contrast, treatment with β-carotene was found to increase significantly the intracellular levels of TGF-β1, a potent growth inhibitor of epithelial cells, in cervical epithelial cells of patients with cervical intra-epithelial neoplasia [60]. In addition, recent evidence shows that exposure of HUVEC cells to β-carotene can modulate the expression of bFGF and VEGF, two key proteins in endothelial cell maturation and vascular repair after injury [61].

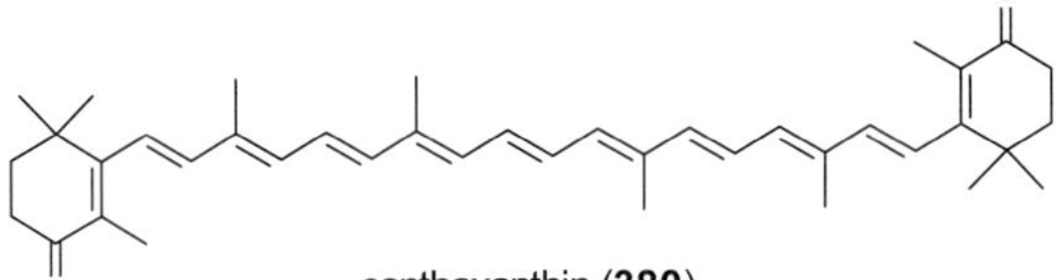

canthaxanthin (**380**)

5. Modulation of cell differentiation

The induction of differentiation may represent an important mechanism for chemoprevention of chronic diseases. Lycopene has been shown to induce differentiation of HL-60 promyelocytic leukaemia cells [21]. Similar effects were also induced by other carotenoids, including β-carotene and lutein [62,63]. The effects of lycopene were associated with an increased expression of several differentiation-related proteins, including cell-surface antigen CD14, oxygen burst oxidase and chemotactic peptide receptors [21]. Moreover, recent studies report the ability of lycopene to stimulate the activity of the differentiation marker alkaline phosphatase in SaOS-2 osteoblasts [64]; this effect depended on the stage of cell differentiation. Although the mechanism by which lycopene affects cell differentiation is not clear, a reasonable hypothesis is that the carotenoid may activate the expression of nuclear hormone and retinoid receptors [65].

6. Modulation of retinoid receptors

Several studies have shown that most, if not all, of the biological activity of retinoic acid is due to its ability to alter gene transcription through changes in nuclear receptors, namely retinoic acid receptors (RARs) and retinoid X receptors (RXRs) (see *Chapter 18*). It has been suggested that carotenoids may regulate several cell functions through the modulation of these transcription factors. In particular, β-carotene has been reported to act as a chemopreventive agent through an up-regulation of retinoid receptors in mouse skin [66]. In lung of ferrets, a combination of smoke and high concentrations of β-carotene has been reported to reduce retinoic acid levels and expression of RARβ, but not of RARα and RARγ [67]. This effect was not observed when the carotenoid was given at low doses [68], which suggests that a pharmacological dose of the carotenoid, in association with smoke, was needed to modify the expression of retinoid receptors. The decreased expression of RARβ was accompanied by increased cell proliferation [67]. The effects of β-carotene on expression of RARβ isoforms were also evaluated in an AJ-mouse model of carcinogenesis induced by the tobacco smoke carcinogen 4-(*N*-methyl-*N*-nitrosamino)-1-(3-pyridyl)-1-butanone (NNK) [69], which reduced the expression of all RAR isoforms. β-Carotene alone, in non-initiated mice, tended to increase expression of RARβ, especially RARβ2 and RARβ4. In the groups initiated with NNK and supplemented with β-carotene, however, the suppressing effect of NNK dominated and β-carotene was not able to restore RARβ expression. In addition, the modulation of retinoic acid-responsive gene expression by NNK and/or β-carotene was not predictive for later tumour development [69].

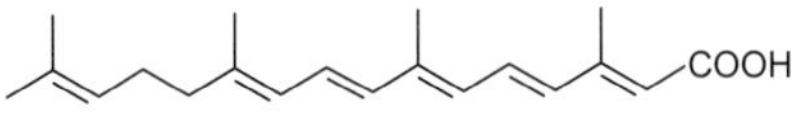

acycloretinoic acid (*3*)

In a similar manner, lycopene metabolites may also act as ligands for a nuclear receptor (see *Chapter 18*). The inhibition of human mammary MCF-7 cancer cell growth and the transactivation of the RAR reporter gene by synthetic acycloretinoic acid (*3*), the acyclic analogue of retinoic acid, was compared to that obtained with lycopene and retinoic acid in the same system [69,70]. The acycloretinoic acid was remarkably less potent than retinoic acid in activating the retinoic acid response element [69]. Lycopene exhibited only very modest activity in this system. In contrast, acycloretinoic acid, retinoic acid, and lycopene each inhibited cell growth with a similar potency, suggesting that the effects of acycloretinoic acid are not entirely mediated by the retinoic acid receptor. Similar results were obtained in other studies [71], which demonstrated that retinoic acid is much more potent than acycloretinoic acid in the transactivation of the retinoic acid responsive promoters of the receptor RAR-β2.

7. Redox-related modulation of transcription factors

a) NF-κB

The nuclear transcription factor NF-κB plays a key role in regulating immune response (see *Chapter 17*). Its chronic expression or activation is associated with inflammation, cancer, *etc.* NF-κB is responsive to oxidative stress [72,73]. Treatment of cells with hydrogen peroxide can activate the NF-κB pathway [74]. The fact that ROS may act as important mediators of NF-κB activation is further supported by the observation that well-known inducers of NF-κB activity, including tumour necrosis factor TNF-α, interleukin IL-1, lipopolysaccharide (LPS), phorbol myristate acetate (PMA), UV and ionizing radiation, are able also to induce an increased production of intracellular ROS.

In tumour cell lines, including HL-60 cells [36] and colon adenocarcinoma cells [75] exposed to β-carotene, a significant increase has been demonstrated in ROS production and/or in glutathione content, accompanied by a sustained elevation of NF-κB and by a significant inhibition of cell growth [75]. α-Tocopherol and N-acetylcysteine could reverse the effects of the carotenoid on cell growth and apoptosis in the cell lines analysed, and were found to prevent the β-carotene-induced increased expression of c-myc. These findings support the hypothesis that a redox regulation of NF-κB is involved in the growth-inhibitory and pro-apoptotic effects of the carotenoid in tumour cells [75]. In contrast, a protective role of β-carotene in cells exposed to oxidative stress has been reported in other studies. In such models, the carotenoid was able to suppress efficiently the activation of NF-κB and the production of pro-inflammatory cytokines such as interleukin IL-6 and TNF-α [76]. In addition, it has been reported recently that lycopene significantly inhibited MMP-9 levels and the binding abilities of NF-κB and Stimulatory protein 1 (Sp1) in the human hepatoma cell line SK-Hep-1, thereby suppressing the invasive ability of these cells [77]. This effect was accompanied by a decrease in the expression of the insulin-like growth factor-1 receptor (IGF-1R) and in the intracellular level of reactive oxygen species [77]. Recent data suggest that carotenoid molecules may represent non-toxic agents for the control of pro-inflammatory genes through a mechanism involving NF-κB. In fact, lycopene prevented macrophage activation induced by gliadin and interferon IF-γ through an inhibition of the activation of NF-κB, IRF-1 (interferon regulatory factor-1) and STAT-1α (signal transducer and activator of transcription-1α) and lowered the levels of both nitric oxide synthase and Cox-2 [78].

b) AP-1

Another well known redox-sensitive transcription factor, AP-1, is implicated in the regulation of cell growth [79]. Recently, it has been reported that β-carotene and its cleavage products are able to inhibit the activation of this transcription factor in mammary tumour cell lines [23]. Moreover, treatment of mammary cancer cells with lycopene inhibited AP-1 binding and reduced induction of the insulin-like growth factor-1 (IGF-1), implying an inhibitory effect of lycopene on mammary cancer cell growth [53]. Some studies *in vivo* also suggest that carotenoids may affect cell growth through their ability to modulate AP-1 activity. Administration of β-carotene to ferrets exposed to tobacco smoke induced increased expression of the AP-1 proteins c-fos and c-jun [67] when given at pharmacological [68], but not at the much lower physiological [68] concentrations. This activation of AP-1 only at high doses of the carotenoid [67] could, at least partially, explain the increased risk of lung cancer among smokers and asbestos workers, as observed in some β-carotene clinical trials [7,8].

c) Nrf2 and phase II enzymes

A group of enzymes known as phase II enzymes provide an important system for detoxifying and combating foreign substances (xenobiotics) including potential carcinogens. These enzymes can conjugate reactive electrophiles and act as indirect antioxidants, and may thus represent potential means of achieving protection against a variety of carcinogens in animals and humans. The expression of such enzymes at the transcriptional level is mediated, at least in part, by the antioxidant response element (ARE) which is found in the regulatory region of their genes. The transcription factor Nrf2, which binds to ARE, appears to be essential for the induction of phase II enzymes such as glutathione S-transferases (GSTs), NAD(P)H quinone oxidoreductase (NQO1) [80], haem oxygenase-1 (HO-1) and the thiol-containing reducing factor, thioredoxin [81]. Several studies have shown that antioxidants present in the diet, such as terpenoids, flavonoids and isothiocyanates, may act as anti-tumour agents by activating this transcription system [82-84]. Carotenoids have been shown to modulate tumour growth by acting as potent inducers of these enzymes [85]. In particular, β-carotene has been shown to modulate the expression of HO-1, either by decreasing it, as observed in cultured FEK4 cells [86] or fibroblasts [87] exposed to UVA, or by increasing it, as observed in another study of human skin fibroblasts enriched with the carotenoid and exposed to UV-light [88]. The pro-oxidant effects of β-carotene were totally suppressed by vitamin E, but only moderately by vitamin C [88]. The modulation of this enzyme may occur through an activation of mitogen-activated protein kinase (MAPK) leading to induction of ARE, as suggested for other dietary chemopreventive compounds [89]. An alternative mechanism to explain the regulation of HO-1 expression by β-carotene has been suggested recently [55]. In this study, with fibroblasts exposed to cigarette smoke condensate, the carotenoid controlled the expression of HO-1 through the induction of Bach1, which is known to act as a HO-1 repressor [90]. It has also been demonstrated that canthaxanthin and astaxanthin, but not lutein or lycopene, were able to induce enzymes of phase II metabolism, namely UDP-glucuronosyl transferase and NQO1, in rats [91]. There is evidence that lycopene may act as an inducer of the activity and/or of the expression of phase II enzymes in healthy animals [92] as well as in animals bearing tumours, including gastric [93] and hamster buccal pouch [94] tumours induced by 7,12-dimethylbenz[a]anthracene (DMBA). Concomitantly to the induction of antioxidant enzymes [93], a reduction of lipid peroxidation was observed following carotenoid treatment [93,94].

The results of these and other studies demonstrate that carotenoids can induce phase II enzymes in various animal systems, but a direct activation of ARE by carotenoids has not been demonstrated. Further studies are necessary.

8. Modulation of hormone action

Prostatic growth and development are controlled by steroid hormones *via* the androgen receptor system. Androgens are also implicated in prostatic neoplasia, including benign prostatic hyperplasia and prostate cancer. 5α-Dihydrotestosterone has been suggested to be the principal androgen responsible for regulating both normal and hyperplastic growth of the prostate gland. This hormone is produced from testosterone by the enzyme steroid 5α-reductase. Some recent evidence shows that lycopene is able to reduce the expression of 5α-reductase I in prostate tumours in the rat MatLyLu Dunning prostate cancer model [54] and, consequently, to down-regulate drastically several androgen target genes, such as those coding for the cystatin-related proteins 1 and 2, prostatic spermine-binding protein, the prostatic steroid-binding protein C1, C2 and C3 chain, and probasin [54]. In addition, lycopene modulated androgen signalling in normal prostatic tissues from young rats [57]. In a recent study to determine whether carotenoids are able to inhibit signalling by steroidal oestrogen and phytoestrogen in breast (T47D and MCF-7) and in endometrial (ECC-1) cancer cells, lycopene, phytoene (**44**) and phytofluene (**42**) have been found to inhibit the oestrogen-induced transactivation of the oestrogen response element ERE that was mediated by the oestrogen receptors ERα and ERβ. These data suggest that these carotenoids may be possible candidates to inhibit the deleterious effect of both 17β-oestradiol and genistein in hormone-dependent mammary and endometrial malignancies [95].

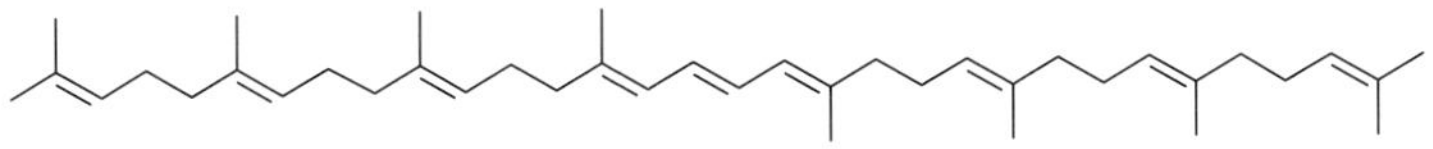

phytoene (**44**)

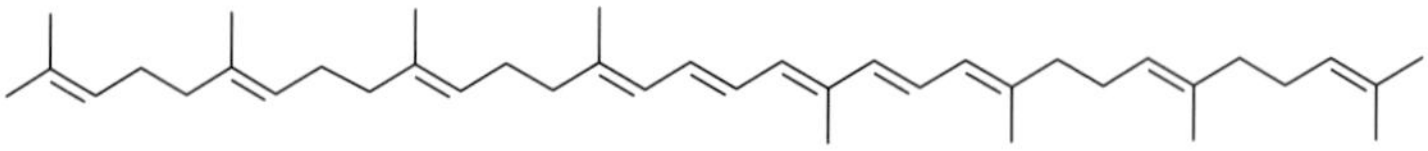

phytofluene (**42**)

9. Modulation of peroxisome-proliferator activated receptors

Peroxisome-proliferator activated receptors (PPARs) are lipid-activated transcription factors that exert several functions in development and metabolism. PPARα is implicated in the regulation of lipid metabolism, lipoprotein synthesis and inflammatory response in liver and other tissues. PPARγ plays an important role in the regulation of proliferation and differentiation in several cell types. The physiological role of PPARδ is still under debate; treatment of obese animals by specific PPARδ agonists resulted in restoration of metabolic parameters to normal and reduction of adiposity. The presence of PPARγ receptors in various cancer cells and their activation by fatty acids, prostaglandins and related hydrophobic agents make these ligand-dependent transcription factors an interesting target for carotenoid derivatives. Moreover, in the nucleus, PPARγ is always found as a dimer with RXR. A recent study [96] of the efficacy of several carotenoids in transactivation of PPAR response element (PPARE) indicated that lycopene, phytoene, phytofluene and β-carotene are able to transactivate PPARE in MCF-7 cells co-transfected with PPARγ. Recently, it has been reported that fucoxanthin (**369**) enhanced the anti-proliferative effect of a PPARγ ligand, troglitazone, in CaCo-2 colon cancer cells [97]. Moreover, fucoxanthin and fucoxanthinol (**368**) inhibited the adipocyte differentiation of 3T3-L1 cells through down-regulation of PPARγ [98].

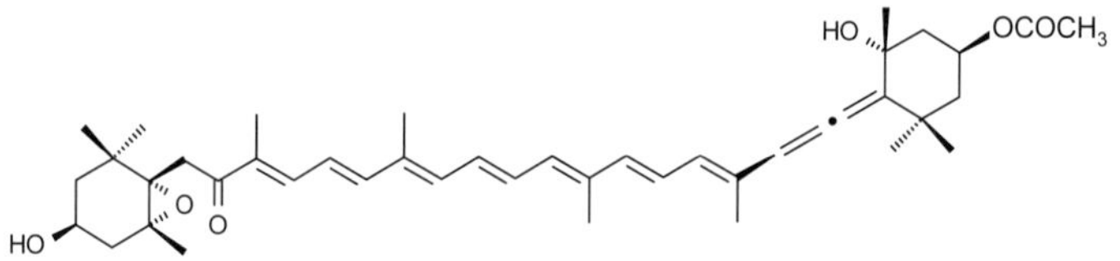

fucoxanthin (**369**)

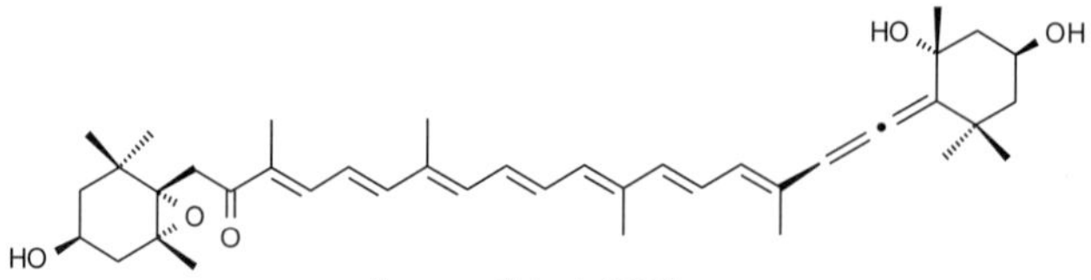

fucoxanthinol (**368**)

Increased PPARγ mRNA and protein levels have been implicated, in association with an increased production of ROS, in the apoptotic effects of β-carotene in MCF-7 cancer cells [99]. Addition of 2-chloro-5-nitro-*N*-phenylbenzamide (GW9662), an irreversible PPARγ antagonist, partly attenuated the cell death caused by the carotenoid in this cell line [99]. Recently, it has been shown also that 14′-apo-β-caroten-14′-al (**513**), but not other structurally related apocarotenals, repressed PPARγ and PPARα responses [100].

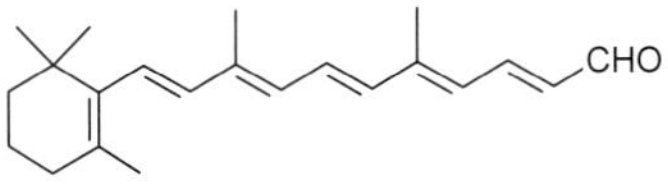

14'-apo-β-caroten-14'-al (**513**)

10. Modulation of xenobiotic and other orphan nuclear receptors

Orphan receptors are structurally related to nuclear hormone receptors but lack known physiological ligands. Xenobiotic receptors represent a family of orphan receptors and make up part of the defence mechanism against foreign lipophilic chemicals (xenobiotics). The family includes the steroid and xenobiotic receptor/pregnane X receptor (SXR/PXR), constitutive androstane receptor (CAR) and the aryl hydrocarbon receptor (AhR). These receptors respond to a wide variety of drugs, environmental pollutants, carcinogens, dietary and endogenous compounds, by regulating the expression of the cytochrome P450 (CYP) enzymes, conjugating enzymes and transporters that are involved in the oxidative metabolism and elimination of foreign substances (xenobiotics). It has been reported that carotenoids may modulate the expression of detoxication enzymes [101-103]. Recently, it has been shown that β-carotene can act as an inducer of several carcinogen-metabolizing enzymes, including the cytochrome P450 forms CYP1A1/2, CYP3A, CYP2B1 and CYP2A, in the lung of Sprague-Dawley rats [103]. Such inductions have been associated with an overgeneration of reactive oxygen-centred radicals [103]. In addition, many procarcinogens found in tobacco smoke are themselves CYP inducers and could act in a synergistic way with β-carotene or with some of its oxidation products, such as 8′-apo-β-caroten-8′-al (**482**), further contributing to the overall risk of carcinogenesis [91].

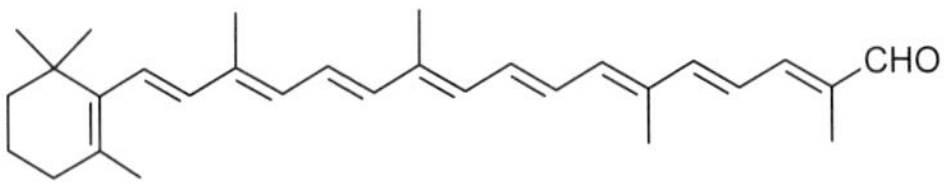

8'-apo-β-caroten-8'-al (**482**)

Moreover, induction of transformation by benzo[a]pyrene and cigarette smoke condensate in BALB/c 3T3 cells was markedly enhanced by the presence of β-carotene in either acute or chronic treatment [101]. Such an enhancement has been related to the boosting effect of the carotenoid on the cytochrome P450 apparatus [101]. β-Carotene has also been reported to enhance the hepatotoxicity of ethanol in both rodents and non-human primates by induction of CYP2E1 and CYP4A1 [104].

Other carotenoids, such as canthaxanthin and astaxanthin, have been recognized as potent inducers of CYP1A1 and 1A2 in mouse liver [105]. The administration of lycopene to rats was shown to induce liver CYP types 1A1/2, 2B1/2 and 3A in a dose-dependent manner [92]. The observation that these enzymic activities were induced at very low lycopene plasma levels suggests that modulation of drug metabolizing enzymes by carotenoids may be relevant to humans [92].

Recently, β-carotene has been shown to act as an activator of phase I enzymes in the human liver *via* a PXR-mediated mechanism [106].

11. Modulation of adhesion molecules and cytokines

Several large epidemiological studies have shown a correlation between higher plasma carotenoid levels and decreased risk of cardiovascular disease (CVD) (see *Chapter 14*). One of the possible mechanisms proposed to explain the beneficial effect of carotenoids is through the functional modulation of potentially atherogenic processes associated with the vascular endothelium. It has been reported recently that β-carotene can modulate the expression of several proteins involved in cell-cell adhesion (VCAM-1, SELP, CD-24), cadherins (CELSR1) and catenins (CTNNA1L, CTNNB1) [61]. Carotenoids, including β-carotene, induce decreased expression of VCAM-1, ICAM and selectin E in human aortic endothelium stimulated with interleukin IL-1β. This decrease was suggested to be responsible for a modulation of inflammatory response and may reflect the anti-inflammatory, protective effect of β-carotene on endothelium matrix proteins and proteases which regulate cell-matrix interaction. β-Carotene, lutein and lycopene were shown to reduce significantly the expression of ICAM-1 [107]. Moreover, both the IL-1β-stimulated and spontaneous adhesion of human amniotic epithelial cells (HAEC) to U937 monocytic cells was found to be decreased by lycopene, but not by other carotenoids [107]. It has been demonstrated recently that lycopene-treated dendritic cells were poor stimulators of naïve allogeneic T-cell proliferation and induced lower levels of interleukin-2 in responding T cells (see *Chapter 17*). The lycopene-treated cells also exhibited impaired interleukin-12 production [108]. Supplementation with pure lycopene reduced the expression of IL-6 in the MatLyLu Dunning rat prostate tumour model [54].

Increased plasma carotenoid concentrations after consumption of vegetable juice are accompanied by a time-delayed modulation of cytokines, including IL-2, IL-4, and TNFα in healthy men who previously consumed a low-carotenoid diet [109].

Further information on the influence of carotenoids on various aspects of the immune system is given in *Chapter 17*.

12. Modulation of gap junction communication

Loss of gap junctional communication between cells may play an important role in cell malignant transformation, and restoration of the communication may reverse the malignancy process [110]. Carotenoids can induce synthesis of the protein connexin 43, a component of the gap junction structure, and increase gap-junctional cell communication (GJC) [111,112]. Both retinoids and carotenoids have been reported to increase the expression of connexin 43, and this result was found to correlate with the suppression of carcinogen-induced transformation in 10T1/2 cells. It seems that the molecular mechanisms for this up-regulation are not the same for provitamin A carotenoids and non-provitamin A carotenoids [113,114]. The retinoic acid receptor antagonist Ro 41-5253 suppressed the expression of connexin 43 induced by retinoids, but not that induced by the non-provitamin A carotenoid astaxanthin. Connexin 43 induction by astaxanthin, but not by an RAR-specific retinoid, was inhibited by GW9662, a PPAR-γ antagonist [114].

Although the influence of lycopene on proliferation of carcinoma cells appears not to be limited to its ability to modulate connexin 43 expression, lycopene and its oxidation products have been reported to enhance GJC in cultured cells [71,115]. Recent data indicate that lycopene may indeed increase connexin 43 expression in human oral cancer cells [115].

D. Towards a Better Understanding of the Regulation of Cell Signalling by Carotenoids

Several studies *in vitro* and *in vivo* in which carotenoids have been shown to influence cell signalling at both protein and transcriptional levels have been summarized here. Despite these promising reports, it is difficult at present to relate the available experimental data directly to human pathophysiology. This is due, amongst other things, to the lack of adequate methods of solubilizing and delivering carotenoids to cells *in vitro* as well as to the difficulty of determining sensitive markers of long-term health effects in *vivo* at an early stage.

1. Delivery of carotenoids to cell cultures

In work with cultured cells, there is a need for methods for carotenoid delivery that are closer to physiological processes. The high hydrophobicity of carotenoids makes them insoluble in aqueous systems and therefore poorly available for cell cultures. In most current studies *in vitro*, carotenoids are provided as water-dispersible beadlets, detergent solutions, or in dilute solution in various water-miscible solvents, such as alcohols, dimethyl sulphoxide, tetrahydrofuran (THF) or, alternatively, in hexane. These methods have allowed potential effects of the pigments to be evaluated, but unspecific uptake and problems of miscibility, aggregation/crystallization and toxicity could give rise to misleading conclusions about the physiological significance of the observed phenomena. For example, THF, which is commonly used as a solvent for carotenoid solubilization, can be toxic to some cell lines and may have other disadvantages, in relation to the stability of carotenoids in solution [116]. Other methods have utilized various types of liposomes [117-119] and niosomes (liposomes made with non-ionic surfactants) [120] as carriers to improve the delivery of β-carotene, but toxic effects of various liposome and niosome constituents have been observed. Micelles incorporating β-carotene and other carotenoids, including lutein, have been used to deliver these carotenoids to cells [118-125]. Although carotenoids have been shown to be stable in these preparations, micelles are a physiological vehicle for carotenoid uptake from the gut, but not for other cell types.

2. Understanding effects and identifying biomarkers

Determining which molecular processes and specific signal transduction pathways may be modulated *in vivo* by carotenoids is also not a simple task. However, 'functional genomics', could provide a technological solution. Functional genomics studies a population of molecules, mRNAs, proteins or metabolites, rather than individual molecules. A large number of parameters can be analysed at the same time, and this not only enhances the chance of finding something but also helps in understanding if these effects are connected. DNA microarrays can be used to analyse simultaneously the response of thousands of genes to the exposure to different carotenoid levels. Analysis of the data by the bioinformatic and statistical tools now available makes it possible to identify what molecular and physiological pathways are affected. This provides a starting point for understanding the molecular effects of carotenoid exposure and thus for identifying sensitive, early biomarkers.

The multitude of possible interactions between carotenoids or other nutrients and the intracellular signal pathways makes the challenge even more daunting. Finally, inter-individual polymorphisms can mask the response to carotenoids and thereby complicate this undertaking to an even greater extent. Nevertheless, understanding the effects of carotenoids on cell signalling is fundamental to improving knowledge and strategies for the prevention of chronic diseases.

References

[1] S. T. Mayne, *FASEB J.*, **10**, 690 (1996).

[2] R. G. Ziegler, S. T. Mayne and C. A. Swanson, *Cancer Causes Control*, **7**, 157 (1996).

[3] R. Peto, R. Doll, J. D. Buckley and M. B. Sporn, *Nature*, **290**, 201 (1981).

[4] P. Palozza and N. I. Krinsky, *Meth. Enzymol.*, **213**, 403 (1992).

[5] N. I. Krinsky, *Annu. Rev. Nutr.*, **13**, 561 (1993).

[6] H. Sies and W. Stahl, *Am. J. Clin. Nutr.*, **62**, 1315S (1995).
[7] The Alpha-Tocopherol, Beta-Carotene Cancer Prevention Study Group, *New Engl. J. Med.*, **330**, 1029 (1994).
[8] G. S. Omenn, G. E. Goodman, M. D. Thornquist, J. Balmes, M. R. Cullen, A. Glass, J. P. Keogh, F. L. Meyskens, B. Valanis, J. H. Williams, S. Barnhart and S. Hammar, *New Engl. J. Med.*, **334**, 1150 (1996).
[9] S. D. Mark, W. Wang, J. F. Fraumeni Jr., J. Y. Li, P. R. Taylor, G. Q. Wang, W. Guo, S. M. Dawsey, B. Li and W. J. Blot, *Am. J. Epidemiol.*, **143**, 658 (1996).
[10] P. Palozza, *Curr. Pharmacol.*, **2**, 35 (2004).
[11] H. J. Palmer and K. E. Paulson, *Nutr. Rev.*, **55**, 353 (1997).
[12] V. J. Thannickal and B. L. Fanburg, *Am. J. Physiol. Lung Cell. Mol. Physiol.*, **279**, L1005 (2000).
[13] C. J. Lowenstein, J. L. Dinerman and S. H. Snyder, *Ann. Intern. Med.*, **120**, 227 (1994).
[14] P. Storz, *Front. Biosci.*, **10**, 1881 (2005).
[15] G. W. Burton and K. U. Ingold, *Science*, **224**, 569 (1984).
[16] P. Palozza, *Nutr. Rev.*, **56**, 257 (1998).
[17] C. A. Rice-Evans, J. Sampson, P. M. Bramley and D. E. Holloway, *Free Radic. Res.*, **26**, 381 (1997).
[18] P. Palozza, C. Luberto, G. Calviello, P. Ricci and G. M. Bartoli, *Free Radic. Biol. Med.*, **22**, 1065 (1997).
[19] T. A. Kennedy and D. E. Liebler, *J. Biol. Chem.*, **267**, 4658 (1992).
[20] A. Nahum, K. Hirsch, M. Danilenko, C. K. Watts, O. W. Prall, J. Levy and Y. Sharoni, *Oncogene*, **20**, 3428 (2001).
[21] H. Amir, M. Karas, J. Giat, M. Danilenko, R. Levy, T. Yermiahu, J. Levy and Y. Sharoni, *Nutr. Cancer*, **33**, 105 (1999).
[22] P. Palozza, S. Serini, N. Maggiano, M. Angelini, F. Di Nicuolo, F. O. Ranelletti and G. Calviello, *Carcinogenesis*, **23**, 11 (2002).
[23] E. C. Tibaduiza, J. C. Fleet, R. M. Russell and N. I. Krinsky, *J. Nutr.*, **132**, 1368 (2002).
[24] N. I. Ivanov, S. P. Cowell, P. Brown, P. S. Rennie, E. S. Guns and M. E. Cox, *Clin. Nutr.*, **26**, 252 (2007).
[25] P. Palozza, S. Serini, A. Boninsegna, D. Bellovini, M. Lucarini, G. Monastra and S. Gaetani, *Br. J. Nutr.*, **98**, 789 (2007).
[26] H. C. Cheng, H. Chien, C. H. Liao, Y. Y. Yang and S. Y. Huang, *J. Nutr. Biochem.*, **18**, 667 (2007).
[27] P. Palozza, S. Serini, F. Di Nicuolo and G. Calviello, *Arch. Biochem. Biophys.*, **430**, 104 (2004).
[28] T. Sacha, M. Zawada, J. Hartwich, Z. Lach, A. Polus, M. Szostek, E. Zdzi Owska, M. Libura, M. Bodzioch, A. Dembinska-Kiec, A. B. Skotnicki , R. Goralczyk, K. Wertz, G. Riss, C. Moele, T. Langmann and G. Schmitz, *Biochim. Biophys. Acta*, **1740**, 206 (2005).
[29] M. Bodzioch, A. Dembinska-Kiec, J. Hartwich, K. Lapicka-Bodzioch, A. Banas, A. Polus, J. Grzybowska, I. Wybranska, J. Dulinska, D. Gil, P. Laidler, W. Placha, M. Zawada, A. Balana-Nowak, T. Sacha, B. Kiec-Wilk, A. Skotnicki, C. Moehle, T. Langmann and G. Schmitz, *Nutr. Cancer*, **51**, 226 (2005).
[30] B. P. Chew, C. M. Brown, J. S. Park and P. F. Mixter, *Anticancer Res.*, **23**, 3333 (2003).
[31] A. Wang and L. Zhang, *Wei Sheng Yan Jin*, **36**, 575 (2007).
[32] H. Zhang, E. Kotake-Nara, H. Ono and A. Nagao, *Free Radic. Biol. Med.*, **35**, 1653 (2003).
[33] P. Palozza, *Biochim. Biophys. Acta*, **1740**, 215 (2005).
[34] F. Esposito, L. Russo, T. Russo and F. Cimino, *FEBS Lett.*, **470**, 211 (2000).
[35] P. Palozza, S. Serini, A. Torsello, A. Boninsegna, V. Covacci, N. Maggiano, F. O. Ranelletti, F. I. Wolf and G. Calviello, *Int. J. Cancer*, **97**, 593 (2002).
[36] P. Palozza, S. Serini, F. Di Nicuolo, A. Boninsegna, A. Torsello, N. Maggiano, F. O. Ranelletti, F. I. Wolf, G. Calviello and A. Cittadini, *Carcinogenesis*, **25**, 1315 (2004).
[37] S. Zhuang, M. C. Lynch and I. E. Kochevar, *Exp. Cell Res.*, **250**, 203 (1999).
[38] K. Chlichlia, M. E. Peter, M. Rocha, C. Scaffidi, M. Bucur, P. H. Krammer, V. Schirrmacher and V. Umansky, *Blood*, **91**, 4311 (1998).

[39] P. Palozza, S. Serini, A. Torsello, F. Di Nicuolo, N. Maggiano, F. O. Ranelletti, F. I. Wolf and G. Calviello, *Nutr. Cancer*, **47**, 76 (2003).
[40] H. L. Hantz, L. F. Young and K. R. Martin, *Exp. Biol. Med.*, **230**, 171 (2005).
[41] M. Terasaki, A. Asai, H. Zhang and A. Nagao, *Mol. Cell. Biochem.*, **300**, 227 (2007).
[42] D. J. Kane, T. A. Sarafian, R. Anton, H. Hahn, E. B. Gralla, J. S. Valentie, T. Ord and D. E. Bredesen, *Science*, **262**, 1274 (1993).
[43] T. Toba, Y. Shidoji, J. Fujii, H. Moriwaki, Y. Muto, T. Suzuki, N. Ohishi and K. Yagi, *Life Sci.*, **61**, 839 (1997).
[44] P. Palozza, A. Sheriff, S. Serini, A. Boninsegna, N. Maggiano, F. O. Ranelletti, G. Calviello and A. Cittadini, *Apoptosis*, **10**, 1445 (2005).
[45] H. Sano, Y. Kawahito, R. L. Wilder, A. Hashiramoto, S. Mukai, K. Asai, S. Kimura, H. Kato, M. Kondo and T. Hla, *Cancer Res.*, **55**, 3785 (1995).
[46] P. Palozza, S. Serini, N. Maggiano, G. Tringali, P. Navarra, F. O. Ranelletti and G. Calviello, *J. Nutr.*, **135**, 129 (2005).
[47] C. Liu, F. Lian, D. E. Smith, R. M. Russell and X. D. Wang, *Cancer Res.*, **63**, 3138 (2003).
[48] Y. Muto, J. Fujii, Y. Shidoji, H. Moriwaki, T. Kawaguchi and T. Noda, *Am. J. Clin. Nutr.*, **62**, 1535S (1995).
[49] M. Y. Khalil, J. R. Grandis and D. M. Shin, *Expert Rev. Anticancer Ther.*, **3**, 367 (2003).
[50] S. E. Hankinson, W. C. Willett, G. A. Colditz, D. J. Hunter, D. S. Michaud, B. Deroo, B. Rosner, F. E. Speizer and M. Pollak, *Lancet*, **351**, 1393 (1998).
[51] J. Ma, M. N. Pollak, E. Giovannucci, J. M. Chan, Y. Tao, C. H. Hennekens and M. J. Stampfer, *J. Natl. Cancer Inst.*, **91**, 620 (1999).
[52] M. Pollak, *Epidemiol. Rev.*, **23**, 59 (2001).
[53] M. Karas, H. Amir, D. Fishman, M. Danilenko, S. Segal, A. Nahum, A. Koifmann, Y. Giat, J. Levy and Y. Sharoni, *Nutr. Cancer*, **36**, 101 (2000).
[54] U. Siler, L. Barella, V. Spitzer, J. Schnorr, M. Lein, R. Goralczyk and K. Wertz, *FASEB J.*, **18**, 1019 (2004).
[55] O. Kucuk, F. H. Sarkar, W. Sakr, Z. Djuric, M. N. Pollak, F. Khachik, Y. W. Li, M. Banerjee, D. Grignon, J. S. Bertram, J. D. Crissman, E. J. Pontes and D. P. Wood Jr., *Cancer Epidemiol. Biomarkers Prev.*, **10**, 861 (2001).
[56] L. A. Mucci, R. Tamimi, P. Lagiou, A. Trichopoulou, V. Benetou, E. Spanos and D. Trichopoulos, *BJU Int.*, **87**, 814 (2001).
[57] A. Herzog, U. Siler, V. Spitzer, N. Seifert, A. Denelavas, P. B. Hunziker, W. Hunziker, R. Goralczyk and K. Wertz, *FASEB J.*, **19**, 272 (2005).
[58] S. Uchiyama and M. Yamaguchi, *Int. J. Mol. Med.*, **15**, 675 (2005).
[59] J. L. Schwartz and G. Shklar, *Nutr. Cancer*, **27**, 192 (1997).
[60] J. T. Comerci Jr., C. D. Runowicz, A. L. Fields, S. L. Romney, P. R. Palan, A. S. Kadish and G. L. Goldberg, *Clin. Cancer Res.*, **3**, 157 (1997).
[61] A. Dembinska-Kiec, A. Polus, B. Kiec-Wilk, J. Grzybowska, M. Mikolajczyk, J. Hartwich, U. Razny, K. Szumilas, A. Banas, M. Bodzioch, J. Stachura, G. Dyduch, P. Laidler, J. Zagajewski, T. Langman and G. Schmitz, *Biochim. Biophys. Acta*, **1740**, 222 (2005).
[62] Y. Liu, R. L. Chang, X. X. Cui, H. L. Newmark and A. H. Conney, *Oncol. Res.*, **9**, 19 (1997).
[63] J. A. Sokoloski, W. F. Hodnick, S. T. Mayne, C. Cinquina, C. S. Kim and A. C. Sartorelli, *Leukemia*, **11**, 1546 (1997).
[64] L. Kim, A. V. Rao and L. G. Rao, *J. Med. Food.*, **6**, 79 (2003).
[65] Y. Sharoni, M. Danilenko, S. Walfisch, H. Amir, A. Nahum, A. Ben-Dor, K. Hirsch, M. Khanin, M. Steiner, L. Agemy, G. Zango and J. Levy, *Pure Appl. Chem.*, **74**, 1469 (2002).
[66] R. M. Ponnamperuma, Y. Shimizu, S. M. Kirchhof and L. M. De Luca, *Nutr. Cancer*, **37**, 82 (2000).

[67] X. D. Wang, C. Liu, R. T. Bronson, D. E. Smith, N. I. Krinsky and R. M. Russell, *J. Natl. Cancer Inst.*, **91**, 60 (1999).

[68] C. Liu, X. D. Wang, R. T. Bronson, D. E. Smith, N. I. Krinsky and R. M. Russell, *Carcinogenesis*, **21**, 2245 (2000).

[69] R. Goralczyk, K. Wertz, B. Lenz, G. Riss, P. Buchwald, B. Hunziker, C. Geatrix, P. Aebischer and H. Bachmann, *Biochim. Biophys. Acta*, **1740**, 179 (2005).

[70] A. Ben-Dor, A. Nahum, M. Danilenko, Y. Giat, W. Stahl, H. D. Martin, T. Emmerich, N. Noy, J. Levy and Y. Sharoni, *Arch. Biochem. Biophys.*, **391**, 295 (2001).

[71] W. Stahl, J. von Laar, H. D. Martin, T. Emmerich and H. Sies, *Arch. Biochem. Biophys.*, **373**, 271 (2000).

[72] K. Schulze-Osthoff, M. Los and P. A. Baeuerle, *Biochem. Pharmacol.*, **50**, 735 (1995).

[73] A. Bowie and L. A. J. O'Neill, *Biochem. Pharmacol.*, **59**, 13 (2000).

[74] F. A. La Rosa, J. W. Pierce and G. E. Sonenshein, *Mol. Cell. Biol.*, **14**, 1039 (1994).

[75] P. Palozza, S. Serini, A. Torsello, F. Di Nicuolo, E. Piccioni, V. Ubaldi, C. Pioli, F. I. Wolf and G. Calviello, *J. Nutr.*, **133**, 381 (2003).

[76] J. Y. Seo, H. Kim, J. T. Seo and K. H. Kim, *Pharmacology*, **64**, 63 (2002).

[77] C. S. Huang, Y. E. Fan, C. Y. Lin and M. L. Hu, *J. Nutr. Biochem.*, **18**, 449 (2007).

[78] D. De Stefano, M. C. Maiuri, V. Simeon, G. Grassia, A. Soscia, M. P. Cinelli and R. Carnucci, *Eur. J. Pharmacol.*, **566**, 192 (2007).

[79] T. Dorai and B. B. Aggarwal, *Cancer Lett.*, **215**, 129 (2004).

[80] M. Ramos-Gomez, M. K. Kwak, P. M. Dolan, K. Itoh, M. Yamamoto, P. Talalay and T. W. Kensler, *Proc. Natl. Acad. Sci. USA*, **98**, 3410 (2001).

[81] Y. C. Kim, H. Masutani, Y. Yamaguchi, K. Itoh, M. Yamamoto and J. Yodoi, *J. Biol. Chem.*, **276**, 18399 (2001).

[82] M. K. Kwak, P. A. Egner, P. M. Dolan, M. Ramos-Gomez, J. D. Groopman, K. Itoh, M. Yamamoto and T. W. Kensler, *Mutat. Res.*, **480-481**, 305 (2001).

[83] A. N. Kong, E. Owuor, R. Yu, V. Hebbar, C. Chen, R. Hu and S. Mandlekar, *Drug Metab. Rev.*, **33**, 255 (2001).

[84] D. Martin, A. I. Rojo, M. Salinas, R. Diaz, G. Gallardo, J. Alam, C. M. De Galarreta and A. Cuadrado, *J. Biol. Chem.*, **279**, 8919 (2004).

[85] A. Ben-Dor, M. Steiner, L. Gheber, M. Danilenko, N. Dubi, K. Linnewiel, A. Zick, Y. Sharoni and J. Levy, *Mol. Cancer Ther.*, **4**, 86 (2005).

[86] M. C. Trekli, G. Riss, R. Goralczyk and R. M. Tyrrell, *Free Radic. Biol. Med.*, **34**, 456 (2003).

[87] E. A. Offord, J. C. Gautier, O. Avanti, C. Scaletta, F. Runge, K. Kramer and L. A. Applegate, *Free Radic. Biol. Med.*, **32**, 1293 (2002).

[88] U. C. Obermüller-Jevic, B. Schlegel, A. Flaccus and H. K. Biesalski, *FEBS Lett.*, **509**, 186 (2001).

[89] E. D. Owuor and A. N. Kong, *Biochem. Pharmacol.*, **64**, 765 (2002).

[90] P. Palozza, S. Serini, D. Currò, G .Calviello, K. Igarashi and C. Mancuso, *Antiox. Redox Signal.*, **8**, 1069 (2006).

[91] S. Gradelet, P. Astorg, J. Leclerc, J. Chevalier, M. F. Vernevaut and M. H. Siess, *Xenobiotica*, **26**, 49 (1996).

[92] V. Breinholt, S. T. Lauridsen, B. Daneshvar and J. Jakobsen, *Cancer Lett.*, **154**, 201 (2000).

[93] B. Velmurugan, V. Bhuvaneswari, U. K. Burra and S. Nasini, *Eur. J. Cancer Prev.*, **11**, 19 (2002).

[94] V. Bhuvaneswari, B. Velmurugan and S. Nasini, *J. Biochem. Mol. Biol. Biophys.*, **6**, 257 (2002).

[95] K. Hirsch, A. Atzmon, M. Danilenko, J. Levy and Y. Sharoni, *Breast Cancer Res. Treat.*, **104**, 221 (2007).

[96] Y. Sharoni, R. Agbaria, H. Amir, A. Ben-Dor, I. Bobilev, N. Doubi, Y. Giat, K. Hirsh, G. Izumchenko, M. Khanin, E. Kirilov, R. Krimer, A. Nahum, M. Steiner, Y. Walfisch, S. Walfisch, G. Zango, M. Danilenko and J. Levy, *Mol. Aspects Med.*, **24**, 371 (2003).

[97] M. Hosokawa, M. Kudo, H. Maeda, H. Kohno, T. Tanaka and K. Miyashita, *Biochim. Biophys. Acta*, **1675**, 113 (2004).
[98] H. Maeda, M. Hosokawa, T. Sashima, N. Takahashi, T. Kawada and K. Miyashita, *Int. J. Mol. Med.*, **18**, 147 (2006).
[99] Y. Cui, Z. Lu, L. Bai, Z. Shi, W. E. Zhao and B. Zhao, *Eur. J. Cancer*, **43**, 2590 (2007).
[100] O. Zionzenkova and J. Plutzky, *FEBS Lett.*, **582**, 32 (2008).
[101] P. Perocco, M. Paolini, M. Mazzullo, G. L. Biagi and G. Cantelli-Forti, *Mutat. Res.*, **440**, 83 (1999).
[102] M. Paolini, G. Cantelli-Forti, P. Perocco, G. F. Pedulli, S. Z. Abdel-Rahma and M. S. Legator, *Nature*, **398**, 760 (1999).
[103] M. Paolini, A. Antelli, L. Pozzetti, D. Spetlova, P. Perocco, L. Valgimigli, G. F. Pedulli and G. Cantelli-Forti, *Carcinogenesis*, **22**, 1483 (2001).
[104] I. G. Kessova, M. A. Leo and C. S. Lieber, *Alcohol Clin. Exp. Res.*, **25**, 1368 (2001).
[105] P. Astorg, S. Gradelet, J. Leclerc and M. H. Siess, *Nutr. Cancer*, **27**, 245 (1997).
[106] R. Ruhl, *Biochim. Biophys. Acta*, **1740**, 162 (2005).
[107] K. R. Martin, D. Wu and M. Meydani, *Atherosclerosis*, **150**, 265 (2000).
[108] G. Y. Kim, J. H. Kim, S. C. Ahn, H. J. Lee, D. O. Moon, C. M. Lee and Y. M. Park, *Immunology*, **113**, 203 (2004).
[109] B. Watzl, A. Bub, K. Briviba and G. Rechkemmer, *Ann. Nutr. Metab.*, **47**, 255 (2003).
[110] A. Hotz-Wagenblatt and D. Shalloway, *Crit. Rev. Oncog.*, **4**, 541 (1993).
[111] L. X. Zhang, R. V. Cooney and J. S. Bertram, *Carcinogenesis*, **12**, 2109 (1991).
[112] L. X. Zhang, R. V. Cooney and J. S. Bertram, *Cancer Res.*, **52**, 5707 (1992).
[113] A. L. Vine and J. S. Bertram, *Nutr. Cancer*, **52**, 105 (2005).
[114] J. S. Bertram and A. L. Vine, *Biochim. Biophys. Acta*, **1740**, 170 (2005).
[115] O. Livny, I. Kaplan, R. Reifen, S. Polak-Charcon , Z. Madar and B. Schwartz, *J. Nutr.*, **132**, 3754 (2002).
[116] S. Shahrzad, E. Cadenas, A. Sevanian and L. Packer, *Biofactors*, **16**, 83 (2002).
[117] S. M.Anderson and N. I. Krinsky, *Photochem. Photobiol.*, **18**, 403 (1973).
[118] P. Grolier, V. Azais-Braesco, L. Zelmire and H. Fessi, *Biochim. Biophys. Acta*, **1111**, 135 (1992).
[119] L. A. Stivala, M. Savio, O. Cazzalini, R. Pizzala, L. Rehak, L. Bianchi, V. Vannini and E. Prosperi, *Carcinogenesis*, **17**, 2395 (1996).
[120] P. Palozza, R. Muzzalupo, S. Trombino, A. Valdannini and N. Picci, *Chem. Phys. Lipids*, **139**, 32 (2006).
[121] M. I. El-Gorab, B. A. Underwood and J. D. Loerch, *Biochim. Biophys. Acta*, **401**, 265 (1975).
[122] E. Reboul, L. Abou, C. Mikail, O. Ghiringhelli, M. Andre, H. Portugal, D. Jourdheuil-Rahmani, M. J. Amiot, D. Lairon and P. Borel, *Biochem. J.*, **387**, 455 (2005).
[123] K. R. Martin, M. L. Failla and J. C. Smith Jr., *J. Nutr.*, **126**, 2098 (1996).
[124] A. During, M. M. Hussain, D. W. Morel and E. H. Harrison, *J. Lipid Res.*, **43**, 1086 (2002).
[125] C. Chitchumroonchokchai, S. J. Schwartz and M. L. Failla, *J. Nutr.*, **134**, 2280 (2004).

Carotenoids
Volume 5: Nutrition and Health

Chapter 12

Antioxidant/Pro-oxidant Actions of Carotenoids

Kyung-Jin Yeum, Giancarlo Aldini, Robert M. Russell and Norman I. Krinsky

A. Introduction

In recent years numerous reviews have discussed in detail the antioxidant action of carotenoids [1-5]. The existence of an antioxidant effect has been questioned by some, however [6]. In addition, there is the complication that, under some circumstances, carotenoids exhibit a pro-oxidant effect [7,8], although some authors do not believe that this occurs *in vivo* [9]. The fundamental chemistry of carotenoid radicals and radical ions, as a basis for understanding mechanisms of antioxidant/pro-oxidant actions, is presented in *Volume 4, Chapter 7.*

Reactive oxygen species (ROS) are continuously generated by normal metabolism in the body [10] and these ROS have various physiological effects [11]. Cellular production of ROS such as superoxide anion ($O_2^{\bullet -}$), hydroxyl radical ($HO^{\bullet}$), peroxyl radical ($ROO^{\bullet}$) and alkoxyl radical ($RO^{\bullet}$), occurs from both enzymic and non-enzymic reactions. Mitochondria appear to be the most important subcellular site of ROS production, in particular of $O_2^{\bullet -}$ and H_2O_2 in mammalian organs. The electron transfer system of the mitochondrial inner membrane is a major source of superoxide production when molecular oxygen is reduced by a single electron. Superoxide can then dismutate to form hydrogen peroxide (H_2O_2). This species can further react to form the hydroxyl radical ($HO^{\bullet}$) and ultimately water, as shown in Scheme 1 [12].

$$O_2 \xrightarrow{1\,e^-} O_2^{\bullet -} \xrightarrow{1\,e^-} H_2O_2 \xrightarrow{1\,e^-} HO^{\bullet} \xrightarrow{1\,e^-} H_2O$$

Scheme 1

In addition to intracellular membrane-associated oxidases, soluble enzymes such as xanthine oxidase, aldehyde oxidase, dihydroorotate dehydrogenase, flavoprotein dehydrogen-ase and tryptophan dioxygenase can generate ROS during catalytic cycling. Auto-oxidation of small molecules such as dopamine, adrenaline (epinephrine), flavins, and quinols can be an important source of intracellular ROS production as well. In most cases, the direct product of such auto-oxidation reactions is superoxide anion [13].

An imbalance between oxidant production and antioxidants may produce excess ROS that can cause oxidative damage in vulnerable targets such as unsaturated fatty acyl chains in membranes, thiol groups in proteins and nucleic acid bases in DNA [14]. Such a state of 'oxidative stress' is thought to contribute to the pathogenesis of a number of human diseases [13], but it is still not clear what kinds of ROS play a role in such pathogenesis or where the major sites of ROS action occur. There is, however, convincing evidence that lipid peroxidation is related to human pathology such as that observed in atherosclerosis [15].

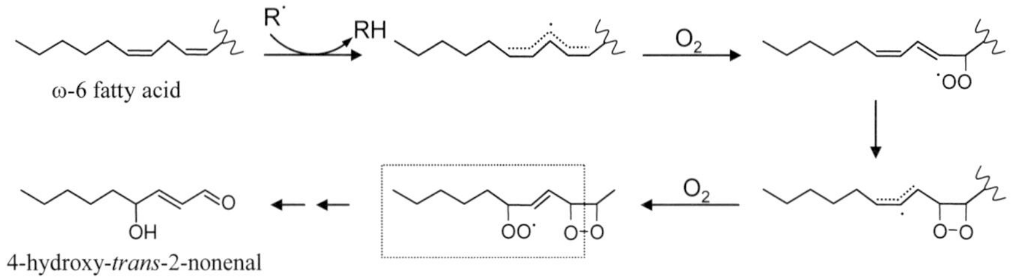

Fig. 1. Scheme for peroxidation of lipids containing ω-6 polyunsaturated fatty acid chains, illustrating the formation of 4-hydroxy-*trans*-nonenal [(*trans*)-4-hydroxynon-2-enal] (HNE).

A simplified pathway for peroxidation of lipids containing ω-6 polyunsaturated fatty acid chains (arachidonic and linoleic acid) and the subsequent formation of (*trans*)-4-hydroxynon-2-enal (HNE) is shown in Fig. 1. The ω-6 polyunsaturated acyl chains are susceptible to free-radical attack to form a free radical intermediate, which further reacts with molecular oxygen to generate first a peroxyl radical and then hydroperoxide derivatives such as (9*Z*,11*E*)-(13*S*)-13-hydroperoxyoctadeca-9,11-dienoic acid (13*S*-HPODE). The products of lipid peroxidation further react to produce HNE. It is important to underline that a peroxyl radical is capable of abstracting a H atom from another lipid molecule leading to the propagation stage of lipid peroxidation. The carbon radical formed can react with O_2 to form another peroxyl radical, and so the chain reaction of lipid peroxidation can continue [16].

The actions of antioxidants in biological systems depend on the nature of oxidants or ROS imposed on the systems, and the activities and amounts of antioxidants and their cooperative/synergistic interactions in these systems. The antioxidant actions of ascorbic acid (vitamin C) and tocopherols (vitamin A) and their interactions *in vitro* are well known [17,18], although the biological significance of those well characterized antioxidants is still not clear [19].

This chapter will evaluate the evidence for an antioxidant action of carotenoids *in vitro*, *ex vivo* or *in vivo*, and will also consider briefly the evidence concerning pro-oxidation. Methods to determine antioxidant/prooxidant actions of carotenoids, factors that affect the efficiency of these actions, and interactions of carotenoids are also discussed.

B. Analytical Methods to Determine Antioxidant/Pro-oxidant Actions of Carotenoids in Biological Samples

Any compound that can inhibit oxidation that is induced either spontaneously or by means of external oxidants is considered to be an antioxidant. This is a relatively simple definition but, at times, it becomes very difficult to evaluate whether a compound actually has an antioxidant action, particularly *in vivo*.

Several methods to measure the antioxidant and pro-oxidant effect of carotenoids have been proposed and applied *in vitro*, *ex vivo* and *in vivo*. Some methods measure the intermediate or final products of the oxidative damage. Because carotenoids are lipid-soluble compounds and act as inhibitors of the lipid peroxidation process, many of the methods are based on the measurement of the consequences of lipid peroxidation, including intermediate (hydroperoxides, conjugated dienes) and/or final breakdown compounds, such as thiobarbituric acid reactive substances (TBARS) as shown in Section **B**.2. The methods must be sensitive and, more importantly, highly specific, because to evaluate the real antioxidant/pro-oxidant significance of carotenoids *in vivo*, they need to be applicable in complex biological matrices such as tissue preparations, plasma or cells, where the synergistic/cooperative effect of carotenoids with hydrophilic/lipophilic compounds takes place. Other methods are based on assays that measure the effect of carotenoids to modulate the oxidative resistance and total antioxidant activity of the biological matrix.

1. Total antioxidant capacity

An assay to measure total antioxidant capacity in a biological sample such as plasma must consider factors such as the heterogeneity of the sample, which consists of both hydrophilic and lipophilic compartments that contain water-soluble and fat-soluble antioxidants, respectively, as shown in Fig. 2. The cooperative/synergistic interactions among antioxidants in biological samples cannot be overlooked.

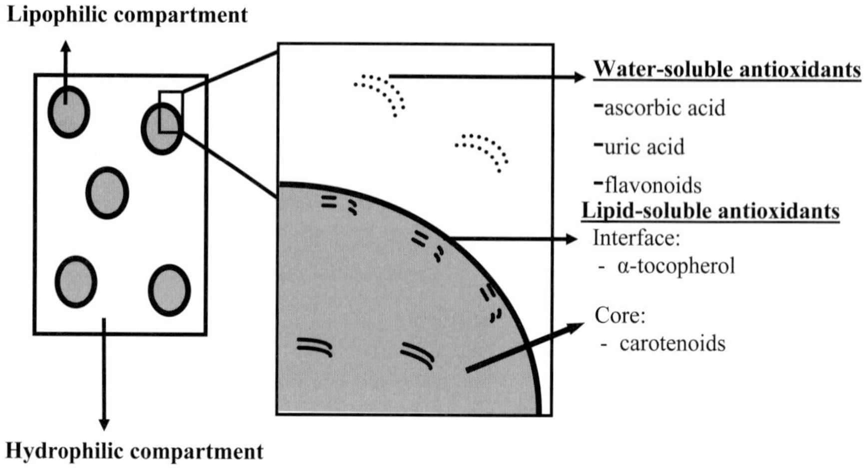

Fig. 2. Schematic representation of the hydrophilic and lipophilic compartments of plasma, indicating the antioxidants that may be present in each compartment.

Azo-initiators are a class of radical inducers (which contain the –N=N– group) widely used in experiments *in vitro* to generate radical species. The azo-initiators decompose at a temperature-controlled rate to give carbon-centred radicals which react rapidly with O_2 to give the peroxyl radical ($ROO^{\bullet}$) (Scheme 2).

$$R–N=N–R \rightarrow N_2 + 2R^{\bullet}$$
$$R^{\bullet} + O_2 \rightarrow ROO^{\bullet}$$

Scheme 2

Peroxyl radicals derived from azo-initiators can induce the lipid peroxidation cascade and can also damage proteins. Depending on the lipophilicity of the azo-initiators (AAPH is water-soluble whereas AMVN and MeO-AMVN are lipophilic), the peroxyl radicals are generated in the aqueous or lipid phase of the sample. The choice of the site of radical generation is of great importance since the activities of antioxidants present in both the lipid and aqueous compartments depend on the localization of the attacking radical species [20].

Table 1 lists assays that are used to determine antioxidant capacity in hydrophilic and lipophilic environments in biological samples such as plasma. Those assays [15,21] that use only hydrophilic radical initiators and probes are not sufficient to determine the antioxidant activity of carotenoids, which are deeply embedded in the lipoprotein core of biological samples. Attempts to determine the activity of fat-soluble antioxidants by measuring the antioxidant activity of lipid extracts dissolved in an organic solvent [22] cannot measure the possible interactions between the fat-soluble and water-soluble antioxidants. The alternative approach of producing radicals in the lipid compartment of whole plasma and monitoring lipid peroxidation by a lipophilic probe [23] allows measurement of the actual 'total' antioxidant activity including possible interactions among antioxidants located in the hydrophilic and lipophilic compartments, because the interference of large amounts of protein in the hydrophilic compartment can be overcome by this approach.

Table 1. Assays that are used to determine antioxidant capacity in hydrophilic and lipophilic environments in plasma and other biological samples.

Assay	Medium	Radical inducer	Oxidizable substrate (probe)	Calculation	Ref.
Hydrophilic					
TRAP	Diluted (x250) plasma	AAPH	DCFH R-Phycoerythrin	Lag time	[15] [24]
ORAC	Diluted (x150) plasma	AAPH	R-Phycoerythrin	AUC	[25]
TEAC	Diluted plasma	ABTS		Absorbance	[26,27]
FRAP	Diluted (x100) plasma	Fe^{3+}		Absorbance	[28]
Crocin bleaching	Diluted (x250) plasma	ABAP		Absorbance	[29,30]
Lipophilic					
Lipophilic ORAC	Plasma extracts dissolved in solvent	AAPH	Fluorescein	AUC	[22]
Lipophilic antioxidant activity	Diluted (~x3) labelled plasma (requires 12 hr for labelling with DPHPC)	AAPH	DPHPC	Lag time	[31]
TAP	Diluted (x5) plasma	MeO-AMVN	BODIPY 581/591	AUC	[23]

TRAP: Total radical-trapping antioxidant parameter
ORAC: Oxygen radical-absorbing capacity
TEAC: Trolox equivalent antioxidant capacity
FRAP: Ferric-reducing ability of plasma
TAP: Total antioxidant performance
AAPH, ABAP: 2,2′-Azobis-(2-amidinopropane) dihydrochloride
ABTS: 2,2′-Azinobis-(3-ethylbenzothiazoline 6-sulphonate)
AUC: Area under the curve
MeO-AMVN: 2,2′-Azobis-(4-methoxy-2,4-dimethylvaleronitrile)
DCFH: 2′,7′-Dichlorodihydrofluorescein
DPHPC: 1-Palmitoyl-2-((2-(4-(6-phenyl-*trans*-1,3,5-hexatrienyl)phenyl)ethyl)-carbonyl-*sn*-glycero-3-phosphocholine
BODIPY 581/591: 4,4-Difluoro-5-(4-phenyl-1,3-butadienyl)-4-bora-3a,4a-diaza-*s*-indacene-3-undecanoic acid

Attempts to show a direct correlation between the consumption of dietary carotenoids and subsequent changes in antioxidant capacity in humans have failed consistently [32,33]. It has even been suggested that carotenoids may not act as antioxidants *in vivo* [6]. However, these suggestions are derived from the lack of an adequate analytical method for measuring antioxidant capacity. Inasmuch as conventional methods such as TRAP, ORAC, *etc*., use primarily hydrophilic radical generators and measure primarily antioxidant capacity in the aqueous compartment of plasma, these methods are unable to determine the antioxidant capacity of the lipid compartment [34,35]. Therefore, it is not surprising that most of the methods used to measure purported 'total antioxidant capacity' of plasma show no effects of lipophilic antioxidants, such as vitamin E and carotenoids [32,33,35]. This can be explained by considering that plasma carotenoids, being deeply embedded in the core of lipoproteins, are not available for reaction with aqueous radical species or ferric complexes used in these assays. It has been reported that the activities of antioxidant nutrients present in the lipid and aqueous compartments depend on the localization of the attacking radical species, and can be increased synergistically by interactions [20].

When the hydrophilic assays were applied, the majority of the antioxidant capacity of plasma could be accounted for by protein (10-28%), uric acid (7-58%), and ascorbic acid (3-27%), whilst the effect of vitamin E (<10%) was minimal [15,28,29,36,37]. These assays measure the antioxidant capacity of the aqueous compartment only, since the radical inducers and probes are all hydrophilic. α-Tocopherol, which has its chroman head group oriented towards the lipoprotein membrane, may participate somewhat in the antioxidant action through interaction with water-soluble antioxidants such as ascorbic acid. However, it is clear that carotenoids that are deeply embedded in the lipid core could not participate in the antioxidant effect under these experimental conditions. The lack of contribution of fat-soluble antioxidants can also be ascribed to the relatively lower amount of fat-soluble antioxidants than of water-soluble antioxidants in plasma, although it should be recognized that the antioxidant activity of fat-soluble antioxidants can be greatly enhanced by synergistic interactions with water-soluble antioxidants and with other fat-soluble antioxidants.

Thus, foods such as green tea [38,39], cocoa [40], red wine [29,41], coffee [42], and strawberries [43], that contain considerable amounts of water-soluble polyphenols, significantly increase plasma antioxidant capacity as determined by these hydrophilic assays. On the other hand, diets rich in carotenoids, *e.g.* lycopene (**31**) or β-carotene (**3**) do not affect antioxidant capacity as measured by the hydrophilic TRAP, FRAP or ORAC assays [32,44,45]. In spite of the consistent failure to show the modification of antioxidant capacity by consumption of a high carotenoid diet [32] or supplementation with carotenoids in humans [20], it is noteworthy that there is considerable and consistent evidence for antioxidant actions of carotenoids [5,46], including *Z* isomers [47], tested in solvent systems *in vitro*.

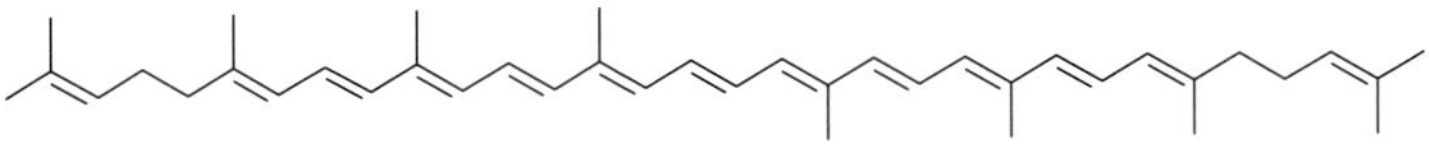

lycopene (**31**)

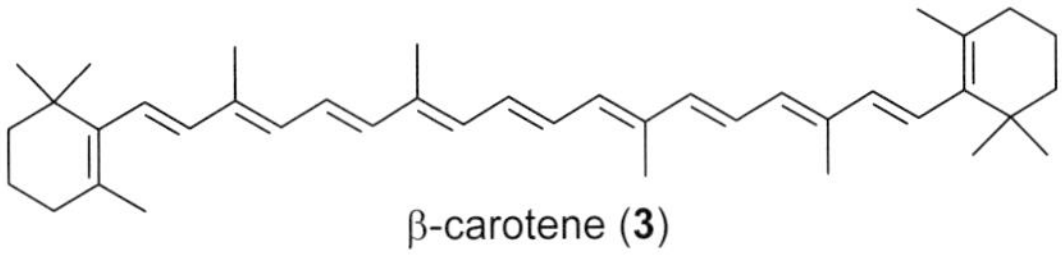

β-carotene (**3**)

2. Lipid peroxidation

In mammalian tissues, malondialdehyde (MDA) originates from the oxidative degradation of polyunsaturated fatty acids (PUFAs) with more than two unconjugated double bonds. The main precursors of MDA are arachidonic acid (20:4) and docosahexaenoic acid (22:6). In certain tissues MDA can also be formed by enzymic processes, for example by human platelet thromboxane synthase from prostaglandins (PGH_2, PGH_3 and PGG_2) or by renal polyamine oxidase from spermine [48].

The thiobarbituric acid (TBA) analysis is one of the most frequently used assays for measuring MDA in biological matrices. The reaction (Fig. 3) is carried out in acid where two moles of TBA react with one mole of MDA, to form a pink reaction product (λ_{max} = 532 nm), which is readily extractable into organic solvents such as butan-1-ol [49].

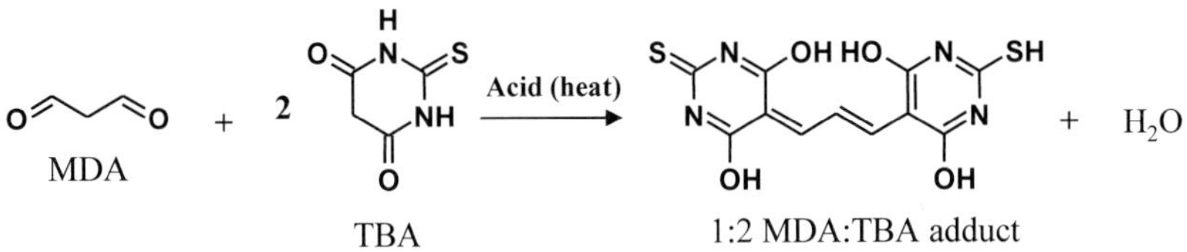

Fig. 3. The 'TBARS' reaction between malondialdehyde (MDA) and thiobarbituric acid (TBA). The 1:2 MDA:TBA adduct that is generated can be determined quantitatively by its light absorbance at 532 nm.

For many years, determination of TBARS such as MDA was assumed to be a valid measure of lipid peroxidation, but it is, in fact, a somewhat unspecific biomarker. Nevertheless, changes in MDA levels have been used to evaluate the effects of enhancing or depleting dietary or supplementary nutrients such as carotenoids in conditions where an oxidative stress might arise [50]. However, the fact that β-carotene interferes with this type of assay because its breakdown yields products that give a positive TBARS reaction should be carefully considered, especially in studies of the pro-oxidant effect of carotenoids at high concentrations [51].

Measurement of prostaglandin F_2-like compounds (F_2-isoprostanes), which are produced *in vivo* by non-enzymic free radical-catalysed peroxidation of arachidonic acid, has emerged as one of the most reliable approaches to assessing oxidative stress status [52,53]. It is generally accepted that F_2-isoprostanes more accurately reflect lipid peroxidation *in vivo* than do TBARS [54] and that F_2-isoprostane concentrations can be lowered by dietary antioxidant supplements [55]. Problems, such as the complicated technique to measure isoprostanes and their instability in biological samples, still need to be overcome, however. It has been reported that supplementary β-carotene, even when given at high doses (50 mg/day) for many years (median 4 years), does not have pro-oxidant effects in either smokers or non-smokers, as measured by urinary excretion of F_2-isoprostanes [56]. Recent efforts to determine plasma isoprostanes may help to eliminate the experimental error introduced by the complicated sample preparation steps, including derivatization, needed for the traditional GC-MS assays.

3. Oxidation of low-density lipoprotein (LDL)

The antioxidant properties of bioactive components present in food, including vitamins E and C, polyphenols and carotenoids, against low-density lipoprotein (LDL) oxidation have been reviewed extensively [57]. The general approach to measure the antioxidant capacity in the lipid compartment of plasma is to determine the susceptibility of isolated LDL to oxidation by hydrophilic radical inducers (AAPH, transition metal ions) or lipophilic radical inducers such as 2,2′-azobis-(2,4-dimethylvaleronitrile) (AMVN). Lipid peroxidation can be estimated by measuring the UV absorbance of conjugated dienes at 234 nm [58], oxidation of 2′,7′-dichlorodihydrofluorescein (DCFH) or oxidation of diphenyl-1-pyrenylphosphine (DPPP) to produce the fluorescent product, DPPP oxide [59]. When the lipophilic AMVN is used as radical initiator and luminol as an oxidizable substrate, the contribution of the fat-soluble antioxidants to the antioxidant capacity of isolated LDL has been shown to be greater than 70% (tocopherol, 73%; ubiquinol-10, 2.5%) [60]. This approach is limited, however, because it does not take into account the potential interaction between water-soluble and fat-soluble antioxidants, a synergism that may greatly increase the total antioxidant activity. It is generally accepted that α-tocopherol can act as a pro-oxidant to initiate lipid peroxidation in isolated LDL [61]. This tocopherol-mediated lipid peroxidation can be prevented, however, by ascorbic acid, which can regenerate α-tocopherol from α-tocopheroxyl radical [62]. In addition, a recent report, which indicates markedly different LDL oxidation kinetics depending on the concentration of copper ion added into LDL, implies possible misinterpretation of LDL oxidation data when the LDL oxidation is calculated on the basis of the inhibition period, *i.e.* the lag-time [63].

4. DNA damage

The single-cell microgel electrophoresis technique, named the 'comet' assay, was developed to detect single or double strand breaks in DNA. The broken DNA fragments show greater migration in electrophoresis than the undegraded DNA, giving rise to a diffuse DNA substance area which, after staining, resembles a comet tail [64] as shown in Fig. 4.

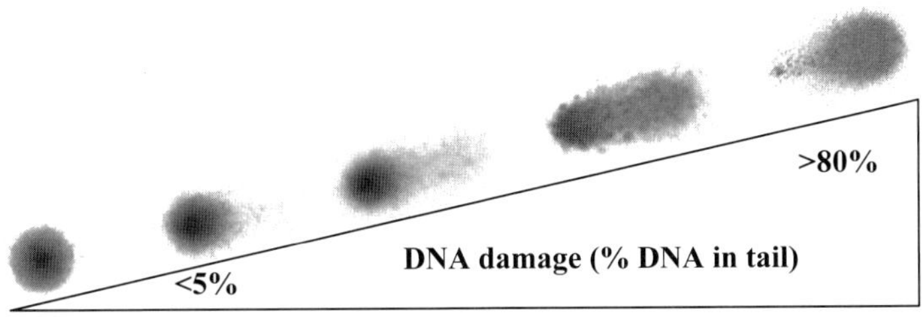

Fig. 4. Illustration of the effects of increasing damage (strand breaks) to DNA, revealed by electrophoresis in the comet assay.

Endogenous strand breaks, as well as the resistance of DNA to oxidative stress caused by treating lymphocytes with hydrogen peroxide (H_2O_2), can be evaluated by the comet assay. Much of the recent material relating to the effects of carotenoids on DNA damage, as well as effects on DNA synthesis and proliferation, has been reviewed [65], leading to the conclusion that, in cell cultures, carotenoids can inhibit DNA synthesis and proliferation, change gene expression, decrease micronucleus frequency and inhibit transformation *via* inhibition of gap-junction proteins. In humans, a diet containing carotenoid-rich foods has been shown to reduce lymphocyte DNA damage, suggesting that the carotenoids may be acting as antioxidants *in vivo* [20]. However, it is not yet known whether oxidative DNA damage in blood cells reflects similar damage in other target tissues.

The product of oxidative damage to DNA most commonly measured in urine and/or blood is 8-hydroxy-2-deoxyguanosine (8-OHdG), even though there is still some question about the validity of this marker for evaluating DNA damage, because of the lack of baseline value standardization and reliability [66].

5. Other assays for biomarkers

a) Pulse radiolysis

The effectiveness of individual carotenoids as antioxidants can be determined by pulse radiolysis. Free-radical forms of several carotenoids, if they are sufficiently long-lived for their reduction potentials to be measured, have been detected by this technique. In addition, the rates of free radical scavenging by carotenoids have been reported [67]. Even though these experiments do not reproduce biological conditions, they contribute to better understanding of the underlying chemistry of carotenoid activity, as discussed in *Volume 4, Chapter 7.*

b) HPLC/mass spectrometry

A sensitive, selective, specific and rapid method, HPLC linked to electrospray ionization tandem mass spectrometry (LC/ESI-MS/MS) was developed and validated for the simultaneous determination of the Michael adducts between (*trans*)-4-hydroxynon-2-enal (HNE), one of the most reactive unsaturated aldehyde products of lipid peroxidation, and endogenous peptides containing histidine and cysteine [68]. The electrophilic nature of α,β-unsaturated aldehydes and ketones makes these compounds highly reactive with cellular nucleophiles and in particular with sulphydryl groups, the imidazolic nitrogen of histidine and the ε-amino group of lysine through the formation of Michael adducts [69]. As shown in Fig. 5 (left), C(3) of an α,β-unsaturated aldehyde is a strong electrophilic centre.

Histidine is one of the most reactive nucleophilic residues in proteins and is a primary reaction site for HNE addition [70] as shown in Fig. 5 (right).

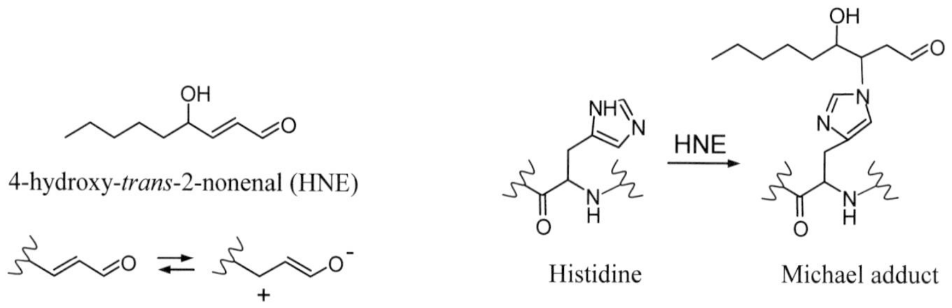

Fig. 5. Left: Tautomeric equilibrium of (*trans*)-4-hydroxynon-2-enal (HNE), illustrating the electrophilic site susceptible to nucleophilic attack by a histidine residue. Right: structure of the Michael adduct formed by reaction between HNE and histidine.

Most of the biological effects of reactive carbonyl-containing intermediate species, mainly α,β-unsaturated aldehydes, arise from the capacity of these compounds to react with the nucleophilic sites of proteins, to form advanced lipoxidation end-products [70,71], or react with DNA bases. Among the oxidation products of β-carotene, several α,β-unsaturated aldehydes and ketones such as trimethylcyclohexenone, β-cyclocitral and β-ionone, have been identified.

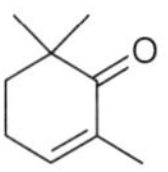

trimethylcyclohexenone (*1*)

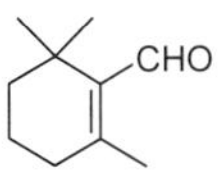

β-cyclocitral (*2*)

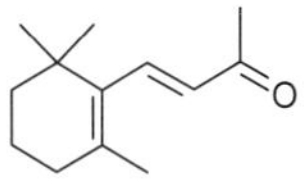

β-ionone (*3*)

Oxidation of β-carotene gives products that react with thiobarbituric acid to give a pink condensation product, suggesting the presence of reactive dicarbonyl derivatives, structurally related to malondialdehyde, among the oxidation products [51]. It is reasonable to hypothesize that some β-carotene oxidation products containing α,β-unsaturated aldehyde or ketone groups could react with nucleophilic biological targets such as cysteine, histidine or lysine residues in proteins, or with DNA bases, to give Michael and Schiff base adducts. This could explain some of the biological effects that have been reported recently for oxidation products of β-carotene and lycopene, such as inhibition of cell proliferation [72], pro-oxidant effects [73], enzyme inhibition [74] and DNA damage [75] (see *Chapter 18*). Therefore, a specific and sensitive assay able to identify the target macromolecules of the reaction products of carotenoids would be useful to elucidate the mechanism by which carotenoids and, in particular, the corresponding oxidative breakdown products, affect the biological response.

C. Studies of Antioxidant/Pro-oxidant Actions of Carotenoids

There have been many reports concerning the relative antioxidant efficacy of carotenoids, with varying results. Part of the problem is that different systems have been used to dissolve the carotenoids, initiate oxidative stress, and then evaluate efficacy. There is probably no single system that can accurately determine the antioxidant/pro-oxidant activity of carotenoids.

1. Studies *in vitro*

There are at least three kinds of reaction of carotenoids with radical species, namely radical addition, electron transfer to the radical, or allylic hydrogen abstraction. The radical addition/adduct formation mechanism [76] suggested that a lipid peroxyl radical ($ROO^{\bullet}$) might add to some positions on the carotenoid (CAR) polyene chain, resulting in the formation of a carbon-centred radical ($ROO\text{-}CAR^{\bullet}$). This resonance-stabilized radical would interfere with the propagating step in lipid peroxidation and would explain the many examples of the antioxidant effects reported for carotenoids in solution [5] (see Section **D**.2).

Electron transfer reactions have been reported that result in the formation of either the radical cation $CAR^{\bullet+}$, the radical anion $CAR^{\bullet-}$, or the neutral alkyl radical $CAR^{\bullet}$. The carotenoid radical cation is frequently detected by very fast spectroscopic techniques such as laser flash photolysis. This radical has been observed in studies of photosynthesis, and it has been proposed to play a role in photoprotection in photosystem 2 [77] (*Volume 4, Chapter 14*).

Hydrogen abstraction processes have been suggested [78] following the detection of 4-methoxy-β-carotene (*4*) and 4,4′-dimethoxy-β-carotene (*5*) when β-carotene is treated with either AIBN or AMVN in the presence of small amounts of methanol. Exposure of β-carotene to cigarette smoke resulted in the production of 4-nitro-β-carotene (*6*), also presumably *via* hydrogen abstraction at the allylic C(4) position [79].

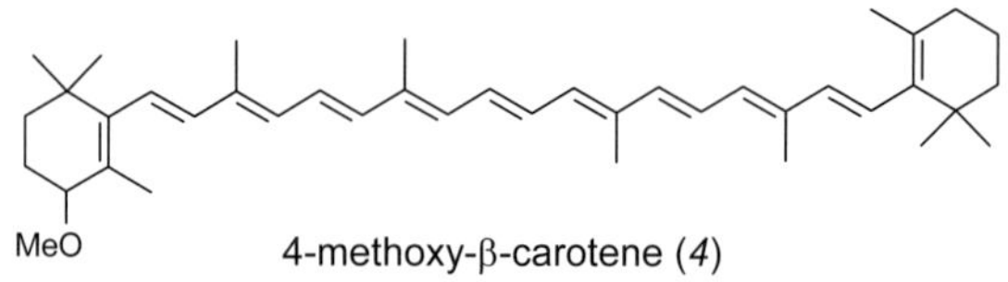

4-methoxy-β-carotene (*4*)

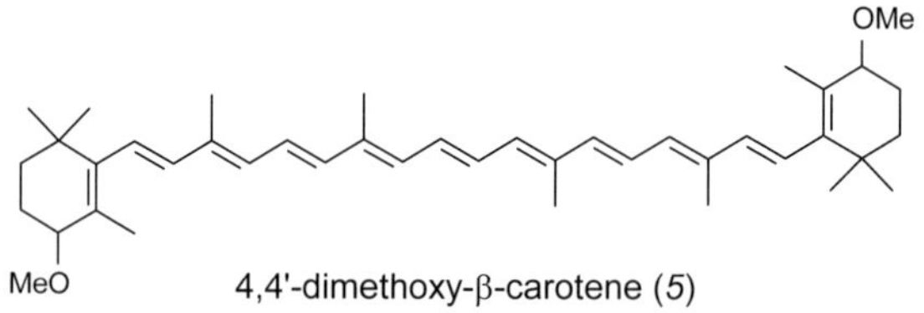

4,4'-dimethoxy-β-carotene (*5*)

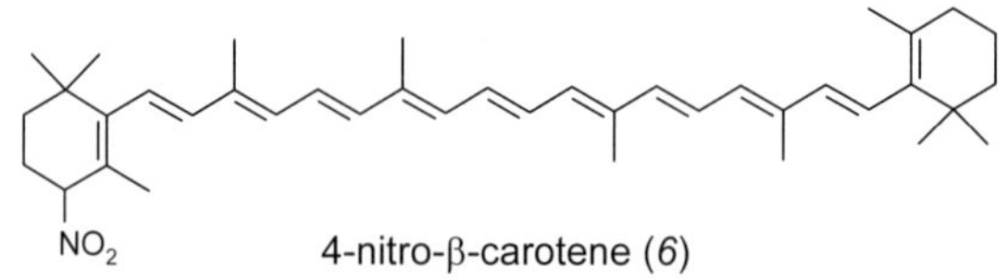

4-nitro-β-carotene (*6*)

Studies on the antioxidant actions of carotenoids have been carried out in artificial membranes (liposomes, micelles) [18,80], isolated LDL [81] and tissue homogenates [82] in an attempt to mimic biological systems. It should be noted that the nature of the interaction between the carotenoids and the matrix in which they are studied dictates the effect, *e.g.* the antioxidant activity of carotenoids depends on the incorporation of carotenoids in the lipid bilayer [83]. In addition, a recent study [84] indicated that certain carotenoids such as astaxanthin (**404-406**), which can preserve membrane structure, exhibited significant antioxidant action.

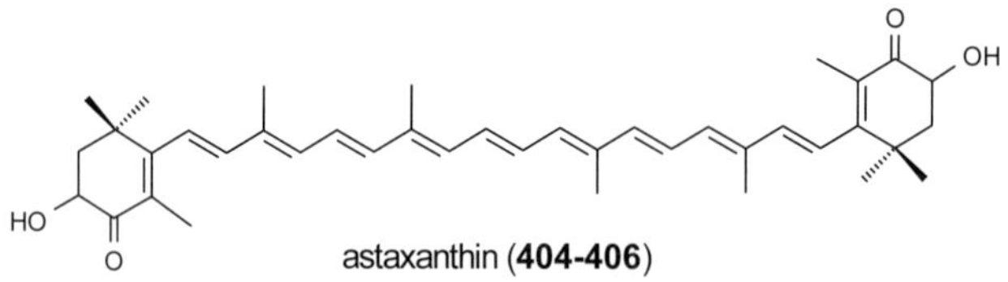

astaxanthin (**404-406**)

Many of the investigations *in vitro* used the development of thiobarbituric acid-related substances (TBARS) as an index of lipid peroxidation [85], but this assay is quite unspecific, as previously discussed. The oxidation of β-carotene itself by either nitrogen dioxide or oxygen results in measurable TBARS activity [51]. More direct effects have been reported; DNA in human promyelocytic leukaemia HL-60 cells exposed to a source of peroxynitrous acid was protected better by the prior administration of β-carotene than of either ascorbate or the water-soluble α-tocopherol analogue Trolox (6-hydroxy-2,5,7,8-tetramethylchroman-2-carboxylic acid [86].

2. Studies *ex vivo*

The ability to detect and/or monitor antioxidant action of any type of molecule *in vivo* is limited by the availability of adequate biomarkers. There is no single method yet that can assess the oxidant stress response or total antioxidant capacity in animals or humans.

Lipoprotein particles from animals or humans that have ingested carotenoids, either as part of their diet or *via* supplementation, can be isolated and evaluated. In the past few years, more investigators are using this approach, which is assumed to insert the carotenoids 'appropriately' in the LDL particle or target membrane. The oxidation of LDL particles was reported to be lower when they were enriched with carotenoids through dietary intervention with fruits and vegetables or by supplementation with carotenoids [87-89]. Supplementation with green vegetables did not protect LDL in either smokers or non-smokers, whereas supplementation with red vegetables was protective, but only in non-smokers [90]. Lycopene-containing, tomato-based products were reported to be effective against LDL oxidation [91] whereas pure lycopene was ineffective [87]. The variability of these results might be attributed to differences in the length of time for which the diet was supplied, in degrees of changes in the plasma carotenoid levels and, certainly, in study populations. Also, when fruits and vegetables are added to the diet, vitamin C and other potential antioxidants such as polyphenols and flavonoids may increase as well as plasma carotenoids, and could be responsible for any changes observed in LDL oxidation. Conflicting results have also been reported in studies with other carotenoids. In some cases, lycopene (**31**), α-carotene (**7**), β-cryptoxanthin (**55**), zeaxanthin (**119**) and lutein (**133**) were reported to be effective as antioxidants [92] but, in some studies in which β-carotene was effective, the addition of either lutein or lycopene actually increased LDL oxidation [93]. A 12-week period of daily supplementation with either 13 mg lycopene or 112 mg β-carotene resulted in an increase in carotenoids in LDL, but no change in LDL oxidizability [81]. Determination of antioxidant nutrient concentrations in LDL, and of LDL resistance as expressed by lag time to oxidation, led to the conclusion that LDL composition did not predict resistance to Cu-stimulated oxidation [94].

Whether dietary or supplemental carotenoids have any protective effect against LDL oxidation, therefore, remains unresolved.

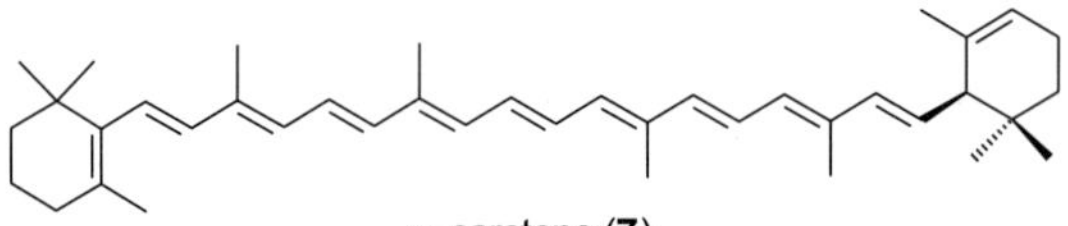

α-carotene (**7**)

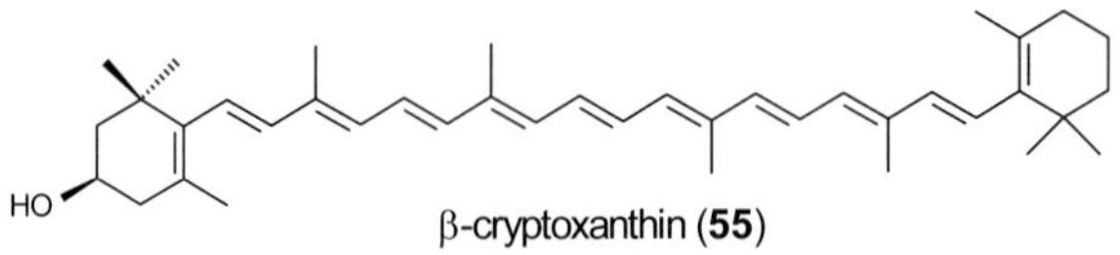

β-cryptoxanthin (**55**)

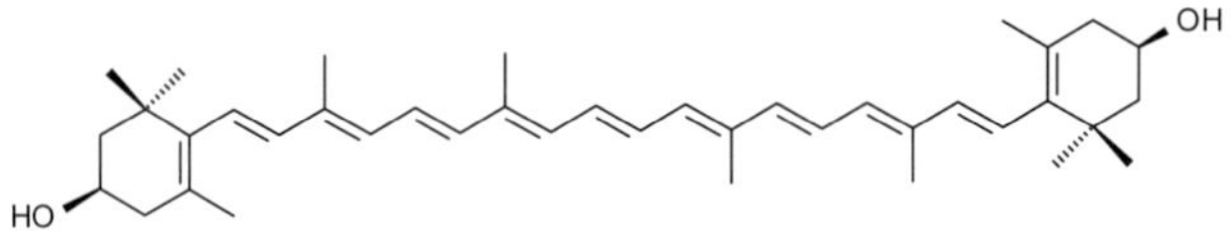

zeaxanthin (**119**)

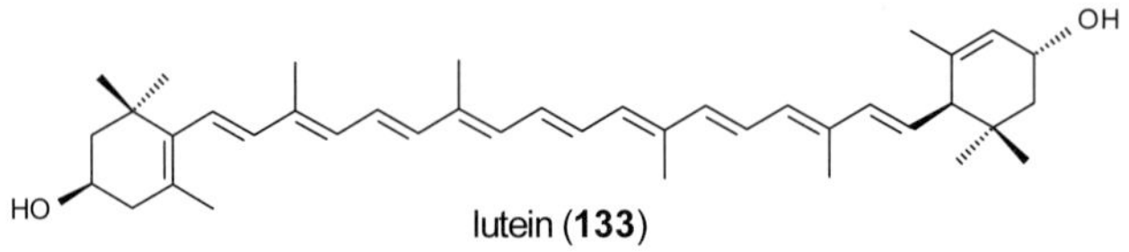

lutein (**133**)

3. Studies *in vivo*

The lack of appropriate biomarkers to determine oxidative stress *in vivo* makes it difficult to determine whether dietary carotenoids alter the oxidative stress in humans. Various animal species, including ferrets [95], gerbils [96] and pre-ruminant calves [97], have been used to study carotenoid absorption. Most experimental animals, however, require large, pharmacological doses of carotenoids, because their ability to absorb carotenoids is low.

For many years, the rather unspecific biomarker TBARS was used to determine lipid peroxidation. It should be noted, however, that any conjugated dialdehyde in the plasma can react with thiobarbituric acid resulting in increased TBARS values, which are usually attributed to increased MDA concentrations. Women on carotenoid-deficient diets showed increased plasma MDA levels [50] but this effect could be reversed when the diets were supplemented with a mixture of carotenoids, strongly supporting the idea that dietary carotenoids can serve to decrease oxidative stress in humans. A recent report [80] found significantly decreased plasma oxidizability in subjects given a high fruit and vegetable diet, and this was followed by a significant increase in oxidizability after transfer to a low fruit and vegetable diet. This, therefore, supports the attribution of antioxidant activity *in vivo* to some component(s) in the fruit and vegetables, which could include carotenoids.

There are a few studies on the effect of carotenoids on the enzymic antioxidant defence systems. For example, no difference was found in superoxide dismutase (SOD) activity in haemolysates of washed erythrocytes from HIV patients who had been given 30 mg/day β-carotene for 1 year [98].

When carotenoids are administered at fairly high doses, they can accumulate in the skin. This phenomenon has been the subject of many investigations to determine if this accumulation can lead to sun protection, which may or may not be related to an antioxidant action (see *Chapter 16*). When a group of volunteers were supplemented for 24 weeks with natural carotene (99% β-carotene), starting at 30 mg/day and increasing to 90 mg/day by the end of the experiment, modest protective effects against sunlight were observed, but no significant dose-dependent inhibition was seen in a commercial assay for lipid peroxidation [99]. It is not clear whether the increased tissue concentrations of carotenoids are directly associated with antioxidant activity. It has been suggested, however, that the carotenoids found in the eye, *e.g.* in the ciliary body, in the retinal pigment epithelium and the choroid, may be acting as antioxidants [100] (see *Chapter 15*).

It has been suggested that oxidative stress plays a role in the early stages of the pathophysiological processes of many chronic diseases. Significantly elevated basal DNA damage, as revealed by the comet assay, was reported in patients suffering from coronary artery disease (53 cases, 28-68 years, 42 controls, 30-67 years) [101], breast cancer (70 cases, 70 controls, 53 years) [102] (40 cases, 60 controls) [103], and head and neck squamous cell cancer (38 cases, 13-78 years, 44 controls, 44-78 years) [104] compared to the level of DNA damage in healthy subjects.

The relationship between DNA damage and the consumption of fruits and vegetables has been suggested by the observation of lower DNA damage in the summer than in the winter, corresponding to the difference in the seasonal intake of dietary antioxidants [105]. However, intervention trials involving increased fruit and vegetable intake have shown mixed results. On the one hand, decreased oxidative DNA damage has been reported [106] with 12 servings/day of fruit and vegetables for 14 days, whereas, in another study, daily consumption of 600 g of fruit and vegetables for 4 weeks showed no effect on DNA damage and repair [107], as determined by urinary and blood 8-hydroxydeoxyguanosine.

Several short-term intervention studies involving carotenoid-rich diets and assay of lymphocyte DNA damage by single-cell gel electrophoresis (comet assay) have been reported. Cross-over studies with healthy female subjects fed 600 mL/day of orange juice for 21 days (16 subjects, 20-27 years) [108], 25g of tomato puree for 14 days [109], or for 21 days (10 subjects, mean age 23.1 years) [110] and/or 150 g of spinach per day for 21 days (9 subjects, mean age 25.2 years) [111] resulted in reduced oxidative DNA damage. A similar reduction was observed in a group of 26 men (mean age 25.4 years) and women (mean age 26 years) treated with 250 mL/day of tomato extract drink for 26 days [112]. Also, a 14-day intervention with tomato juice, carrot juice or dried spinach powder (23 non-smoking male subjects, age 27-40 years) [113] or a polyphenol-rich juice [114] was reported to be beneficial against basal DNA damage in healthy men. In addition, dietary interventions for 3 weeks with tomato sauce, providing 30 mg lycopene per day, resulted in significantly decreased oxidative DNA damage in prostate cancer patients [115,116]. On the other hand, a recent study [117] showed no difference in DNA damage between intervention groups of healthy, well-nourished non-smoking men who received two, five or eight servings/day of fruit and vegetables.

Assuming that, in humans, the bioavailabilities of lycopene and lutein supplements are similar to those of pureed and oil-containing tomato-based foods [118] and green leafy vegetables [119], respectively, (see *Chapter 7*) similar biological actions of pure forms could be expected.

Table 2. Studies that have been reported on the effects of antioxidants on the damage and repair of lymphocytes.

Subjects	**Intervention**	**DNA damage/ repair**	**Ref**
Male (50-59 yr) smokers (n=50) non-smokers (n=50)	Vit C 100 mg, vit E 280 mg and β-carotene 25 mg, daily for 20 wks Placebo controlled	Decreased DNA damage	[120]
Male and female (n=40, 25-45 yr)	α/β-Carotene, lutein or lycopene 15 mg, 12 weeks Placebo controlled	No effect Inverse correlation between serum carotenoids and oxidized pyrimidines	[123]
Male (n=5) female (n=3) (24-34 yr)	β-Carotene, lutein or lycopene consecutively, 15 mg/day for 7 days, with 3wk wash-out	Increased DNA repair by β-carotene and lycopene not by lutein	[124]
Female (n=37, 50-70 yr)	12 mg of either lutein, β-carotene, or lycopene or 4 mg each of lutein, β-carotene, and lycopene, daily for 57 days Placebo controlled	Decreased DNA damage	[125]
Male (n=64, 18-50 yr)	Vitamin C 60 mg daily for 21 days Placebo controlled	No effect	[127]
Male (non-smokers) (n=64, 18-50 yr)	8.2 mg β-Carotene, 3.7 mg α-carotene and 1.75 mg α-tocopherol, daily for 21 days Placebo controlled	Increased DNA repair	[127]
Male and female (n=77, age ≥40)	0, 6.5, 15, or 30 mg lycopene/day for 57 days Placebo controlled	Decreased DNA damage in 30mg/day group only	[126]

Intervention studies with a combination of antioxidant supplements have consistently shown protective effects against DNA damage (Table 2). A combination of micronutrients in a relatively high dose (*i.e.* 100 mg of vitamin C, 280 mg of vitamin E and 25 mg of β-carotene) per day for 20 weeks resulted in a significant decrease in basal DNA damage in adult men aged 50-59 years [120]. Supplementation for 21 days with a combination of carotenoids (8 mg lycopene, 0.5 mg β-carotene) and vitamin C (11 mg) in younger females (mean age 25.2 years) [121] or a combination of carotenoids (8.2 mg β-carotene, 3.7 mg α-carotene) and α-tocopherol (1.75 mg) in males, 18-50 years, has been reported to be protective against oxidative DNA damage [122]. However, it is not clear whether there is any effect of a single carotenoid against DNA damage. No effect on endogenous DNA damage was reported following supplementation with 15 mg/day of either α-carotene and β-carotene, lutein or lycopene for 12 weeks in men and women (25-45 years) in a placebo-controlled parallel study design [123]. This study did, however, reveal an inverse correlation between total serum carotenoids and oxidized pyrimidines. On the other hand, another study using the same amount of lutein, β-carotene or lycopene supplementation (15 mg/day) successively, each for 1 week, showed significant increases in DNA repair in younger men and women (24-34 years) [124]. Further, a recent study [125] reported that there was a significant decrease in basal DNA damage after 15 days supplementation with either 12 mg of a single carotenoid or 12 mg of a combination of carotenoids (4 mg each of lutein, β-carotene and lycopene) in elderly women (50-70 years). The protective effect was maintained throughout the study period of 57 days. A dose-response study (0, 6.5, 15, or 30 mg/day lycopene) indicated that supplementation with 30mg/day lycopene for 8 weeks resulted in significantly decreased DNA damage in healthy adults [126].

In general, carotenoid-rich diets and a combination of carotenoids show a stronger protective effect against DNA damage than does a single carotenoid. Further studies are needed to establish any effect of physiological doses of carotenoids in combination with other antioxidants contained in fruits and vegetables on oxidative DNA damage, and to support the role of a diet rich in fruit and vegetables in the prevention of chronic diseases such as cardiovascular diseases and cancer.

D. Factors that Affect Antioxidant/Pro-oxidant Actions of Carotenoids

Carotenoids can exhibit pro-oxidant as well as antioxidant behaviour, as first described for the auto-oxidation of β-carotene [128].

An impetus to the study of the factors that determine whether carotenoids, in particular β-carotene, show antioxidant or pro-oxidant behaviour was the release of the findings of the Alpha-Tocopherol, Beta-Carotene Cancer Prevention Study (ATBC) [129], an intervention trial of α-tocopherol and β-carotene for the primary prevention of lung cancer. This study showed that, in Finnish male smokers, those who received supplemental β-carotene had an 18% increase in lung cancer incidence. Then the U.S. Carotene and Retinol Efficacy Trial (CARET) was terminated nearly 2 years early because the group receiving the combination of β-carotene plus retinol had a 28% increase in lung cancer incidence [130]. Various theories were put forward to explain why β-carotene may exhibit such pro-carcinogenic activity under these conditions. Because one of the features of both these intervention studies was the high dose of β-carotene given as supplement, several studies *in vitro* addressed the question of whether carotenoids can become harmful at high concentrations. Moreover, because the increased cancer incidence was localized in lung, a tissue exposed to a high partial pressure of oxygen (tracheal or bronchial air 150 torr, alveolar air 105 torr) relative to that of venous blood (40 torr) and the surface of the alveolar cell (20 torr), the effect of oxygen was also considered to be a parameter that could affect the behaviour of β-carotene. In a recent prospective study with 59,910 French women [131], β-carotene intake was inversely associated with the risk of tobacco-related cancers among non-smokers and there was a statistically significant dose-dependent relationship, whereas high β-carotene intake was directly associated with risk among smokers, after a median follow-up of 7.4 years. In a recent systematic review, six randomized clinical trials that examined the efficacy of β-carotene supplements and 25 prospective observational studies that assessed the associations between carotenoids and lung cancer were subjected to random-effects meta-analysis. An inverse association between carotenoids and lung cancer was detected, but the decreases in risk were generally small and not statistically significant. It was concluded that the inverse association may be due to the carotenoid measurements being a marker of a healthier lifestyle *i.e.* higher fruit and vegetable consumption, or to residual confounding by smoking [132].

Factors that may determine the switch from antioxidant to pro-oxidant behaviour are discussed in the following sections.

1. Concentration of carotenoids

The ability of supplementary carotenoids at different concentrations to protect cells against oxidatively induced DNA damage (as measured by the comet assay), and membrane integrity (as measured by ethidium bromide uptake) has been studied [133]. Either lycopene or β-carotene afforded protection against DNA damage only at relatively low concentrations (1-3 μM) whereas, at higher concentrations (4-10 μM), the ability to protect the cell against such oxidative damage was rapidly lost and, indeed, the presence of carotenoids increased the extent of the DNA damage. Similar data were obtained when protection against membrane damage was studied. An increased intracellular level of reactive oxygen species (ROS) was found in adenocarcinoma cells treated with high concentrations of β-carotene, leading to growth inhibition and apoptosis [134]. The pro-oxidant effect of β-carotene was also observed *in vivo*. Excess dietary β-carotene enhanced lipid peroxidation in animals exposed to different conditions such as methyl mercuric chloride intoxication or tocopherol depletion [8].

In recent reviews [9,135], several hypotheses have been proposed for the molecular mechanism involved in the pro-oxidant effect of high concentrations of β-carotene. The main hypotheses are that high concentrations of β-carotene can lead to one or more effects.

(i) A more favourable formation of β-carotene peroxyl radical and/or a faster rate of β-carotene auto-oxidation, leading to the formation of $O_2^{\bullet-}$.
(ii) Modification of iron concentrations, increasing the production of ROS through a Fenton reaction. This can be exacerbated by carotenoids that are characterized by a low oxidation potential (β-carotene has the lowest oxidation potential) which recycle Fe^{2+} by reducing Fe^{3+} ions, thus inducing a carotenoid-driven Fenton reaction [136].
(iii) ROS formation *via* induction of various cytochrome P450 isoforms.
(iv) Formation of aggregates that crystallize out of solution. Such carotenoid aggregates (*Volume 4, Chapter 5*) have been directly observed in membranes, and their presence is thought to have a profound effect on the properties of the membrane itself, by leading to an increase in membrane fluidity, which could result in a pro-oxidant effect.
(v) Formation of oxidation products that exert a pro-oxidant effect.

2. Oxygen tension

Oxygen tension greatly affects the switch between antioxidant and pro-oxidant effects of β-carotene [8]. The pro-oxidant effect of β-carotene induced by a high partial pressure of oxygen has been demonstrated in several experimental models such as rat liver microsomes [137], isolated DNA, and cells [138,139]. The effect of oxygen on the antioxidant/pro-oxidant character of β-carotene has been attributed mainly to a proposed equilibrium between carotenoid radicals and oxygen (Equations 1-6, Scheme 3) [2,140].

$$\mathrm{ROO^{\bullet} + CAR \rightarrow ROO{-}CAR^{\bullet}} \quad (1)$$
$$\mathrm{ROO{-}CAR^{\bullet} + O_2 \leftrightarrow ROO{-}CAROO^{\bullet}} \quad (2)$$
$$\mathrm{ROO{-}CAROO^{\bullet} + CAR \rightarrow ROO{-}CAROOH + CAR^{\bullet}} \quad (3)$$
$$\mathrm{ROO{-}CAROO^{\bullet} + LH \rightarrow ROO{-}CAROOH + L^{\bullet}} \quad (4)$$
$$\mathrm{CAR^{\bullet} + O_2 \leftrightarrow CAROO^{\bullet}} \quad (5)$$
$$\mathrm{L^{\bullet} + O_2 \leftrightarrow LOO^{\bullet}} \quad (6)$$

Scheme 3

According to this scheme, the peroxyl radical-carotenoid adduct (ROO-CAR$^{\bullet}$) is generated by reaction of a carotenoid (CAR) with peroxyl radical (Eq. 1). This adduct then reacts reversibly with oxygen to form the corresponding peroxyl radical (ROO-CAROO$^{\bullet}$) (Eq. 2). At high oxygen concentration, the equilibrium shifts in favour of formation of ROO-CAROO$^{\bullet}$, which can then either react with CAR, inducing auto-oxidation and the formation of the radical CAR$^{\bullet}$ (Eq. 3), or with lipids, thus perpetuating lipid peroxidation (Eq. 4). The resultant radicals, CAR$^{\bullet}$ and L$^{\bullet}$ can also react reversibly with O_2 to form peroxyl radicals (Eq. 5 and 6). At low oxygen concentrations, however, the equilibrium (Eq. 2) shifts toward ROO-CAR$^{\bullet}$, which can react with another ROO$^•$ or form an epoxide *via* alkoxyl radical elimination, thus acting as an antioxidant.

The effect of oxygen partial pressure on the antioxidant activity of β-carotene has been studied [76] by measuring the initial rate of oxidation of tetralin and methyl linoleate in chlorobenzene. This study gave evidence that β-carotene exhibits good radical-trapping antioxidant behaviour only at low partial pressures of oxygen (less than 150 torr) and that at higher oxygen pressures (760 torr), β-carotene loses its antioxidant activity and shows an autocatalytic, pro-oxidant effect, particularly at relatively high concentrations. Changes in the pO_2 alter the ability of β-carotene to inhibit lipid peroxidation in a membrane model, rat liver microsomes [137]. At 150 torr pO_2, β-carotene acted as an antioxidant, inhibiting lipid peroxidation whereas, at 760 torr pO_2, it acted as a pro-oxidant, increasing MDA formation. In cultures of human lung cells treated with AAPH to induce oxidative stress, β-carotene was found to be an antioxidant at both low (0 torr) and normal (143 torr) oxygen tension conditions but, under high oxygen conditions (722 torr; 97% oxygen), the antioxidant effects were decreased, and a pro-oxidant effect was observed, as measured by isoprostane formation [139]. More recently, the formation of lipid peroxidation products (measured as conjugated dienes and TBARS) was studied in rat lung microsomal membranes enriched *in vitro* with varying β-carotene concentrations (from 1 to 10 nmol/mg protein) and then incubated with tar (6-25 μg/ml) under different pO_2. The exposure of the microsomal membranes to tar induced a dose-dependent increment of lipid peroxidation, which progressively increased as a function of pO_2. Under a low pO_2 (15 torr), β-carotene clearly acted as an antioxidant, inhibiting tar-induced lipid peroxidation. The carotenoid progressively lost its antioxidant efficiency as pO_2 increased (50-100 torr), however, and acted as a pro-oxidant at pO_2 from 100 to 760 torr in a dose-dependent manner [141].

3. Exposure to ultraviolet light

The antioxidant/pro-oxidant effect of β-carotene can also be greatly affected by exposure to UV radiation. β-Carotene is an efficient quencher of $^1O_2$, and this allows it to show protective effects against UVA-dependent matrix metalloprotease (MMP) expression in cell culture [142] and against lipid peroxidation in mouse skin [143]. The relationship between carotenoids and skin health is discussed in *Chapter 16*. Other studies, however, have demonstrated a pro-oxidant effect of β-carotene in cell cultures and a carcinogenic effect in mice on UVA exposure. In particular, a pro-oxidant effect of β-carotene in the 0.5-5 μM range was found in skin fibroblasts exposed to suberythaemal doses of UVA light (20 J/cm^2), as deduced from the induction of haem oxygenase-1 [144] and expression of interleukin-6 as markers of oxidative stress [145]. It has also been reported [146] that UVA irradiation of human skin fibroblasts led to a 10-fold to 15-fold rise in matrix metalloprotease-1 (MMP-1) mRNA, and that this increase was suppressed in the presence of low μM concentrations of vitamin E, vitamin C, or carnosic acid but not by β-carotene or lycopene (prepared in a special nanoparticle formulation). Indeed, in the presence of 0.5-1.0 μM β-carotene or lycopene, the UVA-induced MMP-1 mRNA was further increased by 1.5 to 2-fold. This increase was totally suppressed when vitamin E was included in the nanoparticle formulation.

In human dermal fibroblasts, the presence of β-carotene or lycopene (0.5-1.0 μM) led to a 1.5-fold rise in the UVA-induced haem oxygenase-1 mRNA levels. A pro-oxidant effect of β-carotene was reported in mouse fibroblasts under UVA irradiation; β-carotene (20 μM) enhanced DNA strand breaks, an effect which was significantly suppressed by co-incubation with flavonoids such as naringenin, rutin or quercetin [147]. Relatively stable cyclic mono- and diendoperoxides have been identified as first products of the reaction of β-carotene with singlet oxygen [148]. These products remain reactive in the dark and cause auto-oxidation of β-carotene. The cyclic endoperoxides have been considered as candidates to explain the unforeseen pro-oxidant activity of β-carotene.

4. Oxidative stress

Recently an interesting hypothesis has been proposed for the mechanism responsible for the switching between antioxidant/pro-oxidant effects of β-carotene [73,149,150]. According to this, under conditions of moderate oxidative stress, the antioxidant effects of β-carotene predominate, but under heavy oxidative stress, β-carotene undergoes an oxidative breakdown leading to the formation of reactive breakdown products which are responsible for the pro-oxidant activity and harmful effects of β-carotene. These β-carotene breakdown products include reactive aldehydes such as 8′-apo-β-caroten-8′-al (**482**), 10′-apo-β-caroten-10′-al (**499**), 12′-apo-β-caroten-12′-al (**507**), 14′-apo-β-caroten-14′-al (**513**), retinal (*7*), and short-chain products such as β-cyclocitral (*2*), β-ionone (*3*), 5,6-epoxy-β-ionone (*8*) and 4-oxo-β-ionone (*9*) [149].

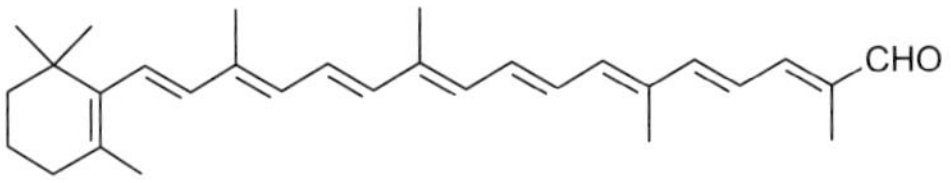

8'-apo-β-caroten-8'-al (**482**)

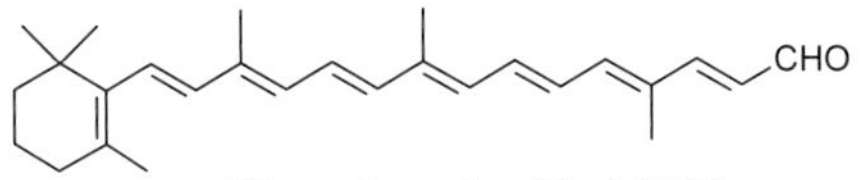

10'-apo-β-caroten-10'-al (**499**)

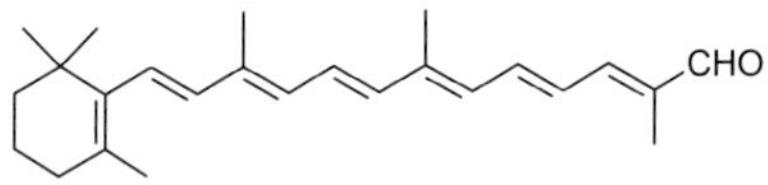

12'-apo-β-caroten-12'-al (**507**)

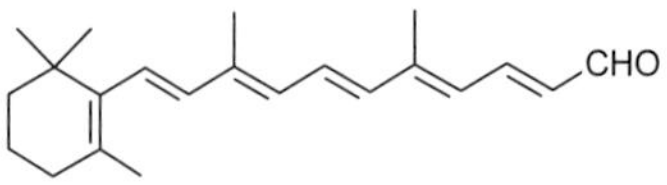

14'-apo-β-caroten-14'-al (**513**)

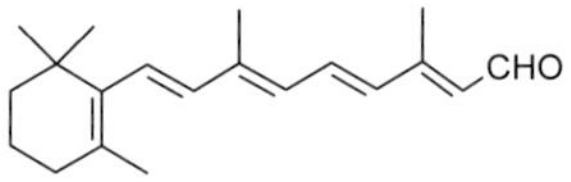

retinal (*7*)

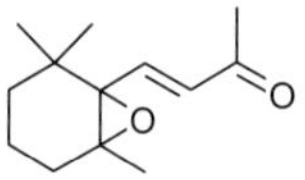

5,6-epoxy-β-ionone (*8*)

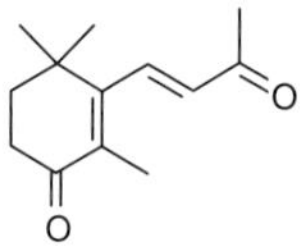

4-oxo-β-ionone (*9*)

The chemical reactivity of most of these compounds with other biomolecules is still unknown although it has been proposed that the reactivity and biological effects may be similar to those induced by the reactive aldehydes from lipid peroxidation, such as 4-hydroxynonenal (HNE) and MDA [149]. These β-carotene breakdown products were found to exert several damaging effects such as (i) inhibiting state 3 respiration in isolated rat liver mitochondria, which is accompanied by increased oxidative stress in the mitochondria, as reflected by a decrease in glutathione and protein SH groups and an increase of MDA, and (ii) genotoxic effects (micronuclei and chromosomal aberrations) at sub-micromolar concentrations [151]. This hypothesis could explain the pro-oxidant effects induced by a series of different factors such as O_2, UV, and general oxidants including smoke or hypochlorous acid (HOCl), that are known to switch on the pro-oxidant effects of β-carotene and are characterized by the induction of oxidative breakdown of carotenoids. The cytotoxic effects of aldehydes derived from breakdown of β-carotene, lutein and zeaxanthin on human retinal pigment epithelial cells (ARPE-19) have been examined. A significant increase of oxidative stress and ROS generation accompanied by an increased number of apoptotic cells was observed following treatment with the aldehydes [152]. A mixture of β-carotene breakdown products in primary hepatocytes showed a genotoxic potential at concentrations in the range100 nM and 1 μM, [153] and significantly enhanced the genotoxic effects of oxidative stress exposure [154].

The question of whether some of the reported beneficial effects of carotenoids may, in fact, be mediated by some of these oxidation products is treated in *Chapter 18*.

5. Interaction with membranes

Studies with model membranes enriched with polyunsaturated fatty acids have indicated that interaction with the membrane is a critical influence on the antioxidant/pro-oxidant activity of carotenoids. Differential effects of carotenoids on lipid peroxidation rates were partially attributed to their orientation and location, as determined by small angle X-ray diffraction (see *Volume 4, Chapters 5* and *10*). The apolar carotenoids lycopene and β-carotene, which disorder the membrane bilayer, show a potent pro-oxidant effect whereas astaxanthin preserves membrane structure and exhibits significant antioxidant activity [84,155].

6. Up-regulation of the receptor for advanced glycation endproducts (RAGE)

Receptors for advanced glycation endproducts (RAGE) have recently been implicated as promoters and/or amplifiers of oxidant-mediated cell death induced by diverse agents. Increased RAGE expression is observed in conditions that are associated with unbalanced production of reactive species, such as atherosclerosis and neurodegeneration. It was proposed that supplementation with retinol increases RAGE protein expression in cultured Sertoli cells, and that co-treatment with antioxidant reversed this effect. Moreover, the retinol-increased RAGE expression was observed only at concentrations that induce production of intracellular reactive species, as assessed by the DCFH assay [156].

E. Interactions of Carotenoids

The synergistic/cooperative interactions of hydrophilic and lipophilic antioxidants have been studied *in vitro* in various systems such as homogeneous phase solvent systems [17], liposomes [18,80], micelles, isolated LDL [157], cells [158] and tissue preparations [82].

1. Interactions between carotenoids

Possible interactions between carotenoids in terms of competition for incorporation into micelles, carotenoid exchange between lipoproteins, and inhibition of provitamin A cleavage have been reviewed [159]. It has been reported that β-carotene supplementation, which results in a moderate increase in the serum β-carotene concentration, does not significantly affect the serum concentrations of other carotenoids [160]. In the past few years, intervention studies with a combination of carotenoids have been reported. Supplementation with 24 mg/day of β-carotene for 12 weeks or an equal amount of a carotenoid mixture, containing lutein, β-carotene and lycopene, ameliorated UV-induced erythaema in humans [161]. In addition, a protective effect of a mixed carotenoid supplement [β-carotene, α-carotene, lycopene, lutein, bixin (**533**) and mixed paprika carotenoids] against LDL oxidation induced by fish oil, has been reported [162]. Furthermore, a recent study comparing the effect of individual carotenoids (12 mg each of lutein, β-carotene or lycopene) with that of an equal amount of mixed carotenoids (4 mg each of lutein, β-carotene and lycopene) against lymphocyte DNA damage clearly indicated synergistic interaction between carotenoids *in vivo* [105].

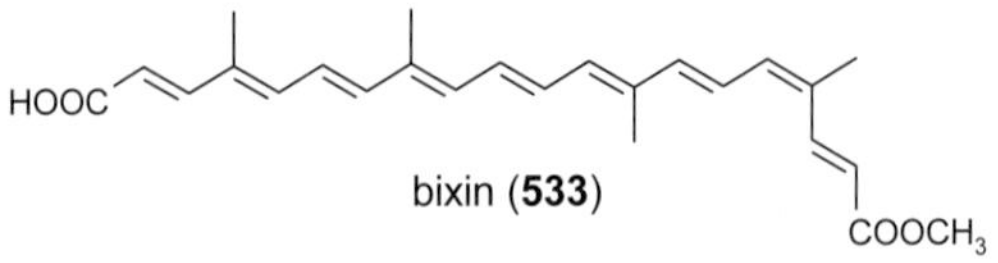

bixin (**533**)

2. Interactions of carotenoids with other antioxidants

Interactions of different antioxidants in plasma have been studied extensively in the past decade. In particular, work has focused both on the interactions between hydrophilic and lipophilic antioxidants, such as ascorbic acid and α-tocopherol [163], or carotenoids and ascorbic acid [164] and between lipophilic antioxidants (carotenoids and α-tocopherol) [82,165].

It has been reported that β-carotene, which is located in the lipophilic core of the membrane bilayer, can directly interact with water-soluble antioxidants. By scavenging radical species in a heterogeneous micellar environment, β-carotene can be converted into its radical cation $CAR^{\bullet+}$ or peroxyl radical cation $CAR\text{-}OO^{\bullet+}$ [166], which are more polar than β-carotene itself, and can be reoriented towards the hydrophilic compartment, allowing ascorbic acid to repair the β-carotene radical [167]. Other work [164] has also shown an interaction between β-carotene radical cations and ascorbic acid.

The combination of α-tocopherol and β-carotene has been reported to act cooperatively to slow down MDA formation initiated by the aqueous peroxyl radical generator, AAPH, in a liver microsomal membrane preparation [82]. β-Carotene added to preformed lipid bilayers produced much less of an antioxidant effect than β-carotene incorporated in the liposomes during bilayer formation [83]. It is possible that α-tocopherol reduces β-carotene peroxyl radicals (LOO-β-C-$OO^{\bullet}$) as well as β-carotene radical cations (β-$C^{\bullet+}$), as has been shown in a homogeneous solution [165]. In addition, β-carotene may recycle α-tocopherol from the α-tocopheroxyl radical (α-$TO^{\bullet}$) through electron transfer [168], although this possible mechanism of action should be studied further, because the reduction potential of β-carotene is reported to be lower than that of α-tocopherol [169,170]. In addition, a synergistic antioxidant activity of lycopene in combination with vitamin E in a liposome system has been reported [171].

It is interesting to note that daily supplementation with moderate doses of combined antioxidants (100 mg vitamin C, 100 mg vitamin E, 6 mg β-carotene and 50 μg selenium) significantly increased plasma antioxidant capacity and decreased chromosome aberrations in lymphocytes [172]. On the other hand, a meta-analysis of randomized trials with antioxidant supplements suggested that high doses of β-carotene [173] or α-tocopherol [174] led to significant increases in mortality due to all causes and no effect against coronary heart disease risk [175,176].

It is likely that physiological doses of a combination of water-soluble and fat-soluble antioxidant nutrients are required to establish an effective antioxidant network *in vivo*.

F. Conclusions: Possible Biological Relevance of Antioxidant/Pro-oxidant Actions of Carotenoids

Epidemiological evidence (see *Chapter 10*) continues to accumulate that diets high in fruits and vegetables [177-181] and carotenoids [182,183] are associated with a reduced risk of chronic diseases such as cardiovascular disease (*Chapter 14*). The evidence for a protective effect against cancer (*Chapter 13*) is much less clear [178], possibly, in part, because of measurement error [184]. Carotenoids may be among the group of antioxidants in fruits and vegetables that help to prevent damage caused by harmful reactive oxygen species, which are continuously produced in the body during normal cellular functioning and are introduced from exogenous sources [10]. It is believed that dietary supplementation with antioxidants, including carotenoids, can be a part of a protective strategy to minimize the oxidative damage in vulnerable populations, such as the elderly. Carotenoids have antioxidant activity *in vitro* at physiological oxygen tensions [139]. However the significance of the antioxidant effect of carotenoids *in vivo* remains controversial and difficult to demonstrate [122,185]. It should be pointed out that the metabolism and functions of carotenoids *in vivo* and *in vitro* may not be the same. For example, antioxidant nutrients can interact with each other during gastrointestinal absorption and metabolism [186-189]. Considering that the antioxidant system *in vivo*, which is finely balanced, and requires the right amounts, possibly an optimal range, of both the hydrophilic and lipophilic antioxidants, then carotenoids located in the core of a lipophilic compartment may be necessary for the antioxidant network to function properly in biological systems.

The benefits of carotenoids for eye health have been of great interest recently (see *Chapter 15*). It is generally accepted that lutein and zeaxanthin are associated with protection of the retina and retinal pigment epithelium from damage by light and oxygen [190,191]. For example, individuals ($n = 356$ Age-related Macular Degeneration (AMD) cases, $n = 520$ controls) in the highest quintile of carotenoid intake had a 43% lower risk for AMD compared with those in the lowest quintile [192], and among the specific carotenoids, lutein and zeaxanthin were most strongly associated with a reduced risk for AMD ($p = 0.001$). Laboratory studies, which had identified the macular pigments as lutein and zeaxanthin, are supportive of the epidemiological observations [193-195]. A recent prospective 18-year follow-up study of 71,494 women and 41,564 men, aged ≥50 and with no diagnosis of AMD, showed no association between lutein/zeaxanthin intake and neovascular AMD risk [196]. Further, a double blind randomized study indicated no evidence of an effect of 9 or 18 months of daily supplementation with a lutein-based nutritional supplement on visual function in healthy subjects [197].

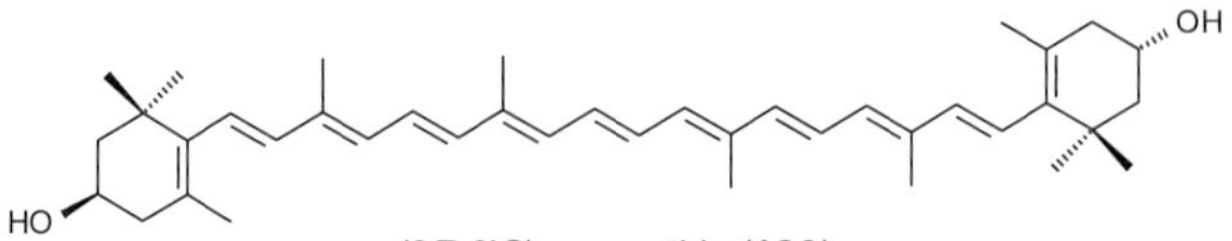

(3*R*,3'*S*)-zeaxanthin (**120**)

In a high fruit and vegetable diet, the intake of lutein is more than five times higher than that of zeaxanthin. It has been proposed that lutein and zeaxanthin in blood are taken up by the retina, where some of the lutein is then converted into (3*R*,3'*S*)-zeaxanthin (*meso*-zeaxanthin, **120**) [198]. It is possible that zeaxanthin may be more effective than lutein as an antioxidant in the central macula [193]. Lutein circulates in the blood at higher concentrations than zeaxanthin, but the concentration of zeaxanthin in the central macula is higher than that of lutein. However, increased consumption of dietary sources of lutein and zeaxanthin has been shown to increase macular pigment in some, but not all individuals. It has been reported that there was 27% prevalence of non-responders in terms of macular pigment density after the consumption of lutein/zeaxanthin-rich foods such as spinach [199]. It is not yet known what factors affect individual responses to lutein supplementation. Further research is required in an effort to determine the biological function of lutein and zeaxanthin in relation to eye health. A recent prospective study of 39,876 female health professionals [200] and a cross-sectional study (n = 1443) conducted in North India [201] reported that higher dietary intakes of lutein/zeaxanthin and vitamin E from food and supplements were significantly associated with decreased risks of cataract. In addition, astaxanthin has been reported to protect porcine lens crystallins from oxidative damage [202]. Even though its antioxidant activity [203] is still questionable [204], the beneficial effect of astaxanthin to improve vascular elastin and arterial wall thickness in hypertensive rats by modulating oxidative conditions has been reported [205].

The antioxidant property of lycopene may be one of the mechanisms for its putative effect against coronary heart disease (*Chapter 14*). It has been reported that lycopene inhibits LDL oxidation synergistically in combination with vitamin E or flavonoids [206]. In addition, in healthy women, increased tomato consumption resulted in reduced susceptibility of LDL to oxidation [207]. However, knowledge of the mechanism of action of lycopene as well as well-designed clinical studies are required to provide stronger evidence for a direct role of lycopene in coronary heart disease prevention [208].

In addition to its anti-atherogenic effect, the possible anti-carcinogenic effect of lycopene has been studied. Increased lycopene consumption resulted in significantly reduced leukocyte

DNA damage in prostate cancer patients [116]. Recently, in a pilot study involving supplementation with 15 mg lycopene/day for 6 months [209], patients with histologically proven benign prostate hyperplasia but free of prostate cancer showed significantly decreased prostate-specific antigen (PSA) levels and no change in the prostate, whereas the placebo group showed progression of prostate enlargement. An anti-carcinogenic property of lycopene has been demonstrated at the molecular level in an animal model, the ferret [210-212]. It has been suggested that lycopene may have protective effects against smoke-induced lung carcinogenesis through up-regulating insulin-like growth factor binding protein-3 (IGFBP-3) [213] as well as against smoke-induced gastric carcinogenesis through changing the gastric mucosal p53 phosphorylation [212] in ferrets (see *Chapter 11*). Whether this is related to any antioxidant action of lycopene remains to be determined.

Evidence has accumulated that high fruit and vegetable intake is associated with lower risk of chronic diseases such as cardiovascular diseases and age-related macular degeneration. It is possible that antioxidant nutrients such as carotenoids in fruits and vegetables can prevent or reduce damage from harmful free radicals that are produced in the body. However, intervention studies have failed to show a consistent beneficial effect of high doses of antioxidant supplementation against chronic diseases. One possible explanation for these apparently contradictory results between observational studies and intervention trials is that the antioxidant system *in vivo*, which is finely balanced, requires the right amount, possibly an optimal range, of both the hydrophilic and lipophilic antioxidants to be working properly. The optimal ranges of antioxidants might be achieved best by a balanced dietary fruit and vegetable intake but not by a high dose of only one or a limited mixture of antioxidant supplements, which could cause an imbalance of the antioxidant machinery leading in some cases to a pro-oxidant effect. In addition, other phytonutrients abundant in fruits and vegetables may not only exert unique biological functions but may also interact synergistically with well recognized antioxidants to promote antioxidant effects.

Various biomarkers have been developed to determine the biological functions of carotenoids and effects on genomic stability. It seems, however, that there is no single system that accurately determines the biological actions of carotenoids, due to the limitations of analytical techniques in relation to the lipophilicity of carotenoids and the model systems used to evaluate the effects. In addition, there are various factors such as carotenoid concentration, oxygen tension, UV exposure and oxidative stress that can affect the antioxidant activity of carotenoids. A synergistic/cooperative interaction between carotenoids and with other antioxidants such as tocopherols, ascorbic acid, and flavonoids, appears to play an important role in the biological antioxidant network. Therefore, an important future direction of research is to elucidate how best to improve our body defence systems against oxidative damage, which in turn might reduce the risk of chronic diseases, by means of dietary modification rather than by taking large amounts of antioxidant supplements.

References

[1] N. I. Krinsky, *Ann. NY Acad. Sci.*, **854**, 443 (1998).

[2] N. I. Krinsky, *Nutrition*, **17**, 815 (2001).

[3] N. I. Krinsky and K. J. Yeum, *Biochem. Biophys. Res. Commun.*, **305**, 754 (2003).

[4] S. A. Paiva and R. M. Russell, *J. Am. Coll. Nutr.*, **18**, 426 (1999).

[5] P. Palozza and N. I. Krinsky, *Meth. Enzymol.*, **213**, 403 (1992).

[6] C. A. Rice-Evans, J. Sampson, P. M. Bramley and D. E. Holloway, *Free Radic. Res.*, **26**, 381 (1997).

[7] R. Edge and T. G. Truscott, *Nutrition*, **13**, 992 (1997).

[8] P. Palozza, *Nutr. Rev.*, **56**, 257 (1998).

[9] A. J. Young and G. M. Lowe, *Arch. Biochem. Biophys.*, **385**, 20 (2001).

[10] L. Gate, J. Paul, G. N. Ba, K. D. Tew and H. Tapiero, *Biomed. Pharmacother.*, **53**, 169 (1999).

[11] K. Hensley and R. A. Floyd, *Arch. Biochem. Biophys.*, **397**, 377 (2002).

[12] E. Cadenas and K. J. Davies, *Free Radic. Biol. Med.*, **29**, 222 (2000).

[13] V. J. Thannickal and B. L. Fanburg, *Am. J. Physiol. Lung Cell. Mol. Physiol.*, **279**, L1005 (2000).

[14] C. Ceconi, A. Boraso, A. Cargnoni and R. Ferrari, *Arch. Biochem. Biophys.*, **420**, 217 (2003).

[15] M. Valkonen and T. Kuusi, *J. Lipid Res.*, **38**, 823 (1997).

[16] K. Uchida, M. Shiraishi, Y. Naito, Y. Torii, Y. Nakamura and T. Osawa, *J. Biol. Chem.*, **274**, 2234 (1999).

[17] E. Niki, T. Saito, A. Kawakami and Y. Kamiya, *J. Biol. Chem.*, **259**, 4177 (1984).

[18] K. Fukuzawa, K. Matsuura, A. Tokumura, A. Suzuki and J. Terao, *Free Radic. Biol. Med.*, **22**, 923 (1997).

[19] I. M. Lee, N. R. Cook, J. M. Gaziano, D. Gordon, P. M. Ridker, J. E. Manson, C. H. Hennekens and J. E. Buring, *JAMA*, **294**, 56 (2005).

[20] K. J. Yeum, G. Aldini, H. Y. Chung, N. I. Krinsky and R. M. Russell, *J. Nutr.*, **133**, 2688 (2003).

[21] G. Cao, H. M. Alessio and R. G. Cutler, *Free Radic. Biol. Med.*, **14**, 303 (1993).

[22] R. L. Prior, H. Hoang, L. Gu, X. Wu, M. Bacchiocca, L. Howard, M. Hampsch-Woodill, D. Huang, B. Ou and R. Jacob, *J. Agric. Food Chem.*, **51**, 3273 (2003).

[23] G. Aldini, K. J. Yeum, R. M. Russell and N. I. Krinsky, *Free Radic. Biol. Med.*, **31**, 1043 (2001).

[24] A. Ghiselli, M. Serafini, G. Maiani, E. Azzini and A. Ferro-Luzzi, *Free Radic. Biol. Med.*, **18**, 29 (1995).

[25] G. Cao, C. P. Verdon, A. H. Wu, H. Wang and R. L. Prior, *Clin. Chem.*, **41**, 1738 (1995).

[26] N. J. Miller, C. Rice-Evans, M. J. Davies, V. Gopinathan and A. Milner, *Clin. Sci.*, **84**, 407 (1993).

[27] R. Re, N. Pellegrini, A. Proteggente, A. Pannala, M. Yang and C. Rice-Evans, *Free Radic. Biol. Med.*, **26**, 1231 (1999).

[28] I. F. Benzie and J. J. Strain, *Anal. Biochem.*, **239**, 70 (1996).

[29] F. Tubaro, A. Ghiselli, P. Rapuzzi, M. Maiorino and F. Ursini, *Free Radic. Biol. Med.*, **24**, 1228 (1998).

[30] M. Kampa, A. Nistikaki, V. Tsaousis, N. Maliaraki, G. Notas and E. Castanas, *BMC Clin. Pathol.*, **2**, 3 (2002).

[31] B. Mayer, M. Schumacher, H. Brandstatter, F. S. Wagner and A. Hermetter, *Anal. Biochem,*. **297**, 144 (2001).

[32] N. Pellegrini, P. Riso and M. Porrini, *Nutrition*, **16**, 268 (2000).

[33] J. J. Castenmiller, S. T. Lauridsen, L. O. Dragsted, K. H. van het Hof, J. P. Linssen and C. E. West, *J. Nutr.*, **129**, 2162 (1999).

[34] S. Lussignoli, M. Fraccaroli, G. Andrioli, G. Brocco and P. Bellavite, *Anal. Biochem.*, **269**, 38 (1999).

[35] G. Cao, S. L. Booth, J. A. Sadowski and R. L. Prior, *Am. J. Clin. Nutr.*, **68**, 1081 (1998).

[36] D. D. Wayner, G. W. Burton, K. U. Ingold, L. R. Barclay and S. J. Locke, *Biochim. Biophys. Acta*, **924**, 408 (1987).

[37] G. Cao and R. L. Prior, *Clin. Chem.*, **44**, 1309 (1998).

[38] M. Serafini, A. Ghiselli and A. Ferro-Luzzi, *Eur. J. Clin. Nutr.*, **50**, 28 (1996).
[39] I. F. Benzie and Y. T. Szeto, *J. Agric. Food. Chem.*, **47**, 633 (1999).
[40] D. Rein, T. G. Paglieroni, D. A. Pearson, T. Wun, H. H. Schmitz, R. Gosselin and C. L. Keen, *J. Nutr.*, **130**, 2120S (2000).
[41] M. Serafini, G. Maiani and A. Ferro-Luzzi, *J. Nutr.*, **128**, 1003 (1998).
[42] F. Natella, M. Nardini, I. Giannetti, C. Dattilo and C. Scaccini, *J. Agric. Food Chem.*, **50**, 6211 (2002).
[43] G. Cao, R. M. Russell, N. Lischner and R. L. Prior, *J. Nutr.*, **128**, 2383 (1998).
[44] V. Böhm and R. Bitsch, *Eur. J. Nutr.*, **38**, 118 (1999).
[45] A. Bub, B. Watzl, L. Abrahamse, H. Delincee, S. Adam, J. Wever, H. Müller and G. Rechkemmer, *J. Nutr.*, **130**, 2200 (2000).
[46] N. J. Miller, J. Sampson, L. P. Candeias, P. M. Bramley and C. A. Rice-Evans, *FEBS Lett.*, **384**, 240 (1996).
[47] V. Böhm, N. L. Puspitasari-Nienaber, M. G. Ferruzzi and S. J. Schwartz, *J. Agric. Food Chem.*, **50**, 221 (2002).
[48] G. Aldini, I. Dalle-Donne, R. M. Facino, A. Milzani and M. Carini, *Med. Res. Rev.*, (2006).
[49] D. R. Janero, *Free Radic. Biol. Med.*, **9**, 515 (1990).
[50] Z. R. Dixon, F. S. Shie, B. A. Warden, B. J. Burri and T. R. Neidlinger, *J. Am. Coll. Nutr.*, **17**, 54 (1998).
[51] K. Kikugawa, K. Hiramoto and A. Hirama, *Free Radic. Res.*, **31**, 517 (1999).
[52] J. D. Morrow, B. Frei, A. W. Longmire, J. M. Gaziano, S. M. Lynch, Y. Shyr, W. E. Strauss, J. A. Oates and L. J. Roberts, 2nd, *New Engl. J. Med.*, **332**, 1198 (1995).
[53] P. Montuschi, P. J. Barnes and L. J. Roberts, 2nd, *FASEB J.*, 18, 1791 (2004).
[54] A. W. Longmire, L. L. Swift, L. J. Roberts, 2nd, J. A. Awad, R. F. Burk and J. D. Morrow, *Biochem. Pharmacol.*, **47**, 1173 (1994).
[55] L. J. Roberts, 2nd and J. D. Morrow, *Biochim. Biophys. Acta*, **1345**, 121 (1997).
[56] S. T. Mayne, M. Walter, B. Cartmel, W. J. Goodwin, Jr. and J. Blumberg, *Nutr. Cancer*, **49**, 1 (2004).
[57] A. C. Kaliora, G. V. Z. Dedoussis and H. Schmidt, *Atherosclerosis*, **187**, 1 (2006).
[58] L. Chancharme, P. Therond, F. Nigon, S. Zarev, A. Mallet, E. Bruckert and M. J. Chapman, *J. Lipid Res.*, **43**, 453 (2002).
[59] M. Takaku, Y. Wada, K. Jinnouchi, M. Takeya, K. Takahashi, H. Usuda, M. Naito, H. Kurihara, Y. Yazaki, Y. Kumazawa, Y. Okimoto, M. Umetani, N. Noguchi, E. Niki, T. Hamakubo and T. Kodama, *Arterioscler. Thromb. Vasc. Biol.*, **19**, 2330 (1999).
[60] S. Voutilainen, J. D. Morrow, L. J. Roberts, 2nd, G. Alfthan, H. Alho, K. Nyyssonen and J. T. Salonen, *Arterioscler. Thromb. Vasc. Biol.*, **19**, 1263 (1999).
[61] K. U. Ingold, V. W. Bowry, R. Stocker and C. Walling, *Proc. Natl. Acad. Sci. USA*, **90**, 45 (1993).
[62] H. Yasuda, N. Noguchi, M. Miki, W. Morinobu, K. Hirano, T. Ogihara, T. Tanabe, M. Mino, K. Terao and E. Niki, *Chem. Biol. Interact.*, **97**, 11 (1995).
[63] O. Ziouzenkova, A. Sevanian, P. M. Abuja, P. Ramos and H. Esterbauer, *Free Radic. Biol. Med.*, **24**, 607 (1998).
[64] A. Collins, M. Dusinska, M. Franklin, M. Somorovska, H. Petrovska, S. Duthie, L. Fillion, M. Panayiotidis, K. Raslova and N. Vaughan, *Environ. Mol. Mutagen.*, **30**, 139 (1997).
[65] A. R. Collins, *Mutation Res.*, **475**, 21 (2001).
[66] A. R. Proteggente, A. Rehman, B. Halliwell and C. A. Rice-Evans, *Biochem. Biophys. Res. Commun.*, **277**, 535 (2000).
[67] A. Mortensen, L. H. Skibsted, J. Sampson, C. Rice-Evans and S. A. Everett, *FEBS Lett.*, **418**, 91 (1997).
[68] G. Aldini, M. Orioli, M. Carini and R. M. Facino, *J. Mass Spectrom.*, **39**, 1417 (2004).
[69] M. Carini, G. Aldini and R. M. Facino, *Mass Spectrom. Rev.*, **23**, 281 (2004).
[70] K. Uchida, *Free Radic. Biol. Med.*, **28**, 1685 (2000).
[71] G. Poli and R. J. Schaur, *IUBMB Life*, **50**, 315 (2000).

[72] E. Nara, H. Hayashi, M. Kotake, K. Miyashita and A. Nagao, *Nutr. Cancer,* **39**, 273 (2001).
[73] W. Siems, O. Sommerburg, L. Schild, W. Augustin, C. D. Langhans and I. Wiswedel, *FASEB J.*, **16**, 1289 (2002).
[74] W. G. Siems, O. Sommerburg, J. S. Hurst and F. J. van Kuijk, *Free Radic. Res.*, **33**, 427 (2000).
[75] S. L. Yeh and M. L. Hu, *Free Radic. Res.*, **35**, 203 (2001).
[76] G. W. Burton and K. U. Ingold, *Science*, **224**, 569 (1984).
[77] C. A. Tracewell, A. Cua, D. H. Stewart, D. F. Bocian and G. W. Brudvig, *Biochemistry*, **40**, 193 (2001).
[78] A. A. Woodall, S. W. Lee, R. J. Weesie, M. J. Jackson and G. Britton, *Biochim. Biophys. Acta*, **1336**, 33 (1997).
[79] D. L. Baker, E. S. Krol, N. Jacobsen and D. C. Liebler, *Chem. Res. Toxicol.*, **12**, 535 (1999).
[80] A. A. Woodall, G. Britton and M. J. Jackson, *Biochem. Soc. Trans.*, **23**, 133S (1995).
[81] Y. L. Carroll, B. M. Corridan and P. A. Morrissey, *Eur. J. Clin. Nutr.*, **54**, 500 (2000).
[82] P. Palozza and N. I. Krinsky, *Arch. Biochem. Biophys.*, **297**, 184 (1992).
[83] D. C. Liebler, S. P. Stratton and K. L. Kaysen, *Arch. Biochem. Biophys.*, **338**, 244 (1997).
[84] H. P. McNulty, J. Byun, S. F. Lockwood, R. F. Jacob and R. P. Mason, *Biochim. Biophys. Acta*, **1768**, 167 (2007).
[85] P. Palozza, G. Agostara, E. Piccioni and G. M. Bartoli, *Arch. Biochem. Biophys.*, **312**, 88 (1994).
[86] K. Hiramoto, S. Tomiyama and K. Kikugawa, *Free Radic. Res.*, **30**, 21 (1999).
[87] T. R. Dugas, D. W. Morel and E. H. Harrison, *Free Radic. Biol. Med.*, **26**, 1238 (1999).
[88] Y. Levy, M. Kaplan, A. Ben-Amotz and M. Aviram, *Isr. J. Med. Sci*,. **32**, 473 (1996).
[89] Y. Lin, B. J. Burri, T. R. Neidlinger, H. G. Muller, S. R. Dueker and A. J. Clifford, *Am. J. Clin. Nutr.*, **67**, 837 (1998).
[90] M. Chopra, M. E. O'Neill, N. Keogh, G. Wortley, S. Southon and D. I. Thurnham, *Clin. Chem.*, **46**, 1818 (2000).
[91] S. Agarwal and A. V. Rao, *Lipids*, **33**, 981 (1998).
[92] O. M. Panasenko, V. S. Sharov, K. Briviba and H. Sies, *Arch. Biochem. Biophys.*, **373**, 302 (2000).
[93] T. R. Dugas, D. W. Morel and E. H. Harrison, *J. Lipid Res.*, **39**, 999 (1998).
[94] A. J. Wright, S. Southon, M. Chopra, A. Meyer-Wenger, U. Moser, F. Granado, B. Olmedilla, B. Corridan, I. Hinninger, A. M. Roussel, H. van den Berg and D. I. Thurnham, *Br. J. Nutr.*, **87**, 21 (2002).
[95] X. D. Wang, N. I. Krinsky, R. P. Marini, G. Tang, J. Yu, R. Hurley, J. G. Fox and R. M. Russell, *Am. J. Physiol.*, **263**, G480 (1992).
[96] J. Pollack, J. M. Campbell, S. M. Potter and J. W. Erdman Jr., *J. Nutr.*, **124**, 869 (1994).
[97] C. L. Poor, T. L. Bierer, N. R. Merchen, G. C. Fahey Jr., M. R. Murphy and J. W. Erdman Jr., *J. Nutr.*, **122**, 262 (1992).
[98] M. C. Delmas-Beauvieux, E. Peuchant, A. Couchouron, J. Constans, C. Sergeant, M. Simonoff, J. L. Pellegrin, B. Leng, C. Conri and M. Clerc, *Am. J. Clin. Nutr.*, **64**, 101 (1996).
[99] A. Lee, D. I. Thurnham and M. Chopra, *Free Radic. Biol. Med.*, **29**, 1051 (2000).
[100] P. S. Bernstein, F. Khachik, L. S. Carvalho, G. J. Muir, D.-Y. Zhao and N. B. Katz, *Exp. Eye Res.*, **72**, 215 (2001).
[101] R. Demirbag, R. Yilmaz and A. Kocyigit, *Mutation Res.*, **570**, 197 (2005).
[102] P. Sanchez, R. Penarroja, F. Gallegos, J. L. Bravo, E. Rojas and L. Benitez-Bribiesca, *Arch. Med. Res.*, **35**, 480 (2004).
[103] T. R. Smith, M. S. Miller, K. K. Lohman, L. D. Case and J. J. Hu, *Carcinogenesis*, **24**, 883 (2003).
[104] O. Palyvoda, J. Polanska, A. Wygoda and J. Rzeszowska-Wolny, *Acta Biochim. Pol.*, **50**, 181 (2003).
[105] M. Dusinska, B. Vallova, M. Ursinyova, V. Hladikova, B. Smolkova, L. Wsolova, K. Raslova and A. R. Collins, *Food Chem. Toxicol.*, **40**, 1119 (2002).
[106] H. J. Thompson, J. Heimendinger, A. Haegele, S. M. Sedlacek, C. Gillette, C. O'Neill, P. Wolfe and C. Conry, *Carcinogenesis*, **20**, 2261 (1999).

[107] P. Moller, U. Vogel, A. Pedersen, L. O. Dragsted, B. Sandstrom and S. Loft, *Cancer Epidemiol. Biomarkers. Prev.*, **12**, 1016 (2003).
[108] P. Riso, F. Visioli, C. Gardana, S. Grande, A. Brusamolino, F. Galvano, G. Galvano and M. Porrini, *J. Agric. Food Chem.*, **53**, 941 (2005).
[109] M. Porrini and P. Riso, *J. Nutr.*, **130**, 189 (2000).
[110] P. Riso, A. Pinder, A. Santangelo and M. Porrini, *Am. J. Clin. Nutr.*, **69**, 712 (1999).
[111] M. Porrini, P. Riso and G. Oriani, *Eur. J. Nutr.*, **41**, 95 (2002).
[112] M. Porrini, P. Riso, A. Brusamolino, C. Berti, S. Guarnieri and F. Visioli, *Br. J. Nutr.*, **93**, 93 (2005).
[113] B. L. Pool-Zobel, A. Bub, H. Muller, I. Wollowski and G. Rechkemmer, *Carcinogenesis*, **18**, 1847 (1997).
[114] A. Bub, B. Watzl, M. Blockhaus, K. Briviba, U. Liegibel, H. Muller, B. L. Pool-Zobel and G. Rechkemmer, *J. Nutr. Biochem.*, **14**, 90 (2003).
[115] L. Chen, M. Stacewicz-Sapuntzakis, C. Duncan, R. Sharifi, L. Ghosh, R. van Breemen, D. Ashton and P. E. Bowen, *J. Natl. Cancer Inst.*, **93**, 1872 (2001).
[116] P. Bowen, L. Chen, M. Stacewicz-Sapuntzakis, C. Duncan, R. Sharifi, L. Ghosh, H. S. Kim, K. Christov-Tzelkov and R. van Breemen, *Exp. Biol. Med.*, **227**, 886 (2002).
[117] K. Briviba, A. Bub, J. Moseneder, T. Schwerdtle, A. Hartwig, S. Kulling and B. Watzel, *Nutr. Cancer*, **60**, 164 (2008).
[118] P. P. Hoppe, K. Krämer, H. van den Berg, G. Steenge and T. van Vliet, *Eur. J. Nutr.*, **42**, 272 (2003).
[119] P. Riso, A. Brusamolino, S. Ciappellano and M. Porrini, *Int. J. Vitam. Nutr. Res.*, **73**, 201 (2003).
[120] S. J. Duthie, A. Ma, M. A. Ross and A. R. Collins, *Cancer Res.*, **56**, 1291 (1996).
[121] P. Riso, F. Visioli, D. Erba, G. Testolin and M. Porrini, *Eur. J. Clin. Nutr.*, **58**, 1350 (2004).
[122] S. B. Astley, R. M. Elliott, D. B. Archer and S. Southon, *Br. J. Nutr.*, **91**, 63 (2004).
[123] A. R. Collins, B. Olmedilla, S. Southon, F. Granado and S. J. Duthie, *Carcinogenesis*, **19**, 2159 (1998).
[124] A. C. Torbergsen and A. R. Collins, *Eur. J. Nutr.*, **39**, 80 (2000).
[125] X. Zhao, G. Aldini, E. J. Johnson, H. Rasmussen, K. Kraemer, H. Woolf, N. R. Musaeus, N. I. Krinsky, R. M. Russell and K.-J. Yeum, *Am. J. Clin. Nutr.*, **83**, 163 (2006).
[126] S. Devaraj, S. Mathur, A. Basu, H. H. Aung, V. T. Vasu. S. Meyers and I. Jialal, *J. Am. Coll. Nutr.*, **27**, 267 (2008).
[127] S. B. Astley, D. A. Hughes, A. J. Wright, R. M. Elliott and S. Southon, *Br. J. Nutr.*, **91**, 53 (2004).
[128] H. S. Olcovich and H. A. Mattill, *J. Biol. Chem.*, **91**, 105 (1931).
[129] The Alpha-Tocopherol, Beta-Carotene Cancer Prevention Study Group, *New Engl. J. Med.*, **330**, 1029 (1994).
[130] G. S. Omenn, G. E. Goodman, M.D. Thornquist, J. Balmes, M. R. Cullen, A. Glass, J. Keogh, F.L. Meyskens Jr., B. Valanis, J.H. Williams Jr., S. Barnhart and S. Hammar, *New Engl. J. Med.*, **334**,1150 (1996)
[131] M. Touvier, E. Kesse, F. Clavel-Chapelon and M. C. Boutron-Ruault, *J. Natl. Cancer Inst.*, **97**, 1338 (2005).
[132] L. Gallicchio, K. Boyd, G. Matanoski, X. G. Tao, L. Chen, T. K. Lam, M. Shiels, E. Hammond, K. A. Robinson, L. E. Caulfield, J. G. Herman, E. Guallar and A. J. Alberg, *Am. J. Clin. Nutr.*, **88**, 372 (2008).
[133] G. M. Lowe, R. F. Bilton, I. G. Davies, T. C. Ford, D. Billington and A. J. Young, *Ann. Clin. Biochem.*, **36**, 323 (1999).
[134] P. Palozza, G. Calviello, S. Serini, N. Maggiano, P. Lanza, F. O. Ranelletti and G. M. Bartoli, *Free Radic. Biol. Med.*, **30**, 1000 (2001).
[135] P. Palozza, *Biochim. Biophys. Acta*, **1740**, 215 (2005).
[136] N. E. Polyakov, T. V. Leshina, T. A. Konovalova and L. D. Kispert, *Free Radic. Biol. Med.*, **31**, 398 (2001).
[137] P. Palozza, G. Calviello and G. M. Bartoli, *Free Radic. Biol. Med.*, **19**, 887 (1995).

[138] P. Palozza, C. Luberto, G. Calviello, P. Ricci and G. M. Bartoli, *Free Radic. Biol. Med.*, **22**, 1065 (1997).

[139] P. Zhang and S. T. Omaye, *Toxicol. In Vitro*, **15**, 13 (2001).

[140] A. El-Agamey, G. M. Lowe, D. J. McGarvey, A. Mortensen, D. M. Phillip, T. G. Truscott and A. J. Young, *Arch. Biochem. Biophys.*, **430**, 37 (2004).

[141] P. Palozza, S. Serini, S. Trombino, L. Lauriola, F. O. Ranelletti and G. Calviello, *Carcinogenesis*, **27**, 2383 (2006).

[142] K. Wertz, N. Seifert, P. B. Hunziker, G. Riss, A. Wyss, C. Lankin and R. Goralczyk, *Free Radic. Biol. Med.*, **37**, 654 (2004).

[143] N. Bando, H. Hayashi, S. Wakamatsu, T. Inakuma, M. Miyoshi, A. Nagao, R. Yamauchi and J. Terao, *Free Radic. Biol. Med.*, **37**, 1854 (2004).

[144] U. C. Obermüller-Jevic, P. I. Francz, J. Frank, A. Flaccus and H. K. Biesalski, *FEBS Lett.*, **460**, 212 (1999).

[145] U. C. Obermüller-Jevic, B. Schlegel, A. Flaccus and H. K. Biesalski, *FEBS Lett.*, **509**, 186 (2001).

[146] E. A. Offord, J. C. Gautier, O. Avanti, C. Scaletta, F. Runge, K. Krämer and L. A. Applegate, *Free Radic. Biol. Med.*, **32**, 1293 (2002).

[147] S. L. Yeh, W. Y. Wang, C. H. Huang and M. L. Hu, *J. Nutr. Biochem.*, **16**, 729 (2005).

[148] J. Fiedor, L. Fiedor, R. Haessner and H. Scheer, *Biochim. Biophys. Acta*, **1709**, 1 (2005).

[149] W. Siems, I. Wiswedel, C. Salerno, C. Crifo, W. Augustin, L. Schild, C. D. Langhans and O. Sommerburg, *J. Nutr. Biochem.*, **16**, 385 (2005).

[150] O. Sommerburg, C. D. Langhans, J. Arnhold, M. Leichsenring, C. Salerno, C. Crifo, G. F. Hoffmann, K. M. Debatin and W. G. Siems, *Free Radic. Biol. Med.*, **35**, 1480 (2003).

[151] A. J. Alija, N. Bresgen, O. Sommerburg, W. Siems and P. M. Eckl, *Carcinogenesis*, **25**, 827 (2004).

[152] N. M. Kalariya, K. V. Ramana, S. K. Srivastava and F. J. Van Kuijk, *Exp. Eye Res.*, **86**, 70 (2008).

[153] A. J. Alija, N. Bresgen, O. Sommerburg, C. D. Langhans, W. Siems and P. M. Eckl, *Biofactors*, **24**, 159 (2005).

[154] A. J. Alija, N. Bresgen, O. Sommerburg, C. D. Langhans, W. Siems and P. M. Eckl, *Carcinogenesis*, **27**, 1128 (2006).

[155] H. McNulty, R. F. Jacob and R. P. Mason, *Am. J. Cardiol.*, **101**, 20D (2008).

[156] D. P. Gelain, M. A. de Bittencourt Pasquali, F. F. Caregnato, A. Zanotto-Filho and J. C. Moreira, *Toxicol. In Vitro*, **22**, 1123 (2008).

[157] A. C. Carr, M. R. McCall and B. Frei, *Arterioscler. Thromb. Vasc. Biol.*, **20**, 1716 (2000).

[158] P. Palozza, S. Serini, F. Di Nicuolo, A. Boninsegna, A. Torsello, N. Maggiano, F. O. Ranelletti, F. I. Wolf, G. Calviello and A. Cittadini, *Carcinogenesis*, **25**, 1315 (2004).

[159] H. van den Berg, *Nutr. Rev.*, **57**, 1 (1999).

[160] D. W. Nierenberg, B. J. Dain, L. A. Mott, J. A. Baron and E. R. Greenberg, *Am. J. Clin. Nutr.*, **66**, 315 (1997).

[161] U. Heinrich, C. Gärtner, M. Wiebusch, O. Eichler, H. Sies, H. Tronnier and W. Stahl, *J. Nutr.*, **133**, 98 (2003).

[162] S. Kiokias and M. H. Gordon, *Eur. J. Clin. Nutr.*, **57**, 1135 (2003).

[163] E. Niki, N. Noguchi, H. Tsuchihashi and N. Gotoh, *Am. J. Clin. Nutr.*, **62**, 1322S (1995).

[164] M. Burke, R. Edge, E. J. Land and T. G. Truscott, *J. Photochem. Photobiol. B*, **60**, 1 (2001).

[165] A. Mortensen and L. H. Skibsted, *FEBS Lett.*, **417**, 261 (1997).

[166] T. J. Hill, E. J. Land, D. J. McGarvey, W. Schalch, J. H. Tinkler and T. G. Truscott, *J. Am. Chem. Soc.*, **117**, 8322 (1995).

[167] A. El-Agamey, A. Cantrell, E. J. Land, D. J. McGarvey and T. G. Truscott, *Photochem. Photobiol. Sci.*, **3**, 802 (2004).

[168] F. Bohm, R. Edge, E. J. Land and T. G. Truscott, *J. Am. Chem. Soc.*, **119**, 621 (1997).

[169] G. R. Buettner, *Arch. Biochem. Biophys.*, **300**, 535 (1993).

[170] R. Edge, E. J. Land, D. J. McGarvey, M. Burke and T. G. Truscott, *FEBS Lett.*, **471**, 125 (2000).
[171] J. Shi, Y. Kakuda and D. Yeung, *Biofactors*, **21**, 203 (2004).
[172] K. Volkovova, M. Barancokova, A. Kazimirova, A. Collins, K. Raslova, B. Smolkova, A. Horska, L. Wsolova and M. Dusinska, *Free Radic. Res.*, **39**, 659 (2005).
[173] D. P. Vivekananthan, M. S. Penn, S. K. Sapp, A. Hsu and E. J. Topol, *Lancet*, **361**, 2017 (2003).
[174] E. R. Miller 3rd, R. Pastor-Barriuso, D. Dalal, R. A. Riemersma, L. J. Appel and E. Guallar, *Ann. Intern. Med.*, **142**, 37 (2005).
[175] P. Knekt, J. Ritz, M. A. Pereira, E. J. O'Reilly, K. Augustsson, G. E. Fraser, U. Goldbourt, B. L. Heitmann, G. Hallmans, S. Liu, P. Pietinen, D. Spiegelman, J. Stevens, J. Virtamo, W. C. Willett, E. B. Rimm and A. Ascherio, *Am. J. Clin. Nutr.*, **80**, 1508 (2004).
[176] R. S. Eidelman, D. Hollar, P. R. Hebert, G. A. Lamas and C. H. Hennekens, *Arch. Intern. Med.*, **164**, 1552 (2004).
[177] S. K. Osganian, M. J. Stampfer, E. Rimm, D. Spiegelman, J. E. Manson and W. C. Willett, *Am. J. Clin. Nutr.*, **77**, 1390 (2003).
[178] H. C. Hung, K. J. Joshipura, R. Jiang, F. B. Hu, D. Hunter, S. A. Smith-Warner, G. A. Colditz, B. Rosner, D. Spiegelman and W. C. Willett, *J. Natl. Cancer Inst.*, **96**, 1577 (2004).
[179] S. T. Mayne, *J. Nutr.*, **133**, 933S (2003).
[180] F. B. Hu, *Am J. Clin. Nutr.*, **78**, 544S (2003).
[181] J. M. Gaziano, J. E. Manson, L. G. Branch, G. A. Colditz, W. C. Willett and J. E. Buring, *Ann. Epidemiol.*, **5**, 255 (1995).
[182] B. Buijsse, E. J. Feskens, L. Kwape, F. J. Kok and D. Kromhout, *J. Nutr.*, **138**, 344 (2008).
[183] G. Riccioni, T. Bucciarelli, N. D'Orazio, N. Palumbo, E. di Ilio, F. Corradi, A. Pennelli and L. A. Bazzano, *Ann. Nutr. Metab.*, **53**, 86 (2008).
[184] A. Schatzkin and V. Kipnis, *J. Natl. Cancer Inst.*, **96**, 1564 (2004).
[185] I. A. Hininger, A. Meyer-Wenger, U. Moser, A. Wright, S. Southon, D. Thurnham, M. Chopra, H. Van Den Berg, B. Olmedilla, A. E. Favier and A. M. Roussel, *J. Am. Coll. Nutr.*, **20**, 232 (2001).
[186] H. van den Berg and T. van Vliet, *Am. J. Clin. Nutr.*, **68**, 82 (1998).
[187] I. Paetau, H. Chen, N. M. Goh and W. S. White, *Am. J. Clin. Nutr.*, **66**, 1133 (1997).
[188] D. Kostic, W. S. White and J. A. Olson, *Am. J. Clin. Nutr.*, **62**, 604 (1995).
[189] W. S. White, M. Stacewicz-Sapuntzakis, J. W. Erdman Jr. and P. E. Bowen, *J. Am. Coll. Nutr.*, **13**, 665 (1994).
[190] B. R. Hammond Jr., J. Curran-Celentano, S. Judd, K. Fuld, N. I. Krinsky, B. R. Wooten and D. M. Snodderly, *Vision Res.*, **36**, 2001 (1996).
[191] B. R. Hammond Jr., K. Fuld and D. M. Snodderly, *Exp. Eye Res.*, **62**, 293 (1996).
[192] J. M. Seddon, U. A. Ajani, R. D. Sperduto, R. Hiller, N. Blair, T. C. Burton, M. D. Farber, E. S. Gragoudas, J. Haller, D. T. Miller, L. A. Yannuzzi and W. Willet, *JAMA*, **272**, 1413 (1994).
[193] J. T. Landrum and R. A. Bone, *Arch. Biochem. Biophys.*, **385**, 28 (2001).
[194] R. A. Bone, J. T. Landrum, G. W. Hime, A. Cains and J. Zamor, *Invest. Ophthalmol. Vis. Sci.*, **34**, 2033 (1993).
[195] G. J. Handelman, E. A. Dratz, C. C. Reay and J. G. van Kuijk, *Invest. Ophthalmol. Vis. Sci.*, **29**, 850 (1988).
[196] E. Cho, S. E. Hankinson, B. Rosner, W. C. Willett and G. A. Colditz, *Am. J. Clin. Nutr.*, **87**, 1837 (2008).
[197] H. E. Bartlett and F. Eperjesi, *Clin. Nutr.*, **27**, 218 (2008).
[198] R. A. Bone, J. T. Landrum, L. M. Friedes, C. M. Gomez, M. D. Kilburn, E. Menendez, I. Vidal and W. Wang, *Exp. Eye Res.*, **64**, 211 (1997).
[199] B. R. Hammond Jr., E. J. Johnson, R. M. Russell, N. I. Krinsky, K. J. Yeum, R. B. Edwards and D. M. Snodderly, *Invest. Ophthalmol. Vis. Sci.*, **38**, 1795 (1997).
[200] W. G. Christen, S. Liu, R. J. Glynn, J. M. Gaziano and J. E. Buring, *Arch. Ophthalmol.*, **126**, 102 (2008).

[201] M. Dherani, G. V. Murthy, S. K. Gupta, I. S. Young, G. Maraini, M. Camparini, G. M. Price, N. John, U. Chakravarthy and A. E. Fletcher, *Invest. Ophthalmol. Vis. Sci.*, **49**, 3328 (2008).
[202] T. H. Wu, J. H. Liao, W. C. Hou, F. Y. Huang, T. J. Maher and C. C. Hu, *J. Agric. Food Chem.*, **54**, 2418 (2006).
[203] M. Santocono, M. Zurria, M. Berrettini, D. Fedeli and G. Falcioni, *J. Photochem. Photobiol. B*, **85**, 205 (2006).
[204] O. Obajimi, K. D. Black, I. Glen and B. M. Ross, *Prostaglandins Leukot. Essent. Fatty Acids*, **76**, 65 (2007).
[205] G. Hussein, H. Goto, S. Oda, U. Sankawa, K. Matsumoto and H. Watanabe, *Biol. Pharm. Bull.*, **29**, 684 (2006).
[206] B. Fuhrman, N. Volkova, M. Rosenblat and M. Aviram, *Antioxid. Redox Signal.*, **2**, 491 (2000).
[207] F. Visioli, P. Riso, S. Grande, C. Galli and M. Porrini, *Eur. J. Nutr.*, **42**, 201 (2003).
[208] A. V. Rao, *Exp. Biol. Med.*, **227**, 908 (2002).
[209] S. Schwarz, U. C. Obermüller-Jevic, E. Hellmis, W. Koch, G. Jacobi and H. K. Biesalski, *J. Nutr.*, **138**, 49 (2008).
[210] C. Liu, R. M. Russell and X. D. Wang, *J. Nutr.*, **133**, 173 (2003).
[211] X. D. Wang, *J. Nutr.*, **135**, 2053S (2005).
[212] C. Liu, R. M. Russell and X. D. Wang, *J. Nutr.*, **136**, 106 (2006).
[213] C. Liu, F. Lian, D. E. Smith, R. M. Russell and X. D. Wang, *Cancer Res.*, **63**, 3138 (2003).

Carotenoids
Volume 5: Nutrition and Health

Chapter 13

Carotenoids and Cancer

Cheryl L. Rock

A. Introduction

Cancer is a major cause of morbidity and mortality around the world, although patterns of cancer, like dietary patterns, are highly variable across regions and countries with different degrees of economic development [1]. Observed associations between dietary patterns and cancer mortality and morbidity have led to hypotheses about cause and effect relationships, which have subsequently been examined more specifically in laboratory studies of biological activities of dietary constituents, case-control and cohort studies within populations, and clinical trials. Food provides nutrients and numerous other bioactive compounds, many of which have been linked specifically to cellular and molecular events and activities that have been identified in the development and progression of cancer [2,3].

Carotenoids are among the bioactive substances that potentially affect risk and progression of cancer, and that have been the focus of numerous investigations. As summarized in comprehensive reviews [4,5], accumulated data on diet and cancer over the past several decades suggest that approximately 30-40% of cancer cases are potentially preventable *via* food choices and the modification of nutritional factors. However, disentangling the effects of various foods, specific dietary constituents, and related lifestyle factors and characteristics (*e.g.* physical activity, obesity) that influence risk and progression of cancer has proved to be very challenging. This challenge is particularly evident when the relationship between carotenoids and cancer is examined, due to the distribution of these compounds in the food supply and the clustering of health-related behaviours. The aim in this *Chapter* is to evaluate the relationship between carotenoids and cancer from data obtained by different experimental approaches. The design, application and interpretation of epidemiology studies are described in *Chapter 10*. Data on the effects of carotenoids on cellular and molecular processes in cultured cells are presented in *Chapter 11* and the influence of carotenoids on the immune response system is discussed in *Chapter 17*.

Cancer is one of the leading causes of death worldwide, accounting for 7.6 million (13%) of all deaths, according to World Health Organization 2006 statistics [6]. Over the past 30 years, improvements have been observed in five-year survival rates for all cancers combined and for several specific cancers, and this has been attributed primarily to improved initial treatments and to increased screening that results in diagnosis at an earlier stage. An increasing population of cancer survivors, *i.e.* individuals with a history of cancer who are thus at risk for recurrence or new cancers, has promoted increased interest in whether dietary factors, including carotenoids, may influence this risk and long-term survival [7].

Examining the evidence linking carotenoids to cancer risk and progression requires an appreciation of the multistage process of carcinogenesis. Cancer results from multiple genetic and epigenetic events involving protooncogenes, tumour suppressor genes and antimetastasis genes throughout progression [8]. Specifically, clinical cancer is not determined by a single molecular event that disrupts normal cellular function or regulation of growth but, instead, results from a series of disruptions across the cancer continuum. This continuum extends from the earliest cellular changes, to a preneoplastic lesion, to a malignant tumour, and finally, to metastasis. Genetic or inherited factors play a role in determining susceptibility to molecular and genetic changes in the process of carcinogenesis, although, notably, nutritional factors appear to influence risk even in the presence of highly penetrant, dominant gene mutations [9].

As reviewed previously [10,11], carotenoids have been shown, in laboratory studies, to exhibit several biological activities that could prevent or slow the progression of cancer. In addition to possible antioxidant activity *in vitro* (see *Chapter 12*), carotenoids have favourable effects on cell growth regulation, such as the inhibition of growth and malignant transformation, and the promotion of apoptosis in transformed cells, similar to the effects of retinoids (see *Chapter 11*). However, demonstrating a specific molecular effect of carotenoids in human cancer, in which a series of genetic and epigenetic changes has occurred over years or decades, is logistically challenging.

Interpretation of results of observational studies (see *Chapter 10*) that identify associations between cancer risk and carotenoid intake or concentrations in the circulation is substantially constrained by the risk for confounding, *i.e.* the misidentification of carotenoid intake or plasma concentration (as a biomarker of carotenoid intake and tissue exposure) as the true protective factor instead of other or associated variables. For example, the majority of carotenoids in the diet are contributed by vegetables and fruit. These foods are complex, containing, in addition to carotenoids, numerous constituents that have biological activities [12]. Also, individuals who report consuming higher intakes of vegetables and fruit (and hence carotenoids), or who have higher plasma carotenoid concentrations, typically are more likely to exhibit other prudent health behaviour, such as limited alcohol intake, not smoking, and increased physical activity [13,14].

In general, interpretation of results from observational studies that rely on self-reported dietary data is particularly constrained by the problem of crude and imprecise methods that are used in the collection of these data, as well as a food content database that is of limited quality. The imprecision of assessing status *via* self-reported intakes, as opposed to biomarkers of intake, can result in the wrong conclusions about associations. For example, self-reported intake of vegetables and fruits, the major sources of dietary carotenoids, was not significantly associated with risk for primary breast cancer in the New York University Women's Health Study [15], whereas serum carotenoid concentrations, which are a biomarker for intake of vegetable and fruit intake, were found to be significantly inversely associated with risk in that same cohort of women [16]. The use of suitable dietary biomarkers of carotenoid intake, rather than reliance only on self-reported dietary intake data, is recognized increasingly as being of value for more accurately characterizing usual patterns of intake and true exposure.

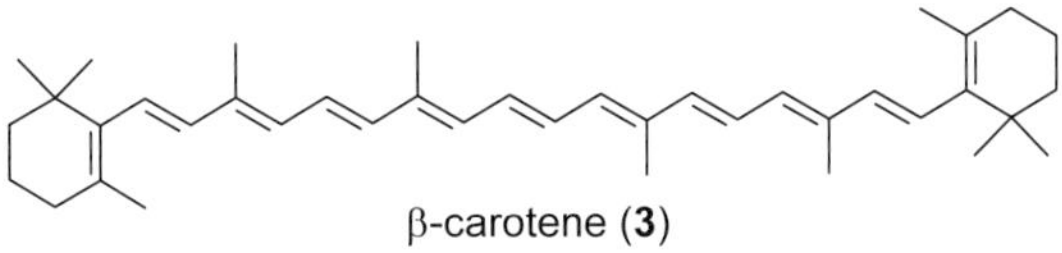

β-carotene (**3**)

Although randomized clinical trials involving carotenoid supplements would seem to be a preferable approach to testing the specific effect of carotenoids on cancer risk and progression, interpretation of the results of studies reported to date is difficult. Most of these studies have involved individuals who are at very high risk for cancer or who have already been diagnosed with precursor lesions or a primary cancer. For example, trials with supplements of β-carotene (**3**) to test an effect on risk for cervical cancer have all involved women who had already been diagnosed with cervical dysplasia, a precursor neoplastic lesion [17]. If the mechanism by which carotenoids may reduce the risk for cervical cancer, as is suggested by observational epidemiological studies, is a favourable effect on the immune system response to exposure to human papillomavirus (the primary cause of the cancer), it is unlikely that a protective effect of carotenoids would be achieved or observed when the infection is already established. Similarly, β-carotene supplement trials to test the effect of carotenoids on risk for colon cancer have focused on only one stage in the development and progression of this cancer, examining whether β-carotene supplementation, or increased vegetable and fruit intake, over 2-5 years, can affect the risk for adenoma recurrence and growth in individuals with a history of adenomatous polyps [18]. The rationale for using recurrence of colorectal adenomas as the primary end point is that adenomatous polyps are considered precursors of most cancers of the large bowel. A general clinical trial testing the effect of carotenoid intake on incident colon cancer would require a large sample, a very long follow-up period, and considerable resources. However, it must be recognized that, without specific knowledge of the critical points at which carotenoids may affect colon carcinogenesis, the focus of intervention studies to date may not have tested appropriately the effects of carotenoids on risk for colon cancer. A finding of no effect in a study of this type does not address the possibility that a lifetime of high or low carotenoid intake, or differential carotenoid intakes at another point in the colon cancer continuum, might affect risk for colon cancer.

Within these recognized constraints on interpretation and conclusions, available data and current evidence suggest a possible role for carotenoids or at least the major food sources of these compounds, *i.e.* vegetables and fruit, in the aetiology of cancer. The weight of the evidence suggests that carotenoids may influence the risk and progression of the following cancers: lung, breast, prostate, large bowel (colon and rectum), head and neck (oral cavity, pharynx and larynx), cervix and ovary.

B. Lung Cancer

In the U.S. and worldwide, lung cancer is the most common cause of cancer death in both men (31%) and women (26%) and is the second most commonly diagnosed invasive cancer for both gender subgroups (13% and 12% of cases for men and women, respectively) [1,6,7]. Smoking behaviour is an established major environmental factor in the aetiology of lung cancer, but higher intakes of carotenoids and their major food sources, vegetables and fruit, have been associated quite consistently with reduced risk for lung cancer in observational epidemiological studies [19-30]. Although early observational studies mainly revealed an inverse relationship between risk for lung cancer and intakes of total vitamin A (a dietary variable that includes preformed vitamin A as well as provitamin A carotenoids), or of vegetables (especially yellow-orange and dark green vegetables), and higher serum or plasma β-carotene concentration, more recent epidemiological studies have suggested a protective association with other carotenoids. These more recent studies have identified inverse relationships between lung cancer risk and intakes or blood concentrations of α-carotene (**7**), lutein (**133**), lycopene (**31**) and β-cryptoxanthin (**55**), as well as β-carotene.

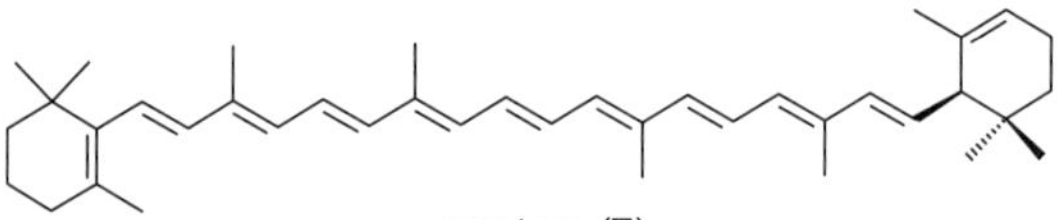

α-carotene (**7**)

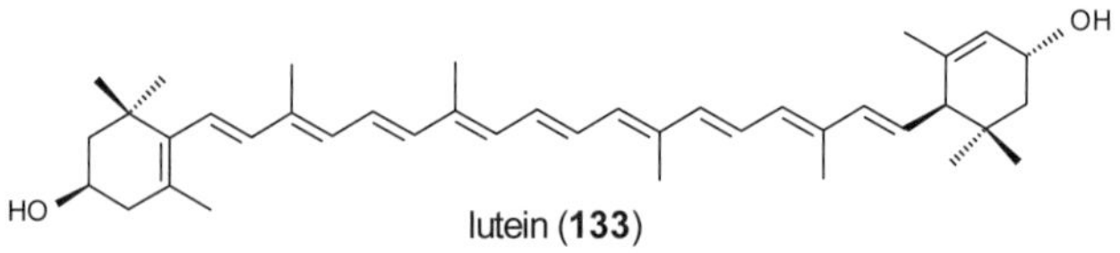

lutein (**133**)

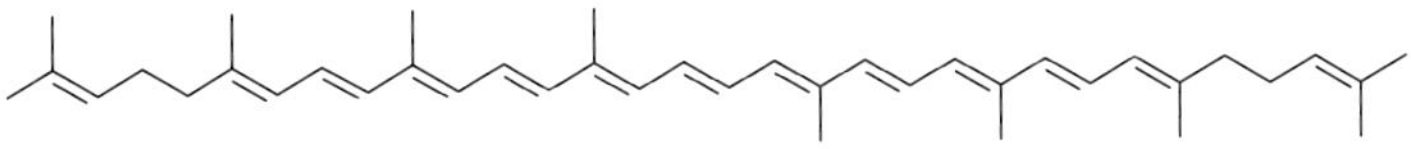

lycopene (**31**)

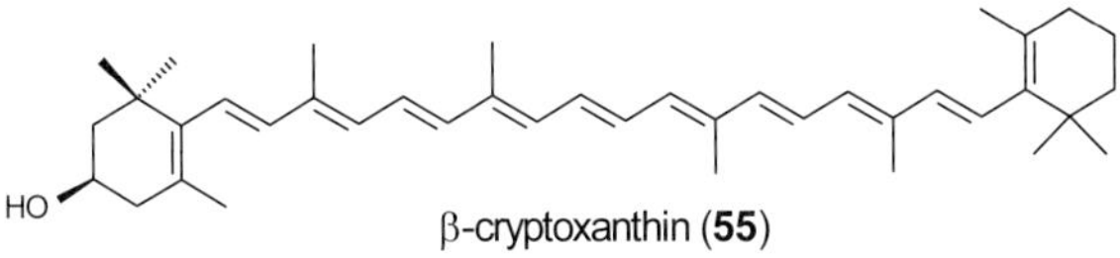

β-cryptoxanthin (**55**)

For example, findings from a multi-centre, case-control study of diet and lung cancer among non-smokers (506 cases, 1,045 controls) found protective effects for intakes of several carotenoids and carotenoid-rich food sources, including tomatoes (odds ratio [OR] 0.5, 95% confidence interval [CI] 0.4, 0.6 for highest *versus* lowest tertile, P = 0.01 for trend), lettuce (OR 0.6, 95% CI 0.3-1.2, P = 0.02 for trend), carrots (OR 0.8, 95% CI 0.5, 1.1, not significant [NS]), total carotenoids (OR 0.8, 95% CI 0.6, 1.0, NS), and β-carotene (OR 0.8, 95% CI 0.6, 1.1, NS) [29]. In another large case-control study (1,000 cases, 1,500 controls) [30], intakes of carrots (relative risk [RR] 0.49, 95% CI 0.31, 0.78 for more than weekly *versus* never), tomato sauce (RR 0.69, 95% CI 0.55, 0.87 for a few times per week or more *versus* never), and tomatoes (RR 0.74, 95% CI 0.57, 0.96 for >45 g/day *versus* <16 g/day) were found to be significantly inversely associated with lung cancer risk.

Note that several ways may be used to express differences between two groups, *e.g.* treated *versus* control. The odds ratio (OR) shows whether the probability of an event is the same in two groups. For example, an OR of 1.0 means that something is equally likely to occur in both groups, an OR of 0.5 means that something is half as likely to occur as it would in a reference group, and an OR of 1.5 means that something is 50% more likely to occur than it would in a reference group. A 95% CI (confidence interval) for a value is the range of values for which one is 95% confident that the true value lies within that range. Relative risk (RR) is the ratio of incidence of disease in the exposed group to incidence of disease in the unexposed group, or the probability of one event divided by the probability of the other event. Hazard ratio (HR) is the ratio of hazard of disease in the exposed group to the hazard of disease in the unexposed or reference group.

In four large placebo-controlled clinical trials, the effect of β-carotene supplementation on lung cancer incidence or mortality was examined [31-34]. Surprisingly, incidence of lung cancer actually increased in response to β-carotene supplementation in two of the studies, particularly among smokers and in subjects who consumed larger amounts of alcohol [35-37]. Results from subsequent studies with laboratory animals suggest that high doses but not low doses (*i.e.* equivalent to 30 mg/day *versus* 6 mg/day in humans) of β-carotene and the resulting cleavage products deplete tissue retinoic acid and interfere with normal retinoid signalling [38,39]. In tissue culture studies, in which the aim is typically to achieve in the cells physiological concentrations, which are considerably lower than the concentrations achieved in most of the supplement trials, β-carotene has been shown to induce both qualitative and quantitative beneficial changes in lung cancer cells [40]. Clearly, the unregulated uptake by peripheral tissues and demonstrable biological effects on major cell regulatory systems set the stage for potentially adverse, in addition to beneficial, effects of carotenoids on risk for lung cancer.

Integration of current evidence across all types of research suggests that the relationship between carotenoids and lung cancer risk may resemble a bell-shaped curve, with smoking behaviour and alcohol consumption acting as modifiers. At physiological concentrations achieved with doses obtainable from the food supply, carotenoids exhibit biological activities that may be protective against lung cancer; at high concentrations, carotenoids and their cleavage products have adverse effects on cellular function that may be of a magnitude high enough to increase lung cancer risk. It appears that intakes of several carotenoids, and not just β-carotene, may reduce risk for lung cancer, but at doses that would be provided by a diet that includes sufficient amounts of foods that are good sources of these compounds, *i.e.* vegetables and fruit. As noted above, it is unclear at this point whether it is the carotenoids obtained from these foods that are the sole or major protective constituents; laboratory evidence, however, suggests that carotenoids are, at least, among the constituents of these foods that are potentially beneficial in reducing risk for lung cancer.

C. Breast and Ovarian Cancers

Carcinomas of the breast and ovary are hormone-related cancers that have biological similarities. In the U.S., 31% of newly diagnosed invasive cancers are breast cancer and, although five-year survival rates are improving, breast cancer is expected to account for 15% of cancer deaths in women in the U.S. in 2006 [7]. Ovarian cancer is much less common than breast cancer but is more likely to have a worse prognosis. Ovarian cancer accounts for 3% of incident cancers in women in the U.S. but 6% of cancer deaths [7].

1. Breast cancer

As with lung cancer, the vast majority of the early observational studies on diet and breast cancer risk examined intake of β-carotene but not of other carotenoids. Also relevant to associations between risk and carotenoid intake are epidemiological studies of vegetable and fruit intake and breast cancer risk, because a protective effect associated with those foods would indicate that carotenoids might be implicated in the effect. Results from these studies are somewhat supportive but not entirely consistent [1,15,41,42]. In general, results from case-control studies, more than those from cohort studies, suggest that carotenoids may be protective against breast cancer. Few studies have examined the relationship between plasma or serum concentrations of carotenoids and risk for breast cancer. In the largest prospective study of the relationship between serum carotenoids and risk for breast cancer [16], the odds ratio for the lowest *versus* highest quartile of total serum carotenoids was 2.31 (95% CI 1.35, 3.96). Serum concentrations of β-carotene (OR 2.21, 95% CI 1.29, 3.79), α-carotene (OR 1.99, 95% CI 1.18, 3.34), and lutein (OR 2.08, 95% CI 1.11, 3.90) were all also inversely associated with risk. Other small observational studies that also relied on prediagnostic serum carotenoid concentrations in the analysis of breast cancer risk found some protective associations for serum β-carotene, lycopene, and total carotenoids [43,44], when adjusted for other influencing factors.

Five of the eight epidemiological studies that examined the relationship between intakes of carotenoids, or their major food sources (vegetables and fruit), and survival suggest a possible modest protective effect on prognosis in women who have been diagnosed with breast cancer [45]. Recently, a cohort study involving 1,511 women previously diagnosed and treated for breast cancer, who were followed for an average of seven years, revealed that women in the highest quartile of plasma total carotenoid concentration had an estimated 43% reduction in risk for a new breast cancer event (recurrence or new primary) compared to those in the lowest quintile [46].

Although no clinical trials involving carotenoid supplements and breast cancer risk have been conducted, one recently completed study, the Women's Healthy Eating and Living (WHEL) Study, examined the effect of substantially increased intakes of vegetables and fruits on risk for recurrence and survival, in women who have been diagnosed with breast cancer [47]. In that study, women who were randomly assigned to receive intensive treatment (diet counselling) reported increased carotenoid intakes and exhibited increased plasma concentrations of α-carotene (+223%), β-carotene (+87%), lutein (+29%), and lycopene (+17%) [47], and these levels appear to be fairly well maintained at four-year follow-up. At the end of the study, this was not shown to be associated with reduced risk for additional breast cancer events or mortality during a median 7.3-year follow-up period [48]. However, higher biological exposure to carotenoids, when assessed over the time frame of the study, was associated with greater likelihood of breast cancer-free survival regardless of study group assignment [49].

Compared with the lack of consistency in the epidemiological studies, especially those relying on self-reported dietary data, the laboratory evidence suggesting the feasibility that carotenoids are protective against breast cancer is consistent and convincing. Cell culture studies strongly suggest specific beneficial effects of both provitamin A and non-provitamin A carotenoids on the development and progression of breast cancer [50-52], and the effects are generally retinoid-like effects on cell growth regulation. At this point, the evidence would suggest that carotenoids may reduce risk and progression of breast cancer, although more epidemiological research involving data on serum carotenoid concentrations might better reconcile the observations from human studies with the evidence from cell culture studies.

2. Ovarian cancer

Few epidemiological studies have addressed associations between dietary factors, including carotenoids, and risk for ovarian cancer. Of the case-control studies in which the relationship between risk for ovarian cancer and carotenoids, or their major food sources, vegetables and fruit, have been examined, six studies found protective effects associated with vegetable and fruit intake [53-58] and four studies found protective effects associated with carotenoid intake [54,59-61]. The potential importance of examining intakes over the whole time course of ovarian carcinogenesis is suggested by one large cohort study, in which intake of vegetables and fruit during adolescence, but not intake during adulthood, was found to be protective (RR 0.54, 95% CI 0.29, 1.03 for women who consumed at least 2.5 servings/day *versus* those with lower intakes) [62]. No relationship between serum carotenoid concentrations and risk for ovarian cancer was observed in the sole, very small, prospective study in which serum carotenoids were examined [63]. In one observational study of the effects of various dietary factors on survival in women who had been diagnosed with ovarian cancer, women who reported higher intake of vegetables had significantly greater likelihood of survival than those with lower intake (hazard ratio [HR] 0.75, 95% CI 0.57, 0.99 for highest *versus* lowest tertile) [64].

Clinical trials have not tested whether carotenoid supplementation or dietary modification can influence the risk and progression of ovarian cancer. At this point, data relating ovarian cancer risk to carotenoid intake are limited but suggestive of a possible favourable relationship.

D. Prostate Cancer

Cancer of the prostate is the most commonly diagnosed invasive cancer among men in most developed countries [1]. In the U.S., it accounts for 33% of new cases [7]. Approximately 9% of cancer deaths among men in the U.S. in 2006 may be attributable to prostate cancer. As recently reviewed [65], evidence from the numerous epidemiological studies that have examined whether carotenoids, or foods that are rich sources of these compounds, are associated with risk for prostate cancer has been suggestive but not consistent. Much attention has been focused on the potential protective effect of lycopene or tomato products, the richest source of lycopene in a typical diet in many developed countries. In a meta-analysis of observational studies (11 case-control and 10 cohort studies), high *versus* low intake of tomatoes was associated with an approximate 10-20% reduction in risk for prostate cancer [66]. In general, results from case-control studies have not supported a protective effect of lycopene or tomato products on prostate cancer risk, whereas cohort studies of dietary factors and incident prostate cancer, especially the largest of this type of epidemiological study, have been more supportive [67]. For example, in the Health Professionals Follow-Up Study, in which dietary intakes and incident prostate cancer were examined in a cohort of 47,365 men, reduced risk of prostate cancer was observed in association with higher lycopene intake (RR 0.84, 95% CI 0.73, 0.96 for high *versus* low quintiles), and a higher level of tomato sauce consumption (RR 0.77, 95% CI 0.66, 0.90 for two or more servings/week *versus* less than one serving/month), controlled for total vegetable and fruit consumption and other influencing factors [68].

Similar to the situation with epidemiological studies of breast cancer, few studies involving serological data indicative of carotenoid status and prostate cancer risk have been reported. The largest of these studies involved blood samples collected from a cohort of 14,916 men, in whom 578 cases were identified, and a statistically significantly lower risk of aggressive prostate cancer (RR 0.56, 95% CI 0.34, 0.92) but not total prostate cancer (RR 0.75, 95% CI 0.54, 1.06) was observed in association with higher plasma lycopene concentrations [69]. Two other smaller cohort studies in which serum lycopene was examined did not identify a protective association [67,70].

Much of the interest in the potential effect of lycopene and food sources of this carotenoid on prostate cancer risk arises from the accumulating laboratory evidence and small clinical studies in which biological markers of prostate cancer have been measured [70]. Because many of these studies tested the effects of tomato products rather than isolated lycopene, it cannot be assumed that the beneficial effects can be attributed to lycopene *per se*. As noted above, studies of the relationship between total vegetable and fruit intake and prostate cancer risk have been inconsistent, as recently reviewed [71,72].

Within one of the placebo-controlled clinical trials testing the effect of β-carotene supplementation on risk for lung cancer, a beneficial effect of β-carotene supplementation on prostate cancer risk was observed [73]. Men in the lowest, *versus* highest, quartile for plasma β-carotene at baseline had a marginally significant ($P = 0.07$) increased risk for prostate cancer over the trial period. When the men in the lowest quartile for baseline plasma β-carotene concentration were compared based on study group assignment, a significant beneficial effect on risk for prostate cancer was observed in those administrated β-carotene *versus* placebo (RR 0.68, 95% CI 0.46, 0.99). The effect of β-carotene supplementation (50 mg every other day, with or without vitamin C, vitamin E, and a daily multiple vitamin formulation) on prostate cancer will continue to be investigated in 15,000 U.S. men aged 55 years and older in the Physicians' Health Study (PHS) II [74].

Thus, current evidence is suggestive of the possibility that carotenoids may reduce risk for prostate cancer, although more data are clearly needed before the relationship can be considered established. Results from the PHS II Study should provide important insight into the effect of β-carotene supplementation. Although a possible beneficial effect of lycopene on prostate cancer risk remains of great interest, randomized clinical trials testing an effect on a cancer outcome have not been reported. Further, the effects of lycopene specifically, as opposed to a complex food source (tomatoes) are difficult to disentangle in the interpretation of results of epidemiological and many of the laboratory studies that have been reported.

E. Colorectal Cancer

Cancer of the colon is the fourth most commonly diagnosed cancer worldwide, and incidence rates have been increasing steadily, especially in developed countries [1]. In the U.S., colorectal cancer accounts for 10% and 11% of the incident cancer cases in men and women, respectively, and 10% of cancer deaths in both gender subgroups [7]. Results from ecological and migrant studies have long suggested that diet is an important environmental factor that influences the risk and progression of colon cancer. Colon and rectal cancers have a well-established and defined continuum of cellular changes and associated lesions that appear to occur in the stepwise process of developing an invasive tumour.

Numerous observational epidemiological studies have examined associations between intakes of carotenoids or their major food sources, vegetables and fruit, and the risk for colon cancer [1,75]. Intakes of carotenoids and/or vegetables and fruit, and serum concentrations of carotenoids, have been inversely associated with colon cancer risk in the majority of the case-control studies and in studies based on pre-diagnosis serum carotenoid concentrations. Results from the more recent studies, in which intakes of a variety of provitamin A and non-provitamin A carotenoids were examined, have suggested beneficial relationships beyond those previously attributed to β-carotene or vegetable and fruit intake. For example, lutein intake was inversely associated with colon cancer risk in both men and women (OR 0.83, 95% CI 0.66, 1.04 for highest *versus* lowest quintile, P = 0.04 for trend) in a large case-control study (1,993 cases, 2,410 controls), whilst associations with the other carotenoids were not significant [76]. In contrast to results from case-control studies, recent cohort studies have not been as supportive of a protective effect of vegetables and fruit on colon cancer risk [77].

In two large, randomized, controlled trials on the risk for recurrence of adenomatous polyps [78,79], a significant beneficial effect of β-carotene supplementation was not observed. Adenomatous polyps are considered to be preneoplastic lesions, although most adenomas do not progress to carcinomas. The precise time course of a progression, should this occur, is not understood, although clinical evidence suggests that malignancy in an adenoma develops over 20 years or more. The effect of increased intake of vegetables and fruit, aiming for 5-8 servings/day, concurrent with reduced fat intake (20% energy from fat), on adenoma recurrence at four years following randomization was the focus of another large randomized trial, the Polyp Prevention Trial (PPT) [80]. The PPT involved 2,079 study participants, and the absolute difference between the self-reported daily vegetable and fruit intake of the intervention and control groups over the four-year period was 1.1 servings per 1000 kcal/day. No effects on adenoma recurrence were observed. Notably, the intervention group exhibited only a minimal increase (approximately 5%) in serum total carotenoid concentration, despite reporting substantially increased carotenoid intakes in association with reported increased intake of vegetables and fruit. Thus, the PPT did not really test the effects of carotenoids on risk for adenoma recurrence, in view of the minimally increased tissue concentrations that were achieved. However, results of a secondary analysis of a sub-cohort of PPT study participants (n = 701) are in agreement with prior observational studies: average serum α-carotene, β-carotene, lutein and total carotenoid concentrations at four time points during the study were found to be associated with decreased risk of polyp recurrence (OR 0.71, 0.76, 0.67, and 0.61, respectively, $P < 0.05$) [81].

In relation to the divergent evidence from epidemiological and clinical studies, biological evidence from cell culture studies involving colon cancer cell lines supports the possibility that carotenoids may affect colon cancer risk and progression, and the mechanisms appear to involve both antioxidant and cell growth regulatory activities [82,83]. As noted above, the inherent challenges in collecting and interpreting dietary data in the observational studies, and the limited target and scope of the intervention studies, may explain the different conclusions about the relationship between carotenoids and colorectal cancer risk that can be drawn from the epidemiological, clinical and laboratory findings.

F. Other Cancers

1. Cancer of the oral cavity, pharynx, and larynx (head and neck)

In the U.S., cancers of the oral cavity, larynx and pharynx account for approximately 2% of the incident cancers diagnosed yearly and approximately 1% of cancer deaths [7]. Notably, survival rate remains low for these cancers, compared to other cancers, with an estimated five-year survival rate of 56%.

Results of numerous observational epidemiological studies, including both case-control and cohort studies, quite consistently suggest that the intakes of carotenoids (especially β-carotene), vegetables and fruit are associated with reduced risk for these cancers, as previously reviewed [84]. Further, several studies have tested the effect of β-carotene supplementation on intermediate end points and on selected preneoplastic lesions, such as oral leukoplakia. The majority of these clinical trials showed favourable responses and significantly increased remission rates. In laboratory animal models (rodents), β-carotene has been shown to be protective against oral carcinogenesis [22].

One randomized, placebo-controlled study of the effect of β-carotene supplementation on a head and neck cancer outcome has been conducted and reported. In this study, the target was recurrence and survival rather than primary prevention [85], and 264 men and women with a recent history of head and neck cancer were randomly assigned to receive 50 mg β-carotene/day or placebo and were followed for a median of approximately four years. The intervention had no effect on risk for the primary outcome, which was second primary tumours plus local recurrences (RR 0.90, 95% CI 0.56, 1.45), and no effect on total mortality (RR 0.86, 95% CI 0.52, 1.42). The risks of second head-and-neck cancer and lung cancer were also examined, and no effects on these other outcomes were identified.

Thus, epidemiological evidence, which is fairly consistent, and results from laboratory studies, suggest a potential protective effect of β-carotene on risk for head and neck cancers, with the added support from clinical trials that have focused on intermediate endpoints. However, the effects on primary risk and outcome have been addressed only minimally with a clinical trial that focused only on one aspect of the cancer continuum. Similar to the conclusions regarding carotenoids and colon cancer risk, there is clearly a suggestion of benefit across the range of scientific evidence, but the evidence is inadequate to permit a definite conclusion that carotenoids reduce risk or progression of head and neck cancers.

2. Cervical cancer

Cervical cancer is the third most common cancer among women worldwide [1]. Like colon cancer, invasive cervical cancer is known to arise through a progression of epithelial cell changes across a continuum of lesions classified as cervical intra-epithelial neoplasia (CIN) I, II, III and carcinoma *in situ*, which are earlier stages of this disease. The primary aetiological factor for cervical cancer is known to be human papilloma virus (HPV), although numerous influencing factors, including dietary factors, appear to be among the determinants of whether the HPV persists, disrupts cellular function, and enables progression of disease in the exposed individual.

Studies of dietary intakes of carotenoid-rich foods, or plasma or serum concentrations of carotenoids, linked these compounds inversely to risk and progression of cervical dysplasia quite consistently [86]. In general, evidence for a protective effect on risk for cervical cancer is more consistent in the studies that use serum or plasma carotenoids, compared to case-control observational studies based on self-reported dietary intake of carotenoids. In one small study of the relationship between serum carotenoids and persistence of HPV infection [87], adjusted mean concentrations of serum β-carotene, β-cryptoxanthin, and lutein were, on average, 24% lower ($P < 0.05$) among women who were HPV positive at two time points, compared with those who were HPV negative at both time points or positive at only one time point. In another study that examined this same point in the cervical cancer continuum, higher levels of vegetable consumption were found to be associated with a 54% decreased risk of HPV persistence (adjusted OR 0.46, 95% CI 0.21, 0.97 for highest *versus* lowest tertile, $P = 0.033$ for trend) [88]. Also, plasma concentration of (*Z*)-lycopene was associated with reduced risk for HPV persistence in that study (adjusted OR 0.44, 95% CI 0.19, 1.01 for highest *versus* lowest tertile, $P = 0.046$ for trend).

Five randomized controlled trials to date have tested the effect of β-carotene supplementation on the progression or regression of cervical dysplasia [89]. None of these studies found a beneficial effect compared with placebo. Additionally, a small randomized clinical trial was conducted to test whether consuming a carotenoid-rich diet, high in vegetables and fruit, would promote increased regression of cervical dysplasia in women who had been diagnosed with that preneoplastic lesion. The diet intervention was successful in promoting increased plasma carotenoid concentrations [90]; however, women assigned to the control group also reported improvements in their diet, and differences in regression rate by study group assignment (diet intervention or control group) were not observed. Overall, consumption of carotenoid-rich foods was associated with increased likelihood of the neoplastic lesion regressing to normal in a year, which is in agreement with the observational studies. Testing the effect of chemoprevention on cervical cancer involving any agent, including carotenoids, is substantially hampered by the facts that spontaneous regression rates are often large and vary a great deal across studies, and that establishing the response, *i.e.* whether regression or progression occurs, in subjects under study is challenging and not readily standardized [89]. The majority of the supplement trials directed towards CIN, as well as the small carotenoid-rich diet intervention study, lacked the statistical power to identify a beneficial effect in this target group at this stage of the cervical cancer continuum.

Cell culture studies have demonstrated that carotenoids can induce growth retardation in cervical dysplasia cell lines and apoptosis in HPV-infected cells [91], thus supporting the evidence from the observational epidemiological studies. Clinical trial results, which have focused solely on the persistence or regression of cervical dysplasia, do not support a protective effect of β-carotene supplementation or a diet high in vegetables and fruit, but the design and limited target of these studies may make these results not applicable to the bigger picture of whether carotenoids affect risk for cervical cancer.

Currently available data suggest that carotenoids may influence the risk and progression of cervical cancer, but increased knowledge of the mechanism would be useful because it would better indicate the time point at which intervention should be directed.

3. Other clinical trials with cancer outcomes

Clinical trials involving β-carotene supplementation have been conducted with a focus on a few additional, although less common, cancer outcomes. Several dietary factors, including β-carotene, have been inversely associated rather consistently with risk for oesophageal and stomach cancer in early epidemiological studies, and two randomized placebo-controlled studies that included β-carotene in a supplement regimen have been conducted in China. In one of these studies, the effect of a supplement that provided 15 mg β-carotene/day plus multivitamins and minerals (*versus* placebo) was tested in 3,318 men and women with oesophageal dysplasia, a precancerous condition [92], and no benefits were observed. In contrast, providing 15 mg β-carotene/day plus vitamin E and selenium was observed to reduce significantly risk for stomach cancer (RR 0.79, 95% CI 0.64, 0.99) in a population-based study conducted in the same region [31]. Results from another randomized controlled trial of β-carotene supplementation, in this case with or without vitamin C, in individuals with a confirmed precancerous gastric lesion (multifocal non-metaplastic atrophy or intestinal metaplasia) suggest improved regression rates in association with β-carotene supplementation, in both conditions [93]. A single randomized trial tested the effect of β-carotene supplementation (50 mg/day) on recurrent non-melanoma skin cancer over a five-year follow-up period, but significant differences from the placebo group were not observed [94].

As noted above, the interpretation of results of placebo-controlled trials is constrained by the timing in the cancer continuum as well as the follow-up period under study. Nonetheless, the evidence for a protective effect of carotenoids in these other cancers is limited, and an increased knowledge base relating to the core biological issues, such as mechanism, would better inform any future clinical trials.

G. Conclusions

Integrating the findings from the numerous studies that are relevant to defining the relationship between carotenoids and the risk and progression of cancer, across the laboratory and epidemiological studies and clinical trials, involves a critical evaluation of the nature of the data examined in these studies. Although cell-culture studies generally provide strong evidence that these compounds can influence the biochemical, biological and molecular processes involved in the development of cell growth dysregulation and carcinogenesis, these systems are not analogous to the complex intact biological system. Laboratory animal models may not be comparable to the human system in many key biological features that are relevant to carotenoid metabolism and the development of human cancer. Epidemiological studies, especially those based on self-reported dietary data, have numerous constraints and, because they are generally based on carotenoid-rich foods rather than pure carotenoids, attributing beneficial associations and responses to carotenoids may be unwarranted. Even more difficult to interpret in the broader picture of cancer risk and progression are the findings from clinical trials conducted to date. The time frame, target groups and limited scope of these trials have not expanded the knowledge base in many instances, and it is questionable whether the results in most cases are relevant to the long process and multi-factorial nature of human cancer.

The most scientifically supportable conclusion, based on currently available data, is that the weight of the evidence suggests a beneficial effect, although for some cancers more than others, and at doses achievable from the food supply. A diet that includes a sufficient amount of vegetables and fruits, including those that are rich in carotenoids, is a scientifically supportable low-risk strategy that would enable the potential beneficial effects of carotenoids on the risk and progression of cancer to be realized.

References

[1] World Cancer Research Fund/American Institute for Cancer Research, *Food, Nutrition and the Prevention of Cancer: A Global Perspective*. American Institute for Cancer Research, Washington, D.C. (1997).

[2] J. A. Milner, *J. Nutr.*, **134**, 2492S (2004).

[3] P. M. Kris-Etherton, K. D. Hecker, A. Bonanome, S. M. Coval, A. E. Binkowski, K. F. Hilpert, A. E. Griel and T. D. Etherton, *Am. J. Med.*, **113**, 71S (2002).

[4] W. C. Willett, *CA Cancer J. Clin.*, **49**, 331 (1999).

[5] W. C. Willett, *The Oncologist*, **5**, 393 (2000).

[6] *World Health Organization Fact Sheet No. 297, February 2006.* http://www.who.int/mediacentre/factsheets/fs297/en/print.html

[7] *Cancer Statistics 2006.* American Cancer Society, Atlanta (2006).

[8] C. C. Harris, *Cancer Res.*, **51 (suppl.)**, 5023S (1991).

[9] J. Kotsopoulos, O. I. Olopado, P. Ghadrian, J. Lubinski, H. T. Lynch, C. Isaacs, B. Weber, C. Kim-Sing, P. Ainsworth, W. D. Foulkes, A. Eisen, P. Sun and S. A. Narod, *Breast Cancer Res.*, **7**, R833 (2005).

[10] N. I. Krinsky, *Ann. NY Acad. Sci.*, **854**, 443 (1998).

[11] J. S. Bertram, *Nutr. Rev.*, **57**, 182 (1999).

[12] K. A. Steinmetz and J. D. Potter, *J. Am. Diet. Assoc.*, **96**, 1027 (1996).

[13] A. K. Kant, *J. Am. Diet. Assoc.*, **104**, 615 (2004).

[14] W. E. Brady, J. A. Mares-Perlman, P. Bowen and M. Stacewicz-Sapuntzakis, *J. Nutr.*, **126**, 129 (1996).

[15] S. A. Smith-Warner, D. Spiegelman, S. S. Yaun, H. O. Adami, W. L. Beeson, P. A. van den Brandt, A. R. Folsom, G. E. Fraser, J. L. Freudenheim, R. A. Goldbohm, S. Graham, A. B. Miller, J. D. Potter, T. E. Rohan, F. E. Speizer, P. Toniolo, W. C. Willett, A. Wolk, A. Zelniuch-Jacquotte and D. J. Hunter, *JAMA.*, **285**, 769 (2001).

[16] P. Toniolo, A. L. Van Kappel, A. Akhmedkhanov, P. Ferrari, I. Kato, R. E. Shore and E. Riboli, *Am. J. Epidemiol.*, **153**, 1142 (2001).

[17] C. L. Rock, C. W. Michael, R. K. Reynolds and M. T. Ruffin, *Crit. Rev. Oncol./Hematol.*, **33**, 169 (2000).

[18] C. L. Rock, *Recent Results in Cancer Research: Cancer Prevention* (ed. P. M. Schlag and H. J. Senn), **Vol. 174**, p. 171, Springer, Berlin (2007).

[19] R. G. Ziegler, S. T. Mayne and C. A. Swanson, *Cancer Causes Control*, **7**, 157 (1996).

[20] D. Albanes, *Am. J. Clin. Nutr.*, **69 (suppl.)**, 1345S (1999).

[21] S. T. Mayne, in *Nutrition in the Prevention and Treatment of Disease* (ed. A. M. Coulston, C. L. Rock and E. R. Monsen), p. 387, Academic Press, San Diego (2001).

[22] International Agency for Research on Cancer. *IARC Handbooks of Cancer Prevention, Vol. 2, Carotenoids*. Oxford University Press, Carey, North Carolina (1998).

[23] L. E. Voorrips, R. A. Goldbohm, H. A. M. Brants, G. A. F. C. Van Poppell, F. Sturmans, J. J. Hermus and P. A. Van den Brandt, *Cancer Epidemiol. Biomarkers Prev.*, **9**, 357 (2000).

[24] D. Ratnasinghe, M. R. Forman, J. A. Tangrea, Y. Qiao, S. X. Yao, E. W. Gunter, M. J. Barrett, C. A. Giffen, Y. Erozan, M. S. Tockman and P. R. Taylor, *Alcohol Alcohol.*, **35**, 355 (2000).

[25] D. S. Michaud, D. Feskanich, E. B. Rimm, G. A. Colditz, F. E. Speizer, W. C. Willett and E. Giovannucci, *Am. J. Clin. Nutr.*, **72**, 990 (2000).

[26] J. M. Yuan, R. K. Ross, X. D. Chu, Y. T. Gao and M. C. Yu, *Cancer Epidemiol. Biomarkers Prev.*, **10**, 767 (2001).

[27] T. E. Rohan, M. Jain, G. R. Howe and A. B. Miller, *Cancer Causes Control*, **13**, 231 (2002).

[28] C. N. Holick, D. S. Michaud, R. Stolzenberg-Solomon, S. T. Mayne, P. Pietinen, P. R. Taylor, J. Virtamo and D. Albanes, *Am. J. Epidemiol.*, **156**, 536 (2002).

[29] P. Brennan, C. Fortes, J. Butler, A. Agudo, S. Benhamou, S. Darby, M. Gerken, K. H. Jockel, M. Kreuzer, S. Mallone, F. Nyberg, H. Pohlabeln, G. Ferro and P. Bottetta, *Cancer Causes Control*, **11**, 49 (2000).

[30] S. Darby, E. Whitley, R. Doll, T. Key and P. Silcocks, *Br. J. Cancer*, **84**, 728 (2001).

[31] W. J. Blot, J. Y. Li, P. R. Taylor, W. Guo, S. Dawsey, G. Q. Want, C. S. Yang, S. F. Zheng, M. Gail, G. Y. Li, Y. Yu, B. Liu, J. Tangrea, Y. Sun, F. Liu, J. F. Fraumeni Jr., Y. H. Zhang and B. Li, *J. Natl. Cancer. Inst.*, **85**, 1483 (1993).

[32] The Alpha-Tocopherol, Beta Carotene Cancer Prevention Study Group, *New Engl. J. Med.*, **330**, 1029 (1994).

[33] G. S. Omenn, G. E. Goodman, M. D. Thornquist, J. Balmes, M. R. Cullen, A. Glass, J. P. Keogh, F. L. Meyskens, B. Valanis, J. H. Williams, S. Barnhart and S. Hammar, *New Engl. J. Med.*, **334**, 1150 (1996).

[34] C. H. Hennekens, J. E. Buring, J. E. Manson, M. Stampfer, B. Rosner, N. R. Cook, C. Belanger, F. LaMotte, J. M. Gaziano, P. M. Ridker, W. Willett and R. Peto, *New Engl. J. Med.*, **334**, 1145 (1996).

[35] D. Albanes, O. P. Heinonen, P. R. Taylor, J. Virtamo, B. K. Edwards, M. Rautalahti, A. M. Hartman, J. Palmgren, L. S. Freedman, J. Haapakoski, M. J. Barrett, P. Pietinen, N. Malila, E. Tala, K. Liipo, E. R. Salomaa, J. A. Tangrea, L. Teppo, F. B. Askin, E. Taskinen, Y. Erozan, P. Greenwald and J. K. Huttunen, *J. Natl. Cancer Inst.*, **88**, 1560 (1996).

[36] G. S. Omenn, G. E. Goodman, M. D. Thornquist, J. Balmes, M. R. Cullen, A. Glass, J. P. Keogh, F. L. Meyskens, B. Valanis, J. H. Williams, S. Barnhard, M. G. Cherniack, C. A. Brodkin and S. Hammar, *J. Natl. Cancer Inst.*, **88**, 1550 (1996).

[37] N. R. Cook, I. M. Lee, J. E. Manson, J. E. Buring and C. H. Hennekens, *Cancer Causes Control*, **11**, 617 (2000).

[38] X. D. Wang, C. Liu, R. T. Bronson, D. E. Smith, N. I. Krinsky and R. M. Russell, *J. Natl. Cancer Inst.*, **91**, 60 (1999).

[39] R. M. Russell, *Pure Appl. Chem.*, **74**, 1461 (2002).

[40] P. Prakash, T. G. Manfredi, C. L. Jackson and L. E. Gerber, *J. Nutr.*, **132**, 121 (2002).

[41] S. Gandini, H. Merzenich, C. Robertson and P. Boyle, *Eur. J. Cancer*, **36**, 636 (2000).

[42] P. Terry, M. Jain, A. B. Miller, G. R. Howe and T. E. Rohan, *Am. J. Clin. Nutr.*, **76**, 833 (2002).

[43] R. Sato, K. J. Helzlsouer, A. J. Alberg, S. C. Hoffman, E. P. Norkus and G. W. Comstock, *Cancer Epidemiol. Biomarkers Prev.*, **11**, 451 (2002).

[44] S. Ching, D. Ingram, R. Hahnel, J. Belby and E. Rossi, *J. Nutr.*, **132**, 303 (2002).

[45] C. L. Rock and W. Demark-Wahnefried, *J. Clin. Oncol.*, **20**, 3302 (2002).

[46] C. L. Rock, S. W. Flatt, L. Natarajan, C. A. Thomson, W. A. Bardwell, V. A. Newman, K. A. Hollenbach, L. Jones, B. J. Caan and J. P. Pierce, *J. Clin. Oncol.*, **23**, 6631 (2005).

[47] J. P. Pierce, V. A. Newmann, S. W. Flatt, S. Faerber, C. L. Rock, L. Natarjan, B. J. Caan, E. B. Gold, K. A. Hollenbach, L. Wasserman, L. Jones, C. Ritenbaugh, M. L. Stefanick, C. A.Thomson and S. Kealey, *J. Nutr.*, **134**, 452 (2004).

[48] J. P. Pierce, L. Natarajan, B. J. Caan, B. A. Parker, E. R. Greenberg, S. W. Flatt, C. L. Rock, S. Kealey, W. K. Al-Delaimy, W. A. Bardwell, R. W. Carlson, J. A. Emond, S. Faerbeer, E. B. Gold, R. A. Hajek, K. Hollenbach, L. A. Jones, N. Karanja, L. Madlensky, J. Marshall, V. A. Newman, C. Ritenbaugh, C. A. Thomson, L. Wasserman, and M. L. Stefanick, *JAMA*, **298**, 289 (2007).

[49] C. L. Rock, L. Natarajan, M. Pu, C. A. Thomson, S. W. Flatt, B. J. Caan, E. B. Gold, W. K. Al-Delaimy, V. A. Newman, R. A. Hajek, M. L. Stefanick, and J. P. Pierce, *Cancer Epidemiol. Biomarkers Prev.*, **18**, 486 (2009).

[50] V. N. Sumantran, R. Zhang, D. S. Lee and M. S. Wicha, *Cancer Epidemiol. Biomarkers Prev.*, **9**, 257 (2000).

[51] C. L. Rock, R. A. Kusluski, M. M. Galvez and S. P. Ethier, *Nutr. Cancer*, **23**, 319 (1995).

[52] P. Prakash, N. I. Krinsky and R. M. Russell, *Nutr. Rev.*, **58**, 170 (2000).

[53] X. O. Shu, Y. T. Gao, J. M. Yuan, R. G. Ziegler and L. A. Brinton, *Br. J. Cancer*, **59**, 92 (1989).

[54] A. Engle, J. E. Muscat and R.E. Harris, *Nutr. Cancer*, **15**, 239 (1991).

[55] H. A. Risch, M. Jain, L. D. Marrett and G. R. Howe, *J. Natl. Cancer Inst.*, **86**, 1409 (1994).

[56] L. H. Kushi, P. J. Mink, A. R. Folsom, K. E. Anderston, W. Zheng, D. Lazovich and T. A. Sellers, *Am. J. Epidemiol.*, **149**, 21 (1999).

[57] F. Parazzini, L. Chatenoud, V. Chiantera, G. Benzi, M. Surace and C. La Vecchia, *Eur. J. Cancer*, **36**, 520 (2000).

[58] C. Boseti, E. Negri, S. Franceschi, C. Pelucchi, R. Talamini, M. Montella, E. Conti and C. La Vecchia, *Int. J. Cancer*, **93**, 911 (2001).
[59] T. Byers, J. Marshall, S. Graham, C. Mettlin and M. A. Swanson, *J. Natl. Cancer Inst.*, **71**, 681 (1983).
[60] M. L. Slattery, K. L. Schuman, D. W. West, T. K. French and L. M. Robison, *Am. J. Epidemiol.*, **130**, 497 (1989).
[61] D. W. Cramer, H. Kuper, B. L. Harlow and L. Titus-Ernstoff, *Int. J. Cancer*, **94**, 128 (2001).
[62] K. M. Fairfield, S. E. Hankinson, B. A. Rosner, D. J. Hunter, G. A. Colditz and W. C. Willett, *Cancer*, **92**, 2318 (2001).
[63] K. J. Helzlsouer, A. J. Alberg, E. P. Norkus, J. S. Morris, S. C. Hoffman and G. W. Comstock, *J. Natl. Cancer Inst.*, **88**, 32 (1996).
[64] C. M. Nagle, D. M. Purdie, P. M. Webb, A. Green, P. W. Harvey and C. J. Bain, *Int. J. Cancer*, **106**, 264 (2003).
[65] J. M. Chan, P. H. Gann and E. L. Giovannucci, *J. Clin. Oncol.*, **23**, 8152 (2005).
[66] M. Etminan, B. Takkouche and F. Caamano-Isorna, *Cancer Epidemiol. Biomarkers Prev.*, **13**, 340 (2004).
[67] E. Giovannucci, *Exp. Biol. Med.*, **227**, 852 (2002).
[68] E. Giovannucci, E. B. Rimm, Y. Liu, M. J. Stampfer and W. C. Willett, *J. Natl. Cancer. Inst.*, **94**, 391 (2002).
[69] P. H. Gann, J. Ma, E. Giovannucci, W. Willett, F. M. Sacks, C. H. Hennekens and M. J. Stampfer, *Cancer Res.*, **59**, 1225 (1999).
[70] C. W. Hadley, E. C. Miller, S. J. Schwartz and S. K. Clinton, *Exp. Biol. Med.*, **227**, 869 (2002).
[71] D. Albanes and T. J. Hartman, in *Diet, Nutrition, and Health* (ed. A. M. Papas), p. 497, CRC Press, Boca Raton (1999).
[72] A. R. Kristal and J. H. Cohen, *Am. J. Epidemiol.*, **151**, 124 (2000).
[73] N. R. Cook, M. J. Stampfer, J. Ma, J. E. Manson, F. M. Sacks, J. E. Buring and C. H. Hennekens, *Cancer*, **86**, 1783 (1999).
[74] W. G. Christen, J. M. Gaziano and C. H. Hennekens, *Ann. Epidemiol.*, **10**, 125 (2000).
[75] J. D. Potter, *Cancer Causes Control*, **7**, 127 (1996).
[76] M. L. Slattery, J. Benson, K. Curtin, K. N. Ma, D. Schaeffer and J. D. Potter, *Am. J. Clin. Nutr.*, **71**, 575 (2000).
[77] K. B. Michels, E. Giovannucci, K. J. Joshipura, B. A. Rosner, M. J. Stampfer, C. S. Fuchs, G. A. Colditz, F. E. Speizer and W. C. Willett, *J. Natl. Cancer Inst.*, **92**, 1740 (2000).
[78] E. R. Greenberg, J. A. Baron, T. D. Tosteson, D. H. Freeman, G. J. Beck, J. H. Bond, T. A. Colacchio, J. A. Coller, H. D. Frankl, R. W. Haile, J. S. Mandel, D. W. Nierenberg, R. Rothstein, D. C. Snover, M. M. Stevens, R. W. Summers and R. W. Van Stock, *New Engl. J. Med.*, **331**, 141 (1994).
[79] R. MacLennan, F. Macrae, C. Bain, D. Battistutta, P. Cahpuis, H. Gratten, J. Lambert, R. C. Newland, M. Ngu, A. Russell, M. Ward and M. L. Wahlqvist, *J. Natl. Cancer Inst.*, **87**, 1760 (1995).
[80] A. Schatzkin, E. Lanza, D. Corle, P. Lance, F. Iber, B. Caan, M. Shike, J. Weissfeld, R. Burt, M. R. Cooper, J. W. Kikendall and J. Cahill, *New Engl. J. Med.*, **342**, 1149 (2000).
[81] S. Steck-Scott, E. Lanza, M. Forman, A. Sowell, C. Borkowf, P. Albert and A. Schatzkin, *FASEB J.*, **15**, A62 (2001).
[82] P. Palozza, S. Serini, N. Maggiano, M. Angelini, A. Boninsegna, F. Di Nicuolo, F. R. Ranelletti and G. Calviello, *Carcinogenesis*, **23**, 11 (2002).
[83] P. Palozza, G. Calviello, S. Serini, N. Maggiano, P. Lanza, F. O. Ranelletti and G. M. Bartola, *Free Radic. Biol. Med.*, **30**, 1000 (2001).
[84] S. T. Mayne and S. M. Lippman, in *Principles and Practice of Oncology, 6th Edition* (ed. V. T. DeVita, S. Hellman and S. A. Rosenberg), p. 575. Lippincott Williams and Wilkins, Philadelphia (2001).

[85] S. T. Mayne, B. Cartmel, M. Baum, G. Shor-Posner, B. G. Fallon, K. Briskin, J. Bean, T. Zheng, D. Cooper, C. Friedman and W. J. Goodwin, *Cancer Res.*, **61**, 1457 (2001).
[86] N. Potischman and L. A. Brinton, *Cancer Causes Control*, **7**, 113 (1996).
[87] A. R. Giuliano, M. Papenfuss, M. Nour, L. M. Canfield, A. Schneider and K. Hatch, *Cancer Epidemiol. Biomarkers Prev.*, **6**, 917 (1997).
[88] R. L. Sedjo, D. J. Roe, M. Abrahamsen, R. B. Harris, N. Craft, S. Baldwin and A. R. Giuliano, *Cancer Epidemiol. Biomarkers Prev.*, **11**, 876 (2002).
[89] M. Follen, F. L. Meyskens, N. Atkinson and D. Schotttenfeld, *J. Natl. Cancer Inst.*, **93**, 1293 (2001).
[90] C. L. Rock, A. Moskowitz, B. Huizar, C. C. Saenz, J. T. Clark, T. L. Daly, H. Chin, C. Behling and M. T. Ruffin, *J. Am. Diet. Assoc.*, **101**, 1167 (2001).
[91] Y. Muto, J. Fujii, Y. Shidoji, H. Moriwaki, T. Kawaguchi and T. Noda, *Am. J. Clin. Nutr.*, **62(suppl)**, 1535S (1995).
[92] J. Y. Li, P. R. Taylor, B. Li, S. Dawsey, G. Q. Wang, A. G. Ershow, W. Guo, S. F. Liu, C. S. Yang and Q. Shen, *J. Natl. Cancer Inst.*, **85**, 1492 (1993).
[93] P. Correa, E. T. H. Fontham, J. C. Bravo, L. E. Bravo, B. Ruiz, G. Zarama, J. L. Realpe, G. T. Malcom, D. Li, W. D. Jonson and R. Mera, *J. Natl. Cancer Inst.*, **92**, 1881 (2000).
[94] E. R. Greenberg, J. A. Baron, T. A. Stukel, M. M. Stevens, J. S. Mandel, S. K. Spencer, P. M. Elias, N. Lowe, D. W. Nierenberg, G. Bayrd, J. C. Vance, D. H. Freeman, W. E. Clendenning and T. Kwan, *New Engl. J. Med.*, **323**, 789 (1990).

Carotenoids
Volume 5: Nutrition and Health

Chapter 14

Carotenoids and Coronary Heart Disease

Elizabeth J. Johnson and Norman I. Krinsky

A. Introduction

Coronary heart disease (CHD) is a disease of the heart caused by atherosclerotic narrowing of the coronary arteries and is likely to produce angina pectoris (chest pain). In atherosclerosis, commonly called hardening of the arteries, the artery walls become thick and lose elasticity. Because of their ability to act as antioxidants (see *Chapter 12*), it has been proposed that carotenoids could be protective against CHD. One factor in the development of coronary vascular disease is the oxidation of low-density lipoproteins (LDL). When LDL is oxidized it is readily taken up by foam cells in the vascular endothelium where it contributes to the development of atherosclerotic lesions [1,2]. The facts that LDL is a major transporter of β-carotene (**3**) and lycopene (**31**) in the circulation [3] and that these carotenoids have the capacity to trap peroxyl radicals and quench singlet oxygen lend support to the hypothesis that the carotenoids may have a protective role.

Theoretically, carotenoids can have either a positive, null, or negative effect on CHD in humans. In this *Chapter*, the evidence for a role of carotenoids in CHD is evaluated.

Most of the data describing a potential relationship between carotenoids and CHD have been derived from epidemiological surveys that sought to associate either the intake of various carotenoids, or the serum levels of these carotenoids, with the incidence of CHD. There have also been several large-scale intervention trials with carotenoids, either alone or in combination with other nutrients, to evaluate the efficacy of interventions with carotenoids for CHD prevention.

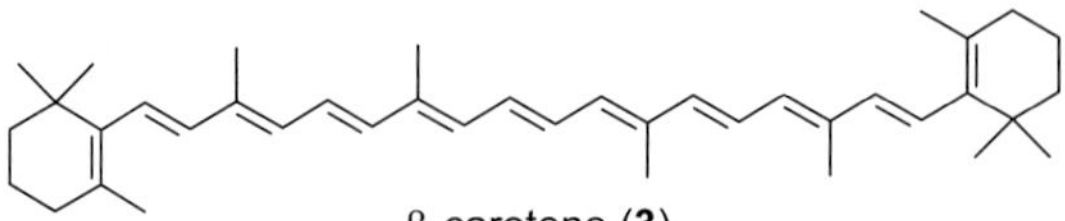

β-carotene (**3**)

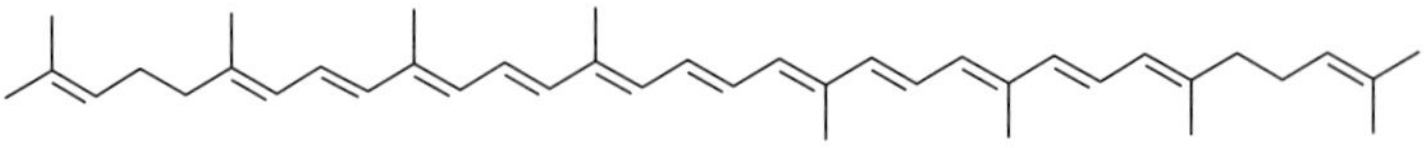

lycopene (**31**)

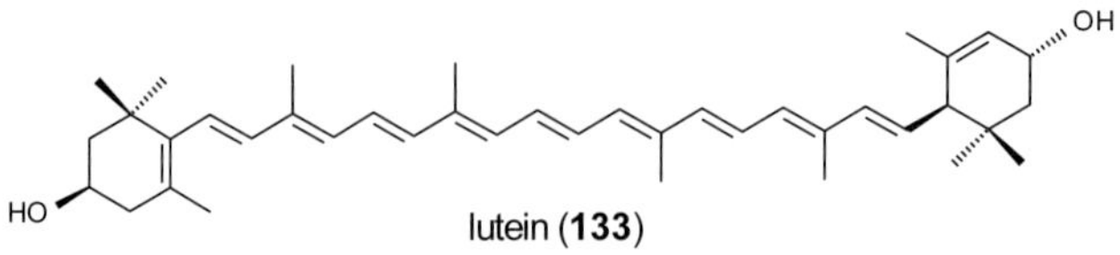

lutein (**133**)

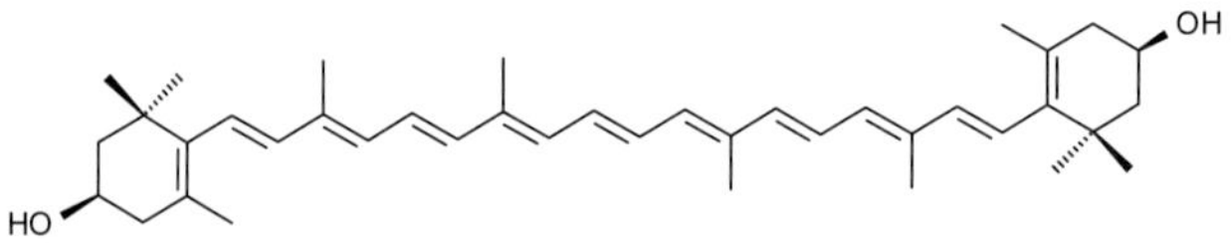

zeaxanthin (**119**)

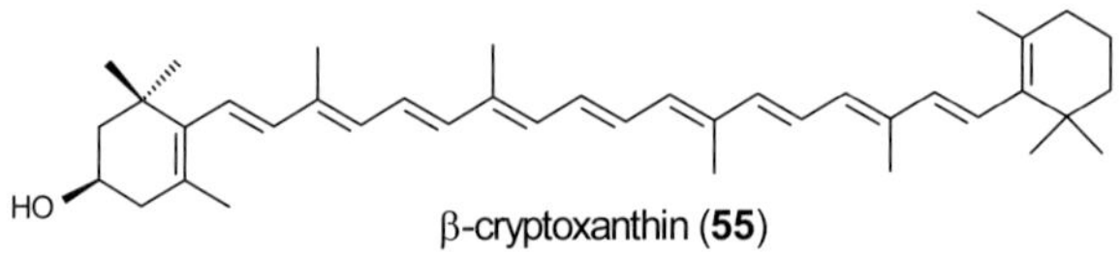

β-cryptoxanthin (**55**)

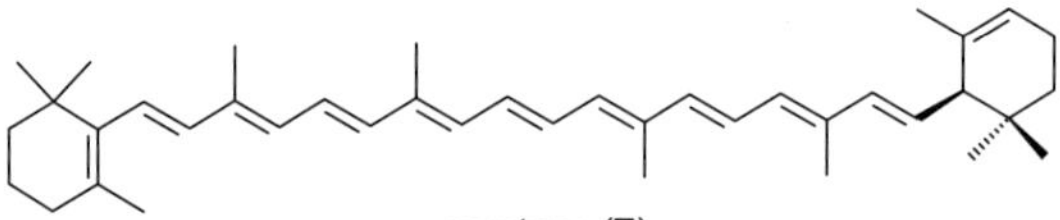

α-carotene (**7**)

B. Observational Epidemiology

1. Case-control studies

The design, application and interpretation of case-control, cohort and other epidemiological studies are described in *Chapter 10*. The hypothesis that carotenoids may decrease the risk of CHD is supported by some, but not all, case-control studies that report relationships between concentrations of carotenoids in diet and serum and risk of CHD [4,5] (Table 1).

Table 1. Case-control studies on intake and plasma/serum or tissue levels of carotenoids and the risk of cardiovascular events.

Cases	Controls	Exposure variables	Effect variables	Outcome	Comments	Ref.
433 ♀	869 ♀	Intake of β-carotene	Non-fatal acute myocardial infarction (AMI)	Significantly lower risk at high *vs* low β-carotene intake	Food frequency by trained interviewers	[6]
760	682	Intake of β-carotene and other carotenoids	AMI	Significantly reduced risk at high *vs* low levels of β-carotene, α-carotene and β-cryptoxanthin	No association with lycopene or lutein+zeaxanthin	[7]
89	50	Plasma carotenoids	Acute coronary syndrome (n = 39); stable CHD (n = 50)	Plasma β-cryptoxanthin and lutein+zeaxanthin significantly reduced in CHD patients	Independent association with natural killer (NK) cells in blood	[8]
52 ♂	52 ♂	Plasma β-carotene	First non-fatal AMI	Significantly reduced risk at high *vs* low levels of β-carotene	Nested case-control study; samples stored at - 80°C	[9]
38	20	Plasma carotene	AMI	Significant association with low levels of plasma β-carotene	Samples obtained within 3 h of AMI onset	[10]
34	40	Plasma β-carotene	Coronary heart disease by angiography	β-Carotene significantly lower in cases	Lipid normalized vitamin levels	[11]
100	249	Serum β-carotene	Coronary artery disease	Significant association with low levels of plasma β-carotene		[12]
483	483	Plasma carotenoids	Cardiovascular disease	Higher plasma lycopene was associated with lower risk of cardiovascular disease	Nested case-control	[13]
28	28	Plasma carotenoids	Acute ischaemic stroke	Plasma lutein, α-carotene, β-carotene and lycopene significantly lower in cases		[14]
297 ♂	297 ♂	Plasma carotenoids	Ischaemic stroke	Baseline α-carotene, β-carotene, and lycopene inversely related to ischaemic stroke; apparent threshold effect	Nested case-control	[15]

Table 1, continued

Cases	Controls	Exposure variables	Effect variables	Outcome	Comments	Ref.
531 ♂	531 ♂	Plasma carotenoids	AMI	No significant association		[16]
123	246	Serum carotenoids	Myocardial infarction	Association with carotenoids only in smokers (RR 0.56 for high *vs* low levels)	Nested case-control design	[17]
662 ♂	717 ♂	Adipose tissue content of carotenoids	AMI	Significantly lower risk with increased lycopene and β-carotene		[18] [19]
125 ♂	430 ♂	Plasma carotene	Angina pectoris	No significant association	Non-fasting samples	[20]
25	200	Plasma carotene	Angina pectoris	No significant association		[21]
23	11	Plasma carotene	Subarachnoid haemorrhage	No significant association	Controls had unruptured cerebral aneurysms	[22]
245 ♂	489 ♂	Serum total carotenoids	Coronary heart disease death and non fatal AMI	No significant association	Nested case-control study; samples stored at -50° to -70°C	[23]

One study reported a significantly lower risk of non-fatal acute myocardial infarction (AMI) with a higher intake of β-carotene [6]. Similarly, in a case-control study involving 760 patients with non-fatal AMI and 682 controls, the risk of AMI decreased with increased intake of β-carotene and of α-carotene (**7**) and β-cryptoxanthin (**55**), but there were no associations with lycopene or lutein (**133**) + zeaxanthin (**119**) [7]. Other studies have reported a significant decrease in AMI [9,10] and CHD [11,12] associated with increased serum concentrations of β-carotene and a decreased risk of cardiovascular disease associated with increased plasma concentrations of lycopene [13].

In a study to evaluate any relationship between carotenoid status and CHD, plasma levels of carotenoids were determined in 39 patients with acute coronary syndrome, 50 patients wih stable CHD and 50 controls [8]. Both patient groups had significantly lower plasma levels of lutein + zeaxanthin than did controls. These levels were independently associated with the proportions of natural killer cells in the blood (see *Chapter 17*). It was concluded that the relationship between natural killer cells and lutein + zeaxanthin may indicate a particular role of some carotenoids in the immunological aspects of coronary heart disease.

Plasma levels of lutein, zeaxanthin, β-cryptoxanthin, lycopene, α-carotene and β-carotene were compared in a small case-control study of 28 subjects with an acute ischaemic stroke and age-matched and sex-matched controls [14]. Plasma levels of lutein, lycopene, α-carotene and β-carotene were significantly lower in patients than in controls. Lower levels of lutein were found in patients with a poor early outcome (functional decline) after ischaemic stroke than in patients who remained functionally stable. It was concluded that the concentrations of most carotenoids in plasma are lowered immediately after an ischaemic stroke, perhaps as a result of increased oxidative stress, as indicated by a concomitant rise in plasma malondialdehyde levels.

The Physician's Health Study was a prospective, nested case-control analysis among male physicians without diagnosed cardiovascular disease; the participants were followed for up to 13 years. Baseline plasma α-carotene, β-carotene, and lycopene level was inversely associated with risk of ischaemic stroke for those in the second to the fifth quintiles of plasma concentrations, suggesting that only those with low levels are at increased risk [15]. No evidence was obtained, however, for a protective effect of increased plasma carotenoids against myocardial infarction [16].

Another nested case-control study reported an inverse association between serum levels of carotenoids and myocardial infarction only in smokers [17]. A study conducted in men with myocardial infarction and matched controls found that higher lycopene concentration in adipose tissue was associated with a decreased risk factor, with an odds ratio (OR) of 0.52 for the contrast between the 10th and the 90th percentiles, after correcting for age, body mass, socioeconomic status, smoking, hypertension and family history [18]. The odds ratio is a way of comparing whether the probability of a certain event is the same for two groups. An odds ratio of 1 indicates that the condition or event under study is equally likely in both groups. An odds ratio greater than 1 indicates that the condition or event is more likely in the first group, and an odds ratio less than 1 indicates that the condition or event is less likely in the first group. In the same study, the age-adjusted and centre-adjusted OR for risk of myocardial infarction in the lowest quintile of β-carotene concentrations in adipose tissue compared with the highest was 2.62 [19]. The increased risk was mainly confined to current smokers. Four other case-control studies, however, that examined plasma or serum carotenoids and risk of cardiovascular events did not find a significant relationship [20-23] (Table 1). For example, in the Multiple Risk Factor Intervention Trial (MRFIT), it was reported that, among 734 men, there were no significant associations between total serum carotenoid concentrations and subsequent risk of coronary disease death or non-fatal myocardial infarction in smokers and non-smokers [23].

2. Cohort studies

Results from prospective epidemiological trials to evaluate the associations between dietary intake and subsequent risk of CHD have not been consistent (Table 2).

Table 2. Cohort studies on intake of carotenoids and the risk for cardiovascular events.

Study [ref]	No. of participants	Follow-up time, yrs	Cardio-vascular events	Exposure variables (dietary)	Effect variables	Relative risk (95% CI) High *vs* low intake level
Nurses Health Study [24]	73,286	12	998	Carotenoids	Non-fatal myocardial infarction	0.74 (0.59-0.93)
Nurses Health Study [24]	73,286	12	998	Carotenoids	Infarction and fatal coronary artery disease	0.80 (0.65-0.99)
Rotterdam Study [25]	4,802	4	124	Carotene	Acute myocardial infarction (AMI)	0.57 (0.37-0.88)
The Nutrition Status Survey [26]	747 (elderly)	9-12	108	Carotenoids	Death from cardiac disorders	0.64 (0.33-1.27)
[27]	39,876 ♀	7.2	719	Tomato products	Cardiovascular disease	0.71 (0.42-1.17)
The Iowa Women's Health Study [28]	34,486 ♀	6	242	Carotenoids	Death from coronary artery disease	1.07 (0.71-1.62)
Western Electric Study [29]	1,556 ♂	25	231 coronary deaths 222 strokes	Carotene	Death from coronary artery disease	0.90 (0.64-1.26)
The Zutphen Study [30]	552 ♂	15	42	β-Carotene	Incidence and death from stroke	0.45 (0.19-1.07)
[31]	18,244 ♂	5.8	245	Carotene	Fatal stroke	0.91 (0.62-1.30)
[32]	293,117	6-11	4647	Carotenoids	Non-fatal myocardial infarction and death from AMI	0.9-0.99
The Health Professional Follow-up Study [33]	39,919 ♂	4	667	Carotene	AMI	0.98 (0.76-1.26)

In the Nurses' Health Study, a prospective study examining the risk factors for coronary artery disease in women, modest but significant inverse associations were observed between the highest quintiles of intake of β-carotene and α-carotene and risk of coronary artery disease, but no significant relation with intakes of lutein/zeaxanthin [24]. Similarly, dietary carotenes were associated with a decreased risk of AMI in older men and women [25]. In a study of 747 elderly men and women [26], dietary intake of carotenoids was not associated with a reduced risk of mortality from heart disease.

In women, dietary lycopene was not strongly associated with risk of cardiovascular disease but a possible inverse association was noted for higher levels of tomato-based products, particularly tomato sauce and pizza, with cardiovascular disease, suggesting that dietary lycopene or other phytochemicals from processed tomato products confer cardiovascular benefits [27]. More recently, it has been reported that higher plasma levels of lycopene are associated with a lower risk of CHD in women [13]. The effects of lycopene on CHD have been reviewed recently [34]. The Iowa Women's Health Study of 34,486 women reported that carotenoid intake was not associated with risk of fatal coronary heart disease [28]. Similar results were reported for studies that included men [29-33].

Cohort studies have also examined the relationships between serum or plasma concentrations of carotenoids and risk for cardiovascular events (Table 3).

Table 3. Cohort studies on serum or plasma carotenoids and the risk for cardiovascular events.

Study [ref]	Participant numbers	Follow-up time, yrs	Cardio-vascular events	Exposure variables	Effect variables	Relative risk (95% CI) High *vs* low plasma/serum or intake level
Basel Prospective Study [35]	2,974 ♂	12	163	Plasma carotene	Death from acute myocardial infarction (AMI)	0.27 (0.19-0.38)
Skin Cancer Prevention Study [36]	1,720	8.2	59	Plasma carotenoids	Death from cardiovascular disease	0.55 (0.33-0.89)
Lipid Research Clinics Coronary Primary Prevention Trial and Follow-up Study [37]	1,899 ♂	13	282	Serum carotenoids	AMI	0.62 (0.44-0.88)
The Nutrition Status Survey [26]	747 (elderly)	9-12	108	Plasma carotenoids	Death from cardiac disorders	1.51 (0.68-3.37)

Three of four studies have reported a decreased risk of AMI or death from cardiovascular disease in men and women with high serum or plasma carotenoid levels [35-37], but one study reported no relationship between plasma carotenoids and death from cardiac disorders in older men and women [26]. It must be emphasized, however, that the plasma/serum values of carotenoids are biomarkers of the consumption of diets rich in yellow, orange, red and green fruits and vegetables, so any such association does not prove that the carotenoids themselves are the active compounds.

C. Randomized Control Trials

1. Carotenoids in the primary prevention of CHD

Four large [38-43] and two smaller randomized trials [36,42] of antioxidant vitamins, which included β-carotene, as dietary supplements for primary prevention have been published, in which severe cardiovascular events were the outcome variable (Table 4).

Table 4. Randomized control trials of dietary supplements of β-carotene in the primary prevention of cardiovascular diseases.

Name of study [ref]	**Participant numbers**	**Cardio-vascular events**	**Intervention**	**Duration, yrs**	**Events reported**	**Relative risk (95% CI) treatment *vs* control**
Linxian Nutrition Intervention Study [38,39]	9 groups with a total of 27,056 in vitamin and placebo groups	523	β-carotene, 15 mg/day α-tocopherol, 30 mg/day selenium, 50 μg/day	5.2	Death from stroke	0.82 (0.62-1.08)
ATBC Study [40,41]	29,133 ♂	1723	β-carotene, 20 mg/day α-tocopherol, 50 mg/day	5-8	Death from all cardiovascular disease non-fatal myocardial infarction	1.11 (1.01-1.23)
CARET Trial [42]	18,314	not available	β-carotene, 30 mg/day + retinol, 25,000 IU/day	4 (mean)	Death from cardiovascular disease	0.81 (0.66-0.99)
Physician's Health Study [43]	22,071 ♂	1939	β-carotene, 50 mg on alternate days	12	Incidence and deaths from all cardiovascular diseases	0.99 (0.91-1.09)
Skin Cancer Prevention Study [36]	1,720	127	β-carotene, 50 mg/day	4.3 (mean)	Death from all cardiovascular diseases	1.29 (0.90-1.09)
[44]	1,203	11	β-carotene, 30 mg/day	4.5 (median)	Death from ischaemic heart disease	0.59 (0.18-1.96)

Two of the trials were performed in high-risk individuals, namely smoking middle-aged men [40], and smokers and asbestos workers [42]. One study was performed in a population at particularly low risk, *i.e.* health professionals [43] and one in a Chinese population in which deficiencies in intake of micronutrients is common [38].

All the six studies failed to show a beneficial effect of β-carotene supplementation on cardiovascular disease, despite the large number of participants and years of observation. Two major intervention trials, one in Linxian County, China, and the other in Finland (the Alpha-Tocopherol, Beta-Carotene study, or ATBC study), showed that supplements containing β-carotene alone or in combination with vitamin E and/or selenium did not reduce the risk of CHD [38,40]. Results of the Physicians' Health Study found that 50 mg β-carotene every other day did not reduce the risk of myocardial infarction, stroke, or cardiovascular death over 12 years, in 22,079 male physicians [43]. The results from the Carotene and Retinol Efficacy Trial (CARET) found that β-carotene (30 mg/day) and vitamin A (25,000 IU/d) combined caused a 26% increase (non-significant) in mortality from cardiovascular disease in a group of smokers, former smokers, and asbestos-exposed individuals [42]. Although this was non-significant, the 95% confidence interval (0.99 to 1.61) suggests that the combination of β-carotene and vitamin A may have had an adverse effect on the risk of death from cardiovascular disease in smokers and workers exposed to asbestos. In the ATBC Study, there was a significant increase in the risk for intracerebral haemorrhage in the group taking β-carotene [45]. In two of the trials, intake of β-carotene supplements was associated with an increased risk of lung cancer in cigarette smokers [40] and asbestos workers [42].

2. Carotenoids in the secondary prevention of CHD

The efficacy of β-carotene supplementation in the secondary prevention in people showing clinical manifestations of cardiovascular disease has been evaluated in four studies (Table 5).

Table 5. Randomized controlled trials of dietary supplementation of β-carotene in secondary prevention in patients with manifest cardiovascular diseases.

Name of study [ref]	**Number**	**Intervention**	**Outcome**
Physician's Health Study [43,46]	333 with ischaemic heart disease included contrary to the study protocol	Factorial design: A. β-carotene, 50 mg alternate days B. aspirin C. A + B D. placebo	Reduction of risk for myocardial infarction by half in the β-carotene group. No infarctions at all in the β-carotene + aspirin group.
[47]	120 patients with intermittent claudication	β-carotene, 3 mg, ascorbic acid, 100 mg, zinc sulphate, 100 mg, nicotinamide, 10 mg, selenium, 1 mg, *vs* placebo	Non-significant reduction of cardiovascular events in the intervention group (event rate average 7.6% *vs* 10.5% per year). No difference in lower limb function.

Table 5, continued

Name of study [ref]	Number	Intervention	Outcome
MRC/BHF Heart Protection Study [48,49]	20,536 men and women, 40-80 yrs, at elevated risk of death from coronary heart disease	Factorial design: A. Antioxidant vitamins (β-carotene, ascorbic acid, α-tocopherol) B. Simvastatin C. A+B D. Placebo	No effects of antioxidant vitamins on any vascular end-point (coronary heart disease, stroke, revascularization or total)
ATBC (substudy) [50]	1,862 with previous AMI, all smokers	Factorial design: A. β-carotene, 20 mg/d B. α-tocopherol, 50 mg/d C. A+B D. placebo	Significant increase in fatal coronary events in carotene groups (adjusted RR = 1.58 - 1.75)

Some beneficial effects were reported in one of these studies in which the risk of myocardial infarction was reduced by half in subjects given β-carotene and no infarctions occurred when β-carotene was given with aspirin [43,46]. Two of the studies reported no significant effects on risk of cardiovascular events [47-49]. In contrast, the ATBC Study reported a significant increase in fatal coronary events in the β-carotene-supplemented groups [50].

3. Intervention trials and CHD biomarkers

Carotid intima-media thickness (IMT) is a marker of early arteriosclerosis and has been used as an endpoint in CHD. IMT is increased in subjects with several risk factors and is a predictor of cardiovascular events and end-organ damage (damage occurring in major organs fed by the circulatory system). Arterial vessel wall changes occur during a presumably long subclinical lag phase characterized by functional disturbances and by gradual thickening of intima-media. The measurement of IMT of large peripheral arteries, especially carotid, has emerged as one of the methods of choice for determining early atherosclerotic changes, the anatomical extent of atherosclerosis and its progression.

In the Atherosclerosis Risk in Communities Study, serum concentrations of carotenoids were correlated with carotid IMT between 231 cases (in which IMT exceeded the 90th percentile) and 231 matched controls (with IMT less than the 75th percentile [51]. After adjustment for the influence of co-variates, an inverse association with IMT was maintained only for lutein/zeaxanthin (OR *per* one SD increase = 0.77, 95% CI = 0.57-1.03), but it was not significant [52].

A cohort of 480 men and women, age 40-60, had their IMT determined over an 18-month period, and the IMT progression declined with increasing quintile of plasma lutein [53]. Since this is an association, it means that foods rich in lutein may have an important role in preventing the increase in carotid artery thickness, a precursor of coronary heart disease.

More recently, carotid IMT was examined in relation to serum lycopene concentration in middle-aged men [54]. The men in the lowest quartile of serum lycopene concentration had a significantly higher mean carotid IMT and maximal carotid IMT than did the other men. The mean and maximal IMT decreased linearly across the quartiles of serum lycopene concentration. This finding suggests that the serum lycopene concentrations may be useful as a biomarker in the early stages of atherosclerosis. The effects of lycopene on CHD are discussed in a recent review [34].

Two clinical studies that measured other intermediate biomarkers of CVD have not demonstrated a benefit from β-carotene. Daily administration of 100 mg/day β-carotene in combination with 500 mg/day vitamin C and 700 IU/day vitamin E did not reduce the rate of restenosis in patients who had undergone angioplasty [55]. Restenosis is the narrowing of a blood vessel following the removal or reduction of a previous narrowing such as in angioplasty. In another study of non-smoking men and women [56], platelet function was not affected by ingestion of 15 mg/day β-carotene.

The effect of tomatoes, a rich source of lycopene, on antioxidant activity in CHD and age-matched controls has been evaluated [57]. At baseline, serum antioxidant enzymes were lower and lipid peroxidation rates were higher in the CHD group. Sixty days of tomato supplementation in the CHD group led to a significant improvement in the levels of serum enzymes involved in antioxidant activity (superoxide dismutase, glutathione peroxidase, glutathione reductase, reduced glutathione). The lipid peroxidation rate was also decreased.

D. Summary and Conclusions

Results of case-control studies of the association between CHD and intakes or serum concentrations of carotenoids have been mixed, and findings from prospective studies have generally found little or no effect [23,26,28]. Some studies found an inverse association [25,36,37,41,58,59], whereas others found no inverse association [17,23,26,28,32,60-62]. In randomized control trials, supplementation with carotenoids has been shown convincingly to have no beneficial effects on the risk of CHD [36,38-44] and, in the case of smokers and asbestos workers, even to have adverse effects [40,42]. Because the supplements in the intervention studies provided much higher β-carotene intakes (20-50 mg/day) than reported in the cohort studies, the results from cohort and intervention studies are not comparable. That leaves open the question of what component(s) in diets rich in fruits and vegetables may be responsible for some of the positive associations. It is emphasized again that plasma/serum values of carotenoids are biomarkers for consumption of diets rich in yellow, orange, red and green fruits and vegetables which contain other potentially bioactive phytochemicals, so any association does not prove that the carotenoids are the active compounds. Suggestions have been made that more attention should be placed on other carotenoids in the diet such as lycopene, lutein and zeaxanthin, and β-cryptoxanthin. Because of their limited distribution in nature, lycopene and β-cryptoxanthin can be increased with relatively simple dietary alterations and may prove of interest with respect to CHD [63].

Carotenoid supplementation above levels that are achievable through diet cannot be recommended in the primary prevention against CHD.

References

[1] S. K. Clinton and P. Libby, *Arch. Pathol. Lab. Med.*, **116**, 1292 (1992).
[2] B. Frei, *Crit. Rev. Food Sci. Nutr.*, **35**, 83 (1995).
[3] B. A. Clevidence and J. G. Bieri, *Meth. Enzymol.*, **214**, 33 (1993).
[4] S. T. Mayne, *FASEB J.*, **10**, 690 (1996).
[5] N. I. Krinsky and E. J. Johnson, *Mol. Asp. Med.*, **26**, 459 (2005).
[6] A. Tavani, E. Negri, G. D'Avanzo and C. La Vecchia, *Eur. J. Epidemiol.*, **13**, 631 (1997).
[7] A. Tavani, S. Gallus, E. Negri, M. Parpinel and C. La Vecchia, *Free Radic. Res.*, **40**, 659 (2006).
[8] C. Lidebjer, P. Leanderson, J. Emerudh and L. Jonasson, *Nutr. Metab. Cardiovasc. Diseases*, **17**, 448 (2007).
[9] M. Bobak, E. Brunner, N. J. Miller, Z. Skovova and M. Marmot, *Eur. J. Clin. Nutr.*, **52**, 632 (1998).
[10] R. Levy, P. Bartha, A. Ben-Amotz, J. G. Brook, G. Dankner, S. Lin and H. Hammerman, *J. Am. Coll. Nutr.*, **17**, 337 (1998).
[11] A. Kontush, T. Spranger, A. Reich, K. Kaum and U. Beisiegel, *Atherosclerosis*, **144**, 117 (1999).
[12] S. Y. Kim, Y. C. Lee-Kim, M. K. Kim, J. H. Suh, E. J. Chung, S. Y. Cho, B. K. Cho and I. Suh, *Biomed. Environ. Sci.*, **9**, 229 (1996).
[13] H. D. Sesso, J. E. Buring, E. D. Norkus and J. M. Gaziano, *Am. J. Clin. Nutr.*, **79**, 47 (2004).
[14] M. C. Polidori, A. Cherubini, W. Stahl, U. Senin, H. Sies and P. Mecocci, *Free Radic. Res.*, **36**, 265 (2002).
[15] A. E. Hak, J. Ma, C. B. Powell, H. Campos, J. M. Gaziano, W. C. Willett and M. J. Stampfer, *Stroke*, **35**, 1584 (2004).
[16] A. E. Hak, M. J. Stamper, H. Campos, H. D. Sesso, J. M. Gaziano, W. C. Willett and J. Ma, *Circulation*, **108**, 802 (2003).
[17] D. A. Street, G. W. Comstock, R. M. Salkeld, W. Schuep and M. J. Klap, *Circulation*, **90**, 1154 (1994).
[18] L. Kohlmeier, J. D. Kark, E. Gomez-Garcia, B. C. Martin, S. E. Steck, A. F. Kardinaal, J. Ringstad, J. Gomez-Aracena, M. Thamm, B. Masaev, R. A. Riemersma, J. M. Martin-Moreno, J. K. Huttunene and F. J. Kok, *Am. J. Epidemiol.*, **146**, 618 (1997).
[19] A. F. M. Kardinaal, F. J. Kok, J. Ringstad, J. Gomez-Aracena, V. P. Mazaev, L. Kohlmeier, B. C. Martin, A. Aro, J. D. Kark and M. Delgado-Rodriguez, *Lancet*, **342**, 1379 (1993).
[20] R. A. Riemersma, D. A. Wood, C. C. A. Macintyre, R. A. Elton, K. F. Gey and M. F. Oliver, *Lancet*, **337**, 1 (1991).
[21] G. G. Duthie, J. A. Beattie, J. R. Arthur, M. Franklin, P. C. Morrice and W. P. James, *Nutrition*, **10**, 313 (1994).
[22] F. Marzatico, P. Baetani, F. Tartara, L. Bertirelli, F. Feletti and D. Adinolfi, *Life Sci.*, **63**, 821 (1998).
[23] R. W. Evans, B. J. Shaten, B. W. Day and L. H. Kutter, *Am. J. Epidemiol.*, **147**, 180 (1998).
[24] S. K. Osganian, M. J. Stamper, E. Rimm, P. Spiegelman, J. E. Manson and W. C. Willett, *Am. J. Clin. Nutr.*, **77**, 1390 (2003).
[25] K. Klipstein-Grobausch, J. M. Geleijnse, J. H. den Breeijen, H. Boeing, H. Hofman, D. E. Grobbee and J. C. Witteman, *Am. J. Clin. Nutr.*, **69**, 261 (1999).
[26] N. R. Sahyoun, P. F. Jacques and R. M. Russell, *Am. J. Epidemiol.*, **144**, 501 (1996).
[27] H. D. Sesso, S. Liu, J. M. Gaziano and J. E. Buring, *J. Nutr.*, **133**, 2336 (2003).

[28] L. H. Kushi, A. R. Folsom, R. J. Prieas, P. J. Monk, Y. Wu and R. M. Bostick, *New Eng. J. Med.*, **334**, 1156 (1996).
[29] M. L. Daviglus, A. J. Orencia, A. R. Dyer, K. Liu, D. K. Morris, V. Persky, V. Chavez, J. Goldberg, M. Drum, R. B. Shekelle and J. Stamler, *Neuroepidemiol.*, **16**, 69 (1997).
[30] S. O. Keli, M. G. L. Hertog, E. J. M. Feskens and D. Kromhout, *Arch. Intern. Med.*, **156**, 637 (1996).
[31] R. K. Ross, J. M. Yuan, B. E. Henderson, J. Park, Y. T. Gao and M. C. Yu, *Circulation*, **96**, 50 (1997).
[32] P. Knekt, A. Reunaned, R. Jarvinen, R. Seppanen, M. Heliovaara and A. Aromaa, *Am. J. Epidemiol.*, **139**, 1180 (1994).
[33] E. B. Rimm, M. J. Stampfer, A. Ascherio, E. Giovannucci, G. A. Colditz and W. C. Willett, *New Eng. J. Med.*, **328**, 1450 (1993).
[34] I. Petr and J. W. Erdman Jr., in *Carotenoids and Retinoids: Molecular Aspects and Health Issues*, (ed. L. Packer, U. Obermüller-Jevic, K. Kraemer and H. Sies), p. 204, AOCS, Champaign (2005).
[35] K. F. Gey, H. B. Stahelin and M. Eichholzer, *Clin. Invest.*, **71**, 3 (1993).
[36] E. R. Greenberg, J. A. Baron, M. R. Karagas, T. A. Stukel, D. W. Nierenberg, M. M. Stevens, J. S. Mandel and R. W. Jaile, *JAMA*, **275**, 699 (1996).
[37] D. L. Morris, S. B. Kritchevsky and C. E. Davis, *JAMA*, **272**, 1439 (1994).
[38] W. J. Blot, J. Y. Li, P. R. Taylor, W. Guo, S. Dawsey, G. Q. Wang, C. S. Yang, S. F. Zheng, M. Bail and G. Y. Li, *J. Natl. Cancer Inst.*, **85**, 1483 (1993).
[39] S. D. Mark, W. Wang, J. F. Fraumeni, J. Y. Li, P. R. Taylor, G. O. Wang, S. M. Dawsey, B. Li and W. J. Blot, *Epidemiology*, **9**, 9 (1998).
[40] The Alpha-Tocopherol Beta-Carotene Cancer Prevention Study Group, *New Engl. J. Med.*, **330**, 1029 (1994).
[41] J. Virtamo, J. M. Rapolo, S. Ripatti, O. P. Heinonen, P. R. Taylor, D. Albanes and J. K. Huttenen, *Arch. Intern. Med.*, **158**, 668 (1998).
[42] G. S. Omenn, G. E. Goodman, M. D. Thornquist, J. Balmes, M. R. Cullen, A. Glass, J. P. Keogh, F. L. Meyskens, B. Valanis, J. H. Williams, S. Barnhart and S. Hammar, *New Engl. J. Med.*, **334**, 1150 (1996).
[43] C. H. Hennekens, J. E. Buring, J. E. Manson, M. J. Stamper, B. Rosner, N. R. Cook, C. Belanger, F. La Motte, J. M. Gaziano, R. Ridker, W. Willett and R. Peto, *New Engl. J. Med.*, **334**, 1483 (1996).
[44] N. H. deKlerk, A. W. Musk, G. L. Ambrosini, J. L. Eccles, J. Hansen, N. Olsen, V. L. Watts, H. G. Lund, S. C. Pang, J. Beilby and M. S. Hobbs, *Int. J. Cancer*, **75**, 362 (1998).
[45] J. M. Leppala, J. Virtamo, R. Fogelholm, J. K. Huttunen, D. Albanes, P. R. Taylor and O. P. Heinonen, *Arterioscler. Thromb. Vasc. Biol.*, **20**, 230 (2000).
[46] C. H. Hennekens and J. M. Gaziano, *Clin. Cardiol.*, **16**, 10 (1993).
[47] G. C. Leng, A. J. Lee, G. R. Fowkes, D. Horrobin, R. G. Jepson, G. D. Lowe, A. Rumley, E. R. Skinner and B. F. Mowat, *Vascular Med.*, **2**, 279 (1997).
[48] Heart Protection Study Collaborative Group, *Eur. Heart J.*, **20**, 725 (1999).
[49] Heart Protection Study Collaborative Group, *Lancet*, **360**, 23 (2002).
[50] J. M. Rapola, J. Virtamo, S. Ripatti, J. K. Huttunen, D. Albanes, P. R. Taylor and O. P. Heinonen, *Lancet*, **349**, 1715 (1997).
[51] C. Iribarren, A. R. Folsom, D. R. Jacobs, M. D. Gross, J. D. Belcher and J. H. Eckfeldt, *Arterioscler. Thromb. Vasc. Biol.*, **17**, 1171 (1997).
[52] S. B. Kritchevsky, G. S. Tell, T. Shimakawa, B. Dennis, R. Li, L. Kohlmeier, E. Steere and G. Heiss, *Am. J. Clin. Nutr.*, **68**, 726 (1998).
[53] J. H. Dwyer, M. Navab, K. M. Dwyer, K. Hassan, P. Sun, A. Shircore, S. Hama-Levy, G. Hough, X. Wang, T. Drake, C. N. B. Merz and A. M. Fogelman, *Circulation*, **103**, 2922 (2001).
[54] T. H. Rissaneen, S. Voutilainen, K. Nyyssonene, R. Salonen, G. A. Kaplan and J. T. Salonen, *Am. J. Clin. Nutr.*, **77**, 133 (2003).

[55] J. C. Tardif, B. Cote, J. Lesperance, M. Bourassa, J. Lambert, S. Doucet, L. Blolodeau, S. Nattel and P. deGuise, *New Eng. J. Med.*, **337**, 365 (1997).
[56] C. Calzada, K. R. Bruckdorfer and C. A. Rice-Evans, *Atherosclerosis*, **128**, 97 (1997).
[57] K. S. C. Bose and B. K. Agrawal, *Singapore Med. J.*, **48**, 415 (2007).
[58] S. Todd, J. Woodward, H. Tunstall-Pedoe and C. Bolton-Smith, *Am. J. Epidemiol.*, **150**, 1073 (1999).
[59] J. M. Gaziano, J. E. Manson, L. G. Branch, G. A. Colditz, W. C. Willett and J. E. Buring, *Ann. Epidemiol.*, **5**, 255 (1995).
[60] A. Mezzetti, G. Zuliani, F. Romano, F. Costantini, S. D. Pieromenico and F. Cuccurullo, *J. Am. Geriatric Soc.*, **49**, 533 (2001).
[61] K. Nyyssonen, M. T. Parviainen, R. Salonen, J. Tuomilehto and J. T. Salonen, *BMJ*, **314**, 634 (1997).
[62] M. Eichholzer, H. B. Stahelin and K. F. Gey, in *Free Radicals and Aging* (ed. I. Emerit and G. Chance), p. 398, Birkhäuser Verlag, Basel (1992).
[63] H. D. Sesso and J. M. Gaziano, in *Carotenoids in Health and Disease* (ed. N. I. Krinsky, S. T. Mayne and H. Sies), p. 473, Marcel Dekker, New York, NY (2005).

Carotenoids
Volume 5: Nutrition and Health

Chapter 15

The Eye

Wolfgang Schalch, John T. Landrum and Richard A. Bone

A. Introduction

The retina of the human and primate eye, especially the macula, the circular 5-6 mm diameter central part, is a target organ for the deposition of carotenoids. This *macula lutea* (yellow spot) has the highest local concentration of carotenoids in the human body. Although the *macula lutea* was well known by the late 18th century, it was considered by many to be a *post-mortem* artefact. It was not accepted to be an anatomical feature *in vivo* until the invention of the ophthalmoscope, when researchers were able to observe the yellow to orange colour of the macula in the living eye by using appropriate, red-free, illumination. A comprehensive article [1] reviews the evolution of ideas in relation to the macular pigment over a period of 200 years. In 1945, the macular yellow pigment was broadly characterized by absorption spectroscopy as being "in all probability lutein or leaf xanthophyll" [2].

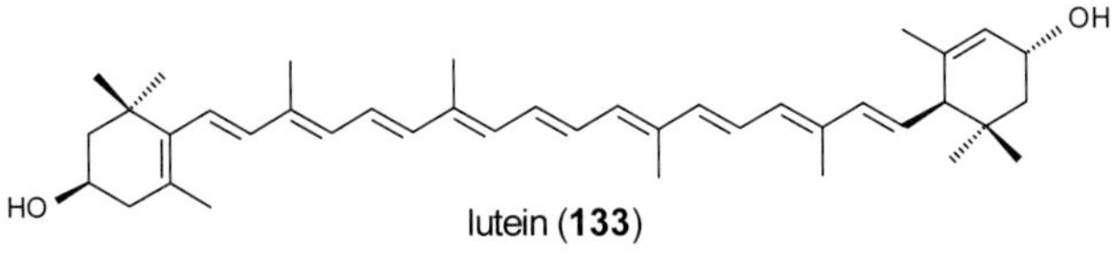

lutein (**133**)

In 1985, lutein (**133**) and zeaxanthin were identified as its main constituents [3] and, in 1993, it was reported that macular zeaxanthin comprises two stereoisomers, (3*R*,3'*R*)-zeaxanthin (**119**) and (3*R*,3'*S*)-zeaxanthin [(*meso*)-zeaxanthin] (**120**) [4].

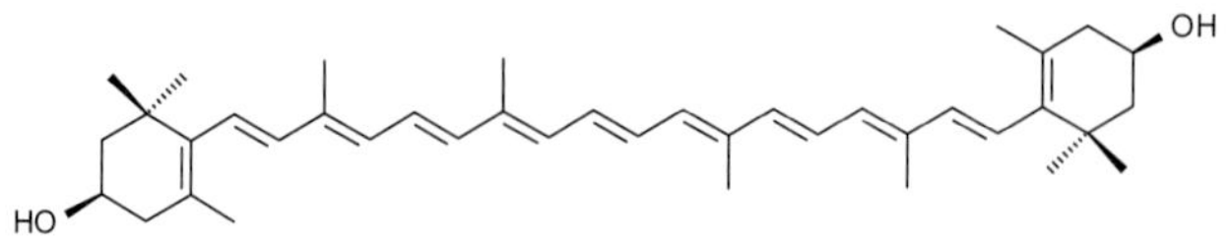

(3*R*,3'*R*)-zeaxanthin (**119**)

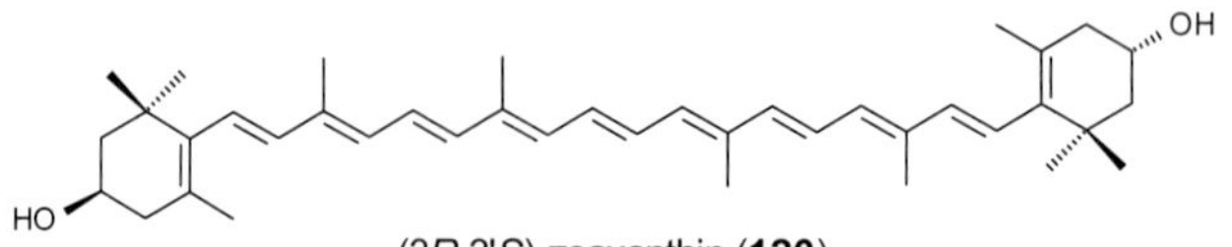

(3*R*,3'*S*)-zeaxanthin (**120**)

Because the macula is responsible for highest visual acuity, and mediates the detailed vision that is necessary for reading and similar high-resolution tasks, the high concentration of carotenoids in this particular area of the eye raises expectations that they must be essential in maintaining the structure and function of this important part of the retina and, therefore, may be important for the health of the eye, in particular of the retina and macula. It is now well established that the ophthalmic disease, age-related macular degeneration (AMD), is a major cause of severe visual impairment in the elderly. The roles of carotenoids in the healthy eye, and their significance in relation to protection not only against AMD but also against cataract, the serious degenerative disease of the lens, are described and evaluated in this *Chapter*.

B. Anatomy of the Eye and Retina

Like those of other vertebrates, eyes of primates and humans are optically quite simple [5]. Light reflected from a scene or object is transmitted successively through several transparent media: the cornea, which is the major refractive structure of the eye, the aqueous humour, the lens that focuses the image, accommodating for distance, and the vitreous body. The vitreous body is adjacent to the neural retina. The light must then pass through the anterior layers of the retina before reaching the photoreceptors, where the initial neural signals of vision are generated. Before these signals are transmitted to the brain *via* the optic nerve, they are pre-processed and amplified by a complex network of other retinal cells. The retina can be compared to a computer chip that handles the output from an array of about 10^8 photo-detectors (the photoreceptors) and prepares it for further and final processing by the brain. The main features of the anatomy of the human eye are shown in Fig. 1.

The photoreceptors of the retina are of two types, cones and rods, allowing human eyes to function not only in bright daylight (cones) but also in crepuscular and dim light conditions (rods). The cone density increases dramatically towards the fovea at the centre of the retina, a region that is devoid of rods. This area provides the highest visual acuity of any region in the retina. It is an area approximately 1.5-2 mm wide, formed as a gradual depression of the retinal surface towards the centre of the macula. The fovea is thinner than the rest of the retina because the overlying retinal layers and blood vessels are pushed aside, thereby reducing the scattering of light that could blur the image and reduce visual acuity. From the centre of the fovea, the axons of the photoreceptors extend radially outwards, like the spokes of a wheel, to form 'Henle's fibre' layer where the majority of the yellow macular pigment is located, producing a visually discernible spot that extends a little beyond the fovea.

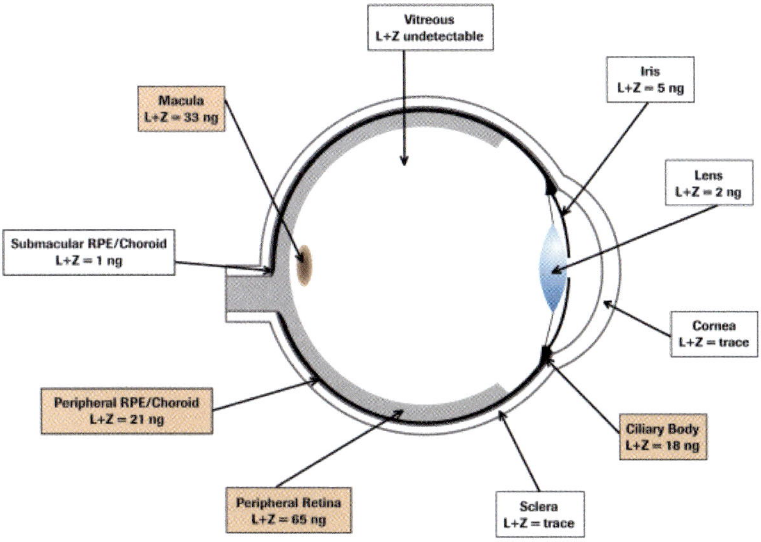

Fig. 1. Anatomy of the human eye. The figure also shows the distribution of the macular xanthophylls lutein (L) and zeaxanthin (Z) in different parts of the eye. (Adapted from [6]).

The importance of the integrity of this central area of the retina for vision cannot be over-emphasized. Photoreceptor density reaches a maximum of around 300,000 per mm^2 in the fovea, so a visual resolution of 1.2 seconds of arc is possible. According to the geometry of the optical system of the eye, a 2.7 cm diameter circle, viewed from a normal reading distance (30 cm), forms a circular image on the retina just covering the 1.5 mm diameter of the fovea. If this retinal area is dysfunctional, reading ability is impaired as is the capacity to perform other tasks that require high visual acuity, such as driving and recognizing facial features.

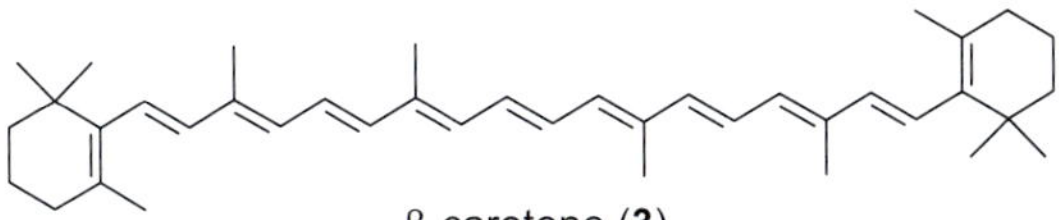

β-carotene (**3**)

C. Occurrence of Carotenoids in the Eye

The principal carotenoids present in the human eye are the xanthophylls lutein and zeaxanthin. They are most highly concentrated in the retina, but are also present in the lens, ciliary body and retinal pigment epithelium (RPE) [6]. The orbital adipose tissue also has measurable quantities of lutein and β-carotene (**3**), and perhaps other carotenoids as minor constituents [7].

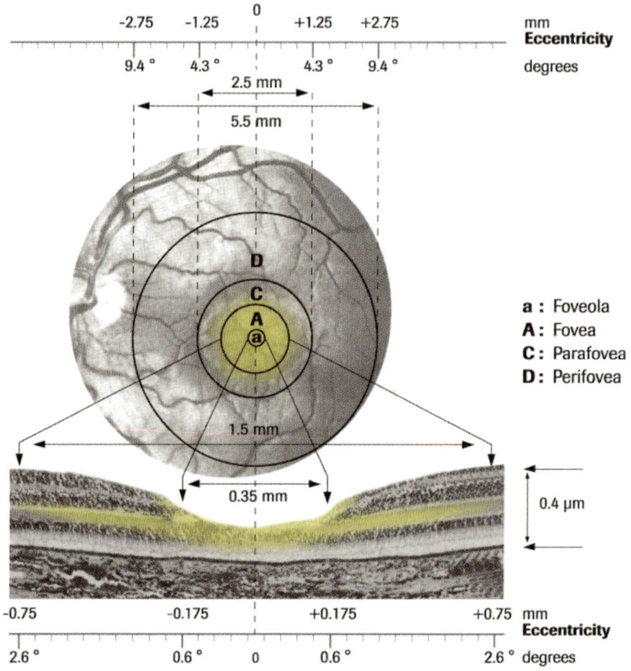

Fig. 2. Retinal topography and the macular pigment. The figure illustrates the location of the macular pigment (marked yellow) in a horizontal map (top) and a vertical section (bottom) of the retina. Most of the macular pigment occurs in the fovea and in Henle's fibre layer that is formed by the axons of the photoreceptors. The rod outer segments also contain macular pigment but at lower concentrations (not marked). The macula comprises the 2.5 mm wide area from "a" to "C". The fovea ("A"), an area only 1.5 mm wide, is the zone of highest visual acuity. Note that approximate retinal eccentricities are given in two scales, *i.e.* in mm and in degrees (°): most publications report only one or the other, not both. Adapted from *Spectroscopic Atlas of Macular Diseases*, (ed. J.D.M. Gass, 1997), with permission of the publisher, Mosby Year Book Inc., St. Louis, MO, USA.

1. Retina

The human macula has an approximately radial symmetry and the central 2 mm is the visually discernible yellow *macula lutea* which marks the fovea (Fig. 2) and contains the highest concentration of the macular carotenoids, lutein and zeaxanthin. The macular carotenoids are not confined to this central area but are also present, albeit at much lower concentrations, in the parafoveal and peripheral retina. In the peripheral retina, the carotenoid concentration is so low that only the greater geometrical area permits collection of sufficiently large amounts of tissue to make detection by HPLC possible.

In retinal cross-sections, the yellow pigmentation can be seen to be concentrated primarily within Henle's fibre layer [8] (Fig. 3). Consequently, light must pass through the yellow macular pigment before reaching the photoreceptor outer segments. As indicated in Fig. 3, lutein and zeaxanthin are also present in the rod outer segments [9], though at low concentrations, and may be present also in cone outer segments. Furthermore, Müller cells have been suggested to be a reservoir of macular xanthophylls [10].

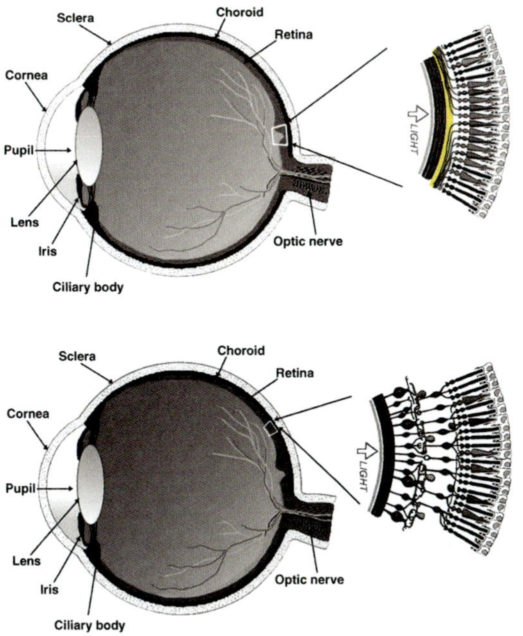

Fig. 3. Schematic representation, not to scale, of the anatomy of the eye and retina in relation to macular pigment. (Adapted from website http://webvision.med.utah.edu/). The location of the macular pigment is marked yellow. Most occurs in Henle's fibre layer (top) and forms an effective pre-receptoral light filter. Its occurrence in the outer segments of cones is still hypothetical, but its occurrence in rod outer segments (bottom) has been confirmed. The retinal anatomy of the fovea (top) and the parafovea (bottom) are fundamentally different. In the fovea, the retina is substantially thinner, because only the photoreceptor axons (Henle's fibre layer) are present in front of the photoreceptors, with the other retinal layers having been pushed aside in order to allow the light a relatively 'barrier-free' access to the area of highest visual acuity, the fovea.

A number of other carotenoids are also present in the retina at low levels, including 3'-epilutein (**137**), lactucaxanthin (**150**), 3′-dehydrolutein (**302**), and ε,ε-carotene-3,3'-dione (**385**). Interestingly, β-carotene, the provitamin A precursor of retinal (*1*), the chromophore of the visual pigments, is not detected in the retina itself.

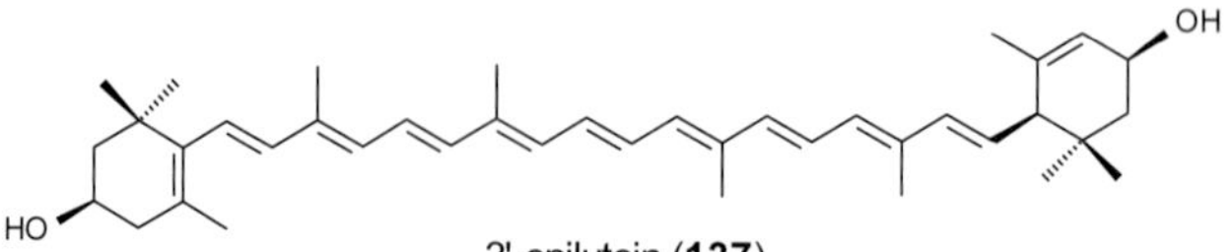

3'-epilutein (**137**)

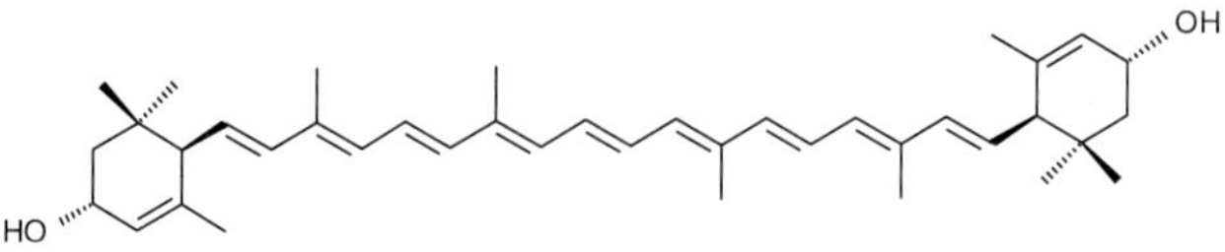

lactucaxanthin (**150**)

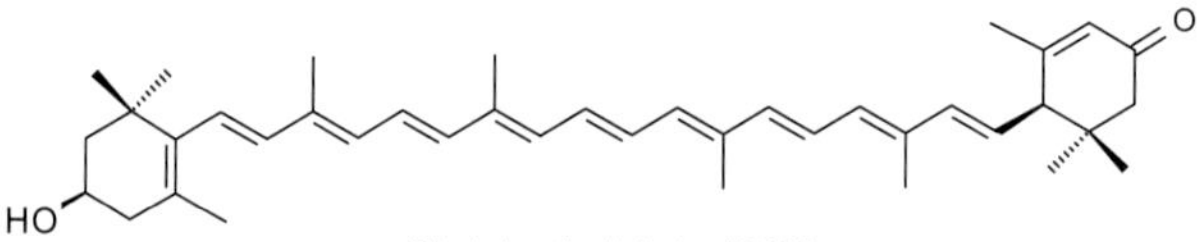

3'-dehydrolutein (**302**)

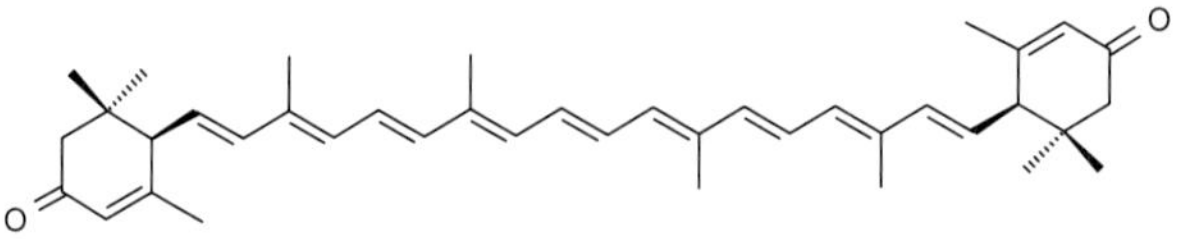

ε,ε-carotene-3,3'-dione (**385**)

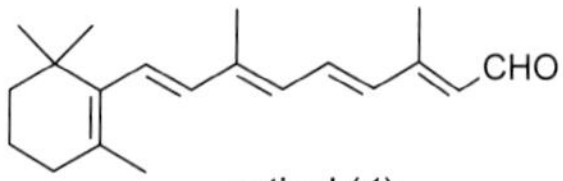

retinal (*1*)

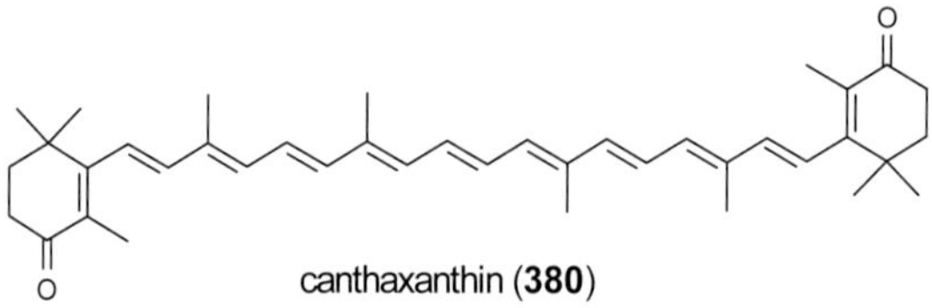

canthaxanthin (**380**)

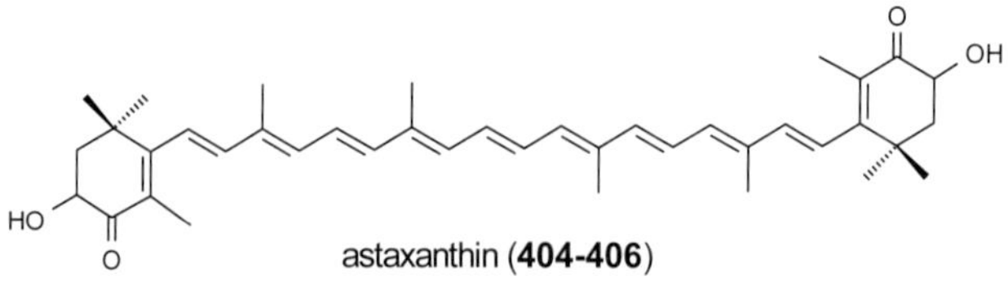

astaxanthin (**404-406**)

Canthaxanthin (**380**) has been used as a treatment for polymorphous light eruptions (which occur, for example, in the rare haematological disease erythropoietic protoporphyria) and as an oral tanning agent. For these applications, high doses of canthaxanthin have been taken for extended periods of time. Canthaxanthin was found to induce the formation of crystalloid deposits in the human retina in these instances. The crystalloid deposits around and within the macular region gave rise to the name 'canthaxanthin retinopathy' for this condition, but this term is inappropriate because the crystalloid deposition has no clinical consequences [11-13]. In spite of its structural similarity to zeaxanthin and canthaxanthin, astaxanthin (**404-406**) has not been identified in retinal tissues.

2. Lens

The human lens contains lutein and zeaxanthin in roughly equal amounts, but no other carotenoids. The average total quantity present is about 4 ng and the average concentration is estimated to be about 10 nM, which is six orders of magnitude less than the concentration in the centre of the retina. Growth of the lens continues throughout life but little is known about the amounts of the carotenoids present at different ages, or their distribution. Approximately 75% of the carotenoid content of the lens appears to be present in the relatively young epithelium/cortex tissue, which comprises about half of the lens.

3. Ciliary body and retinal pigment epithelium

The ciliary body contains the highest quantity of carotenoids, after the retina. Whilst lutein and zeaxanthin are still the dominant carotenoids in non-retinal eye tissue, the additional presence of lycopene (**31**) and β-carotene has been reported in the ciliary body and the retinal pigment epithelium (RPE). In these two tissues, the amount of these other carotenoids is, in total, about equal to that of lutein and zeaxanthin. In the ciliary body, lutein is the most abundant carotenoid followed by lycopene and zeaxanthin.

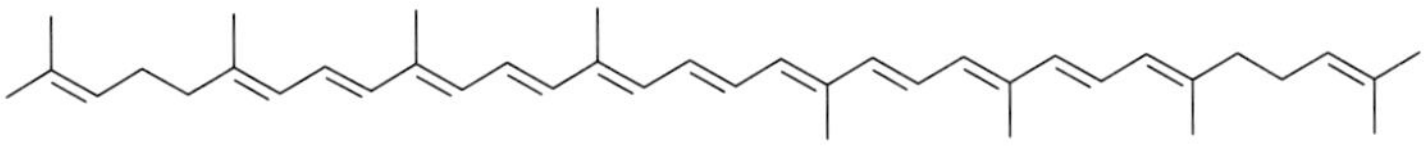

lycopene (**31**)

D. The Macular Xanthophylls

The macular xanthophylls are most highly concentrated in the centre of the macula, but the zeaxanthin to lutein ratio in the retina varies from a maximum exceeding 2:1 in the central fovea to a low of near 1:2 in the peripheral retina (Fig. 4). The existence of the *macula lutea* as a localized feature within the retina suggests both a function for the macular carotenoids and an active mechanism for their accumulation. The variation in the zeaxanthin/lutein ratio across the retina further suggests that the chemical and biochemical influences operating on the carotenoids in the peripheral retina differ in some way from those in the central macula.

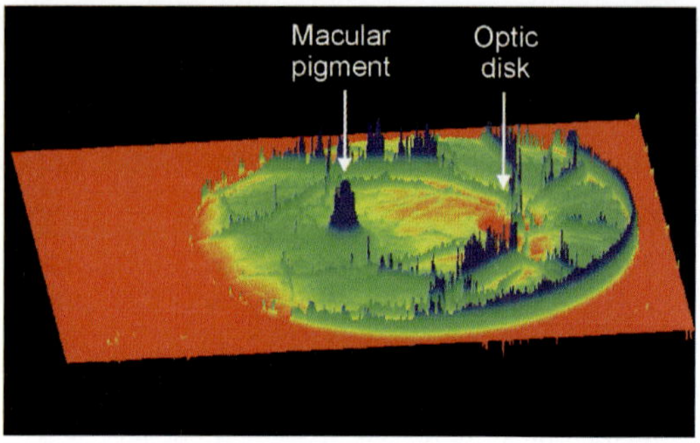

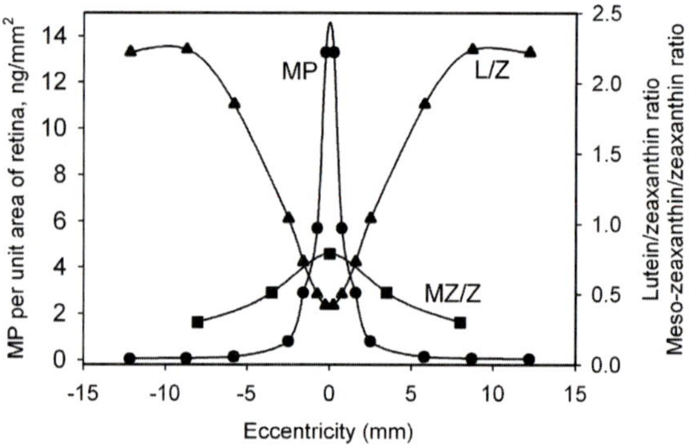

Fig. 4. The distribution of the macular pigment (MP), lutein (L) and zeaxanthin (Z) across the retina. The three-dimensional representation of macular pigment distribution (upper) was obtained by imaging reflectometry (see Section **F**.2.b). The macular pigment appears as a 'hill' whose height is proportional to the peak macular pigment optical density, MPOD. The MPOD distribution is represented differently in the lower graph. The variation in the lutein:zeaxanthin ratio (L/Z) and the (3*R*,3'*S, meso*)-zeaxanthin:(3*R*,3'*R*)-zeaxanthin ratio (MZ/Z) with eccentricity are also shown.

Surprisingly, macular zeaxanthin was found to comprise two stereoisomers, the normal dietary (3*R*,3'*R*)-zeaxanthin (**119**) and (3*R*,3'*S*)-zeaxanthin (**120**) which is not a normal dietary component [4]. The concentration of (3*R*,3'*S*)-zeaxanthin within the retina varies from a maximum within the central fovea to a minimum in the peripheral retina. This distribution inversely reflects the relative abundance of lutein (see Fig. 4) and gave rise to a hypothesis [14] that (3*R*,3'*S*)-zeaxanthin is formed in the retina from lutein. This was confirmed by an experiment in which xanthophyll-depleted monkeys were supplemented with zeaxanthin-free lutein or pure (3*R*,3'*R*)-zeaxanthin [15]. In spite of the relatively high daily doses used, (3*R*,3'*S*)-zeaxanthin was not detected in the plasma of any of the supplemented monkeys, and appears to be a retina-specific metabolite of (3*R*,3'*R*,6'*R*)-lutein, though the mechanism of its formation has not been established.

Another metabolite found in the retina, 3'-dehydrolutein (**302**), may be an intermediate in the conversion. The reduction products 4'-hydroxyechinenone (**296**) and isozeaxanthin (**129**) have been identified in the retinas of primates fed high amounts of canthaxanthin (**380**). The greatest relative concentration of these two compounds is observed in the macula. This indicates that the retina, especially the macula, has the ability to reduce keto groups in the carotenoids [13].

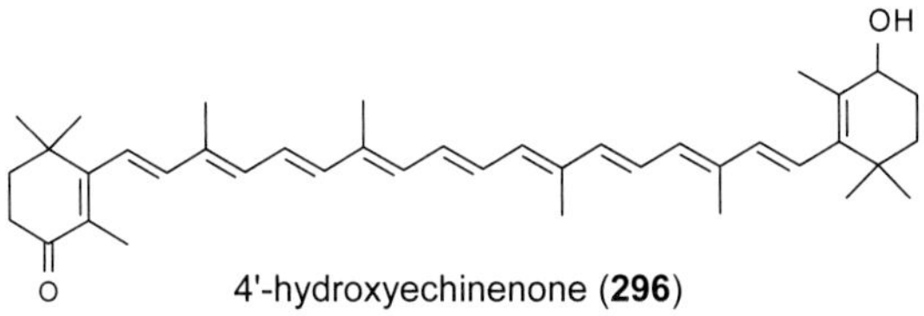

4'-hydroxyechinenone (**296**)

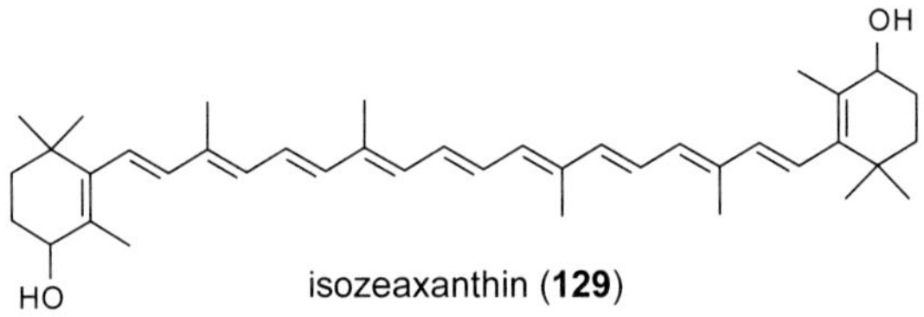

isozeaxanthin (**129**)

Lutein (**133**), (3*R*,3'*R*)-zeaxanthin (**119**), and (3*R*,3'*S*)-zeaxanthin (**120**) are chemically very similar but the small structural differences have consequences for the preferred conformation of the molecule (Fig. 5). The ring-chain orientation of the ε-ring of lutein is particularly distinctive. Studies suggest that zeaxanthin prefers to occupy a membrane-spanning, perpendicular orientation whereas lutein is apparently able to locate itself at both spanning and superficial (parallel oriented) sites [16]. This could account, at least in part, for why lutein and zeaxanthin are functionally different from one another, but it leaves open the question of how the retina distinguishes between them. Protein-binding could be important; a recently identified xanthophyll-binding protein shows a clear binding preference for zeaxanthin over lutein [17]. The structural and conformational differences between the lutein and zeaxanthin molecules may appear minimal, but they are sufficiently significant for a mechanism to have evolved for the formation of (3*R*,3'*S*)-zeaxanthin in the macula.

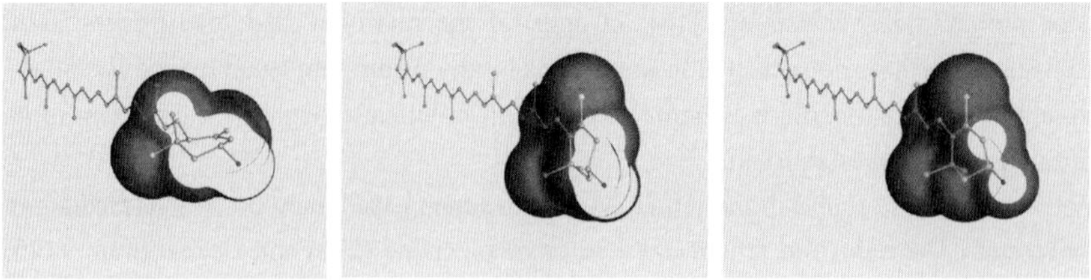

Fig. 5. Space-filling models of the end groups of: left, (3*R*,3′*R*,6′*R*)-lutein (**133**); middle, (3*R*,3′*S*)-zeaxanthin (**120**); right, (3*R*,3′*R*)-zeaxanthin (**119**). The structural differences appear minimal but they have notable consequences for the spatial conformation of the molecules.

E. 'Classical' Features of the Macular Pigment

1. General

The presence of high carotenoid concentrations in a very small area of the retina gives rise to two entoptic phenomena, Maxwell's spot and Haidinger's brushes.

a) Maxwell's spot

Maxwell's spot, first described in 1856 [18], is generally seen as a dark spot approximately 4° in diameter when subjects focus on a uniform blue surface. It owes its presence to the absorption of blue light by the macular pigment in the inner retina. Some subjects describe an annular pattern rather than a spot, implying a lower macular pigment density in the centre of the fovea, surrounded by a region of higher density. This type of distribution has been observed in some subjects when the macular pigment has been mapped by both psychophysical and physical means. However, the distributions that are more often observed in subjects who report a ring-like Maxwell's spot are characterized by a central peak with a lower, concentric, subsidiary maximum at about 0.7° eccentricity. This type of distribution appears to be more prevalent in females [19]. It has been hypothesized that the differences in the individually perceived shapes of Maxwell's spot reflect differences in the distribution profile of the macular pigment. Such differences have been described [20] and linked to differences in retinal architecture [21].

b) Haidinger's brushes

Haidinger's brushes, first described in 1844 [22], are reported as a dark, 'hour-glass' or 'bow-tie' figure which could be likened to a bundle of fibres bound in the middle with the 'brushes' protruding at both ends. This phenomenon appears at the central fixation point when a surface, uniformly illuminated by blue light, is viewed through a plane-polarizing filter. The non-random arrangement of lutein and zeaxanthin in Henle's fibre layer, together with their dichroic properties, are responsible for this phenomenon. (all-*E*)-Lutein and (all-*E*)-zeaxanthin are dichroic, absorbing maximally light that is polarized parallel to the polyene chain. The axis of the observed 'hour-glass' is perpendicular to the electric field vector of the polarized light. A preferential alignment of the lutein and zeaxanthin molecules perpendicular to the radially directed Henle fibres is necessary to account for the brushes, as indicated in Fig. 6. Incorporation of these molecules transversely in the cylindrical membranes of the fibres is one possible structure consistent with the Haidinger's brush phenomenon.

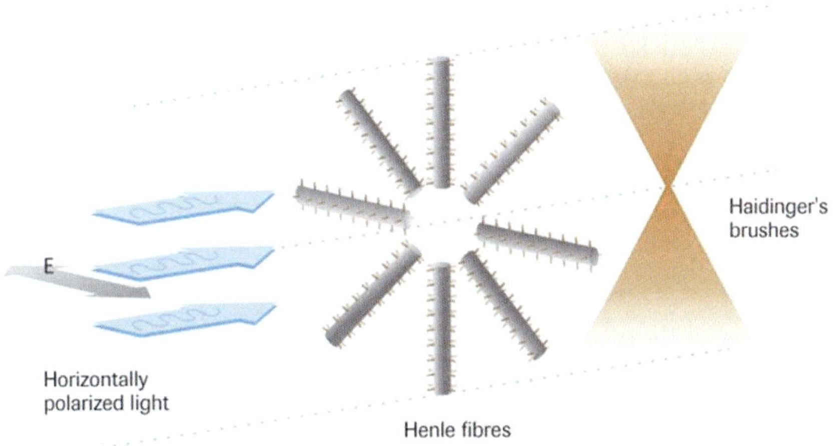

Fig. 6. The formation of Haidinger's brushes. The Henle's fibres (grey) are shown radiating from the centre of the fovea. The orange 'whiskers' on the Henle's fibres represent the polyene chains of the xanthophyll molecules located in the membranes of the fibres. The entoptical phenomenon of Haidinger's brushes is generated by the dichroism of the xanthophyll molecules. Those molecules whose polyene chains are parallel to the plane of polarization (horizontal in this example; the grey double arrow indicates the electric field vector) will absorb the horizontally polarized light (blue arrows) maximally while those whose chains are perpendicular will absorb minimally. The transmitted light intensity will, therefore, be minimum at the 12 o'clock and 6 o'clock positions, and maximum at the 9 o'clock and 3 o'clock positions. The spatial characteristics of the transmitted light, incident on the light-sensitive outer segments of the photoreceptors, give rise to the characteristic hour-glass shaped shadow figure (right) having, in this example, a vertical orientation.

Together with Maxwell's spot, Haidinger's brushes have been used clinically as approximate markers for the position of the fovea, and can indicate qualitatively the absence or presence of the macular pigment. Furthermore, because the visibility of Haidinger's brushes is crucially dependent on the correct orientation of the xanthophyll molecules, any process that disturbs their orientation may lead to a disappearance of the entoptic image, and thus may provide an indication of developing disease even before the photoreceptors themselves are affected [23].

2. Effects of macular pigment on visual performance

Long before the chemical identity of the 'macular yellow' was determined, there were hypotheses about its role in vision. As early as 1861, it was conjectured that the yellow colour of the 'macular yellow' might be physiologically important for human vision. By 1920, before the localization of macular pigment within the retinal structure was determined, 'macular yellow' was believed to be a pre-receptoral light filter which could reduce chromatic aberration (thereby improving visual acuity), reduce blue haze, glare and dazzle (thereby improving comfort), and enhance contrast. Thus, ideas about the functions of macular pigment in the healthy eye originated much earlier than the ideas that it may contribute in risk reduction of age-related macular degeneration (AMD). One reason for this may have been that the prevalence of AMD was much less at that time because of the lower life expectancy.

A recent hypothesis [24] suggests that increased macular pigment could reduce the amount of light in the wavelength range where rhodopsin absorbs. This effect would be most important in the mesopic range when the photoreceptors are adapting from photopic (high light) to scotopic (low light) conditions. At this transition, both rods and cones are active but the quality of the image is degraded by the contribution of rods, with their poor contrast sensitivity and resolving power.

a) Visual acuity and contrast sensitivity

The focal length of the optic media decreases with wavelength, the rate of decrease being greatest at shorter wavelengths [25]. This effect, chromatic aberration, results in an imperfect retinal image fringed with prismatic colours. In other words, if the eye is in focus for green light, the blue parts of an image are focused in front of the retina, while the red parts are focused behind the retina [25]. Because of the dispersive properties of the optic media, the aberration is much stronger for blue light than for the longer wavelengths of the spectrum, and it is in the blue wavelength range that the macular xanthophylls absorb. At 460 nm, the dominant wavelength of sky light and the peak absorption of macular pigment, the magnitude of aberration of blue light quantitatively amounts to –1.2 dioptres [26].

Visual acuity and contrast sensitivity are related parameters that determine the resolving power of the eye. Visual acuity is a measure of the smallest angle between two points subtended at the retina, or between two lines, that can be seen to be separate. The normal visual acuity test, where subjects attempt to read lines of letters of decreasing size, is the most familiar method of assessing visual acuity. In a contrast sensitivity test, the subject views sinusoidal gratings covering a range of spatial frequencies and, for each, adjusts the contrast ratio until the bars can only just be discriminated. The ability of subjects to demonstrate high visual acuity or contrast sensitivity, assuming their refractive errors have been corrected, will depend on a variety of factors such as pupil size, cone density and clarity of the optic media. Not surprisingly, visual acuity and contrast sensitivity in healthy eyes tend to decrease with age. Since carotenoids may be associated with a reduction in the incidence of cataracts [27], and therefore with preservation of the clarity of the lens, supplementation with lutein or zeaxanthin may, from this perspective, assist in the maintenance of visual acuity.

Supplementation with lutein and zeaxanthin was shown to improve contrast acuity, a parameter compounded from visual acuity and contrast sensitivity, in the mesopic range [28]. This was the first controlled trial with lutein and zeaxanthin to study the effect of supplementation on visual performance in healthy subjects. Although the study was small, the results increase the credibility of the classical hypotheses of vision and the macular pigment. In addition to these data for healthy subjects, there is also some limited evidence that lutein supplementation can improve visual acuity in subjects with degenerative retinal diseases [29-31].

b) Glare sensitivity and light scatter

Glare sensitivity and light scatter can lead to deterioration of the image on the retina and thus reduce visual performance. Sensitivity to glare is often exacerbated by increasing age and by diseases of the lens that result in increased light scatter within the eye. Glare sensitivity may be assessed in a subject by measuring contrast sensitivity (see above) in the presence of a nearby glare source, for example a pair of halogen lamps that simulate the headlights of an oncoming car. In 36 healthy non-supplemented subjects, macular pigment optical density (MPOD) and the sensitivity to glare were measured by assessing their photostress recovery time, that is the time span until vision returns after the subects had been "blinded" by a bright glare light. It was found that photostress recovery time was significantly shorter for subjects with higher MPOD levels [32]. These correlational data were later extended by supplementing 40 healthy subjects with a mixture of 10 mg of lutein and 2 mg of zeaxanthin for 6 months and again measuring photostress recovery time. Supplementation increased MPOD levels on average by 35% and, along with this MPOD increase, photostress recovery time was significantly ($p = 0.01$) reduced [33]. Although the study was not placebo-controlled or randomized, when it is taken together with the results of the correlational study, the data strongly support an inverse relationship between MPOD and photostress recovery time. In addition, it is possible that increasing the level of macular pigment would diminish the amount of scattered blue light reaching the photoreceptors, and this might result in lowered sensitivity to glare [26]. Light scatter within the eye has been demonstrated to be independent of wavelength, however [34], so the scattered longer wavelengths would not be removed. This may be the reason why supplementation with lutein, zeaxanthin, or a combination of both carotenoids was consistently shown to reduce intra-ocular light scatter in healthy eyes, though not at a level of statistical significance [28].

On the other hand, light scatter by the atmosphere does depend on wavelength, being greater for shorter wavelengths, and accounts for the blue haze that tends to blur the visibility of distant objects. Higher levels of macular pigment could filter out a significant fraction of atmospherically scattered blue light and thereby improve outdoor visual performance [35].

F. Macular Pigment Optical Density (MPOD) and its Measurement

The MPOD, measured at the peak absorption wavelength (*ca.* 460 nm), is a direct indicator of the concentration of pigment present in the macula. The most reliable and accurate method, HPLC, can only be used to analyse carotenoids in various eye tissues, notably the retina and lens, in *post-mortem* or pathology samples. Non-invasive measurement of carotenoids in the retinas of living, human subjects is crucial not only for studies that evaluate the effectiveness of carotenoids in reducing the risk of eye diseases but also for investigating the role of carotenoids in the healthy retina. Unfortunately, the low concentration of carotenoids in the lens has, so far, precluded the development of a non-invasive measurement in this structure.

1. Analysis of carotenoids in retina and lens *in vitro*

Human eye tissues can be obtained from tissue banks. The use of formalin reduces degradation *post mortem* and does not result in any measurable leaching of carotenoids from the tissue. Quantitative analysis of the lutein and zeaxanthin content is easily accomplished by HPLC on a C_{18} reversed-phase column. With a chiral column, the individual stereoisomers (3*R*,3'*R*)-zeaxanthin and (3*R*,3'*S*)-zeaxanthin may also be resolved [15] (see *Chapter 2*).

2. Non-invasive determination of carotenoids in the retina *in vivo*

a) Quantitative estimation by psychophysical methods

The presence of macular pigment in a layer anterior to the central photoreceptors modifies the observer's photopic luminosity function - a function describing the relative luminosity of different wavelengths. This, and related effects, are the basis of a number of psychophysical procedures for determining a subject's macular pigment density.

i) Heterochromatic flicker photometry. The most commonly used method of assessing the density of the macular pigment *in vivo* is heterochromatic flicker photometry (HFP) [36]. The subject views a small stimulus, subtending a visual angle of about 1°, which alternates between a wavelength that is strongly absorbed by the macular pigment (*e.g.* 460 nm - blue) and another which is minimally absorbed by it (*e.g.* 540 nm - green). The subject seeks to eliminate, or minimize, the sensation of flicker by adjusting the blue brightness until the luminous intensities of the blue and green components of the stimulus are equalized at the level of the photoreceptors (Fig. 7). Subjects with high or low MPOD require more or less blue light, respectively, in order to compensate for absorption in the macular pigment layer. In order to eliminate the influence of absorption by the lens and the subject's individual spectral sensitivity, a second setting is made with the stimulus image directed to the parafovea or perifovea, typically 5° to 8° from the centre of the fovea. There, absorption by the macular pigment is assumed to be minimal. However, there are indications that, with lutein or zeaxanthin supplementation even at moderately high doses, this assumption may not be valid and pigment accumulation at the parafoveal (reference) location can indeed occur [37]. Furthermore, with age, MPOD at the parafovea (around 5.5° eccentricity) may increase [38]. Notwithstanding, the MPOD is given by the log ratio of the foveal to parafoveal blue wavelength intensity settings.

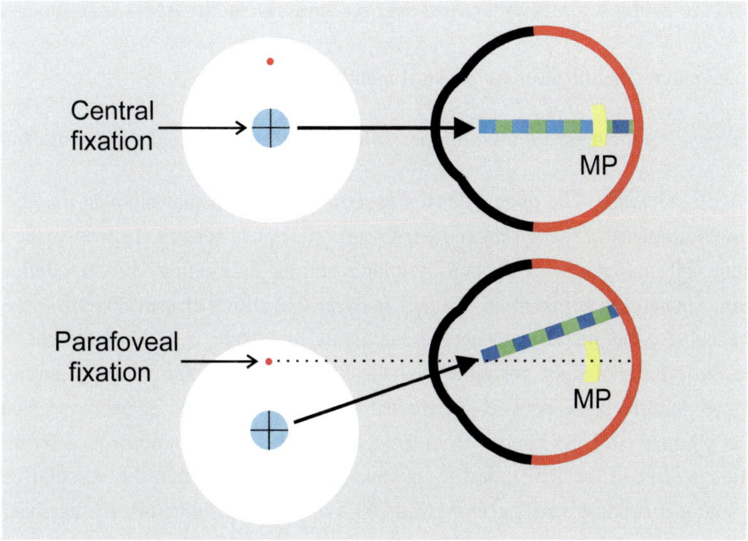

Fig. 7. The principle of heterochromatic flicker photometry. The subject views a stimulus (turquoise circle), with either foveal (upper diagram) or parafoveal (lower diagram) fixation. The stimulus consists alternately of blue (460 nm) or green (540 nm) light, resulting in a flickering, turquoise appearance. With foveal viewing the blue light is partially absorbed upon passing through the macular pigment. If the perceived luminances of the blue and green light are different, the subject observes flicker, and proceeds to adjust the blue luminance until no, or only minimal, flicker is observed (the iso-luminant point). Repeating the procedure with parafoveal fixation provides a reference measurement to which the central measurement is compared. The amount of blue light needed for iso-luminance of the blue and green lights under parafoveal fixation divided by the amount needed under foveal fixation is equal to the transmittance T of the macular pigment at the blue wavelength. The macular pigment optical density (MPOD) is given by the relationship: $MPOD = -\log_{10}T$.

ii) Motion photometry. In this related technique [39], the two colours employed in flicker photometry appear as alternating bars of a grating that move across the field of view. A perception of minimum motion of the stimulus occurs when the luminances of the bars are matched. This psychophysical technique was recently compared to the physical fundus autofluorescence technique (see b.*ii* below) and a reasonably good correlation between the two techniques was reported [40].

iii) Conclusion. The psychophysical methods have the advantage of being very well validated. If a series of test wavelengths are used successively, the absorption spectrum of the macular pigment may be generated. This spectrum is remarkably consistent with that obtained from lutein and zeaxanthin in a lipid environment [41]. In addition, both methods provide the possibility of mapping the macular pigment density distribution in the retina by viewing stimuli at various eccentricities. Disadvantages of these methods are the time taken for testing and the requirement of a certain degree of subject skill. Neither method is likely to prove practical for testing subjects with retinal degenerations, including AMD, that impair vision.

b) Quantitative determination by physical methods

Several physical methods are used for measuring and mapping macular pigment density.

i) Retinal reflectance. The optical density spectrum of the macular pigment may be obtained from measurements of the foveal and perifoveal reflectance spectra. Typically, the fundus of the retina is illuminated with a xenon flash lamp and the light reflected from a defined area of the retina is analysed with a multi-channel analyser that allows all wavelengths to be analysed simultaneously. Before reflectance measurements are made, a bleaching light is used to reduce the effects of light absorption by visual pigments. In the simplest application, the difference in reflectance between the foveal and peripheral sites is attributed solely to the macular pigment [42]. To remove from the spectrum the effects of absorption by the lens, by melanin (present in the RPE), and by oxyhaemoglobin (present in the vascular choroid), a curve-fitting procedure may be employed [43,44]. Imaging reflectometry has been used to determine the spatial distribution of the macular pigment [45] (Fig. 4). Digital images of the fundus are obtained at two wavelengths, such as 462 and 559 nm, for which macular pigment absorption is close to the maximum and zero, respectively. The highest quality images are obtained when a scanning laser ophthalmoscope is used.

The question of whether it is possible to detect lutein and zeaxanthin individually rather than the combined macular pigments was addressed recently [46]; apparatus was designed, based on spectral fundus reflectometry, and was claimed to provide this distinction. If this is confirmed independently, it would be an important tool in xanthophyll supplementation studies.

ii) Autofluorescence of lipofuscin. The fluorescence of lipofuscin, which is situated in the RPE and therefore posterior to the macular pigment layer, provides a means of obtaining the optical density of the macular pigment. Two different exciting wavelengths are provided by a scanning laser ophthalmoscope, *e.g.* 470 and 550 nm, which are differentially absorbed by the macular pigment. The resulting fluorescence is measured at ~710 nm, beyond the absorption range of the macular pigment. In this way, the differential absorbance of the exciting wavelengths by the macular pigment can be obtained, leading to a measurement of its optical density [47]. A simpler technique employs a single exciting wavelength, *e.g.* 470 nm, that excites fluorescence of lipofuscin in the fovea and periphery. Assuming that the fluorescence efficiencies are the same at these two locations, the MPOD can be obtained by a subtraction process. The one-wavelength and two-wavelengths techniques have been compared [48].

iii) Resonance Raman spectroscopy. Lutein and zeaxanthin in the retina produce strong, resonance-enhanced, Raman signals when excited by the 488 or 514.5 nm lines of an argon laser. The resulting photon count depends on the amount of carotenoids present. A recent development has been to image the Raman-scattered light from the retina onto a digital camera [49,50], thereby creating a distribution map of the macular pigment. Some of the maps reveal less macular pigment in the centre of the fovea compared with the surrounding area, consistent with the annular appearance of Maxwell's spot reported by some observers. Raman spectroscopy has been compared with heterochromatic flicker photometry [51].

iv) Conclusion. Despite the advantage of being objective, each method has its drawbacks. Macular pigment densities determined by reflectometry are typically low, probably because of light scatter within the ocular media. Raman spectroscopy provides photon counts rather than optical density, and requires calibration for comparison with other methods. Methods that exploit the autofluorescence of lipofuscin may be of limited use in the case of young subjects and those with AMD, both groups tending to have less lipofuscin in the RPE. However, each method is fast, independent of subject skill and capable of revealing the distribution of macular pigment in the retina.

G. The Determinants of Macular Pigment Optical Density

1. Transport of carotenoids into the retina

Before the macular xanthophylls can reach the eye and be deposited in the macula, an important first step is their absorption from food or supplements into the plasma (*Chapter 7*). The kinetic parameters describing this uptake are well documented [52,53]. In contrast to the hydrocarbons β-carotene and lycopene, which are mainly bound to LDL, lutein and zeaxanthin are predominantly transported in plasma by HDL lipoproteins [54,55]. However, the processes by which they are then deposited in the macula remain unknown. A specific binding protein may be important in this regard, and at least one has been proposed [56]. This protein shows specificity for xanthophylls relative to the carotenes. Recently it was isolated from human retina and described as the Pi isoform of glutathione S-transferase [57]. It exhibits the largest binding affinity towards (3*R*,3′*R*)-zeaxanthin and (3*R*,3′*S*)-zeaxanthin whereas lutein is only weakly bound. The topic of xanthophyll-binding proteins has been reviewed comprehensively [58,59]. The discovery of a zeaxanthin-binding protein has been reported recently [60].

2. Diet

The possibility of modulating MPOD in the macula by dietary means was first investigated in non-human primates [61]. When these animals were raised on a carotenoid-free diet, the carotenoids disappeared quickly from the plasma, but the yellow macular pigment disappeared only very slowly, over several years. The carotenoid deprivation led to distinct ophthalmological consequences that were similar to human age-related maculopathy.

Under special conditions, deprivation of carotenoids can also occur in humans. The human disease cystic fibrosis, a consequence of which is that the absorption of fat-soluble nutrients, including carotenoids, is severely impaired, provides a model for this [62]. Like the carotenoid-deprived monkeys, cystic fibrosis patients had lower plasma lutein and zeaxanthin concentrations, and MPOD levels as much as 50% lower than age-matched and gender-matched healthy subjects. On the other hand, the concentration of macular pigment can be increased by providing subjects with fruits or vegetables rich in lutein and zeaxanthin. While the responses to lutein and zeaxanthin intake from the diet are quite variable between individuals, most subjects respond with increases in both plasma concentration and MPOD, though some show only plasma increases, and a minority show no response in either plasma or MPOD [63]. Compared to the relationship of xanthophyll intake to plasma levels, the correlation between MPOD and plasma or dietary intake levels of lutein and zeaxanthin is weak.

Another factor with an influence on variability of MPOD may be seasonal variation in the dietary intake of lutein and zeaxanthin, though two recent studies could not detect seasonal fluctuations over one year [64,65].

3. Supplementation

A considerable body of information is available on the response of MPOD to intake of lutein and zeaxanthin supplements, yet many details remain unresolved. In an extension of the earlier investigation [61], groups of six carotenoid-depleted rhesus monkeys were supplemented either with pure lutein or pure zeaxanthin at doses of 2.2 mg/kg/day (equivalent to 12-24 mg of carotenoid per day per animal) for 6-12 months [66]. Plasma levels of lutein rose faster, and to higher initial levels, than those of zeaxanthin but, by approximately 16 weeks, both had stabilized at comparable concentrations of around 0.8 μmol/L. This was equivalent to a ten-fold increase compared with plasma levels of normal chow-fed animals. MPOD increased gradually and variably in both groups. However, by 16 months, MPOD had approached levels of only *ca.* 50% of that seen in monkeys that were fed normal monkey chow throughout life. The life-long carotenoid deprivation may, thus, have impaired the retina's natural ability to accumulate xanthophylls to their full extent during the given supplementation period.

Human supplementation studies with lutein and zeaxanthin have yielded a wide range of results, characterized by a substantially larger variability than that of plasma responses; MPOD values can vary by one order of magnitude or more. Differences between measuring techniques, study length, and subject training are all factors that could contribute to this variability, as well as inherent differences between individuals. On the other hand, the use of esterified *versus* unesterified carotenoids has been shown not to contribute much to this variability in terms of bioavailability [67,68]. Both appear to be equally effective. In the first controlled investigation of both plasma and MPOD responses to supplementation of humans with relatively large daily doses, two subjects received 30 mg per day of lutein (as esters, suspended in 2 mL canola oil) for a period of 140 days [69]. Plasma lutein increased rapidly in both subjects, after about three weeks reaching a plateau that represented a roughly ten-fold increase. On cessation of supplementation, plasma lutein concentration decayed exponentially, reaching pre-supplementation levels within a period of 40-60 days. MPOD, measured approximately five times per week by heterochromatic flicker photometry (HFP), began to increase only after about 30-40 days of supplementation. Linear increases in MPOD were observed thereafter throughout the supplementation period, and continued during post-supplementation until plasma lutein had fallen to baseline levels. The maximum MPOD increases from baseline observed were about 20% in one subject and 40% in the other. In addition, MPOD showed no tendency to decrease during a subsequent 200-day period.

In the first double-blind, placebo-controlled and randomized supplementation study with lutein over a period of one year, sixty AMD patients were supplemented with 10 mg of lutein per day. MPOD increases of around 40% were observed, while in the thirty subjects of the placebo group, MPOD had decreased slightly [29]. In a smaller study, participants received 20 mg of lutein esters (equivalent to 10 mg of unesterified lutein) per day for 4-5 months. MPOD increases amounted to 29% and 35% in AMD patients and healthy controls, respectively [70]. In another supplementation study, 108 subjects with early signs of AMD were given daily doses of 12 mg of esterified lutein, and showed a maximum MPOD increase of 48% [71]. About 25 subjects were identified as 'non-responders' who did not exhibit measurable MPOD responses, while plasma concentration of lutein had clearly increased. Other studies of supplementation with lutein, for a few months up to a year, have shown more modest net increases of MPOD, 15% [72], 23% [73], 20% [74], 22% [75], and 14%, after supplementation with lutein or zeaxanthin [76]. On the other hand, it has been reported that no change in MPOD was detected in 29 subjects who were supplemented with 9 mg per day of lutein for only 5 weeks [77].

Another study investigated the response to daily supplementation with 20 mg of unesterified lutein for 6 months in patients with ABCA4-associated retinal degeneration, such as Stargardt disease [78]. A majority of the patients showed increases of lutein concentration in plasma, while only 63% of the subjects responded with significant increases of MPOD. The responding patients tended to be female and to have lower plasma xanthophyll concentrations as well as lower MPOD at baseline.

In summary, it can be concluded that, after supplementation with lutein, for which the most data are available, MPOD will be raised in most but not in all subjects. Longer periods of supplementation (>6 months) and higher dosage (>20 mg/day) may be necessary to cause significant responses and to determine if subjects are 'non-responders' in either plasma or MPOD, or merely respond more slowly than normal.

Knowledge of MPOD responses resulting from supplementation with (3*R*,3′*R*)-zeaxanthin is limited. MPOD levels increased significantly in seven volunteers who, for 3 months, received daily doses of about 20 mg zeaxanthin from the zeaxanthin-rich Chinese Wolfberry 'Gou Qi Xi' [79]. In another study [80], two subjects were supplemented with 30 mg per day of pure zeaxanthin for four months. Significant MPOD increases were observed but these were smaller than those observed with lutein in an earlier study [69]. However, this might be due to differences in formulation of the zeaxanthin and lutein supplements that were used. In another, slightly larger study, eight subjects were supplemented with pure zeaxanthin. MPOD increases could be identified by HFP in five of the subjects but, at the end of supplementation, MPOD values below baseline were reported in the other three [81]. Another study [76] showed increased pigment concentration in the parafoveal region in response to supplementation with zeaxanthin. Such pigment increases in the parafoveal location that HFP uses as reference would cause MPOD to appear to decline, as was observed. That such effects of xanthophyll supplementation on parafoveal pigment concentrations are indeed possible, was recently confirmed for lutein [82] and for zeaxanthin [46].

Only a very small number of studies have reported the responses to supplementation with (3*R*,3′*S*)-zeaxanthin. In one study [83], plasma xanthophyll levels were measured in nineteen subjects who were supplemented for 3 weeks with a mixture of lutein, (3*R*,3′*S*)-zeaxanthin, and (3*R*,3′*R*)-zeaxanthin. It was concluded that (3*R*,3′*S*)-zeaxanthin was less well absorbed than (3*R*,3′*R*)-zeaxanthin from the administered mixture. Only one publication has reported elevated plasma and MPOD levels after supplementation with (3*R*,3′*S*)-zeaxanthin [84].

4. Other factors

Correlations of macular pigment density or plasma xanthophyll levels with a number of factors associated with the risk of AMD have emerged in recent years. These include obesity [85,86], diabetes [87], smoking [88], and female gender [85]. Individuals with these characteristics appear to have reduced MPOD. On the other hand, factors that can increase MPOD, such as consumption of lutein/zeaxanthin-containing vegetables [89] or elevated plasma levels [90], appear to decrease the risk of AMD.

Such potential MPOD determinants have recently been evaluated in a cohort of 698 women, aged 53-86 years [91]. The strongest direct predictors of MPOD in this study were plasma concentrations of lutein and zeaxanthin followed by the dietary intake levels of these carotenoids [92]. Larger waist circumference and the presence of diabetes predicted a decrease of MPOD. In contrast to earlier findings, iris colour was not related to MPOD. No dependence of MPOD on age was revealed but, because of the lower age limit of 53 years, the study had limited power to do so. Recently, reports on the age-dependency of MPOD from 23 published studies have been reviewed [93]. Most indicate either no relationship with age, or a decline. Based on the totality of evidence, the normal MPOD value of a population appears to be nearly constant for all age groups, but individual MPOD values are clearly susceptible to significant change. It is important to note, however, that increases of pigment density in the parafoveal area (5.5° eccentricity) with age have been identified by HFP [38]. If this finding is confirmed, the decline of MPOD with age, as measured by reference to this parafoveal region, would have to be interpreted carefully, because the apparent decline could be accounted for by the increase in parafoveal pigment. On the other hand, this may be an important question to investigate, because it could indicate that, as individuals age, macular pigment may accumulate preferentially in, or redistribute to, the parafoveal and peripheral retina, thus covering a wider area. Earlier precise measurements of macular pigment in the central 3 mm *via* chemical analysis indicate that carotenoid levels are independent of age [92]. This finding is not inconsistent with the idea of parafoveal accumulation as any fovea-parafovea redistribution of pigment would not have been detected in that study.

The age dependence of MPOD may be genetically determined [95]. A genetic linkage may not be the primary determining factor, however. MPOD is open to modulation by diet or supplementation, as demonstrated by earlier work [96] which indicated that monozygotic twins could indeed have different levels of MPOD, depending on differences in their specific environmental and, particularly, dietary factors.

The year 2005 was a landmark year as the genetic basis of AMD began to become clearer. Several research groups independently have identified genes that appear to be strongly linked with the risk for AMD [97]. Genetic factors are certainly important to consider in the context of the macular xanthophylls. Whether and how the presence or absence of these genes is linked to an individual's MPOD remains to be resolved, as does the question of whether intake of the macular xanthophylls, either dietary or as supplements, can influence the course of the disease in subjects who possess one or more genes linked to AMD.

None of the ocular biometric parameters, such as anterior chamber depth, lens thickness, vitreous chamber depth, and ocular dominance, yielded a significant correlation with MPOD or with plasma levels of lutein and zeaxanthin, in 180 healthy subjects [93] after correction for age and height. This is important because it shows that it is not necessary to correct for these parameters when investigating the relationship of MPOD to AMD risk.

H. The Role of Carotenoids in Risk Reduction of Macular Degeneration and Cataract

It is now established that macular degeneration and cataract are two age-related ophthalmic diseases that can lead to very significant visual impairment. In the U.S., they affect 1.5% [98] and 17% [99], respectively, of the population over 40 years of age. Whilst cataract is easily treated and cured, treatment of AMD has had only limited success, though recent medications are being evaluated and appear promising. So, AMD prevention or risk reduction is an important strategy, particularly because the prevalence of this disease is increasing dramatically and, in the U.S. alone, may reach almost 3 million individuals in less than 20 years [98]. With the realization that they have essential functions in the macula, it is reasonable to believe that lutein and zeaxanthin have a key role in maintaining eye health through reduction of risk of AMD and perhaps other ophthalmic diseases such as cataracts.

1. Mechanistic Basis

a) Absorption of blue light

The ability of intense light to cause acute damage to the retina is undisputed. Acceptance that chronic (long term) exposure to less intense light sources is also a cause of retinal injury is still controversial. The action spectrum for acute exposure to light required to induce cellular damage to the retina and other ocular tissues has been well studied. Not surprisingly, the higher energy shorter wavelengths are the most efficient at producing this damage. The cornea and lens are effective at absorbing UV radiation, so little or no UV light reaches the retina. However, UV radiation is presumed to be one of the factors that induce free-radical photo-oxidation in the lens, resulting in irreversible protein damage and ultimately leading to cataract. Nearly 60% of the incident blue light at 460 nm is absorbed by the macular pigment. Over the entire damaging wavelength range from 400 to 500 nm, the average value is about 47%. Thus, the macular pigment acts as a light filter and prevents significant amounts of blue light from reaching the photoreceptors, the RPE, and the vascular choroid. Unlike UV radiation, blue light does not possess sufficient energy to lead directly to the breaking of most covalent chemical bonds, but it can promote the formation of reactive oxygen species (ROS), such as singlet oxygen, which have the potential to cause extensive damage, especially to unsaturated fatty acids which are important and easily oxidized components of the photoreceptors of the retina [100].

b) Protection against photooxidation

With the eye, as with other tissues, it is not known whether carotenoids act as antioxidants *in vivo*. In the retina, where the simultaneous presence of high levels of both light and oxygen are the norm [101], numerous reactive oxygen species (ROS), including singlet oxygen and superoxide, can be generated and cause damage *e.g.* to polyunsaturated fatty acids. Thus they pose a serious risk to the retina in regard to its ability to recycle these essential membrane components [100]. In particular, docosahexaenoic acid, the major lipid constituent of vertebrate photoreceptor outer segments, is highly susceptible to oxidation. Supplementation with antioxidants can lessen this damage, though the contribution of carotenoids remains unclear.

Lipofuscin is a by-product formed from retinal in the RPE during the catabolism of the photoreceptors [102], and increases with age. One component, A2-E, a pyridinium bis-retinoid (*2*), is a fluorophore capable of sensitizing the formation of singlet oxygen when irradiated at wavelengths up to 430 nm, *i.e.* in the blue-light wavelength range. Photochemical damage to DNA is enhanced by the presence of A2-E, and data strongly implicate a role for singlet oxygen in this process. Vitamin E (α-tocopherol), lutein and zeaxanthin have been shown to give some protection against this blue-light-induced photooxidation. In particular, significantly greater protection was observed with zeaxanthin than with lutein, and an even larger protection was seen when zeaxanthin was combined with α-tocopherol. In another study, the amount of A2-E present in cultured rabbit RPE cells could be reduced by addition of carotenoids such as lutein, zeaxanthin or lycopene [103]. A study of photochemical damage in RPE cells [104] showed that, in addition to A2-E, (*trans*)-retinal (*1*) may also be an important photochemical generator of singlet oxygen [105,106]. Zeaxanthin appears to be a better photoprotectant than lutein, and zeaxanthin has a stronger protective effect in the presence of other antioxidants such as ascorbate and α-tocopherol. The evidence indicates that zeaxanthin and lutein can protect against photooxidation by quenching the excited state sensitizer and also by intercepting and quenching singlet oxygen after it is formed. The antioxidant α-tocopherol, in turn, protects the carotenoid from destruction. The different orientation of zeaxanthin and lutein in membrane bilayers may be an important factor in the different efficacy of the two xanthophylls [107,108].

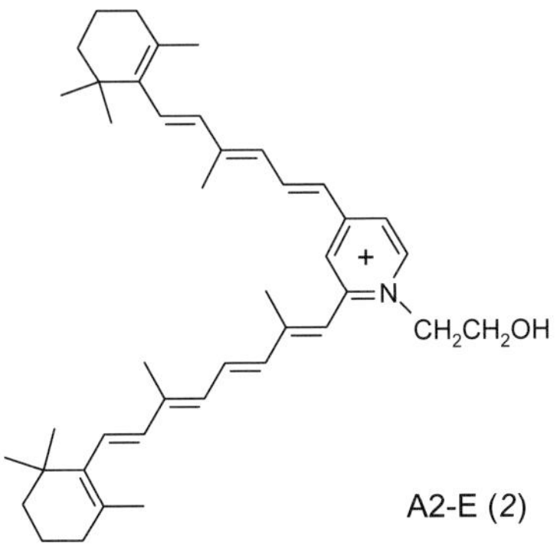

A2-E (2)

c) Other properties

Carotenoids may function in the eye in ways other than those already described. The possibility that the macular carotenoids, or their breakdown products, participate in complex biochemical processes and cell-to-cell communication in the eye, as they do in other tissues (see *Chapters 11* and *19*), cannot be dismissed [109].

2. Evidence obtained from experiments with animals

Whilst numerous animal models for cataract formation and development are available, there is no well-established animal model for AMD. Only primates exhibit the anatomical characteristics of the *macula lutea* and monkeys develop drusen (small yellow or white accumulations of extracellular material) and changes in the macula that are similar to those that characterize human age-related maculopathy [110,111]. Only in a few instances, however, have primates been utilized to study the pathophysiology of macular degeneration in the context of the macular xanthophylls. In a study in 1980 [61], the long-term dietary depletion of carotenoids in monkeys led to the disappearance of lutein and zeaxanthin from the plasma and to loss of the *macula lutea*. Ophthalmological consequences of the carotenoid depletion were demonstrated, including increased numbers of drusen, and RPE defects, both being early indicators of AMD in humans [61]. More recently, in carotenoid-depleted rhesus monkeys [112], a distinct decrease of the density of RPE cells in the fovea was identified; these RPE defects were not present to the same extent in animals after subsequent supplementation. In another set of experiments, the vulnerability of the same monkeys to exposure of the retina to a blue-light laser was investigated [113]. Before supplementation, the sizes of the photochemical lesions induced by controlled laser exposures of foveal and parafoveal areas were comparable. After supplementation and concurrent partial repletion of the macular pigment, lesions produced in the fovea were significantly smaller than parafoveal lesions, confirming a photoprotective effect of supplementation by lutein or zeaxanthin.

Long-term, high-level supplementation of cynomolgous monkeys with lutein or zeaxanthin at doses from 0.2-20 mg/kg/day for 52 weeks [114] produced no adverse effects, supporting the good safety profile of lutein and zeaxanthin, which was later demonstrated in humans [115]. In 2004 the Joint FAO/WHO Expert Committee on Food Additives (JECFA) set a group ADI (Advisory Daily Intake) of 0-2 mg/kg body weight/day for lutein and zeaxanthin taken together (Summary and conclusions of the 63rd meeting of the Joint FAO/WHO Expert Committee on Food Additives (JECFA), June 8-17, 2004, Geneva).

Although birds lack a *macula lutea*, some, such as quail, have a cone-rich retina in which carotenoids accumulate [116,117] and which have the ability to form drusen [118]. Experiments with quail have shown that lutein and zeaxanthin can contribute directly to reduction of photodamage in the retina [116,119].

3. Investigations in humans

a) Observational studies

For hundreds of years, Chinese Wolfberry, 'Gou Qi Zi' (*Lycium barbarum*) which contains high levels of (3*R*,3'*R*)-zeaxanthin (up to 50 mg/kg [120]) and is commonly used in home cooking in China, has been a constituent of traditional Chinese herbal medicine for the treatment of visual disorders. This is probably the first recorded 'medicinal use' of one of the macular xanthophylls. The age-induced decline in sensitivity of short-wavelength-sensitive cones is slower in the central retinal area where yellow macular pigment is present, but the decline is accelerated in older individuals; lower macular pigment concentrations are known to precede the clinical manifestation of AMD and other macular diseases [121]. Older subjects with high levels of macular pigment, however, have a retinal sensitivity similar to that of young subjects [122]. This correlation indicates that macular pigment may act to preserve retinal sensitivity.

A number of changes in the retina due to toxicity and degeneration, such as those caused by the photosensitizing drug chloroquine, show an annular pattern, called bull's-eye maculopathy. A circular ring of structural change, surrounding the macula, can be seen, but the macula itself is not affected. The area spared from the degenerative changes corresponds very closely to that with the highest concentration of macular pigment. In the past, retinal lesions were often caused during eye surgery, such as cataract extraction, by exposure to excessive white light from the operating microscope. When these lesions overlapped the macula they were noted to have a crescent moon appearance resulting from the protection that the macular carotenoids bestow against photodamage.

HPLC analysis of eyes *post mortem* from normal subjects and subjects with AMD showed that the average lutein and zeaxanthin levels were lower in the AMD retinas than in the normal retinas [123]. The difference was greatest in a central disc (0 to 5°) of the retina and somewhat smaller in a surrounding annulus (5 to 19°). This difference was smallest but still significant in an outer annular region (19 to 38°) that is relatively unaffected by the disease. The carotenoid concentrations in this region of AMD and healthy retinas were compared and a risk ratio for AMD was calculated. Those individuals having the highest quartile of carotenoid concentration in the outer annulus had an 82% lower risk for AMD than those in the lowest quartile. Thus, low carotenoid concentrations in the retina can be a risk factor for AMD. Some caution, however, is warranted since a causal relationship cannot necessarily be demonstrated from such statistical observations.

The hypothesis that lutein and zeaxanthin can lower the risk for AMD is also supported by a study in which macular pigment density was measured in healthy subjects and in the healthy eye of patients whose fellow eye had advanced AMD [124]. On average, the macular pigment density of patients where the fellow eye had diagnosed AMD was significantly smaller than that of healthy patients. Bilateral symmetry of the macular pigment is usually very high, though exceptions are known. In a small percentage of the population, macular pigment differences as great as 40% in the central 3 mm have been reported between fellow eyes. When macular pigment density is measured by lipofuscin autofluorescence, a statistically significant reduction of macular pigment is seen in AMD patients who have the late stages of the exudative form of the disease, compared with that in normal controls or their healthy offspring [125]. Measurement of macular pigment levels in patients with and without AMD by Raman spectroscopy [126] gave data consistent with lower macular pigment in AMD eyes, but a more recent study did not confirm these findings [127]. However, none of the techniques employed provides the same confidence level as direct analysis by HPLC [123]. How and to what extent the amount of carotenoids in the macula modulates an individual's AMD risk is still open to debate.

Macular pigment density varies widely among individuals and populations, even if corrected for differences in measuring techniques [128]. In a survey of MPOD across the U.S.A., mean MPODs of around 0.2 were reported for populations in the Indianapolis, Phoenix, and New York areas, whereas values around 0.35 were measured for a population in the North-Eastern U.S.A. [63]. A comparative macular pigment density study of South Pacific populations would be interesting. Fijians, for example, are reported to have dietary lutein intakes of up to 26 mg per day [129] compared with the 1-6 mg per day range of total carotenoid intake in the U.S.A. [130].

b) Epidemiological studies

i) AMD. In one of the first epidemiological studies, plasma levels of lutein and zeaxanthin in 356 subjects with neovascular AMD, compared with those of 520 healthy control subjects, revealed a statistically significant inverse relationship between plasma levels of lutein and zeaxanthin and the risk for neovascular AMD [90], *i.e.* higher plasma levels were correlated with a lower risk ratio for AMD. A later study gave complementary results showing a lower risk for AMD in subjects with a higher dietary intake of lutein [89]. Individuals who consumed approximately 6 mg lutein per day had a 57% reduction in risk for AMD when compared with the group that took only 0.6 mg lutein per day. However, the results of such epidemiological studies are not entirely consistent. The Beaver Dam Eye Study [131,132] examined a largely Caucasian community in South-Central Wisconsin, U.S.A., comparing AMD subjects to normal control subjects. Individual plasma concentrations of lutein and zeaxanthin were slightly lower in subjects with exudative AMD, though not to a statistically significant extent. A recently reported epidemiological study [133] indicated that subjects had a lower risk for AMD if they had higher plasma concentrations of zeaxanthin, confirming earlier results [134]. Data from over 8,000 AMD cases examined in the course of the third NHANES (National Health and Nutrition Examination Survey) did not reveal any overall inverse relationship between intake or serum levels of lutein and/or zeaxanthin and any form of AMD. However, the youngest age group at risk for developing either early or late AMD had a lower risk for developing pigmentary changes, an early sign of AMD, if they had higher levels of lutein or zeaxanthin in the diet. It is emphasized, however, that AMD is a multifactorial disease, so interpretation of the correlation of a single group of supposed risk factors, *e.g.* the combined intake levels of lutein + zeaxanthin and their concentration in plasma and retina, is difficult.

ii) Cataract. The correlation of carotenoid intake with reduction of the risk of cataract has been fairly consistent in several epidemiological studies [135] which revealed a significantly lower risk of cataract in the upper percentiles of intake of lutein and zeaxanthin. In the Beaver Dam Eye Study [136], however, the incidence of age-related nuclear cataract was not reduced in people with higher plasma concentrations of these carotenoids. When considering the data on risk reduction of cataract by carotenoids, it should be borne in mind that the amount of carotenoids in the lens is very small. The presence of lutein and zeaxanthin and absence of β-carotene in the human lens is well documented [6,137-139] but the concentration of lutein and zeaxanthin in the lens is at least six orders of magnitude lower than in the macula. Direct involvement of these carotenoids in prevention of pathogenesis of cataract would be difficult to explain, but indirect effects, associated in some way with elevated plasma levels, cannot be precluded.

c) Supplementation studies (intervention trials)

Epidemiological studies (see *Chapter 10*) cannot provide definitive proof that lutein and zeaxanthin can lower the risk of either AMD or cataract. They provide evidence of the possible correlations but do not establish a causal relationship. The situation is different with intervention studies in which agents are administered as a supplement on a double-blind, placebo-controlled, and randomized basis, and where results are evaluated according to pre-defined efficacy parameters. It is only such studies that are likely to provide a definite answer about an effect of lutein and zeaxanthin on AMD [140] or cataract. Nevertheless, the long-term time-course and nature of these diseases make the design of such trials difficult.

i) AMD. To date, no large, well-controlled intervention trials involving the administration of lutein and/or zeaxanthin to influence disease-specific endpoints have been completed. One reason is that only recently have lutein and zeaxanthin supplements become available for human consumption. In 1992, the National Eye Institute initiated the Age-Related Eye Disease Study (AREDS) of 3600 people [141], and demonstrated that consumption of a dietary antioxidant supplement produces significant lowering in the development of advanced AMD relative to controls. Another major clinical intervention study (AREDS II), which began early in 2007, includes the macular xanthophylls as well as the polyunsaturated fatty acids. From a pilot study with 40 subjects, it was concluded that adding ω-3 long-chain polyunsaturated fatty acids to lutein and zeaxanthin did not change plasma levels of either carotenoid [142].

There have been some recent small-scale intervention studies with lutein. One [30] reported significantly improved visual function in sixteen patients, with congenital retinal degenerations, who were supplemented with 20-40 mg lutein per day for 26 weeks. A case-control study [143,144] found improvements in a number of visual function tests, such as contrast sensitivity, in patients who consumed lutein-rich spinach, delivering 30 mg of lutein per day, for 26 weeks. A larger, 1-year, double-blind, placebo-controlled Lutein Antioxidant Study (LAST), of 90 subjects showed significant improvements in visual function of AMD patients who took either 10 mg/day of lutein alone, or 10 mg/day of lutein in an antioxidant formula, compared with those taking a placebo [29]. In another intervention trial with lutein, 21 patients diagnosed with *retinitis pigmentosa* (a genetic eye disease characterized by a progressive decline in photoreceptor function) and eight normal subjects were supplemented with a daily dose of 20 mg of lutein for a period of 6 months [72]. Plasma lutein concentrations increased in all participants, but MPOD, as measured by heterochromatic flicker photometry, increased significantly only in half of them. These 'retinal responders' had a less severe course of the disease than had the non-responders. Inner retinal thickness, measured by optical coherence tomography, correlated positively with the level of MPOD at 0.5° eccentricity, a relationship that was significant for patients, but not for healthy controls. In contrast, however, results of a recent study indicate that central retinal thickness is indeed directly correlated with MPOD in healthy subjects [145].

ii) Cataract. In 1993, the first cataract intervention trial, with 4000 participants aged 45 to 74 years, was reported [146]. The group receiving a multivitamin combination containing β-carotene had a significant 43% reduction in the prevalence of nuclear cataract. Another supplementation trial (REACT) [147] demonstrated that administration of β-carotene together with vitamins E and C could significantly slow the progression of age-related cataract. Whilst in absolute terms the effect was not large, it was the first direct demonstration of the efficacy of a nutritional type intervention for this disease. In contrast, the AREDS study did not show any effect of the trial treatment on cataract; the subjects in AREDS, however, had more advanced cataract, and more sensitive cataract assessing techniques were employed in the REACT trial. This supports the idea that antioxidant supplementation appears to have rather a preventive than a therapeutic value. Because supplements of mixed antioxidants were given, no effect could be attributed specifically to β-carotene. When 50 mg of lutein per week was provided to five subjects with cataract and five subjects with AMD for an average of 13 months, visual acuity and glare sensitivity improved slightly in some subjects, but only in those with cataract [31].

I. Conclusions

The contributions of lutein and zeaxanthin to reduction in risk of AMD and cataract may be only small or moderate, and will be very difficult, or impossible, to determine quantitatively. Both diseases have long time-courses and a multifactorial pathogenesis. Furthermore, the 'disease-initiating lesions' may have genetic origins (as is the case for Stargardt's disease) or may occur very early in life. Both genetic and environmental factors are underlying causes in most instances. It is clear, however, that lutein and zeaxanthin have a role in the context of health and disease of the eye, especially the macula. Given the excellent safety profile of these xanthophylls, even small benefits, such as those that have been reported in the small number of clinical studies and with limited numbers of patients, justify the view that increased intake of lutein and zeaxanthin, preferably in a well balanced diet, is desirable for most people. The alternative of supplementing the diet *via* carotenoid-fortified foods or by taking a multivitamin/carotenoid type product is also a safe and effective strategy to ensure adequate intake of lutein and zeaxanthin.

There are several important, unresolved questions that arise from all these studies. The presence of Haidinger's brushes proves that a net organizational symmetry exists for the carotenoids present in the central macula. It is important to know whether the carotenoids are bound to proteins within the cellular cytosol or located in the lipid bilayer of membranes, because this will indicate the natural limits to which these carotenoids can be concentrated within the macula. It appears that uptake is a controlled biochemical process involving at least one protein capable of shuttling the carotenoids from the blood to the cellular target. Such a protein, if identified, would provide a marker that could help to explain the location and distribution of the different carotenoids within the eye and the specificity for lutein and zeaxanthin. It would also be important to determine which cell types have receptors for such a transport protein. With the exception of the small amounts located in the rod outer segments, the cellular or sub-cellular localization of the bulk of the macular xanthophylls has yet to be determined. Immunological techniques may help to provide answers, particularly for determining if significant quantities of the xanthophylls are also present in the outer segments of the macular cones.

The mechanism by which (3*R*,3'*S*)-zeaxanthin is formed from lutein remains a mystery. Is it a photochemically driven process, or enzyme-mediated? Do these two very similar molecules, lutein and zeaxanthin, contribute differently or complementarily to risk reduction for AMD?

Measurement of the distribution of individual carotenoids is now receiving attention as the tools for accurate determination of macular pigment *in vivo* are developed. Several phenotypes appear to exist. From the clinical perspective, further improvements in the measurement techniques for MPOD levels in diverse populations, both with and without macular or other ocular diseases, is needed, with a reference technique that can serve as the 'standard'. Only with such an accepted standard will it be possible to compare quantitatively MPOD levels across continents, populations and time periods. The question of whether macular pigment distributions are a genetic characteristic or whether they are predominantly environmental remains unresolved. The contribution of genetic factors to the determination of MPOD and pigment distribution and of the risk of AMD must be defined. Populations in some cultures have diets that afford a dramatically higher intake of carotenoids. They may also have unique genetic characteristics that could provide insight into the effects of these two factors on the accumulation of lutein and zeaxanthin. Essential to field studies in areas that are remote and lacking in modern conveniences will be techniques that use portable equipment and are essentially objective and free from subject bias.

References

[1] J. J. Nussbaum, R. C. Pruett and F. C. Delori, *Retina*, **1**, 296 (1981).
[2] G. L. Wald, *Nature*, **101**, 653 (1945).
[3] R. A. Bone, J. T. Landrum and S. L. Tarsis, *Vision Res*., **25**, 1531 (1985).
[4] R. A. Bone, J. T. Landrum, G. W. Hime, A. Cains and J. Zamor, *Invest. Ophthalmol. Vis. Sci*., **34**, 2033 (1993).
[5] C. W. Oyster, *The Human Eye: Structure and Function*. Sinauer, Sunderland (1999).
[6] P. S. Bernstein, F. Khachik, L. S. Carvalho, G. J. Muir, D. Y. Zhao and N. B. Katz, *Exp. Eye Res*., **72**, 215 (2001).
[7] B. S. Sires, J. C. Saari, G. G. Garwin, J. S. Hurst and F. J. van Kuijk, *Arch. Ophthalmol*., **119**, 868 (2001).
[8] D. M. Snodderly, P. K. Brown, F. C. Delori and J. D. Auran, *Invest. Ophthalmol. Vis. Sci*., **25**, 660 (1984).
[9] L. M. Rapp, S. S. Maple and J. H. Choi, *Invest. Ophthalmol. Vis. Sci*., **41**, 1200 (2000).
[10] J. D. M. Gass, *Arch. Ophthalmol*., **117**, 821 (1999).
[11] G. B. Arden and F. M. Barker, *J. Toxicol. Cutaneous Ocular Toxicol*., **10**, 115 (1991).
[12] W. Koepcke, F. M. Barker and W. Schalch, *J. Toxicol. Cutaneous Ocular Toxicol*., **14**, 89 (1995).
[13] R. Goralczyk, F. M. Barker, S. Buser, H. Liechti and J. Bausch, *Invest. Ophthalmol. Vis. Sci*., **41**, 1513 (2000).
[14] R. A. Bone, J. T. Landrum L. M. Friedes, C. Gomez, M. D. Kilburn, E. Menendez, I. Vidal and W. Wang, *Exp. Eye Res*., **64**, 211 (1997).
[15] E. J. Johnson, M. Neuringer, R. M. Russell, W. Schalch and D. M. Snodderly, *Invest. Ophthalmol. Vis. Sci*., **46**, 692 (2005).
[16] W. Gruszecki, A. Sujak, K. Strzalka, A. Radunz and G. H. Schmid, *Z. Naturforsch*., **54c**, 517 (1999).
[17] H. H. Billsten, P. Bhosale, A. Yemelyanov, P. S. Bernstein and T. Polivka, *Photochem. Photobiol*., **78**, 138 (2003).
[18] J. C. Maxwell, *Rep. British Assocn*., **2**, 12 (1856).
[19] F. C. Delori, D. G. Goger, C. Keilhauer, P. Salvetti and G. Staurenghi, *J. Opt. Soc. Am*., **A 23**, 521 (2006).
[20] T. T. Berendschot and D. van Norren, *Invest. Ophthalmol. Vis. Sci*., **47**, 709 (2006).
[21] J. M. Nolan, J. M. Stringham, S. Beatty and D. M. Snodderly, *Invest. Ophthalmol. Vis. Sci*., **49**, 2134 (2008).
[22] W. K. Haidinger, *Poggendorf Annalen*, **63**, 29 (1844).
[23] W. M. Hart (ed.), *Adler's Physiology of the Eye*, Mosby, St. Louis (1992).
[24] J. Kvansakul, M. Rodriguez-Carmona, D. F. Edgar, F. M. Barker, W. Koepcke, W. Schalch and J. L. Barbur, *Ophthal. Physiol. Opt*., **26**, 362 (2006).
[25] V. M. Reading and R. A. Weale, *J. Opt. Soc. Am*., **64**, 231 (1974).
[26] B. R. Hammond Jr., B. R. Wooten and J. Curran-Celentano, *Arch. Biochem. Biophys*., **385**, 41 (2001).
[27] B. R. Hammond Jr., B. R. Wooten and D. M. Snodderly, *Optom. Vis. Sci*., **74**, 499 (1997).
[28] J. Kvansakul, M. Rodriguez-Carmona, D. F. Edgar, F. M. Barker, W. Koepcke, W. Schalch and J. L. Barbur, *Ophthal. Physiol. Opt*., **26**, 362 (2006).

[29] S. Richer, W. Stiles, L. Statkute, J. Pulido, J. Frankowski, D. Rudy, K. Pei, M. Tsipurski and J. Nyland, *Optometry*, **75**, 216 (2004).

[30] G. Dagnelie, I. S. Zorge and T. M. McDonald, *Optometry*, **71**, 147 (2000).

[31] B. Olmedilla, F. Granado, I. Blanco, M. Vaquero and C. Cajigal, *J. Sci. Food Agric.*, **81**, 904 (2001).

[32] J. M. Stringham and B. R. Hammond Jr., *Optom. Vis. Sci.*, **84**, 859 (2007).

[33] J. M. Stringham and B. R. Hammond, *Optom. Vis. Sci.*, **85**, 82 (2008).

[34] D. Whittaker, R. Steen and D. B. Elliott, *Optom. Vis. Sci.*, **70**, 963 (1993).

[35] B. R. Wooten and B. R. Hammond, *Prog. Retin. Eye Res.*, **21**, 225 (2002).

[36] R. A. Bone and J. T. Landrum, *Arch. Biochem. Biophys.*, **430**, 137 (2004).

[37] M. Rodriguez-Carmona, J. Kvansakul, J. A. Harlow, W. Koepcke, W. Schalch and J. L. Barbur, *Ophthal. Physiol. Opt.*, **26**, 137 (2006).

[38] T. T. Berendschot and D. van Norren, *Exp. Eye Res.*, **81**, 602 (2005).

[39] J. D. Moreland, A. G. Robson, N. Soto-Leon and J. J. Kulikowski, *Vision Res.*, **38**, 3241 (1998).

[40] A. G. Robson, G. Harding and F. J. van Kuijk, *Perception*, **34**, 1029 (2005).

[41] R. A. Bone, J. T. Landrum and A. Cains, *Vision Res.*, **32**, 105 (1992).

[42] D. van Norren and L. F. Tiemeijer, *Vision Res.*, **26**, 313 (1986).

[43] J. van de Kraats, T. T. Berendschot and D. van Norren, *Vision Res.*, **36**, 2229 (1996).

[44] J. van de Kraats, T. Berendschot, T. S. Valen and D. van Norren, *J. Biomed. Opt.*, **11**, 064031 (2006).

[45] P. E. Kilbride, K. R. Alexander, M. Fishman and G. A. Fishman, *Vision Res.*, **29**, 663 (1989).

[46] J. van de Kraats, M. J. Kanis, S. W. Genders and D. van Norren, *Invest. Ophthalmol. Vis. Sci.*, **49**, 5568, (2008).

[47] F. C. Delori, D. G. Goger, B. R. Hammond, D. M. Snodderly and S. A. Burns, *J. Opt. Soc. Am.*, **A 18**, 1212 (2001).

[48] M. Trieschmann, B. Heimes, H. W. Hense and D. Pauleikhoff, *Graefes Arch. Clin. Exp. Ophthalmol.*, **244**, 1565 (2006).

[49] W. Gellerman, I. V. Ermakov, R. W. McClane and P. S. Bernstein, *Optics Lett.*, **27**, 833 (2002).

[50] W. Gellermann and P. S. Bernstein, *J. Biomed. Opt.*, **9**, 75 (2004).

[51] K. Neelam, N. O'Gorman, J. Nolan, O. Donovan, H. B. Wong, K. G. A. Eong and S. Beatty, *Invest. Ophthalmol. Vis. Sci.*, **46**, 1023 (2005).

[52] P. A. Thürmann, W. Schalch, J.-C. Aebischer, U. Tenter and W. Cohn, *Am. J. Clin. Nutr.*, **82**, 88 (2005).

[53] D. Hartmann, P. A. Thürmann, V. Spitzer, W. Schalch, B. Manner and W. Cohn, *Am. J. Clin. Nutr.*, **79**, 410 (2004).

[54] B. A. Clevidence and J. G. Bieri, *Meth. Enzymol.*, **214**, 33 (1993).

[55] W. Wang, S. L. Connor, E. J. Johnson, M. L. Klein, S. Hughes and W. E. Connor, *Am. J. Clin. Nutr.*, **85**, 762 (2007).

[56] A. Y. Yemelyanov, N. B. Katz and P. S. Bernstein, *Exp. Eye Res.*, **72**, 381 (2001).

[57] P. Bhosale, A. J. Larson, J. M. Frederick, K. Southwick, C. D. Thulin and P. S. Bernstein, *J. Biol. Chem.*, **279**, 49447 (2004).

[58] P. Bhosale and P. S. Bernstein, *Arch. Biochem. Biophys.*, **458**, 121 (2007).

[59] E. Loane, J. M. Nolan, O. Donovan, P. Bhosale, P. S. Bernstein and S. Beatty, *Surv. Ophthalmol.*, **53**, 68 (2008).

[60] P. Bhosale and P. S. Bernstein, *Biochim. Biophys. Acta*, **1740**, 116 (2005).

[61] M. R. Malinow, L. Feeney-Burns, L. H. Peterson, M. Klein and M. Neuringer, *Invest. Ophthalmol. Vis. Sci.*, **19**, 857 (1980).

[62] C. Schupp, E. Olano-Martin, C. Gerth, B. M. Morrissey, C. E. Cross and J. S. Werner, *Am. J. Clin. Nutr.*, **79**, 1045 (2004).

[63] B. R. Hammond, E. J. Johnson, R. M. Russell, N. I. Krinsky, K. J. Yeum, R. B. Edwards and D. M. Snodderly, *Invest. Ophthalmol. Vis. Sci.*, **38**, 1795 (1997).

[64] J. M. Nolan, J. Stack, J. Mellerio, M. Godhinio, O. O'Donovan, K. Neelam and S. Beatty, *Curr. Eye Res.*, **31**, 199 (2006).

[65] C. Jahn, C. Brinkmann, A. Mossner, H. Wustemeyer, U. Schnurrbusch and S. Wolf, *Ophthalmologe*, **103**, 136 (2006).

[66] M. Neuringer, E. J. Johnson, D. M. Snodderly, M. M. Sandstrom and W. Schalch, *Invest. Ophthalmol. Vis. Sci., ARVO 2001 Abstracts*, Abstract no. 1209 (2001).

[67] H. Y. Chung, H. M. Rasmussen and E. J. Johnson, *J. Nutr.*, **134**, 1887 (2004).

[68] P. Bowen, S. Herbst-Espinosa, E. Hussain and M. Stacewicz-Sapuntzakis, *J. Nutr.*, **132**, 3668 (2002).

[69] J. T. Landrum, R. A. Bone, H. Joa, M. D. Kilburn, L. L. Moore and K. E. Sprague, *Exp. Eye Res.*, **65**, 57 (1997).

[70] H. H. Koh, I. J. Murray, D. Nolan, D. Carden, J. Feather and S. Beatty, *Exp. Eye Res.*, **79**, 21 (2004).

[71] M. Trieschmann, S. Beatty, J. M. Nolan, W. Hense, B. Heimes, U. Austermann, M. Fobker and D. Pauleikhoff, *Exp. Eye Res.*, **84**, 718 (2007).

[72] T. S. Aleman, J. L. Duncan, M. L. Bieber, E. de Castro, D. A. Marks, L. M. Gardner, J. D. Steinberg, A. V. Cideciyan, M. G. Maguire and S. G. Jacobson, *Invest. Ophthalmol. Vis. Sci.*, **42**, 1873 (2001).

[73] J. L. Duncan, T. S. Aleman, L. M. Gardner, E. De Castro, D. A. Marks, J. M. Emmons, M. L. Bieber, J. D. Steinberg, J. Bennett, E. M. Stone, I. M. Macdonald, A. V. Cideciyan, M. G. Maguire and S. G. Jacobson, *Exp. Eye Res.*, **74**, 371 (2002).

[74] T. T. Berendschot, J. J. Willemse-Assink, M. Bastiaanse, P. T. de Jong and D. van Norren, *Invest. Ophthalmol. Vis. Sci.*, **43**, 1928 (2002).

[75] D. Schweitzer, G. E. Lang, B. Beuermann, H. Remsch, M. Hammer, E. Thamm, C. W. Spraul and G. K. Lang, *Ophthalmologe*, **99**, 270 (2002).

[76] W. Schalch, W. Cohn, F. Barker, W. Köpcke, J. Mellerio, A. C. Bird, A. G. Robson, F. F. Fitzke and F. J. G. M. van Kuijk, *Arch. Biochem. Biophys.*, **458**, 128 (2007).

[77] N. Cardinault, J. Gorrand, V. Tyssandier, P. Grolier, E. Rock and P. Borel, *Exp. Gerontol.*, **38**, 573 (2003).

[78] T. S. Aleman, A. V. Cideciyan, E. A. Windsor, S. B. Schwartz, M. Swider, J. D. Chico, A. Sumaroka, A. Y. Pantelyat, K. G. Duncan, L. M. Gardner, J. M. Emmons, J. D. Steinberg, E. M. Stone and S. G. Jacobson, *Invest. Ophthalmol. Vis. Sci.*, **48**, 1319 (2007).

[79] I. Y. F. Leung, M. O. M. Tso and T. T. Lam, *Invest. Ophthalmol. Vis. Sci.*, **42**, S359; Program#1942 (2001).

[80] R. A. Bone, J. T. Landrum, L. H. Guerra and C. A. Ruiz, *J. Nutr.*, **133**, 992 (2003).

[81] K. M. Garnett, L. H. Guerra, J. D. Lamb, J. L. Epperson, D. L. Greenbury, K. Dorey and N. E. Craft, *ARVO 2002, Abstracts Volume #1*, Program# 2820 (2002).

[82] E. J. Johnson, H. Y. Chung, S. M. Cardarella and D. M. Snodderly, *Am. J. Clin. Nutr.*, **87**, 1521 (2008).

[83] D. I. Thurnham, A. Tremel and A. N. Howard, *Br. J. Nutr.*, **100**, 1307 (2008).

[84] R. A. Bone, J. T. Landrum, Y. Cao, A. N. Howard and F. Alvarez-Calderon, *Nutr. Metab.*, **4**, 1 (2007).

[85] B. R. Hammond Jr., J. Curran Celentano, S. Judd, K. Fuld, N. I. Krinsky, B. R. Wooten and D. M. Snodderly, *Vision Res.*, **36**, 2001 (1996).

[86] E. J. Johnson, *Nutr. Rev.*, **63**, 9 (2005).

[87] L. Brazionis, K. Rowley, C. Itsiopoulos and K. O'Dea, *Br. J. Nutr.*, **101**, 270 (2009).

[88] B. R. Hammond, B. R. Wooten and D. M. Snodderly, *Vision Res.*, **36**, 3003 (1996).

[89] J. M. Seddon, U. A. Ajani, R. D. Sperduto, R. Hiller, N. Blair, T. C. Burton, M. D. Farber, E. S. Gragoudas, J. Haller and D. T. Miller, *JAMA*, **272**, 1413 (1994).

[90] EDCC Study Group, *Arch. Ophthalmol.*, **111**, 104 (1993).

[91] J. A. Mares, T. L. LaRowe, D. M. Snodderly, S. M. Moeller, M. J. Gruber, M. L. Klein, B. R. Wooten, E. J. Johnson and R. J. Chappell, for the CAREDS Macular Pigment Study Group and Investigators, *Am. J. Clin. Nutr.*, **84**, 1107 (2006).

[92] R. A. Bone, J. T. Landrum and B. Brener, *ARVO 2006, Abstract Book on CDROM*, Program#: 3806 (2006).
[93] K. Neelam, J. Nolan, E. Loane, J. Stack, O. O'Donovan, K. G. Au Eong and S. Beatty, *Vision Res.*, **46**, 2149 (2006).
[94] R. A. Bone, J. T. Landrum, L. Fernandez and S. L. Tarsis, *Invest. Ophthalmol. Vis. Sci.*, **29**, 843 (1988).
[95] S. H. Liew, C. E. Gilbert, T. D. Spector, J. Mellerio, J. Marshall, F. J. van Kuijk, S. Beatty, F. Fitzke and C. J. Hammond, *Invest. Ophthalmol. Vis. Sci.*, **46**, 4430 (2005).
[96] B. R. Hammond, K. Fuld and J. Curran-Celentano, *Invest. Ophthalmol. Vis. Sci.*, **36**, 2531 (1995).
[97] J. Marx, *Science*, **24**, 5768 (2006).
[98] D. S. Friedman, B. J. O'Colmain, B. Munoz, S. C. Tomany, C. McCarty, P. T. de Jong, B. Nemesure, P. Mitchell, J. Kempen and E. D. P. R. Group, *Arch. Ophthalmol.*, **122**, 564 (2004).
[99] N. Congdon, J. R. Vingerling, B. E. Klein, S. West, D. S. Friedman, J. Kempen, B. O'Colmain, S. Y. Wu and H. R. Taylor, *Arch. Ophthalmol.*, **122**, 487 (2004).
[100] W. L. Stone, C. C. Farnsworth and E. A. Dratz, *Exp. Eye Res.*, **28**, 387 (1979).
[101] W. Schalch. in *Free Radicals and Aging* (ed. B. Chance), p. 280, Birkhäuser, Basel (1992).
[102] S. R. Kim, K. Nakanishi, Y. Itagaki and J. R. Sparrow, *Exp. Eye Res.*, **82**, 828 (2006).
[103] S. P. Sundelin and S. E. Nilsson, *Free Radic. Biol. Med.*, **31**, 217 (2001).
[104] M. Wrona, M. Rozanowska and T. Sarna, *Free Radic. Biol. Med.*, **36**, 1094 (2004).
[105] A. Pawlak, M. Wrona, M. Rozanowska, M. Zareba, L. E. Lamb, J. E. Roberts, J. D. Simon and T. Sarna, *Photochem. Photobiol.*, **77**, 253 (2003).
[106] M. Rozanowska and T. Sarna, *Photochem. Photobiol.*, **81**, 1305 (2005).
[107] W. I. Gruszecki and K. Strzalka, *Biochim. Biophys. Acta*, **1740**, 108 (2005).
[108] A. Sujak, J. Gabrielska, W. Grudzinski, R. Borc, P. Mazurek and W. I. Gruszecki, *Arch. Biochem. Biophys.*, **371**, 301 (1999).
[109] L. X. Zhang, P. Acevedo, H. Guo and J. S. Bertram, *Mol. Carcinogenesis*, **12**, 50 (1995).
[110] W. A. Monaco and C. M. Wormington, *Optom. Vis. Sci.*, **67**, 532 (1990).
[111] G. M. Hope, W. W. Dawson, H. M. Engel, R. J. Ulshafer, M. J. Kessler and M. B. Sherwood, *Br. J. Ophthalmol.*, **76**, 11 (1992).
[112] I. F. Leung, D. M. Snodderly, M. M. Sandstrom, C. L. Zucker and M. Neuringer, *Invest. Ophthalmol. Vis. Sci., ARVO 2002 Abstracts*, Abstract no. 717 (2002).
[113] F. M. Barker, M. Neuringer, E. J. Johnson, W. Schalch, W. Koepcke and D. M. Snodderly, ARVO 2005: *Abstract Book on CDROM*: Program#: 1770 (2005).
[114] R. Goralczyk, F. M. Barker, O. Froescheis, J.-C. Aebischer, B. Niggemann, U. Korte, J. Schierle, F. Pfannkuch and J. Bausch, *ARVO 2002, Abstracts Volume #1*, Program # 2546 (2002).
[115] W. Schalch and F. M. Barker, *ARVO 2005: Abstract Book on CDROM*: Program#: 1765 (2005).
[116] Y. Toyoda, L. R. Thomson, A. Langner, N. E. Craft, K. M. Garnett, C. R. Nichols, K. M. Cheng and C. K. Dorey, *Invest. Ophthalmol. Vis. Sci.*, **43**, 1210 (2002).
[117] J. K. Bowmaker, J. K. Kovach, A. V. Whitmore and E. R. Loew, *Vision Res.*, **33**, 571 (1993).
[118] K. V. Fite, C. L. Bengston and F. Cousins, *Exp. Eye Res.*, **59**, 417 (1994).
[119] C. K. Dorey, Y. Toyoda, L. Thomson, K. M. Garnett, M. Sapuntzakis, N. Craft, C. Nichols and K. Cheng, *Invest. Ophthalmol. Vis. Sci.*, **38**, S355 (1997).
[120] K. W. Lam and P. But, *Food Chem.*, **67**, 173 (1999).
[121] J. S. Sunness, R. W. Massof, M. A. Johnson, N. M. Bressler, S. B. Bressler and S. L. Fine, *Ophthalmol.*, **96**, 375 (1989).
[122] B. R. Hammond Jr., B. R. Wooten and D. M. Snodderly, *Invest. Ophthalmol. Vis. Sci.*, **39**, 397 (1998).
[123] R. A. Bone, J. T. Landrum, S. T. Mayne, C. M. Gomez, S. E. Tibor and E. E. Twaroska, *Invest. Ophthalmol. Vis. Sci.*, **42**, 235 (2001).

[124] S. Beatty, I. J. Murray, D. B. Henson, D. Carden, H. Koh and M. E. Boulton, *Invest. Ophthalmol. Vis. Sci.*, **42**, 439 (2001).
[125] D. Schweitzer, G. E. Lang, H. Remsch, B. Beuermann, M. Hammer, E. Thamm, C. W. Spraul and G. K. Lang, *Ophthalmologe*, **97**, 84 (2000).
[126] P. S. Bernstein, D. Y. Zhao, S. W. Wintch, I. V. Ermakov, R. W. McClane and W. Gellermann, *Ophthalmology*, **109**, 1780 (2002).
[127] C. Jahn, H. Wustemeyer, C. Brinkmann, S. Trautmann, A. Mossner and S. Wolf, *Graefes Arch. Clin. Exp. Ophthalmol.*, **243**, 222 (2005).
[128] B. R. Hammond, B. R. Wooten and D. M. Snodderly, *J. Opt. Soc. Am.*, **A 14**, 1187 (1997).
[129] L. Le Marchand, J. H. Hankin, F. Bach, L. N. Kolonel, L. R. Wilkens, S. Stacewicz, P. E. Bowen, G. R. Beecher, F. Laudon, P. Baque, R. Daniel, L. Servatu and B. E. Henderson, *Int. J. Cancer*, **63**, 18 (1995).
[130] A. Ascherio, M. J. Stampfer, G. A. Colditz, E. B. Rimm, L. Litin and W. C. Willett, *J. Nutr.*, **122**, 1792 (1992).
[131] J. A. Mares-Perlman, W. E. Brady, R. Klein, B. E. Klein, P. Bowen, M. Stacewicz-Sapuntzakis and M. Palta, *Arch. Ophthalmol.*, **113**, 1518 (1995).
[132] J. A. Mares-Perlman, R. Klein, B. E. Klein, J. L. Greger, W. E. Brady, M. Palta and L. L. Ritter, *Arch. Ophthalmol.*, **114**, 991 (1996).
[133] C. Delcourt, I. Carriere, M. Delage, P. Barberger-Gateau and W. Schalch, *Invest. Ophthalmol. Vis. Sci.*, **47**, 2329 (2006).
[134] C. R. Gale, N. F. Hall, D. I. W. Phillips and C. N. Martyn, *Invest. Ophthalmol. Vis. Sci.*, **44**, 2461 (2003).
[135] P. F. Jacques, L. T. Chylack Jr., S. E. Hankinson, P. M. Khu, G. Rogers, J. Friend, W. Tung, J. K. Wolfe, N. Padhye, W. C. Willett and A. Taylor, *Arch. Ophthalmol.*, **119**, 1009 (2001).
[136] B. J. Lyle, J. A. Mares-Perlman, B. E. Klein, R. Klein and J. L. Greger, *Am. J. Epidemiol.*, **149**, 801 (1999).
[137] K. J. Yeum, A. Taylor, G. Tang and R. M. Russell, *Invest. Ophthalmol. Vis. Sci.*, **36**, 2756 (1995).
[138] C. J. Bates, S. J. Chen, A. Macdonald and R. Holden, *Int. J. Vitam. Nutr. Res.*, **66**, 316 (1996).
[139] J. T. Landrum, R. A. Bone, E. Kenyon, K. Sprague and A. Maya, *Invest. Ophthalmol. Vis. Sci.*, **38**, S1026 (1997).
[140] J. M. Seddon and C. H. Hennekens, *Arch. Ophthalmol*, **112**, 176 (1994).
[141] AREDS Research Group, *Contr. Clin. Trials*, **20**, 573 (1999).
[142] L. L. Huang, H. R. Coleman, J. Kim, F. de Monasterio, W. T. Wong, R. L. Schleicher, F. L. Ferris and E. Y. Chew, *Invest. Ophthalmol. Vis. Sci.*, **49**, 3864 (2008).
[143] S. Richer, *J. Am. Optom. Assoc.*, **70**, 24 (1999).
[144] S. Richer, *Optometry*, **71**, 657 (2000).
[145] S. H. Liew, C. E. Gilbert, T. D. Spector, J. Mellerio, F. J. Kuijk, S. Beatty, F. Fitzke, J. Marshall and C. J. Hammond, *Exp. Eye Res.*, **82**, 915 (2006).
[146] R. D. Sperduto, T. S. Hu, R. C. Milton, J. L. Zhao, D. F. Everett, Q. F. Cheng, W. J. Blot, L. Bing, P. R. Taylor and J. Y. Li, *Arch. Ophthalmol.*, **111**, 1246 (1993).
[147] L. T. Chylack Jr., N. P. Brown, A. Bron, M. Hurst, W. Kopcke, U. Thien and W. Schalch, *Ophthal. Epidemiol.*, **9**, 49 (2002).

Carotenoids
Volume 5: Nutrition and Health

Chapter 16

Skin Photoprotection by Carotenoids

Regina Goralczyk and Karin Wertz

A. Introduction

In Western populations, a lifestyle favouring tanned skin leads to increased exposure to natural and artificial sources of UV-radiation (UVR). To keep the adverse effects of this exposure, such as sunburn, immunosuppression, photoaging and photocarcinogenesis, to a minimum, nutritional manipulation of the basic endogenous protective properties of skin is an attractive target. In this respect, considerable interest has been directed for many years towards the dietary carotenoids, because of their radical scavenging and singlet oxygen quenching properties and thus their putative role in photochemistry, photobiology and photomedicine.

Hypothetically, carotenoids could be involved in several ways to protect skin from sunlight damage, namely by increasing optical density, quenching singlet oxygen ($^1O_2$) or, for provitamin A carotenoids, *via* formation of retinoic acid (*1*), a known topical therapeutic agent against photodermatoses. The role of $^1O_2$ in UVA-induced oxidative stress is well established and has been reviewed extensively [1,2]. Carotenoids can also scavenge other reactive oxygen species [3,4], such as superoxide anions, hydroxyl radicals or hydrogen peroxide. Under certain conditions, however, *i.e.* higher oxygen partial pressure, carotenoids may act as pro-oxidants [5,6] (*Chapter 12*).

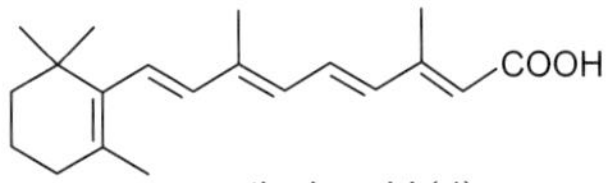

retinoic acid (*1*)

Figure 1 shows the various structural layers of the skin, and the depth of penetration of radiation of different wavelengths; the shorter the wavelength, the greater is the energy of the radiation. Chronic exposure to UV radiation leads to epidermal and dermal damage, such as hyperkeratosis, keratinocyte dysplasia and dermal elastosis, in affected skin areas, clinically presenting as photoaged skin with actinic or solar keratosis. These precancerous lesions show an increased risk for the development of squamous cell carcinoma (SCC).

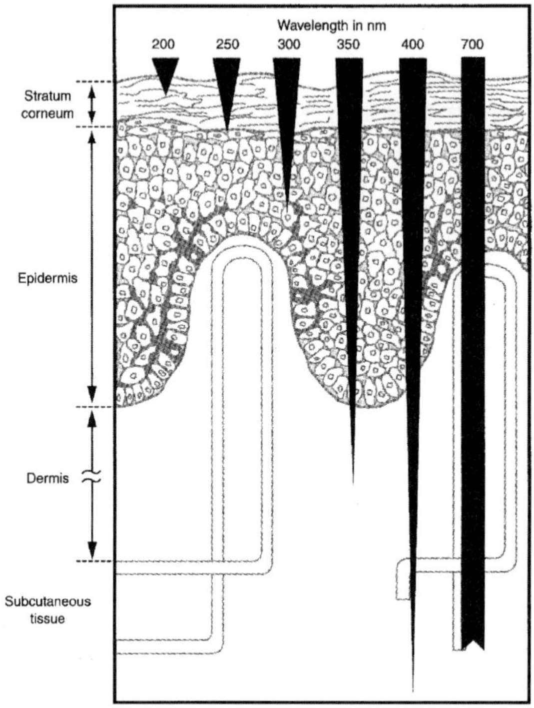

Fig. 1. Scheme of the structural layers of the skin, *i.e.* the stratum corneum, epidermis, dermis and subcutis. The black arrows show the penetration depth of increasing wavelengths [7]. In contrast to UVB (290-320 nm), which only penetrates through the epidermis, UVA (320-400nm) can penetrate deep into the dermis and subcutis.

The molecular mechanisms of skin photodamage and photoaging have been subjects of extensive research [8] (Fig. 2). UV radiation activates a wide range of cell-surface growth factors and cytokine receptors [9]. This ligand-independent receptor activation induces multiple downstream signalling pathways that converge to stimulate the transcription factor AP-1. Among the genes that are up-regulated by AP-1 are several members of the matrix metalloprotease (MMP) family. Increased MMP expression and activity causes enhanced collagen proteolysis and, together with reduced collagen expression, results in skin elastosis and wrinkling [10]. Under chronic UV exposure, the clinical condition is accompanied by dilated and twisted microvasculature, *i.e.* teleangiectasia and hyper-pigmentation (clinical features of photoaging [11]).

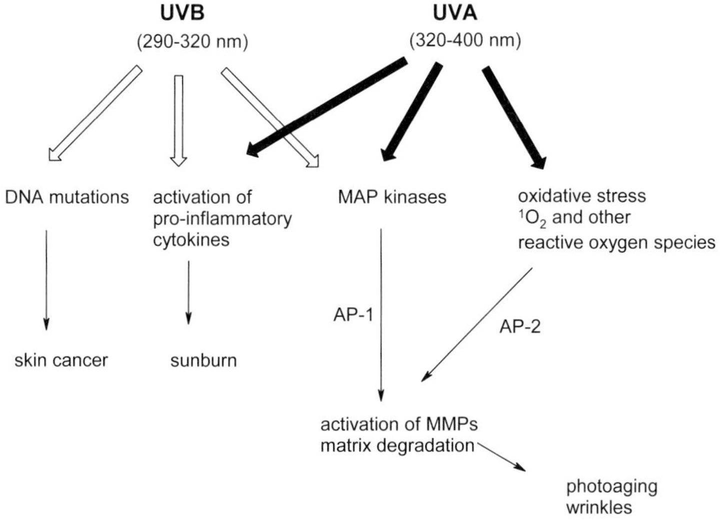

Fig. 2. Major mechanisms for the involvement of UVB and UVA in photocarcinogenesis and photoaging

UVB (290-320 nm) is mainly absorbed by keratinocytes in the epidermis. By direct interaction with the DNA, it causes mutations and skin cancer. UVB also leads to sunburn, which is an erythema resulting from an inflammatory response to the photodamage to the skin.

UVA (320-400 nm) plays a major role in photoaging. UVA can penetrate into the deeper dermis and induces the generation of reactive oxygen species (ROS), which can induce mutations in the mitochondrial DNA, thus leading to losses of enzymes involved in oxidative phosphorylation (see Section **D**.3.) and deficiencies in energy metabolism. The defects in the respiratory chain lead to further inductions of ROS. Singlet oxygen can also induce up-regulation of MMPs directly, independent of the AP-1 pathway (see Section **D**.4.).

In this *Chapter*, photoprotective effects of dietary carotenoids, especially β-carotene (**3**), towards skin damage induced by UVA and UVB are reviewed, underlying molecular mechanisms are discussed, and the availability of carotenoids at the target skin tissue is summarized.

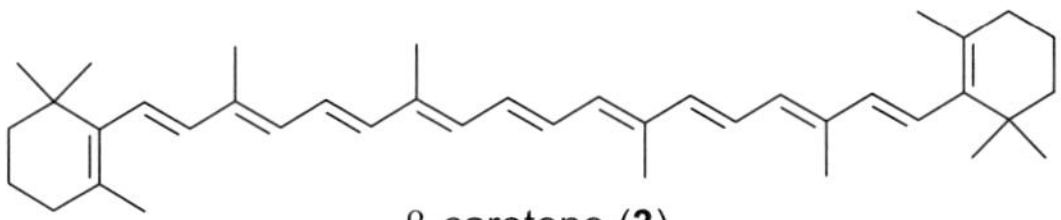

β-carotene (**3**)

B. Uptake and Metabolism of Carotenoids in Skin Cells

1. Humans and mouse models

The effectiveness of photoprotection will largely depend on the local concentration of the carotenoids in the specific skin compartment at the site of UV-induced radical formation.

β-Carotene and other carotenoids are transported to the skin and accumulate mainly in the epidermal layers. High β-carotene concentration in skin leads to increased reflection and scattering of light. Thus, penetration of photons into deeper skin layers is lessened. Reflection of light has also been utilized to measure β-carotene concentration in skin by non-invasive reflection spectroscopy. β-Carotene does not, however, act as an optical UV filter [12], since its main absorption maximum, like that of most carotenoids, is around 460 nm and not in the UVB/UVA range of wavelengths.

The amount of carotenoids deposited in skin correlates with dietary intake and bioavailability from the food source (see *Chapter 7*). After absorption, carotenoids are transported in the bloodstream *via* lipoproteins to the various target tissues [13-16]. Recently, cholesterol transporters such as SR-B1 and CD 36 were shown to mediate a facilitated absorption of carotenoids in the gut [17-19]. It is likely that carotenoids are taken up by these transporters also in the epidermis, which is an active site of cholesterol accumulation for maintenance of permeability barrier function. SR-B1 is expressed in human epidermis [20], predominantly in the basal layers.

Unfortunately, reports on carotenoid concentrations in skin of humans or laboratory animals are rare, many of them old and most referring to β-carotene only. Comparisons across publications are complicated by the fact that different methods were applied, such as simple absorption spectroscopy of skin extracts, non-invasive reflectance spectroscopy or HPLC of skin biopsies. The latter method can be regarded as the most appropriate technique, but it requires analysis of skin biopsies, which are often collected in different ways, *i.e.* blister, scrape or punch, and result in different fractions of dermis/epidermis or even contamination with subcutaneous fat. Furthermore, absorption spectroscopy, non-invasive reflection spectroscopy and earlier HPLC of skin extracts were only able to detect total carotenoids, and did not differentiate between the various carotenes, xanthophylls and their isomers. In addition, efficiency of extraction of skin samples varies, thus leading to large differences in carotenoid recoveries. In general, however, the correlation of skin to plasma carotenoid concentration is very good [21-23]. A compilation of reported β-carotene/carotenoid concentrations in skin is shown in Table 1.

Table 1. β-Carotene or carotenoid concentration in human skin, normal and after dietary supplementation.

Treatment	Analysis Method	Tissue	Value nmol/g	Ref.
Normal skin	Extraction/ absorption spectrum	Scrapings Epidermis Dermis	 0.39 0.01	[24]
β-Carotene (beadlets) 180 mg/day, 10 weeks	Extraction/ absorption spectrum	Epidermis blister	1.7	[24]
Normal skin	Extraction/ absorption spectrum shave biopsy	Epidermis Dermis Subcutis Surface lipid Comedones	4.1 1.3 3.5 10.0 14.5	[25]
Normal skin	HPLC	Whole skin	0.09	[21]
a) Baseline b) 120 mg β-carotene, single dose	HPLC	Whole skin	a) 1.41 b) 1.74	[26]
a) Baseline b) 30 mg/day β-carotene (beadlets), 10 weeks	HPLC	Punch biopsy (Dermis/ Epidermis)	a) 8.3 b) 24.2	[27]
β-Carotene (24 mg/day) from algal extract, 12 weeks	Reflection spectroscopy, total carotenoids	Forehead	1.4	[28]
Tomato paste, 16 mg lycopene, 20 weeks	Reflection spectroscopy, total carotenoids	Hand palm	Control: 0.33- 0.19 Treated: 0.26-0.3	[29]
Combination of vitamin E, β-carotene, lycopene, selenium, proantho-cyanidins (Seresis), 16 weeks	HPLC	Punch biopsy (Dermis/ Epidermis)	β-Carotene: nmol/mg protein baseline: 0.007 56 days: 0.022 112 days: 0.012	[30]
a) β-Carotene, 24 mg/day b) mixed carotenoids from algae, 24 mg/day, 12 weeks	Reflection spectroscopy, total carotenoids	Hand palm	a) ~ 1.1 b) ~ 1.5 controls ~0.5	[31]

Apparently, carotenes are present at higher concentration in the epidermis and in surface lipids than in the dermis, consistent with the distribution of lipid transporters. Physiological levels between 0.09 and 4 nmol/g wet weight are reported. Upon supplementation with β-carotene, reported values vary widely, *i.e.* 1.7 nmol/g (determined by absorption spectrophotometry) [24] in the epidermis after administration of supplements of 180 mg/day over 10 weeks, or 8 nmol/g in punch biopsies at baseline compared to 24 nmol/g after supplementation with 30 mg β-carotene/day over 10 weeks [27]. In contrast, a lower concentration of 1.4 nmol/g was determined by reflectometry after a 12-week supplementation with β-carotene from an algal source [28]. The variability of skin carotenoid concentrations across human studies may be due to differences in the bioavailability of the supplemented product and/or to the use of different analytical methods.

The level of β-carotene in plasma and in epithelial cells (oral mucosa cells, OMC) is dependent on skin-type [32]; individuals with Type I, *i.e.* fair skin and hair, and high UV-sensitivity, have significantly lower β-carotene levels than Type IV individuals, who have strong pigmentation, dark hair and low UV-sensitivity.

Similar large variations in skin β-carotene concentrations have been reported in studies with rodent models. Depending on the protocols used for intervention and the bioavailability of the β-carotene supplement, values in mice vary as extremely as from 0.27 to 8 nmol/g [33-35]. This demonstrates the difficulty of establishing the absolute β-carotene concentration in the target tissue for correlation with its photoprotective effects. Nevertheless, although much higher doses are required, skin levels of β-carotene in mice are in the same order of magnitude as in humans, making the mice relevant models for studying the interactions of carotenoids with UV-induced processes in skin.

There are fewer HPLC data on skin levels of other dietary carotenoids. In normal skin, xanthophylls such as lutein (**133**), zeaxanthin (**119**), 2′,3′-anhydrolutein (**59.1**), and α-cryptoxanthin (**62**) and β-cryptoxanthin (**55**) were detected, as well as low amounts of their monoacyl and diacyl esters [36].

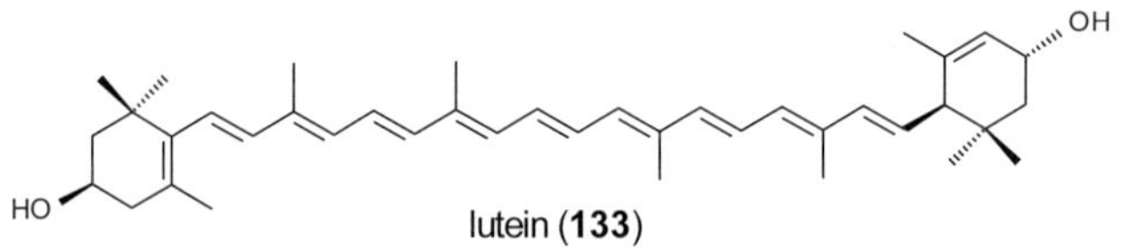

lutein (**133**)

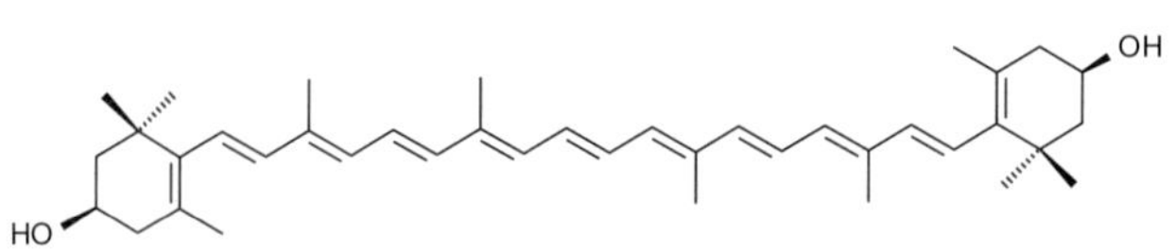

zeaxanthin (**119**)

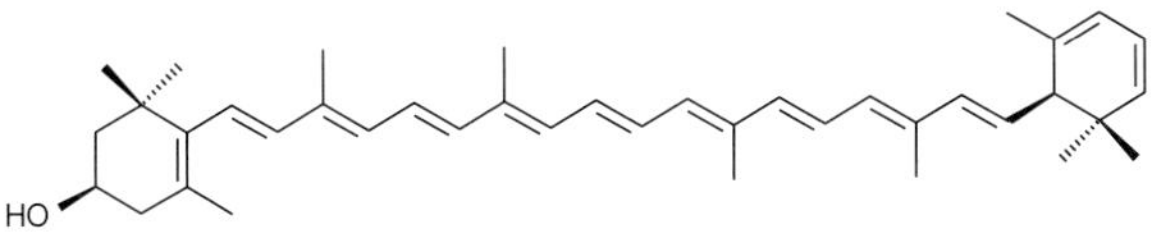

2',3'-anhydrolutein (**59.1**)

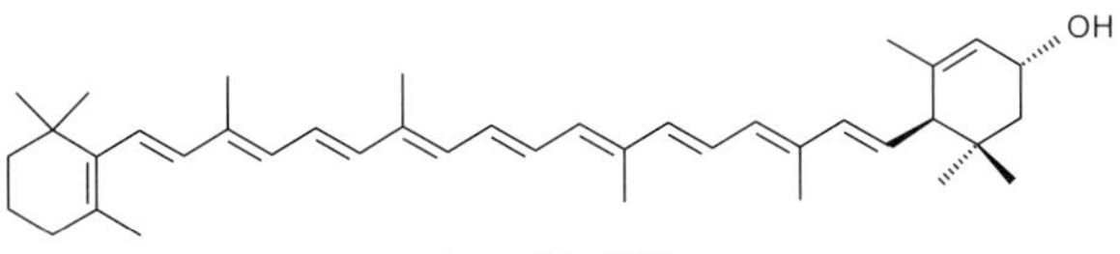

α-cryptoxanthin (**62**)

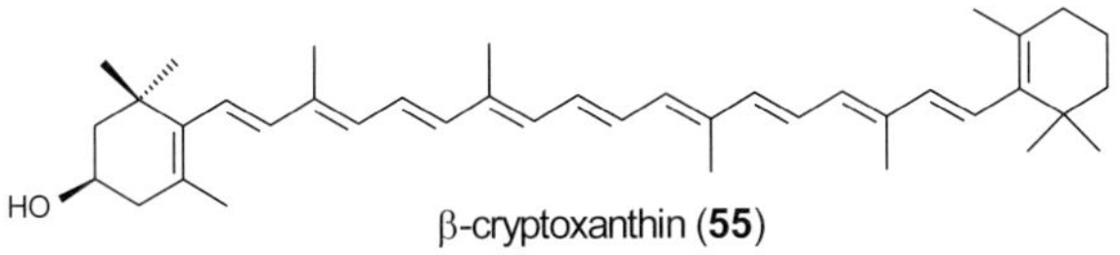

β-cryptoxanthin (**55**)

Supplementation with lycopene-rich products, *i.e.* carrot juice (from the variety 'Nutrired', containing 2.5 mg lycopene and 1.3 mg β-carotene/100 ml), a lycopene supplement from tomato extract, a lycopene-containing drink or a supplement of synthetic lycopene, for 12 weeks led to about 20-40% increases in total skin carotenoid levels as measured by reflection spectroscopy [37]. Daily supplementation with 40 g tomato paste (providing 16 mg lycopene) for 10 weeks, however, did not lead to significant increases in skin total carotenoid levels as determined by reflection spectroscopy [29]. Lycopenodermia, a rare reversible cutaneous condition similar to carotenodermia, can be observed after excessive dietary ingestion of lycopene-containing products [38]. Two oxidative metabolites of lycopene, namely the stereo-isomeric 2,6-cyclolycopene-1,5-diols A and B (**168.1**), which are only present in tomatoes in extremely low concentrations, have been isolated and identified in human skin [39].

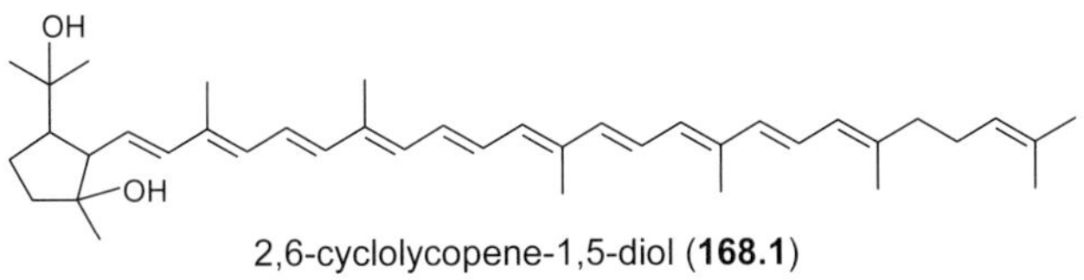

2,6-cyclolycopene-1,5-diol (**168.1**)

Recently, a novel approach for non-invasive, laser optical detection of carotenoid levels in skin by Raman spectroscopy has been established [40]. The Raman scattering method monitors the presence of carotenoids in human skin and is highly reproducible. Evaluation of five anatomical regions demonstrated significant differences in carotenoid concentration by body region, with the highest carotenoid concentration noted in the palm of the hand.

Comparison of carotenoid concentrations in basal cell carcinomas, actinic keratosis, and their peri-lesional skin demonstrate a significantly lower carotenoid concentration than in region-matched skin of healthy subjects. Furthermore, the method reveals that carotenoids are a good indicator of antioxidant status. People with high oxidative stress, *e.g.* smokers, and subjects with high exposure to sunlight, in general, have reduced skin carotenoid levels, independent of their dietary carotenoid consumption. Portable versions of the Raman spectroscopy instruments are now available and could have a broad application in dermatology and cosmetics.

The levels of β-carotene in serum decreased in unsupplemented but not in supplemented individuals on chronic UV exposure [32,41]. Depletion of skin carotenes and retinol after UV irradiation, and restoration by carotene supplementation were also observed in hairless mouse models [35]. When skin is subjected to UV light stress, more lycopene is destroyed than β-carotene, suggesting a role of lycopene as first defence line towards oxidative damage in tissues [26].

In conclusion, carotenoids from dietary intake accumulate in skin, thus allowing them to exert their photoprotective function at the target site. The levels correlate with bioavailability of the supplement, UV-exposure, and genetic factors such as skin type.

2. Carotenoids in skin cell models

a) Culture conditions

A precondition for carotenoid efficacy in photoprotection is that the carotenoids are taken up by the cells. Since the amount of carotenoid accumulated depends on many factors, such as cell line [42], carotenoid concentration in the cell culture media, vehicle used to bring the carotenoid into solution, treatment period *etc.*, it is essential to analyse the uptake and metabolism in cultures, before drawing conclusions on the efficacy of the carotenoid.

The major difficulty concerns the choice of the vehicle to be used to solubilize the highly lipophilic carotenoids without adverse effects on the cells. The vehicle should also prevent oxidative degradation of the carotenoids without affecting the UV response of the cells. Among the vehicles that have been used are organic solvents such as tetrahydrofuran (THF) [43], cyclodextrins [44], liposomes [45] or adsorption on nanoparticles [46]. When carotenoids were supplied in the latter three vehicles, deleterious pro-oxidative rather than protective effects were observed, in particular in the absence of stabilizing antioxidants such as vitamin E. Caution has to be exercised in interpreting such negative results because, for example, cyclodextrins were shown to deplete cholesterol from cells and alter the UV-response [47,48]. In liposomes, carotenoids are soluble only to a limited degree, leading to lower loading of the cells. In addition, enhanced pro-oxidative reactions can occur due to peroxidation of the liposomal lipids.

Use of THF as a vehicle leads to reliable results. It requires, however, removal of peroxides from the solvent by column chromatography on alumina. Carotenoid stock solution should always be prepared fresh before each experiment, to avoid oxidative degradation. Even then, carotenoids degrade rapidly in medium under cell culture conditions, *i.e.* within 24 hours. Thus, the medium must be changed daily to avoid accumulation of degradation products in the cells [43]. The concentration of the carotenoid in media and cells should always be monitored carefully by HPLC.

b) Uptake and metabolism of carotenoids in skin cells

It has been demonstrated that HaCaT keratinocytes take up β-carotene in a time-dependent and dose-dependent manner (Table 2). The HaCaT cells had to be supplemented for at least two days to achieve significant β-carotene accumulation. The cells continued to take up β-carotene thereafter, and maximum β-carotene levels were found after three days of supplementation. If no fresh β-carotene was added, the β-carotene content decreased, demonstrating that a daily supply of fresh β-carotene is crucially required to maintain the cellular β-carotene content.

Table 2. β-Carotene uptake and metabolism in HaCaT skin keratinocytes.
HaCaT cells were treated with 0.5, 1.5 or 3 μM β-carotene for 2 days. Cellular contents of β-carotene and β-carotene metabolites were determined by HPLC. <LOD: below limit of detection. Retinol and retinyl palmitate concentrations were below the limit of detection in all cases.

β-Carotene supplementation (μM)	**(all-*E*)-β-Carotene (pmol/10^6 cells)**	**(*Z*)-β-Carotene (pmol/10^6 cells)**	**Apocarotenals (pmol/10^6 cells)**
Placebo	<LOD	<LOD	<LOD
0.5	9.70 ± 0.09	0.20 ± 0.07	1.18 ± 0.04
1.5	34.30 ± 0.05	0.41 ± 0.02	3.21 ± 0.19
3.0	63.90 ± 0.22	0.82 ± 0.16	5.04 ± 0.11

As a provitamin A, β-carotene may act *via* retinoid pathways through local metabolism to retinol or apocarotenals and further to retinoic acid. Human skin fibroblasts *in vitro* increased their intracellular retinol after β-carotene supplementation [49]. HaCaT keratinocytes expressed β-carotene 15,15′-monooxygenase at low levels, but the retinol content in HaCaT cells was below the HPLC detection limit. Also, only marginal amounts of retinoic acid (RA) were formed from β-carotene, as detected indirectly by the induction of the RA target gene RARβ (Fig. 3, right). In contrast, expression of the β-carotene 9,10-oxygenase was much higher, and apocarotenals were detected in cells [43].

In keratinocytes (Fig. 3, left) [43] and similarly in skin fibroblasts [48], UVA irradiation destroyed β-carotene so that only 13% remained of the content before irradiation. Consistent with this finding, the retinoic acid response element (RARE)-dependent gene activation by β-carotene was reduced if the cells were irradiated with UVA (Fig. 3, right) [43].

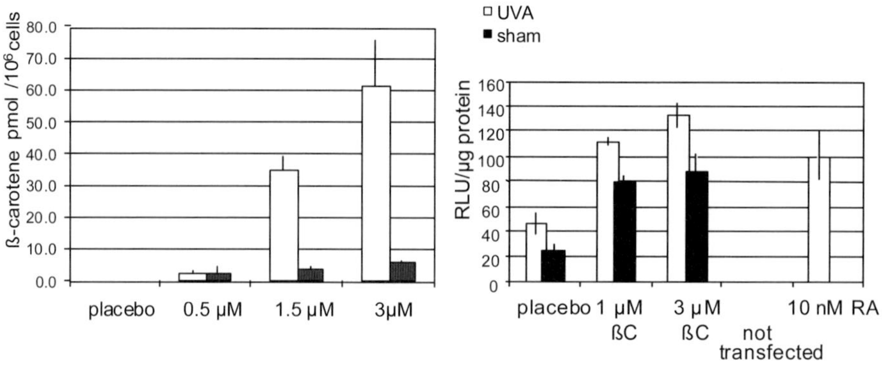

Fig. 3. Left: Dose-dependent uptake of β-carotene in HaCaT skin keratinocytes and depletion of cellular β-carotene stores by UVA irradiation. HaCaT cells were supplemented with 0.5, 1.5 or 3 μM β-carotene for 2 days prior to UVA irradiation (270 kJ/m^2). Cellular β-carotene content was analysed by HPLC. Right: Effect of β-carotene on transactivation of a retinoic acid-dependent reporter construct: HaCaT cells were transiently transfected with a reporter gene construct containing 5 direct repeats of the wild type. Luciferase activity was determined after 40 h treatment with β-carotene. RLU: random luminescence units, RA: retinoic acid.

UVA caused down-regulation of all retinoid receptors about 2-fold, except for RARα, which was not influenced by UVA. Apparently, regulation of RARα and RARγ expression, as well as regulation of RXRα and RXRγ, has a $^1O_2$-dependent component, as UVA irradiation in the presence of D_2O, which is known to extend the lifetime of $^1O_2$, had a significant effect on these transcripts. β-Carotene had no significant effect on the basal or UVA-regulated expression levels of the RARs and RXRs. Of note, β-carotene non-significantly induced RARβ in a dose-dependent manner, an effect observed predominantly in unirradiated cells. It shows that weak retinoid activity is generated from β-carotene in HaCaT cells; this may be attributed to the products of excentric cleavage of β-carotene, apocarotenals, which are present at detectable concentrations in HaCaT cells.

These findings are in agreement with observations *in vivo*, which also show that UVA exposure depleted epidermal vitamin A stores [35,50]. It has been reported [51] that retinoid content and RXRα expression were reduced in UV-irradiated SKH-1 hairless mice, and β-carotene 15,15′-monooxygenase activity was induced in response to this UV-induced depletion.

In conclusion, the observation of a depletion of vitamin A and provitamin A stores by UV light calls for an awareness of an increased requirement for vitamin A and carotenoid in situations of extensive sun exposure, in view of the role of vitamin A in maintaining skin integrity.

C. Photoprotection *in vivo*

1. Photosensitivity disorders

Elucidation of the function of carotenoids in singlet oxygen quenching in photosynthetic plants, algae and bacteria has led to the assumption that similar protection might be relevant in human skin, where UV light in the presence of endogenous photosensitizers can also induce formation of excited triplet species. The accumulation of large amounts of protoporphyrin, an endogenous photosensitizer, in the blood and skin of patients with inherited erythropoietic protoporphyria (EPP) gives rise to symptoms of itching and burning of the skin when patients are exposed to sun light.

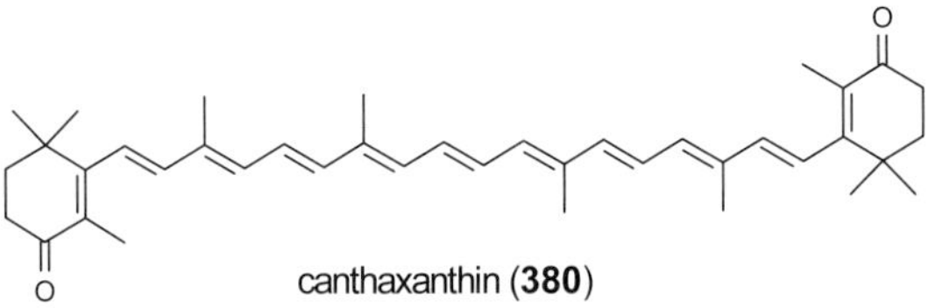

canthaxanthin (**380**)

In particular, β-carotene and canthaxanthin (**380**) have been shown to be beneficial in alleviating the symptoms of erythropoietic protoporphyria and other conditions such as polymorphous light eruptions [52-56]. These findings are mainly based on uncontrolled human studies performed in the 1960s and 1970s, usually with low case numbers. About 84% of the patients responded to successively increasing doses of oral β-carotene (formulated as beadlets) of up to 180-300 mg/day by showing increased tolerance to sunlight exposure. The US Food and Drug Administration approved the use of β-carotene for the treatment of EPP in 1975. This high dose β-carotene treatment did not lead to any adverse side effects other than a discolouration of the skin.

In conclusion, some patients react with improvement of skin symptoms in erythropoietic protoporphyria after oral supplementation with β-carotene, but extremely high doses over several months or years, leading to plasma levels of about 8 μmol/L, are required to achieve an effect.

2. Photocarcinogenesis

The encouraging results with β-carotene on erythropoietic protoporphyria led to further speculation that β-carotene might also have a protective role against photocarcinogenesis. Although several studies in rodent models initially showed promising results with high doses of β-carotene, these effects could not be reproduced later, and even an exacerbation of UVB-induced skin carcinogenesis was observed [57-59]. These contradictory findings remain unexplained; an influence of the specific diet was discussed, however. In humans, subsequent large randomized skin cancer prevention trials did not find a risk reduction in non-melanoma skin cancer by β-carotene (50 mg β-carotene/day for 5 years [60]; 20 mg β-carotene/day for 4.5 years [61]; 50 mg β-carotene every other day for 12 years [62]). A possible explanation of these results could be that supplementation would be necessary to increase carotene content during earlier phases of life, before the initial pathogenic events. Yet, from a mechanistic point of view, it seems rather unlikely that β-carotene is able to interact with the direct mutagenic and carcinogenic actions of UVB. This process involves absorption of short wavelength radiation by DNA, formation of the major types of DNA damage photoproducts, *i.e.* cyclobutane pyrimidine dimers and pyrimidine-6-4-pyrimidone photoproducts which are formed between adjacent pyrimidine nucleotides on the same strand of DNA. The resulting DNA mutations consequently lead to activation of oncogenes or inactivation of tumour suppressor genes.

Observational studies do not support a role of dietary carotenoids in non-melanoma skin cancer risk reduction [63-67]. Results from a prospective nested case control study embedded in the Nambour Skin Cancer Trial in Australia [67] suggested a positive association of basal cell carcinoma development with intake of lutein, but not of other carotenoids, selenium or vitamin E. In another recent observational study within the Isotretinoin-Basal Cell Carcinoma Prevention Trial [68], serum lutein, zeaxanthin and β-cryptoxanthin were positively related to risk of squamous cell carcinomas; risk ratios for subjects in the highest *versus* lowest tertiles were for lutein 1.63 [95% confidence interval (95% CI) 0.88-3.01; P for trend = 0.01], for zeaxanthin 2.40 (95% CI 1.30-4.42; P for trend = 0.01), and for β-cryptoxanthin 2.15 (95% CI 1.21-3.83; P for trend = 0.09), respectively. These observations would imply a detrimental effect of higher carotenoid intake rather than a protective effect.

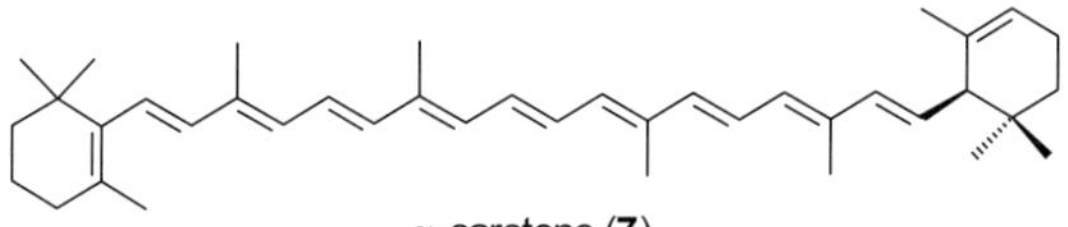

α-carotene (**7**)

A case control study for assessment of melanoma risk found that individuals in high *versus* low quintiles of energy-adjusted vitamin D, α-carotene (**7**), β-carotene, β-cryptoxanthin, lutein, and lycopene had significantly reduced risk for melanoma [Odds Ratios (a measure of the degree of association, *e.g.* the odds of exposure among the cases compared with the odds of exposure among the controls) ≤ 0.67], which remained significant after adjustment for the presence of dysplastic nevi, education, and skin response to repeated sun exposure. Larger prospective population studies would be required to substantiate such a protective effect.

Together, these studies provide only little or no evidence for a role of β-carotene and other dietary carotenoids in prevention of melanoma and non-melanoma skin cancer in humans.

3. Sunburn

Sunburn is the inflammatory reaction of the skin in response to excessive exposure to natural or artificial solar light of UVB wavelength. It is characterized by reddening of the skin, and, depending on the severity, by blister formation and ablation of the epidermis. On a histological level, sunburn cells, *i.e.* keratinocytes undergoing programmed cell death, form within hours after exposure. The minimal dose of UVB required to produce an erythema (MED) is dependent on the skin type. The MED is assessed by chromametry and used routinely to determine the sun protection factor (SPF) of sun screens.

Human dietary intervention studies of the effect of carotenoids on sun erythema formation have recently been reviewed comprehensively [23,69]. The effect of the carotenoid on the endpoint minimal erythema dose was investigated at various doses ranging from 24 to 180 mg per day β-carotene, or mixed carotenoid/micronutrient combinations, or carotenoids supplied as vegetable juices. Supplements were administered for between 3 days and 24 weeks. In eight of the ten studies reviewed, the MED was increased or sun erythema was less pronounced, indicating a protective effect. Two studies, where supplementation was very short, *i.e.* 3 days [70] to 4 weeks [71] showed no protective effect. Recently, another study with 15 mg β-carotene over 8 weeks also showed no effect on MED, but there was also no increase in skin β-carotene levels after the supplementation [72].

The evidence for a protective effect of β-carotene against sunburn was confirmed in a recent meta-analysis of the literature up to June 2007 on human supplementation studies and dietary protection against sunburn by β-carotene [73] (Fig. 4).

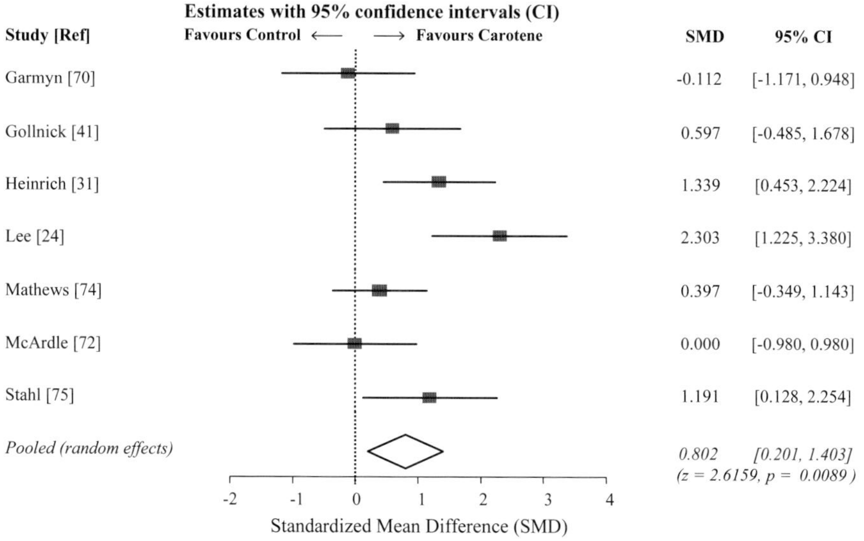

Fig. 4. Results of seven studies of effects of β-carotene *versus* placebo on protection against sunburn evaluated in a meta-analysis. (From [73] with permission).

Seven studies which evaluated the effectiveness of β-carotene in protection against sunburn were identified in Pubmed, ISI Web of Science and EBM Cochrane library [73]. Data were abstracted from these studies by means of a standardized data collection protocol. Although two of the studies considered showed no protective effect of β-carotene, the other five all showed varying levels of protection, with Standardized Mean Difference (SMD) ranging from 0.397 (95% CI -0.349, 1.143) to 2.303 (95% CI 1.225, 3.380). When the results were pooled, this gave an overall SMD of 0.802 (95% CI 0.201, 1.403, $p = 0.0089$). The meta-analysis showed that (i) β-carotene supplementation protects against sunburn and that (ii) the study duration had a significant influence on the size of the effect. Regression plot analysis revealed that protection required a minimum of 10 weeks of supplementation with a mean increase in the protective effect of 0.5 standard deviations with every additional month of supplementation. Thus, dietary supplementation of humans with β-carotene provides protection against sunburn in a time-dependent manner. These studies taken together show that erythema reduction is the photoprotection parameter which is most consistently affected by carotenoids. The effect seems not to be specific for a particular carotenoid, since a mixture of 6 mg each of lutein, lycopene and β-carotene was as effective as 24 mg β-carotene alone. Similarly, a mixture of antioxidants consisting of lycopene, β-carotene (6 mg/d each), vitamin E (10 mg/d) and selenium (75 μg/d) for 7 weeks increased erythema threshold significantly [76].

It should be noted that erythema reduction by carotenoid is mild and correlates with a Sun Protection Factor (SPF) of 2, putting into question the clinical relevance. In no case should oral supplementation with carotenoids replace the use of UV filters. On the other hand, orally supplemented β-carotene was shown to enhance the effectiveness of topical sun lotions [41]. Overall, dietary carotenoids may find their use and are important as part of a basic skin protection, in particular upon occasional sun exposure, when a UV filter is not applied.

4. Photoaging

No large human intervention studies have yet been conducted to address the effects of carotenoids on clinical parameters of premature photoaging, such as wrinkling, pigmentation, teleangiectasia (a widening of the fine capillaries in skin), dryness and inelasticity. The Nambour Skin Cancer Trial in Australia [61] addressed photoaging only to a limited extent. The subjects, 556 adults aged 25-50 years, were randomized in a 2 x 2 factorial trial to a daily sunscreen with Sun Protection Factor (SPF)-15 *vs.* usual (occasional) sunscreen use, and β-carotene (30 mg daily) *vs.* placebo treatment over a period of 4.5 years. Participants were exposed to the natural sunlight during the course of the trial. Silicone impressions of skin texture of the back of the hand were evaluated before and after treatment. There was a significant interaction effect of sunscreen and β-carotene on photoaging. Relative to the placebo group, the adjusted odds ratio (the odds of the occurrence of an event or disease is compared between the unexposed and exposed groups) for photoaging was about two-thirds for those on sunscreen, about one-third for those on β-carotene but slightly increased for those on both treatments. This was taken to suggest independent roles for sunscreen and β-carotene in the prevention of photoaging of the skin in sun-exposed white populations [77]. The negative interaction observed for the combination of sunscreen with β-carotene remained unexplained. Although the study had some limitations with parameter assessments and statistical analyses, it could be considered to provide the first evidence of a preventive effect of β-carotene on clinical photoaging caused by sunlight, including UVA.

In the Seresis study on molecular markers for photoaging, the effect of an antioxidant mixture containing β-carotene and lycopene [vitamin E (10 mg), β-carotene (2.4 mg), standardized tomato extract (25 mg lycopene), selenium yeast (25 mg), and proanthocyanidins from grape seed extract (25 mg)] was addressed [30]. In a 2 x 2 factorial design, 48 volunteers who had received either the antioxidant medication or placebo for 10 weeks were exposed to low dose UVB for 2 weeks and MED measurements taken. Before and after irradiation, the proteins MMP-1 and MMP-9, two major metalloproteases which degrade various collagens and other interstitial matrix proteins and also cleave the cytokine IL1β from its propeptide, were analysed in skin biopsies. After 2 weeks of UVB exposure, MMP-1 was slightly increased in the placebo group ($p<0.03$) and decreased in the Seresis group ($p<0.044$). MMP-9 did not change significantly. The MED was increased in the Seresis-treated group, *i.e.* sunburn induction was reduced by the antioxidant mixture.

Recently, it was demonstrated [78] that long-term supplementation with antioxidant micronutrients was able to improve parameters related to skin structure. Thirteen volunteers per group received a daily supplement consisting of either (i) lycopene (3 mg), lutein (3 mg), β-carotene (4.8 mg), α-tocopherol (10 mg), and selenium (75 μg), or (ii) lycopene (6 mg), β-carotene (4.8 mg), α-tocopherol (10 mg), and selenium (75 μg), or (iii) placebo, for 12 weeks. Skin density and thickness were assessed by ultrasound measurement, and roughness, scaling, smoothness and wrinkling assessed by Surface Evaluation of Living Skin. Both supplement mixtures containing carotenoids, vitamin E and selenium increased skin density and thickness over the treatment period, and skin surface, including roughness and scaling, was significantly improved.

A recently published human study [79] demonstrated that oral supplementation with 10 mg/day lutein and 0.6 mg/day zeaxanthin in combination with a topical treatment (50 ppm lutein, 3 ppm zeaxanthin) for 12 weeks provided better photoprotection and skin hydration, and an increase in superficial skin lipids than did the individual treatments. The reduction in lipid peroxidation following oral supplementation alone was equal to that given by the combined treatment. Skin elasticity was improved significantly by the topical treatment, and to a lesser extent by the combined and oral treatments. These results also show that, in addition to a photoprotective action, carotenoids, in this case the xanthophylls lutein and zeaxanthin, are able to improve physiological cosmetic skin parameters.

Modern optical non-invasive methods were used to investigate the structures of furrows and wrinkles *in vivo* and to correlate them with the concentration of lycopene, analysed by resonance Raman spectroscopy, in the forehead skin of 20 volunteers aged between 40 and 50 years [80]. In a first step, no significant correlation was found between the age of the volunteers and their skin roughness. In a second step, a significant correlation was obtained between the skin roughness and the lycopene concentration ($R = 0.843$, $p<0.01$). The indication from these findings is that higher levels of lycopene in the skin effectively lead to lower levels of skin roughness.

The results of these studies provide the first evidence to support the hypothesis that antioxidant mixtures containing carotenoids can reduce UV-induced molecular markers of premature photoaging in humans and also improve skin structure parameters. Complementary mixtures of low dose micronutrients constituting a synergistic antioxidant network are as effective as moderate to high-dose supplements of a single carotenoid.

5. Photoimmune modulation

UV-radiation has been shown consistently to induce a number of immunological changes to the immune system. Continuous alleviation of photoimmune suppression by protective dietary micronutrients is warranted in vulnerable populations, *i.e.* children and the elderly [81]. Whilst there is evidence from preclinical and clinical studies about general immune modulatory effects of carotenoids [82-84] (*Chapter 17*), few studies in humans and animals have addressed the protection against photoimmune suppression. β-Carotene supplements (30 mg/day) given to healthy young volunteers for four weeks protected against suppression of delayed type hypersensitivity (DHT) induced by UVA [85]. In the Eilath study [41], the same dose of 30 mg/day β-carotene over 10 weeks prevented the UV-exposure-induced loss of Langerhans cell density in the epidermis.

A diet enriched with 0.4% lutein and 0.04% zeaxanthin for 2 weeks decreased significantly the UVB-induced inflammatory oedematous cutaneous response and the hyper-proliferative rebound in female hairless Skh-1 mice [86].

Mice fed dietary lutein demonstrated significant inhibition of ear swelling induced by UVB radiation compared to controls on a standard laboratory diet. Suppression of contact hypersensitivity response by a lower, repeated dose of UVB radiation was also significantly inhibited by feeding lutein. When UVB radiation was given at a single dose of 10,000 J/m^2 to inhibit the induction of contact hypersensitivity at a distant, non-irradiated site, no effect of lutein was seen. Lutein accumulated in the skin of the mice following diet supplementation and was also shown to decrease UVR-induced ROS generation [87].

D. Mechanistic Aspects of Photoprotection by Carotenoids

UV radiation induces reactive oxygen species (ROS) including singlet oxygen, $^1O_2$ [1], which can damage lipids, proteins and DNA. The UVA wavelengths between 320 and 400 nm are considered the part of the light spectrum most likely to cause this oxidation.

Singlet oxygen, induced mainly by UVA, can regulate the expression level of a variety of genes involved in the cell cycle or apoptosis (see *Chapter 11*). Furthermore, genes involved in photoaging (such as MMPs, [88,89], haem oxygenase (HO)-1 [90], and intracellular adhesion molecule 1 [91]) have been reported to be regulated by UVA and/or $^1O_2$. Inhibition or moderation of these molecular events could confer photoprotection on target cells.

1. Inhibition of lipid peroxidation

Several studies have used cultured human or other skin fibroblasts to examine the protective effects of carotenoids on UV-induced lipid peroxidation. β-Carotene prevented UVA-induced membrane damage of human skin fibroblasts [92]. Lycopene, β-carotene and lutein, applied in liposomes as the vehicle, decreased UVB-induced formation of thiobarbituric acid-reactive substances (TBARS, see *Chapter 12*) at 1 hour to levels 40-50% of those of controls free of carotenoids [45]. The amounts of carotenoid needed for optimal protection were 0.05, 0.40 and 0.30 nmol/mg protein for lycopene, β-carotene and lutein, respectively. Further increases of carotenoid content in cells beyond the optimum levels led to pro-oxidant effects. In another study, the depletion of catalase and superoxide dismutase (SOD) by UVA was restored, and TBARS reduced by culturing rat kidney fibroblasts with β-carotene or lutein (1 μM each), or with astaxanthin (**404-406**), which was reported to give superior protective activity at concentrations as low as 10 nM [93]. Cultivation of human skin fibroblasts and melanocytes with pure astaxanthin or an astaxanthin-containing algal extract prevented UVA-induced oxidative DNA damage, and restored also UVA-induced alterations in SOD activity and glutathione content [94].

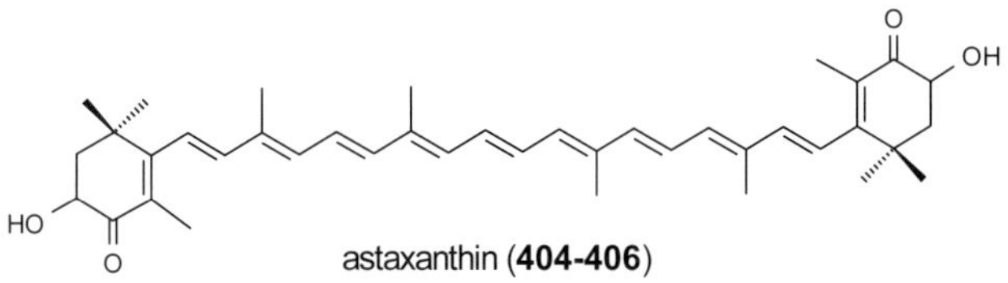

astaxanthin (**404-406**)

In humans, a mixture of antioxidants consisting of lycopene (6 mg), β-carotene (6 mg), vitamin E (10 mg) and selenium (75 μg) per day for 7 weeks reduced lipid peroxide levels, and also improved parameters of the epidermal defence system against UV-induced damage such as sunburn cell formation and pigmentation [76].

Studies in mouse models confirm the prevention by carotenoids of oxidative stress induction by UV irradiation in skin [35]. Baseline TBARS were lower than in controls in hairless mice receiving β-carotene or palm fruit carotenoids (α-carotene 30%, β-carotene 60%, other carotenoids including lycopene 10%) at 0.005% dispersed as emulsions in drinking water. The palm fruit carotenoids accumulated in skin to a higher degree than β-carotene alone. UV irradiation-induced TBARS were decreased by palm fruit carotenes, but not by β-carotene, which may be explained by the differences in the bioavailability of the supplemented products. β-Carotene reduced the degree of lipid peroxidation in UVA-irradiated skin homogenates *ex vivo* from Balb/c mice, which had been supplemented for three weeks with 50 mg β-carotene/100 g diet [95]. β-Carotene 5,8-endoperoxide (*2*), a marker for the $^1O_2$ reaction, increased in the homogenates.

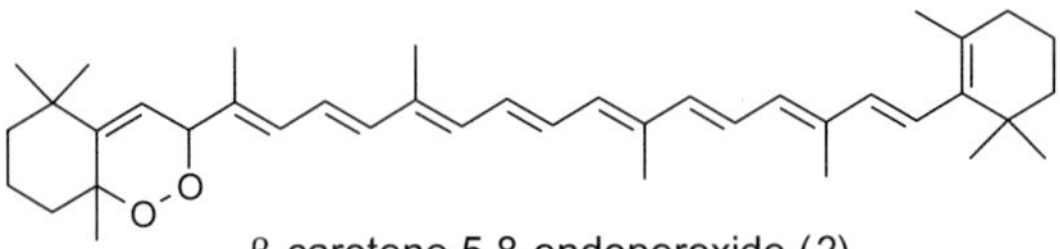

β-carotene 5,8-endoperoxide (*2*)

In healthy volunteers who had been supplemented with 15 mg β-carotene daily for 8 weeks, skin malondialdehyde concentrations after UVR (270-400 nm) were not reduced, whereas the effect of 400 mg vitamin E supplementation was significant [72]. No effects were observed on other indicators of oxidation. The lack of efficacy in this study may be explained by the low skin levels of β-carotene.

Overall, these studies *in vitro* and *in vivo* show that carotenoids can exert their protective antioxidant function when present at sufficiently high concentration in the skin cells.

2. Inhibition of UVA-induced expression of haem oxygenase 1

The human haem oxygenase 1 (HO-1) enzyme catalyses the first and rate-limiting step in haem degradation. The HO-1 gene is strongly activated within the first hours that follow UVA irradiation of normal human dermal fibroblasts and this response is being used as a marker of oxidative stress in cells. It has been shown that the induction of this gene occurs *via* $^1O_2$ produced on interaction of UVA radiation with an as yet undefined cellular chromophore. Carotenoids could be expected to suppress the UVA induced HO-1 gene activation in human cells. Unexpectedly, two studies with skin fibroblasts *in vitro* found an opposite effect. The first study applied β-carotene in cyclodextrins at levels of 0.5 and 5 μM [44]. A significant pro-oxidative effect and enhancement of UVA-induced HO-1 expression were observed. Combined application of β-carotene with vitamin E prevented the pro-oxidative effect, but did not exhibit a protective effect. In the second study, β-carotene or lycopene (0.5-1.0 μM) were prepared in nanoparticle formulations together with vitamin C and/or vitamin E. As in the study above, either β-carotene or lycopene led to a further 1.5-fold rise in the UVA-induced HO-1 mRNA levels [46].

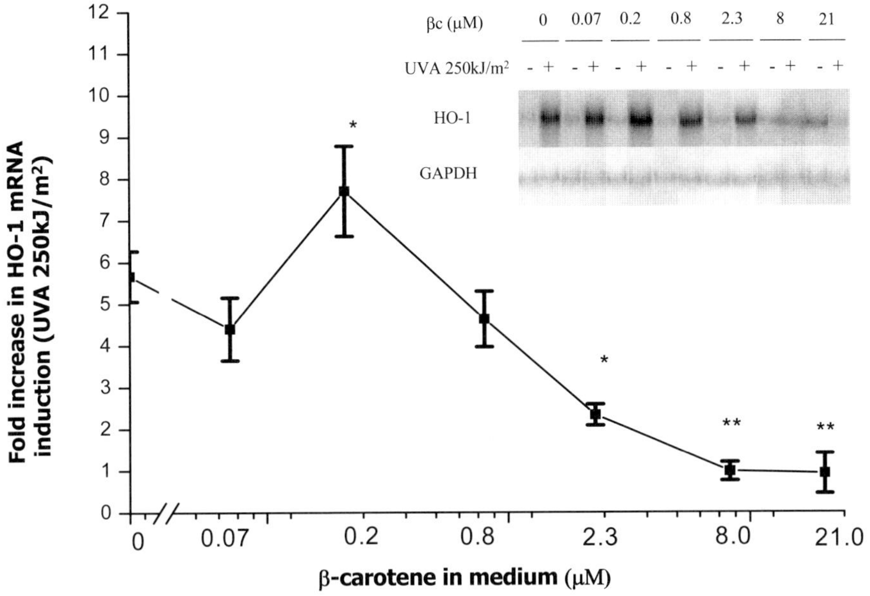

Fig. 5. Main graph: Modulation by β-carotene of UVA-induced haemoxygenase-1 (HO-1) mRNA accumulation [48]. Insert: The modulation, by 0.07, 0.2, 0.8, 2.3, 8.0 and 21 μM β-carotene (in THF), of UVA-induced HO-1 mRNA levels (UVA 250kJ/m^2) normalized over glyceraldehyde-3-phosphate dehydrogenase (GAPDH) mRNA in FEK4 skin fibroblasts, as measured by Northern Blot Analysis.

In another study (Fig. 5), the suppression of UVA-induced levels of HO-1 mRNA was measured after addition of a series of six β-carotene concentrations to the culture medium of FEK4 skin fibroblasts for three days, under the conditions described in Section **B**.2.a.

A concentration-dependent inhibition of UVA-induced transcriptional activation of HO-1 in exponentially growing FEK4 cells by β-carotene was observed, despite a UVA-induced increase of apocarotenals, indicators for oxidative degradation. Inhibition occurred at concentrations observed in human plasma after dietary supplementation with β-carotene.

These results also demonstrate, as mentioned earlier, the importance of culture conditions to avoid secondary influences *in vitro* that may cause altered responsiveness to UV and oxidative stress in cells.

3. Prevention of mitochondrial DNA deletions

Mutations of mitochondrial (mt) DNA have been reported to play a causative role in processes such as carcinogenesis, normal aging and premature photoaging of the skin [96-98].

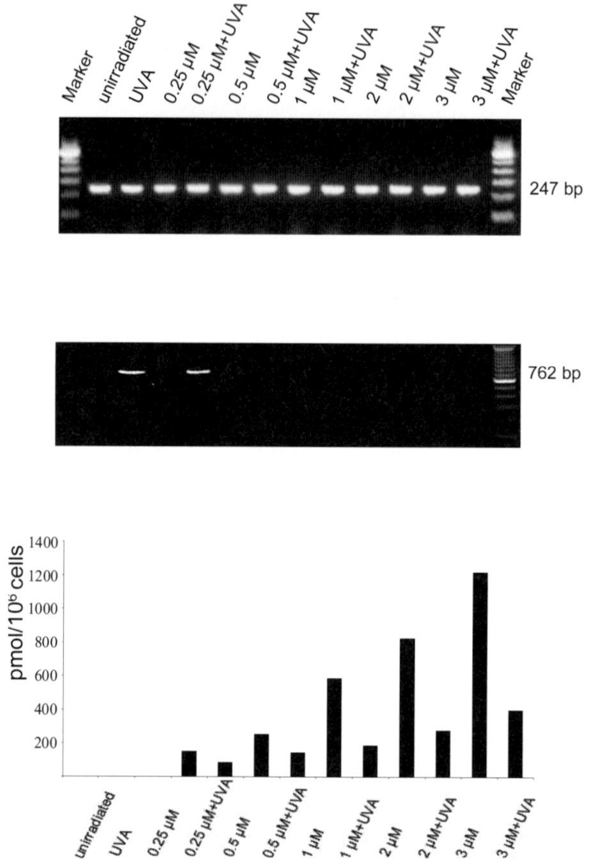

Fig. 6. Protective effect of β-carotene against photoaging-associated mtDNA deletion. Human dermal fibroblasts were repetitively exposed to UVA in the presence or absence of β-carotene at concentrations ranging from 0.25 to 3.0 μM, with HPLC assessment of β-carotene levels and PCR amplification of the common deletion and the reference fragment after each week of irradiation [96].

Top) Representative agarose gel of PCR amplifications of the reference fragment.

Middle) Representative agarose gel of PCR amplifications of the common deletion.

Bottom) Levels of β-carotene (pmol/10^6 cells).

Skin showing clinical signs of photoaging is characterized by an increase of mitochondrial mutations. The most frequent mutation of mtDNA is a 4977bp deletion, also called 'common deletion', which is considered to be a marker for alterations of the mt genome. Repetitive exposure of normal human fibroblasts to sublethal doses of UVA radiation leads to the induction of the common deletion and this is mediated in a singlet oxygen dependent fashion. The ability of β-carotene to protect normal human fibroblasts from the induction of photoaging-associated mtDNA deletions was investigated [99]. (all-*E*)-β-Carotene was tested at doses from 0.25 to 3.0 μM for uptake into cells as well as for its protective capacity. Assessment of cellular uptake of (all-*E*)-β-carotene, measured by HPLC, revealed a dose-dependent increase of intracellular concentration, as well as an increase in oxidative metabolites, *i.e.* apocarotenals and epoxides. UVA exposure led to a decrease of (all-*E*)-β-carotene, its *Z* isomers and oxidative metabolites. Assessment of mtDNA deletions by polymerase chain reaction (PCR) revealed reduced levels of mtDNA mutagenesis in cells incubated with β-carotene at concentrations of 0.5 μM and higher (Fig. 6). Taken together, these results indicate that β-carotene is taken up into the skin fibroblasts in a dose-dependent manner, interacts with UVA radiation in the cell and shows protective properties against the induction of a photoaging-associated mtDNA mutation.

4. Metalloprotease inhibition

Matrix metalloproteases (MMPs) are among the most important photoaging-associated genes induced by $^1O_2$. Investigation of the effect of carotenoids on suppression of UVA-induced MMPs is therefore of major relevance for establishing the protective effects of the carotenoids against photoaging.

In a detailed investigation, HaCaT keratinocytes were precultured with β-carotene at physiological concentrations (0.5, 1.5 and 3.0 μM) prior to UVA exposure from a Hönle solar simulator (270 kJ/m^2) [43]. The lifespan of $^1O_2$ was enhanced by irradiation in the presence of deuterium dioxide (D_2O). Expression levels of target genes such as MMP-1 were determined by TaqMan® Quantitative Real Time RT-PCR (Fig. 7).

β-Carotene suppressed the UVA-induction of MMP-1, MMP-3, and MMP-10, three major MMPs involved in photoaging (Fig. 7). Not only MMP-1, but also MMP-10 regulation was demonstrated to involve $^1O_2$-dependent mechanisms. β-Carotene quenched $^1O_2$-mediated induction of MMP-1 and MMP-10 dose-dependently with an approximately 50% reduction compared to cells treated with vehicle alone without β-carotene.

In contrast to this, in another study [46] an enhancement effect of β-carotene and lycopene on MMP-1 induction by UVA in fibroblasts was observed. As discussed above for HO-1, it is likely that the mode of β-carotene application is responsible for the differences in effects.

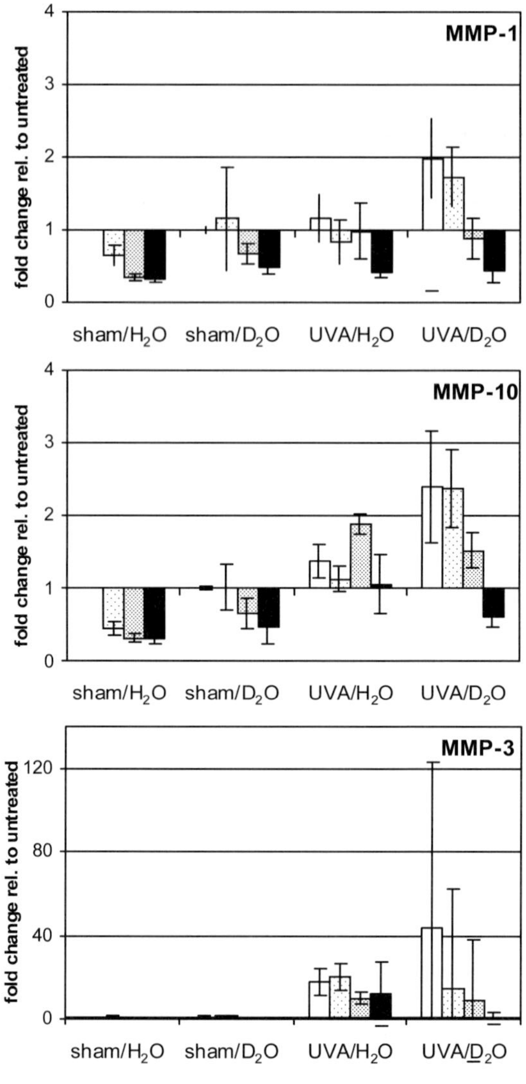

Fig. 7. Effect of β-carotene on ^2H_2O-enhanced UVA induction of (top) MMP-1, (middle) MMP-10, and (bottom) MMP-3. HaCaT cells were pretreated for 2 days with 0.5, 1.5 or 3.0 μM β-carotene. The cells were irradiated with UVA (270 kJ/m^2) in phosphate-buffered saline (PBS) made with ^2H_2O or H_2O, to analyse $^1O_2$ inducibility of genes. Gene expression 5 hours after UVA irradiation was analysd by TaqMan® Quantitative Real Time PCR. Values are geometric means ± standard error from three independent experiments [42].

5. Use of microarray analysis to profile gene expression

The introduction of modern molecular techniques and tools such as gene expression microarrays, proteomics and metabolomics (in nutrition research, termed 'nutrigenomics') created unique opportunities to identify the modes of action of nutritional compounds and study their influences on disease prevention beyond the commonly established functions. Microarrays allow genome-wide monitoring of gene expression in one step in small samples and their clustering to biological pathways. Some technical aspects of the methodology are summarized below.

RNA is extracted from treated cells or from control cells. Biotin-labelled probes are generated from these RNAs and incubated with microarrays. Microarrays carry oligonucleotides which recognize and bind probe molecules corresponding to a specific RNA ('riboprobes'). Riboprobes binding to their respective oligonucleotides are made visible by a fluorophore coupled to streptavidin. Nowadays, microarrays are available that can detect genes of selected pathways or which cover the entire (*ca.* 30 000 genes) of the genome. Gene activity is defined by the number of transcripts derived from a gene. In microarrays, this signal is converted to signal intensity of the fluorophore. Gene regulation by, for example, a carotenoid, is detected by comparing the signal intensity for a given RNA in treated *vs* control cells. Bioinformatics programs are used to analyse the vast amount of data, and translate the regulation of thousands of genes into biological meanings.

To analyse overall gene expression and identify specific processes influenced by β-carotene, Affymetrix® Gene Chip technology was applied in studies similar to those for MMPs (Section **D**.4) [100]. HaCaT cells were pre-cultured with β-carotene at physiological dose levels (0.5, 1.5. and 3.0 μmol/L) before exposure to UVA from a solar light lamp.

The results from Gene Chip hybridizations show that β-carotene altered UVA-induced changes in gene expression, in some cases reducing, in others enhancing the specific UVA effect. Downregulation of growth factor signalling, moderate induction of pro-inflammatory genes, upregulation of immediate early genes including apoptotic regulators, and suppression of cell cycle genes were hallmarks of the UVA effect. Of the 568 genes that were regulated by UVA, β-carotene reduced the UVA effect for 143, enhanced it for 180, and did not alter the UVA effect for 245 genes.

In unirradiated keratinocytes, gene regulations suggested that β-carotene reduced stress signals and extracellular matrix (ECM) degradation, and promoted keratinocyte differentiation. In UVA-irradiated cells, β-carotene inhibited those gene regulations by UVA that promote ECM degradation, suggesting a photoprotective effect of β-carotene. β-Carotene enhanced UVA-induced expression of tanning-associated protease-activated receptor 2, suggesting that β-carotene enhances tanning after UVA exposure. The combination of β-carotene-induced differentiation with the cellular 'UV response' led to a synergistic induction of cell cycle arrest and apoptosis by UVA and β-carotene. The different interaction modes imply that β-carotene/UVA interactions involve multiple mechanisms.

The 'transcriptomics' results, *i.e.* the expression profiles of retinoic acid target genes, confirmed the finding (Section **B**.2.b) that the retinoid-mediated effect of β-carotene in this cell system was minor, indicating that the β-carotene effects reported here were predominantly mediated through vitamin A-independent pathways.

A model of the interactions of β-carotene and UVA is shown in Fig. 8. It is proposed that β-carotene reduced the UVA-induction of genes involved in ECM degradation and inflammation by acting as a $^1O_2$ quencher. The mild photoprotective effect of β-carotene is suggested to be based on inhibition of these $^1O_2$-induced gene regulations, rather than on a physical filter effect, since the absorption maximum of β-carotene, *e.g.* 460 nm, lies outside the UVB/UVA range. β-Carotene, if scavenging ROS other than $^1O_2$, is irreversibly damaged and converted into radicals, if not rescued by other antioxidants. Thus, β-carotene did not inhibit UVA-induced stress signals, and enhanced some. UVA exposure suppressed several retinoic acid target genes. Since HaCaT cells produce marginal amounts of retinoid activity from β-carotene, the provitamin A activity of β-carotene did not translate into restored expression of RA target genes in this system.

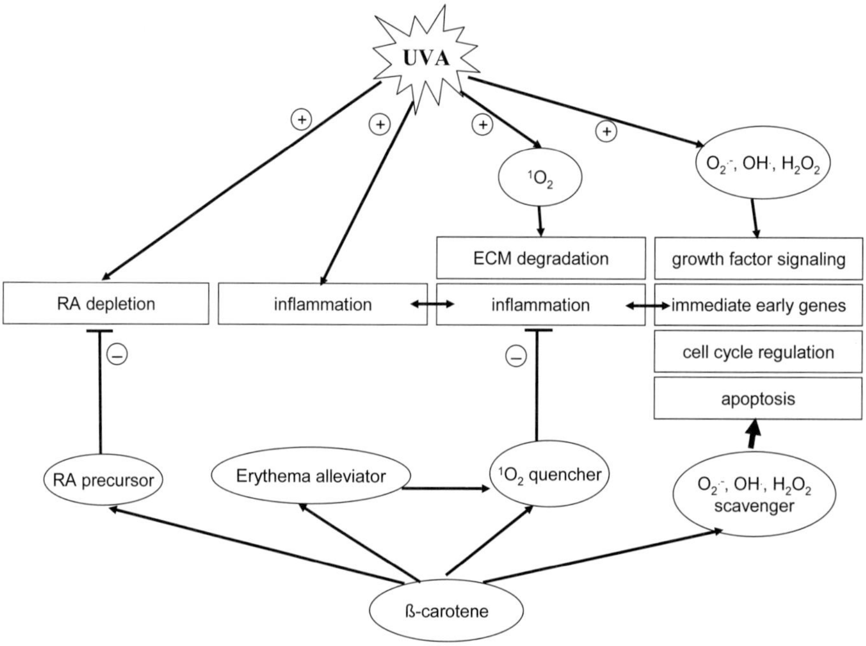

Fig. 8. Proposed relationship of the modes of action of β-carotene to its influence on UVA-induced biological processes. + indicates upregulation, – downregulation of processes.

E. Summary and Conclusion

Besides alleviation of symptoms in photosensitivity disorders by β-carotene, data obtained from human trials with carotenoids consistently show a moderate reduction in the development of sun-induced erythema. Some human studies also point to a possible beneficial effect of single carotenoids or of antioxidant compositions containing carotenoids in reducing the effects of premature skin aging. Chronic supplementation for more than 10 weeks is required to achieve these effects. The required doses, mainly established for β-carotene, lycopene and lutein, are between 10 and 20 mg/day, but can be lowered below 10 mg/day when the carotenoid is applied as part of an antioxidant composition containing mixed carotenoids and/or vitamins E, C and selenium. That the function of carotenoids in skin is strongly linked to their $^1O_2$ quenching properties is supported by studies *in vivo* and *in vitro*. Dietary intake of carotenoids can prevent the UV-induced losses in antioxidant defence systems and stores of skin retinol. Recent research elucidating the molecular modes of action shows that β-carotene can reduce up-regulation of UVA-induced pathways that are strongly involved in photoaging processes.

In conclusion, a considerable body of evidence, mostly from experiments with β-carotene, has emerged over the past 30-40 years on the benefits of carotenoids in photoprotection of human skin. Therefore, nutritional manipulation of carotenoid levels in skin, in conjunction with other antioxidants, has its importance as part of a concept of basic lifetime photoprotection to complement topical sun protection.

References

[1] R. M. Tyrrell, *Biochem. Soc. Symp.*, **61**, 47 (1995).

[2] R. M. Tyrrell, *Meth. Enzymol.*, **319**, 290 (2000).

[3] H. Sies and W. Stahl, *Am. J. Clin. Nutr.*, **62**, 1315S (1995).

[4] R. Edge, D. J. McGarvey and T. G. Truscott, *J. Photochem. Photobiol. B*, **41**, 189 (1997).

[5] R. Edge and T. G. Truscott, *Nutrition*, **13**, 992 (1997).

[6] H. D. Martin, C. Jaeger, C. Ruck, M. Schmidt, R. Walsh and J. Paust, *J. Pract. Chem.*, **341**, 302 (1999).

[7] J. Gray and J. L. M. Hawk, *The Benefits of Lifetime Photoprotection*, Royal Society of Medicine Services, London (1997).

[8] J. Krutmann and A. Morita, in *Hautalterung* (ed. J. Krutmann and T. Diepgen), p. 46, Springer, Berlin (2003).

[9] L. Rittie and G. J. Fisher, *Ageing Res. Rev.*, **1**, 705 (2002).

[10] M. Berneburg, in *Hautalterung* (ed. J. Krutmann and T. Diepgen), p. 14, Springer, Berlin, (2003).

[11] B. A. Gilchrest, *Br. J. Dermatol.*, **135**, 867 (1996).

[12] R. M. Sayre and H. S. Black, *J. Photochem. Photobiol. B*, **12**, 83 (1992).

[13] G. M. Lowe, R. F. Bilton, I. G. Davies, T. C. Ford, D. Billington and A. J. Young, *Ann. Clin. Biochem.*, **36**, 323 (1999).

[14] T. Wingerath, W. Stahl and H. Sies, *Arch. Biochem. Biophys.*, **324**, 385 (1995).

[15] S. Lin, L. Quaroni, W. S. White, T. Cotton and G. Chumanov, *Biopolymers*, **57**, 249 (2000).

[16] K. H. van het Hof, B. C. de Boer, L. B. Tijburg, B. R. Lucius, I. Zijp, C. E. West, J. G. Hautvast and J. A. Weststrate, *J. Nutr.*, **130**, 1189 (2000).
[17] A. van Bennekum, M. Werder, S. T. Thuahnai, C. H. Han, P. Duong, D. L. Williams, P. Wettstein, G. Schulthess, M. C. Phillips and H. Hauser, *Biochemistry*, **44**, 4517 (2005).
[18] E. Reboul, L. Abou, C. Mikail, O. Ghiringhelli, M. Andre, H. Portugal, D. Jourdheuil-Rahmani, M. J. Amiot, D. Lairon and P. Borel, *Biochem. J.*, **387**, 455 (2005).
[19] A. During, H. D. Dawson and E. H. Harrison, *J. Nutr.*, **135**, 2305 (2005).
[20] H. Tsuruoka, W. Khovidhunkit, B. E. Brown, J. W. Fluhr, P. M. Elias and K. R. Feingold, *J. Biol. Chem.*, **277**, 2916 (2002).
[21] Y. M. Peng, Y. S. Peng, Y. Lin, T. Moon and M. Baier, *Cancer Epidemiol. Biomarkers Prev.*, **2**, 145 (1993).
[22] W. Stahl, U. Heinrich, H. Jungmann, H. Tronnier and H. Sies, *Meth. Enzymol.*, **319**, 494 (2000).
[23] H. Sies and W. Stahl, *Annu. Rev. Nutr.*, **24**, 173 (2004).
[24] R. Lee, M. M. Mathews-Roth, M. A. Pathak and J. A. Parrish, *J. Invest. Dermatol.*, **64**, 175 (1975).
[25] A. Vahlquist, E. Stenstrom and H. Torma, *Ups. J. Med. Sci.*, **92**, 253 (1987).
[26] J. D. Ribaya-Mercado, M. Garmyn, B. A. Gilchrest and R. M. Russell, *J. Nutr.*, **125**, 1854 (1995).
[27] H. K. Biesalski, C. Hemmes, W. Hopfenmuller, C. Schmid and H. P. Gollnick, *Free Radic. Res.*, **24**, 215 (1996).
[28] W. Stahl, U. Heinrich, H. Jungmann, J. von Laar, M. Schietzel, H. Sies and H. Tronnier, *J. Nutr.*, **128**, 903 (1998).
[29] W. Stahl, U. Heinrich, S. Wiseman, O. Eichler, H. Sies and H. Tronnier, *J. Nutr.*, **131**, 1449 (2001).
[30] A. K. Greul, J. U. Grundmann, F. Heinrich, I. Pfitzner, J. Bernhardt, A. Ambach, H. K. Biesalski and H. Gollnick, *Skin Pharmacol. Appl. Skin Physiol.*, **15**, 307 (2002).
[31] U. Heinrich, C. Gartner, M. Wiebusch, O. Eichler, H. Sies, H. Tronnier and W. Stahl, *J. Nutr.*, **133**, 98 (2003).
[32] H. P. Gollnick and C. Siebenwirth, *Skin Pharmacol. Appl. Skin Physiol.*, **15**, 360 (2002).
[33] M. Mathews-Roth, D. Hummel and C. Crean, *Nutr. Rep. Int.*, **16**, 419 (1977).
[34] W. Wamer, A. Giles Jr. and A. Kornhauser, *Nutr. Rep. Int.*, **32**, 295 (1977).
[35] K. Someya, Y. Totsuka, M. Murakoshi, H. Kitano and T. Miyazawa, *J. Nutr. Sci. Vitaminol.*, **40**, 303 (1994).
[36] T. Wingerath, H. Sies and W. Stahl, *Arch. Biochem. Biophys.*, **355**, 271 (1998).
[37] W. Stahl, U. Heinrich, O. Aust, H. Tronnier and H. Sies, *Photochem. Photobiol. Sci.*, **5**, 238 (2006).
[38] M. La Placa, M. Pazzaglia and A. Tosti, *J. Eur. Acad. Dermatol. Venereol.*, **14**, 311 (2000).
[39] F. Khachik, L. Carvalho, P. S. Bernstein, G. J. Muir, D. Y. Zhao and N. B. Katz, *Exp. Biol. Med.*, **227**, 845 (2002).
[40] I. V. Ermakov, M. R. Ermakova, W. Gellermann and J. Lademann, *J. Biomed. Opt.*, **9**, 332 (2004).
[41] H. Gollnick, W. Hopfenmuller, C. Hemmes, S. Chun, C. Schmid, K. Sundermeier and H. K. Biesalski, *Eur. J. Dermatol.*, **6**, 200 (1996).
[42] N. L. Franssen-van Hal, J. E. Bunschoten, D. P. Venema, P. C. Hollman, G. Riss and J. Keijer, *Arch. Biochem. Biophys.*, **439**, 32 (2005).
[43] K. Wertz, N. Seifert, P. B. Hunziker, G. Riss, A. Wyss, C. Lankin and R. Goralczyk, *Free Radic. Biol. Med.*, **37**, 654 (2004).
[44] U. C. Obermüller-Jevic, P. I. Francz, J. Frank, A. Flaccus and H. K. Biesalski, *FEBS Lett.*, **460**, 212 (1999).
[45] O. Eichler, H. Sies and W. Stahl, *Photochem. Photobiol.*, **75**, 503 (2002).
[46] E. A. Offord, J. C. Gautier, O. Avanti, C. Scaletta, F. Runge, K. Kramer and L. A. Applegate, *Free Radic. Biol. Med.*, **32**, 1293 (2002).
[47] R. Gniadecki, N. Christoffersen and H. C. Wulf, *J. Invest. Dermatol.*, **118**, 582 (2002).

[48] M. C. Trekli, G. Riss, R. Goralczyk and R. M. Tyrrell, *Free Radic. Biol. Med.*, **34**, 456 (2003).

[49] R. R. Wei, W. G. Wamer, L. A. Lambert and A. Kornhauser, *Nutr. Cancer*, **30**, 53 (1998).

[50] O. Sorg, C. Tran, P. Carraux, L. Didierjean, F. Falson and J. H. Saurat, *J. Invest. Dermatol.*, **118**, 513 (2002).

[51] A. Takeda, T. Morinobu, K. Takitani, M. Kimura and H. Tamai, *J. Nutr. Sci. Vitaminol.*, **49**, 69 (2003).

[52] M. M. Mathews-Roth, *Ann. N Y Acad. Sci.*, **691**, 127 (1993).

[53] M. M. Mathews-Roth, *Br. J. Dermatol.*, **134**, 977 (1996).

[54] M. M. Mathews-Roth, *Semin. Liver. Dis.*, **18**, 425 (1998).

[55] M. M. Mathews-Roth, *Clin. Dermatol.*, **16**, 295 (1998).

[56] M. M. Mathews-Roth, *Meth. Enzymol.*, **319**, 479 (2000).

[57] H. S. Black, *Nutr. Cancer*, **31**, 212 (1998).

[58] H. S. Black, G. Okotie-Eboh and J. Gerguis, *Nutr. Cancer*, **37**, 173 (2000).

[59] M. M. Mathews-Roth and N. I. Krinsky, *Photochem. Photobiol.*, **46**, 507 (1987).

[60] E. R. Greenberg, J. A. Baron, M. M. Stevens, T. A. Stukel, J. S. Mandel, S. K. Spencer, P. M. Elias, N. Lowe, D. N. Nierenberg, G. Bayrd, J. C. Vance and Skin Cancer Prevention Study Group, *Control Clin. Trials*, **10**, 153 (1989).

[61] A. Green, G. Williams, R. Neale, V. Hart, D. Leslie, P. Parsons, G. C. Marks, P. Gaffney, D. Battistutta, C. Frost, C. Lang and A. Russell, *Lancet*, **354**, 723 (1999).

[62] U. M. Frieling, D. A. Schaumberg, T. S. Kupper, J. Muntwyler and C. H. Hennekens, *Arch. Dermatol.*, **136**, 179 (2000).

[63] R. A. Breslow, A. J. Alberg, K. J. Helzlsouer, T. L. Bush, E. P. Norkus, J. S. Morris, V. E. Spate and G. W. Comstock, *Cancer Epidemiol. Biomarkers Prev.*, **4**, 837 (1995).

[64] M. R. Karagas, E. R. Greenberg, D. Nierenberg, T. A. Stukel, J. S. Morris, M. M. Stevens and J. A. Baron, *Cancer Epidemiol. Biomarkers Prev.*, **6**, 25 (1997).

[65] T. T. Fung, D. J. Hunter, D. Spiegelman, G. A. Colditz, F. E. Speizer and W. C. Willett, *Cancer Causes Control*, **13**, 221 (2002).

[66] T. T. Fung, D. Spiegelman, K. M. Egan, E. Giovannucci, D. J. Hunter and W. C. Willett, *Int. J. Cancer*, **103**, 110 (2003).

[67] S. A. McNaughton, G. C. Marks, P. Gaffney, G. Williams and A. C. Green, *Cancer Causes Control*, **16**, 609 (2005).

[68] J. F. Dorgan, N. A. Boakye, T. R. Fears, R. L. Schleicher, W. Helsel, C. Anderson, J. Robinson, J. D. Guin, S. Lessin, L. D. Ratnasinghe and J. A. Tangrea, *Cancer Epidemiol. Biomarkers Prev.*, **13**, 1276 (2004).

[69] H. Sies and W. Stahl, *Photochem. Photobiol. Sci.*, **3**, 749 (2004).

[70] M. Garmyn, J. D. Ribaya-Mercado, R. M. Russell, J. Bhawan and B. A. Gilchrest, *Exp. Dermatol.*, **4**, 104 (1995).

[71] C. Wolf, A. Steiner and H. Honigsmann, *J. Invest. Dermatol.*, **90**, 55 (1988).

[72] F. McArdle, L. E. Rhodes, R. A. Parslew, G. L. Close, C. I. Jack, P. S. Friedmann and M. J. Jackson, *Am. J. Clin. Nutr.*, **80**, 1270 (2004).

[73] W. Köpcke and J. Krutmann, *Photochem. Photobiol.*, **84**, 284 (2008).

[74] M. M. Mathews-Roth, M. A. Pathak, J. A. Parrish, T. B. Fitzpatrick, E. H. Kass, K. Toda and W. Clemens, *J. Invest. Dermatol.*, **59**, 349 (1972).

[75] W. Stahl, U. Heinrich, H. Jungmann, H. Sies and H. Tronnier, *Am. J. Clin. Nutr.*, **71**, 795 (2000).

[76] J. P. Cesarini, L. Michel, J. M. Maurette, H. Adhoute and M. Bejot, *Photodermatol. Photoimmunol. Photomed.*, **19**, 182 (2003).

[77] D. Battistutta, D. L. Williams and A. Green, *Abstr. 13th Int. Congress on Photobiology, July 1-6, 2000*, 58 (2000).

[78] U. Heinrich, H. Tronnier, W. Stahl, M. Bejot and J. M. Maurette, *Skin Pharmacol. Physiol.*, **19**, 224 (2006).
[79] P. Palombo, G. Fabrizi, V. Ruocco, E. Ruocco, J. Fluhr, R. Roberts and P. Morganti, *Skin Pharmacol. Physiol.*, **20**, 199 (2007).
[80] M. Darvin, A. Patzelt, S. Gehse, S. Schanzer, C. Benderoth, W. Sterry and J. Lademann, *Eur. J. Pharm. Biopharm.*, **69**, 943 (2008).
[81] E. Boelsma, H. F. Hendriks and L. Roza, *Am. J. Clin. Nutr.*, **73**, 853 (2001).
[82] D. A. Hughes, *Proc. Nutr. Soc.*, **58**, 713 (1999).
[83] D. A. Hughes, *Nutrition*, **17**, 823 (2001).
[84] B. P. Chew and J. S. Park, *J. Nutr.*, **134**, 257S (2004).
[85] C. J. Fuller, H. Faulkner, A. Bendich, R. S. Parker and D. A. Roe, *Am. J. Clin. Nutr.*, **56**, 684 (1992).
[86] S. Gonzalez, S. Astner, W. An, D. Goukassian and M. A. Pathak, *J. Invest. Dermatol.*, **121**, 399 (2003).
[87] E. H. Lee, D. Faulhaber, K. M. Hanson, W. Ding, S. Peters, S. Kodali and R. D. Granstein, *J. Invest. Dermatol.*, **122**, 510 (2004).
[88] K. Scharffetter-Kochanek, M. Wlaschek, K. Briviba and H. Sies, *FEBS Lett.*, **331**, 304 (1993).
[89] M. Wlaschek, K. Briviba, G. P. Stricklin, H. Sies and K. Scharffetter-Kochanek, *J. Invest. Dermatol.*, **104**, 194 (1995).
[90] R. Tyrrell, *Free Radic. Res.*, **31**, 335 (1999).
[91] J. Krutmann and M. Grewe, *J. Invest. Dermatol.*, **105**, 67S (1995).
[92] M. L. Skoog, K. Ollinger and M. Skogh, *Photodermatol. Photoimmunol. Photomed.*, **13**, 37 (1997).
[93] I. O'Connor and N. O'Brien, *J. Dermatol. Sci.*, **16**, 226 (1998).
[94] N. M. Lyons, and N. M. O'Brien, *J. Dermatol. Sci.*, **30**, 73 (2002).
[95] N. Bando, H. Hayashi, S. Wakamatsu, T. Inakuma, M. Miyoshi, A. Nagao, R. Yamauchi and J. Terao, *Free Radic. Biol. Med.*, **37**, 1854 (2004).
[96] M. Berneburg, S. Grether-Beck, V. Kurten, T. Ruzicka, K. Briviba, H. Sies and J. Krutmann, *J. Biol. Chem.*, **274**, 15345 (1999).
[97] M. Berneburg and J. Krutmann, *Meth. Enzymol.*, **319**, 366 (2000).
[98] M. Berneburg, H. Plettenberg, K. Medve-Konig, A. Pfahlberg, H. Gers-Barlag, O. Gefeller and J. Krutmann, *J. Invest. Dermatol.*, **122**, 1277 (2004).
[99] J. Eicker, V. Kurten, S. Wild, G. Riss, R. Goralczyk, J. Krutmann and M. Berneburg, *Photochem. Photobiol. Sci.*, **2**, 655 (2003).
[100] K. Wertz, P. B. Hunziker, N. Seifert, G. Riss, M. Neeb, G. Steiner, W. Hunziker and R. Goralczyk, *J. Invest. Dermatol.*, **124**, 428 (2005).

Carotenoids
Volume 5: Nutrition and Health

Chapter 17

The Immune System

Boon P. Chew and Jean Soon Park

A. The Immune System and Disease

1. Introduction

The immune system plays an essential role in maintaining the body's overall health and resistance to diseases. It comprises two branches, known as the innate or antigen-nonspecific branch, and the adaptive or antigen-specific branch. A truly effective immune defence is based on a balance of the different arms of the whole immune system. The human immune response system is very complex and carotenoids have been reported to have effects on many different aspects. To understand the significance of this it is necessary to have a working knowledge of the immune system. An outline of the main features and principles is given below, but the non-specialist reader is recommended to consult a modern biology or biochemistry textbook, or an introductory book on immunology.

Stimulation of a particular immune response does not necessarily translate into improved immune defence or health; it must be taken in context with changes in other immune responses and with the physiological state in question. Any immune stimulation must be significant, yet within the range of a normal response; hyperactivity can mean an autoimmune or immune-mediated disease whilst a hypoactive response can result in immune suppression and incompetence, such as in human AIDS and feline immunodeficiency syndrome.

Inflammation, increased temperature and swelling of the tissue, is a localized non-specific respone to infection or injury. The increased temperature and blood flow facilitate the migration in of neutrophils, monocytes and macrophages to attack the infection. The acute inflammatory response is mediated by the cytokine TNF (tumour necrosis factor).

2. Features of the immune system

a) The innate or antigen-non-specific immune system

The innate immune system is the first line of defence and is composed of barriers such as the skin, gastrointestinal tract and lungs, as well as phagocytic cells and non-cellular components such as lysozymes and complements. If the physical barriers are breached, the body then employs a collection of non-specific cellular and chemical defences that respond to any microbial infection without the need to recognize and identify it. These cells and chemicals circulate throughout the body in the blood and the lymphatic system. The most important of the cells are several forms of white blood cells (leukocytes), which fight invading microorganisms in different ways. In this process, the leukocytes first migrate along a chemical gradient toward the microorganism that has been opsonized (coated with immunoglobulins (Ig) or complements), adhere to it, and attack and destroy it.

Macrophages are large irregular cells that engulf the invader by phagocytosis and kill it with oxygen free radicals produced by lysosomal enzymes. At the site of infection, undifferentiated leukocytes (monocytes) are transformed into additional macrophages. Neutrophils, the most abundant circulating leukocytes, also ingest and kill bacteria by phagocytosis but, in addition, they release oxidizing chemicals that kill other bacteria in the neighbourhood. Eosinophils defend against invading parasites. Natural killer cells do not attack invading microbes directly, but kill cells of the body that have been infected. They kill by creating a hole in the plasma membrane of the target cell, not by phagocytosis. Natural killer cells also attack cancer cells, often before a detectable tumour develops.

These cellular defences are enhanced by the complement system, which consists of approximately twenty different proteins that circulate in the blood. These proteins aggregate into a complex that inserts into the membrane of the marked foreign cell, and forms a pore through which fluids can enter the cell. The complement proteins have various other effects. They can amplify the inflammatory response, attract neutrophils to the site of infection, or coat the surface of the foreign cell to facilitate the attachment of phagocytes. Other important proteins are the interferons (IFN). These are secreted by body cells that have been infected with a virus, and protect neighbouring cells from being infected.

Cell-to-cell adhesion occurs during leukocyte trafficking. The contact, through a pair consisting of intracellular adhesion molecule-1 (ICAM-1) and the leukocyte-function-associated antigens-1 (LFA-1) ligand receptor, serves to co-stimulate an immune response, thereby enhancing cell proliferation and cytokine production. The LFA-1 is a β_2-integrin protein expressed on leukocytes and is involved in the migration of lymphocytes, monocytes and neutrophils. LFA-1 binds to ICAM-1 and ICAM-2 expressed on the vascular endothelium, and controls the migration of lymphocytes into inflammatory sites. The endothelial expression of ICAM-1 is inducible, whilst that of ICAM-2 is constitutive.

b) The adaptive or antigen-specific specific immunity

The specific immune system is mediated by two types of cells that circulate in the lymphatic system, known as T cells and B cells, which direct the cell-mediated and humoural responses, respectively. These lymphocytes, also sometimes referred to as splenocytes, are not themselves phagocytic.

T cells originate in the bone marrow and then migrate to the thymus, where they develop the ability to identify microorganisms and viruses by the antigen molecules exposed on the surface of the invaders. There are different subsets of T cells. Inducer T cells mediate the development and maturation of other T cells in the thymus. Helper T cells (Th) detect infection and initiate both T cell and B cell responses. The Th cells can be sub-divided into Th1 cells, that are important in response to bacterial infection, and Th2 cells, that are important in response to parasite infection. Cytotoxic T cells (Tc) lyse cells that have been infected by viruses. Suppressor T cells terminate the immune response. Unlike T cells, B cells complete their maturation in the bone marrow and do not migrate to the thymus. B cells are specialized to recognize particular foreign antigens. When activated, a B cell becomes a plasma cell that produces specific antibodies.

c) Cell-mediated immune response

When an invading foreign particle, *e.g.* a virus, is taken into a body cell, it is partially digested and the viral antigens thus produced are processed and moved to the surface of the cell, which thus becomes an antigen-presenting cell (APC). At the membrane of the APC the processed antigens are complexed with major histocompatibility complex proteins MHC-II. MHC-II is found only on macrophages, B cells and helper T cells, also known as CD4+ T cells because they have the CD4 surface co-receptor, which interacts only with the MHC-II proteins of another lymphocyte. Cytotoxic T cells (CD8+) have the co-receptor CD8 and can interact only with the MHC-I proteins of an infected cell. The human form of MHC is also known as the human leukocyte-associated antigen (HLA) complex. The T-cell antigen receptor (TCR) recognizes a peptide antigen in conjunction with the MHCII molecule. Dendritic cells, derived from the bone marrow, are the most potent of the antigen-presenting cells. Immature dendritic cells in peripheral tissues capture and process antigens; the maturing dendritic cells then migrate to lymphoid organs where they stimulate naïve T cells *via* TCRs which recognize peptide antigens in conjunction with MHC-II molecules. The degree of immune response is proportional to the number of PACs that possess MHC-II molecules and the density of the latter on the cell surface. Dendritic cells are also highly responsive to inflammatory cytokines such as TNF-α or to bacterial products that induce phenotypic and functional changes.

Activation of a Th cell by such an APC is mediated by soluble regulatory proteins known as cytokines. The most important of these are the interleukins. Interleukin-1 (IL-1) is secreted by macrophages and signals Th cells to bind to the antigen-MHC protein complex. The Th cells then release IL-2, which stimulates the multiplication of Tc cells that are specific for the antigen. Cytotoxic T cells can only destroy infected cells that display the foreign antigen together with their MHC-I proteins. Interleukin-4 (IL-4), secreted by T cells, stimulates the proliferation of B cells, and thus the humoural response. Cytotoxic T cells will attack any cells recognized as carrying a foreign version of MHC-I. This includes transplanted cells from another individual, and cancer cells that reveal abnormal surface antigens.

d) The humoural immune response

The B cells of the humoural immune system also respond to Th cells activated by IL-1. B cells recognize invading microbes but do not attack them; they mark the pathogen for destruction by macrophages and natural killer cells. The B cells recognize antigens and divide to produce plasma cells and memory B cells, resulting in the circulation of high titres of antibodies against those antigens. Antibodies are immunoglobulin (Ig) proteins, of which there are several subclasses with different structures and functions, namely IgM, IgG, IgD, IgA and IgE. IgM antibodies are produced first and they activate the complement system. Following this, large amounts of IgG antibodies are produced, and these bind to antigens on an infected cell thereby serving as markers that stimulate phagocytosis by macrophages; antibodies do not kill invading pathogens directly.

3. Nutritional intervention

Some physiological or environmental insults can weaken the immune system, resulting in increased risk of infection and disease. Under these conditions, nutritional intervention can be beneficial in modulating the immune response. A variety of immune response tests have thus been developed to assess the effect of nutritional intervention on different aspects of the immune response. These include: (i) gene expression and cell signalling associated with cell-cycle progression and apoptosis (see *Chapter 11*), (ii) lectin-induced lymphocyte proliferation, (iii) NK cell cytotoxic activity, (iv) cytokine production, (v) phenotyping, (vi) Ig production, (vii) delayed type hypersensitivity (DTH), and (viii) phagocytosis and killing ability. The assessment of these immune responses has been aided by the use of techniques associated with flow cytometric analysis, genomics, proteomics and metabolomics.

The interpretation of the results of nutritional intervention studies on immunity must, therefore, consider the whole immune system.

4. Immunity and oxidative stress

Cellular oxidative damage by reactive oxygen species (ROS) has been suggested to be a key factor in numerous chronic diseases (see *Chapter 12*). The ROS destroy cellular membranes, cellular proteins and nucleic acids. Immune cells are particularly sensitive to oxidative stress because their plasma membranes contain a high percentage of polyunsaturated acyl lipids, which easily undergo peroxidation [1] (see *Chapter 12*). Immune cells rely on cell-to-cell communication *via* membrane-bound receptors; peroxidation of the polyunsaturated acyl chains in the cell membrane, therefore, can lead to the loss of membrane integrity and altered membrane fluidity, resulting in impairment of intracellular signalling and overall cell function. Indeed, exposure to ROS leads to decreased expression of membrane receptors [2].

The ROS can arise from several sources. Immune cells are very active cells and, therefore, generate ROS during normal cellular activity, mainly through their mitochondria. Oxidative stress on the membrane of normal healthy cells can also be caused by oxidizing pollutants and many viruses, factors that are capable of inducing excessive production of ROS. A third source of ROS is from the 'respiratory burst' used by phagocytic cells (macrophages and neutrophils) during the killing of invading antigens. In this oxidative bactericidal mechanism, the NADP oxidase system is activated, and a large amount of superoxide anion ($O_2^{\bullet-}$) is produced from molecular oxygen. The $O_2^{\bullet-}$ is rapidly converted into hydrogen peroxide (H_2O_2) by superoxide dismutase. Neutrophils contain myeloperoxidase that converts H_2O_2 into the highly potent bactericidal component, hypochlorite ion (OCl^-), whilst macrophages generate oxygen-derived free radicals such as the hydroxyl radical ($OH^\bullet$). Excess ROS can in turn destroy both the cells that produce them and surrounding cells. Excess ROS can be eliminated by endogenous or dietary antioxidants which together maintain an optimal oxidant:antioxidant balance that is critical for maintaining normal cellular function and health. Tipping this balance in favour of ROS is thought to be a major contributor to several age-related diseases such as cancer, and neurodegenerative, cardiovascular and eye diseases.

Even though ROS are usually portrayed as the villain, research has now demonstrated that they are important signalling molecules involved in the regulation of gene expression, cell growth and cell death. Therefore, the action of antioxidants, including carotenoids, on immune response is anything but straightforward; their actions hang in a delicate balance between the total elimination of toxic ROS on one hand, and the maintenance of an optimal ROS concentration for cell signalling on the other.

B. Carotenoids and the Immune Response

Immune cells are very active cells and, therefore, generate ROS during normal cellular activity. The mitochondrial electron transport system utilizes approximately 85% of the oxygen consumed by the cell to generate ATP, so mitochondria are the most important source of ROS [3]. Unfortunately, the mitochondria are also a target of the ROS.

1. Effects of carotenoids

The localization of carotenoids in the mitochondria is of particular relevance as these carotenoids may serve to protect the subcellular organelles of immune cells against oxidative injury. Optimal function of the subcellular organelles ensures that cellular functions, including apoptosis, cell signalling and gene regulation, are optimal. Studies with cats and dogs have shown significant uptake of orally fed lutein (**133**) [4,5], β-carotene (**3**) [6,7], and astaxanthin (**404-6**) [8] into the mitochondria, nuclei, and microsomes of circulating lymphocytes, with the mitochondria showing high total uptake of these carotenoids. Uptake of β-carotene by human neutrophils [9], and by neutrophils and lymphocytes from calves [10] and pigs [11,12], has similarly been demonstrated.

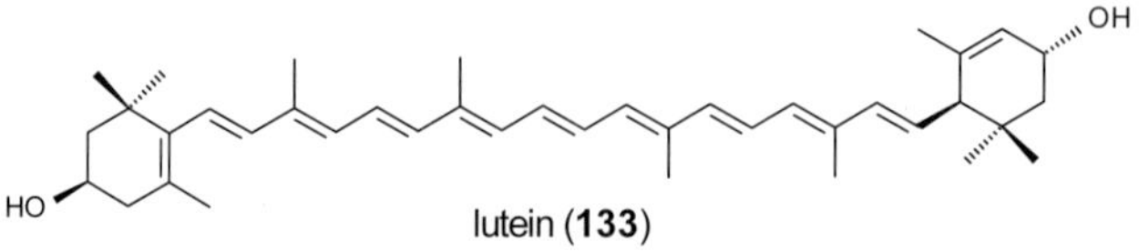

lutein (**133**)

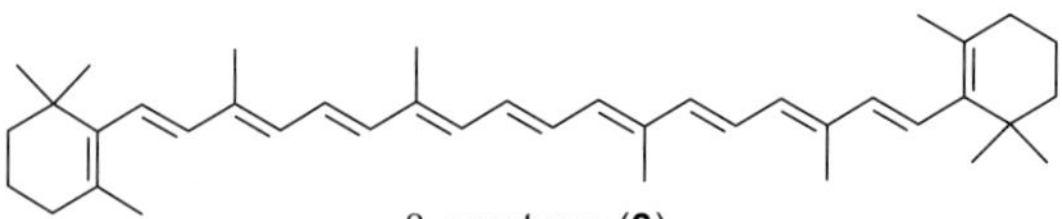

β-carotene (**3**)

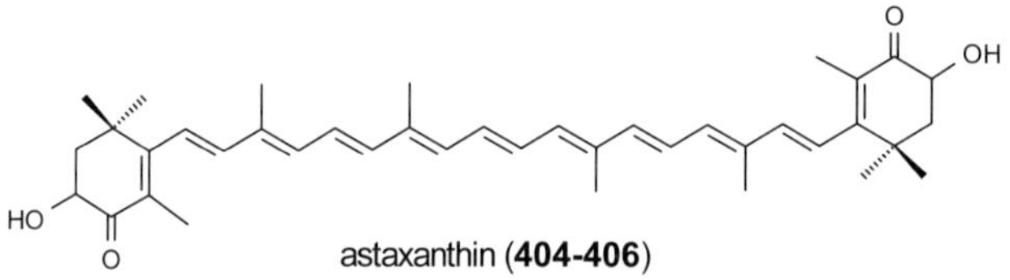

astaxanthin (**404-406**)

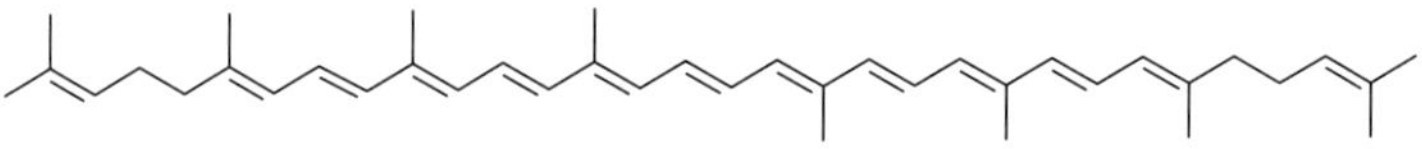

lycopene (**31**)

It is proposed that dietary carotenoids help to protect the immune system from oxidative damage and thereby enhance cell-mediated immune response [13]. Most studies have been with β-carotene, though more now deal with other carotenoids. Earlier reports that lycopene (**31**) prolonged the survival time of bacterially-infected mice [14] and that β-carotene markedly increased the growth of the thymus gland and the number of thymic lymphocytes [15] stimulated the study of the possible immune modulation action of carotenoids.

a) Specific effects

The ability of carotenoids to modulate the embryonic development (ontogenesis) of the immune system begins early during neonatal development. β-Carotene supplementation significantly changed the percentage and total number of splenic CD3+, CD4+ and CD8+ cells, and IgG production, in mice between days 7 and 14 of age [16]. Lycopene also increased the number of splenic T and B cell subsets, but its immune modulation action occurred at a later time point. In humans, the sequence of events in T lymphocyte development starts during embryogenesis, and is thus comparable to that in mice, so it is possible that dietary carotenoids can influence the ontogenesis of the human immune system.

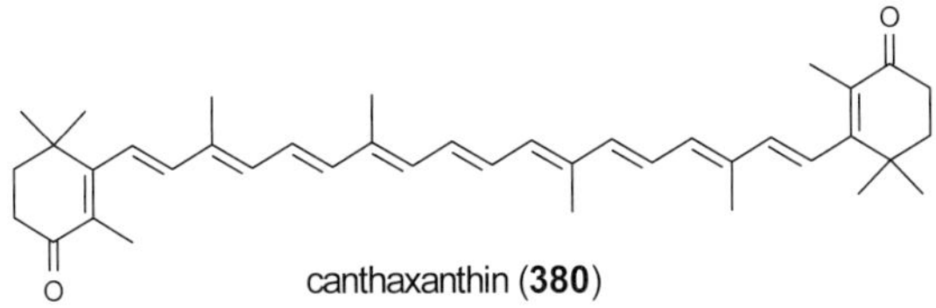

canthaxanthin (**380**)

Numerous studies have reported that high plasma β-carotene status or β-carotene supplementation enhance immune response [17-22]. Supplementation with β-carotene stimulated lymphocyte proliferation in several human intervention studies [19,23,24], and in rats [25], pigs [26], and cattle [27]. In mice, astaxanthin and β-carotene but not canthaxanthin (**380**) stimulated phytohaemagglutinin-induced splenocyte proliferation [28]. Higher lymphocyte proliferation after β-carotene supplementation was accompanied by an increase in specific lymphocyte populations. For example, human adults given β-carotene orally had increased numbers of Th and T inducer lymphocytes [17,29]. Subjects given 60 mg/day β-carotene for 4 weeks showed a slight increase in the number of CD4+ cells [30]. In contrast, another study [31] failed to show significant changes in the number of T cells, lectin-stimulated lymphocyte proliferation, or surface molecule expression, in older subjects (>65 years) who were given β-carotene or lycopene daily for 12 weeks. Short-term supplementation with β-carotene (30 mg/day [32] or 89 mg/day [33]) had no effect on T lymphocyte function in healthy women.

The number of lymphoid cells with surface markers for NK cells and for IL-2 and transferrin receptors also was increased substantially in peripheral blood from individuals after short-term supplementation with β-carotene [17,34]; the NK cells as a percentage of the total increased [17]. The lytic activity of NK cells was increased in elderly but not in middle-aged men on long-term β-carotene supplementation [21]. β-Carotene reversed the age-related decline in NK cell lytic activity in older (65-86 years) male subjects, restoring levels to those of younger (51-64 years) subjects. Similar studies [35] also reported higher NK cell cytotoxicity in human subjects given β-carotene orally. It was concluded that additional low amounts of β-carotene or lycopene are unable to enhance cell-mediated immune response in well-nourished healthy individuals.

Astaxanthin also possesses immune enhancing activity. In mice, astaxanthin stimulated splenocyte proliferation [28]. In a double-blind, placebo-controlled study [36], 2-8 mg astaxanthin given daily to young healthy human female subjects stimulated mitogen-induced lymphoproliferation, and increased NK cell cytotoxic activity. Astaxanthin also increased the number of total T and B cells but did not influence the sub-populations of Th, Tc or NK cells. There was a heightened DTH response and a higher frequency of cells that expressed the marker LFA-1 in subjects given 2 mg astaxanthin. Delayed type hypersensitivity is a local inflammation occurring 24-48 hours after challenge with an antigen against which the person has previously been immunized. In response to inflammation, plasma concentrations of some proteins, known as acute phase proteins, may increase or decrease. Astaxanthin decreased DNA damage and plasma concentrations of acute phase proteins. In mice, lutein and astaxanthin increased the antibody response of splenocytes *ex vivo* to T-cell antigens [37]. *In vitro*, lycopene directly suppressed the antigen-presenting function of lipopolysaccharide-stimulated bone marrow-derived murine myeloid dendritic cells by down-regulating the expression of co-stimulatory molecules CD80 and CD86, and MHC-II molecules [38]. Dendritic cells treated with lycopene were poor stimulators of naïve allogeneic T cell proliferation and they showed impaired IL-12 production in responding T cells [39]. IL-12 is a pro-inflammatory cytokine and its expression is a specific marker of functionally activated dendritic cells [40,41]. Lycopene, therefore, may control chronic immune and/or inflammatory diseases through down-regulation of dendritic cell maturation.

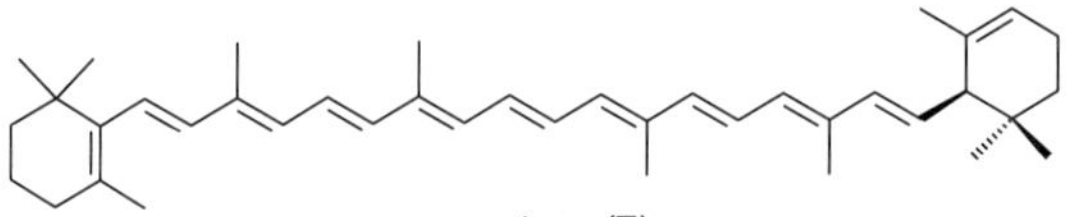

α-carotene (**7**)

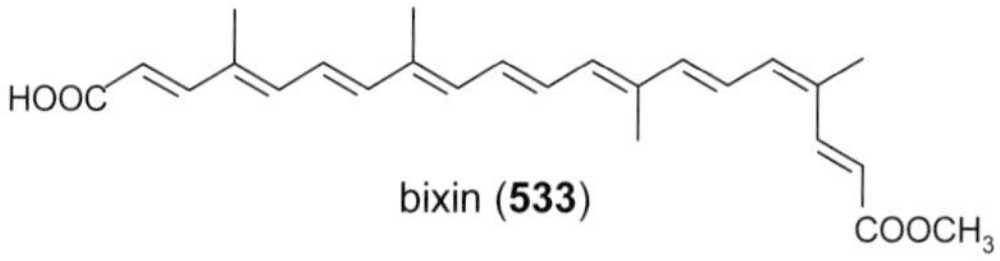

bixin (**533**)

There are also reports that carotenoids can modulate the activity of phagocytic cells, although this has been less well studied. Murine macrophages incubated with canthaxanthin, β-carotene or α-carotene (**7**) had higher cytochrome oxidase and peroxidase activities than did those incubated with (*cis*)-retinoic acid; the highest activity was observed with canthaxanthin [42]. Dietary β-carotene stimulated the phagocytic and killing ability of bovine blood neutrophils [27,43,44]. On the other hand, β-carotene, lutein, bixin (**533**), and canthaxanthin decreased luminol-dependent chemiluminescence generated from rat peritoneal macrophages stimulated by phorbol myristate acetate. This suggests that suppression by carotenoids of the respiratory burst of macrophages represents a way to protect host cells and tissues from the harmful effects of oxygen metabolites that may be overproduced during specific immune response [45].

Studies with dogs and cats have provided direct comparisons of the immune modulation action of several carotenoids. In dogs, dietary β-carotene [46], lutein [47] or astaxanthin [48] stimulated DTH response, the number of CD4+ Th cells, and IgG production, and lutein but not β-carotene enhanced mitogen-induced lymphocyte proliferation [47]. Cats fed astaxanthin [49], β-carotene [50] or lutein [51] also showed heightened DTH response, higher Th and B cell sub-populations, and increased plasma IgG concentrations. These results demonstrated that β-carotene, lutein and astaxanthin can have immune-enhancing activity, especially after antigenic challenge with a vaccine. However, the specific immune response factors modulated may be different for different carotenoids and for different species.

b) Effects of carotenoid-rich foods and extracts

The above studies used pure carotenoids, but the importance of a balance and interaction of different dietary components is recognized. Recent studies, therefore, have used whole foods to provide a more complete array of important carotenoids. Tomatoes and tomato products are rich in lycopene. Tomato intake is inversely related to the risk of diarrhoea and respiratory infections in young children [52]. Tomato juice improved T lymphocyte function in subjects who otherwise consumed low carotenoid diets [53]. In a blinded, randomized crossover study, healthy male subjects on a low carotenoid diet were fed 330 mL/day tomato juice (providing 37 mg/day lycopene) or carrot juice (27 mg/day β-carotene + 13 mg/day α-carotene) for 2 weeks followed by a 2-week depletion period [54]. There was a time-delayed modulation of IL-2, NK cytotoxicity, and lymphocyte proliferation during the depletion period. An earlier study [55] showed no stimulation of cell-mediated immune response in well-nourished elderly men and women fed tomato juice for 8 weeks. Male non-smokers on a low carotenoid diet were given three carotenoid-rich food sources sequentially, each for two weeks, namely first tomato juice (containing 40 mg lycopene + 1.5 mg β-carotene), then carrot juice (22 mg β-carotene + 16 mg α-carotene + 0.5 mg lutein), and finally dried spinach powder (11 mg lutein + 3 mg β-carotene) [56]. Tomato juice, but not carrot juice or spinach powder, enhanced IL-2 and IL-4 secretion. On the other hand, tomato oleoresin, when given to smokers and non-smokers in a double-blind, placebo-controlled randomized study, returned IL-4 production to normal in smokers but had no effect on lymphocyte proliferation, NK cell activity, IL-2 or TNF-α [57]. There was a decrease in DNA strand breaks in both smokers and non-smokers. High levels of circulating IL-4 led to increased susceptibility of smokers to viral or mycobacterial infections [58].

It must be emphasized, however, that these plant materials and extracts would contain a large collection of other phytochemicals, including antioxidants and phytosterols, besides a mixture of carotenoids, so the effects seen cannot safely be attributed to the carotenoids.

c) Model studies of health benefits

Whether the immune regulatory action of carotenoids translates into health benefits has been studied with several biological models such as exercise-induced oxidative damage, photoprotection and *Helicobacter pylori* infection.

i) Exercise-induced oxidative stress. During intense prolonged exercise, oxidative stress (originating from metabolism in the mitochondria, ischaemic-reperfusion injury and phagocytic cell activity) is greatly increased. Intense exercise in sled dogs suppressed T cell and B cell mitogenic response, suppressed the number of MHC-II+ cells, and increased Th cell and B cell populations [59]. Exercise also increased the concentration of acute phase proteins, suggesting a stress-induced response similar to inflammation or infection. Dietary β-carotene, lutein and α-tocopherol, supplemented together, returned to normal the exercise-induced changes in Tc and B cell sub-populations, and the concentration of acute phase proteins. Sled dogs supplemented with antioxidants also had decreased DNA oxidation and increased resistance of blood lipoproteins to oxidation compared to unsupplemented exercised dogs [60].

ii) Exposure to UV light. Exposure to UV light can suppress immune response, but carotenoids have photoprotective properties against this. In young male subjects fed 30 mg/day β-carotene for 28 days before periodic exposure to UV light, no DTH suppression by UV exposure was reported, and the DTH response was inversely proportional to plasma β-carotene concentration [18]. In healthy older males given 30 mg/day β-carotene for 28 days and exposed to UV light, UV-induced DTH suppression was also reduced, but β-carotene was not as protective as it was in younger male subjects [61], perhaps because of lower plasma β-carotene response or higher vitamin E status in the older individuals. Because UV light can inhibit expression of the human MHC protein HLA-DR, and the adhesion molecule ICAM-1 in human cell lines, a carotenoid-induced increase in cell surface molecules may help to explain the ability of β-carotene to prevent a decrease in DTH response after UV exposure.

iii) Helicobacter pylori *infection.* Infection by *Helicobacter pylori* is a major cause of chronic gastritis and is marked by an active inflammatory response due to neutrophilic infiltration. During *H. pylori* infection, the immune response is polarized to a Th1 cell-mediated immune response with the release of IFN-γ which activates phagocytic cells and contributes to mucosal damage [62,63]. Supplementation with astaxanthin led to a decrease in bacterial load and gastric inflammation in infected mice by shifting the T-lymphocyte response from a Th1 response dominated by IFN-γ to a Th1/Th2 response dominated by IFN-γ and IL-4 [64].

C. Carotenoids and Disease

Considerable interest has been generated around the possible use of the immune-enhancing activity of carotenoids in prevention of inflammatory diseases, cancer, and human immunodeficiency disease.

1. Age-related diseases

a) Age-related immunity decline

Overall immune response declines with advancing age, thereby increasing susceptibility to infection and a number of age-related conditions such as inflammatory, cardiovascular and neurodegenerative diseases, and cancer. The 'mitochondria theory of aging' states that oxidative damage to DNA, proteins and lipids accumulates in the mitochondria over the lifespan of the organism [3]. Indeed, numerous studies have reported increased oxidative damage to mitochondrial macromolecules with age, resulting in mitochondrial dysfunction and loss of ATP production. Mitochondrial dysfunction can lead to impaired immune response and to neurodegenerative conditions such as Alzheimer's, Parkinson's, Huntington's diseases and Amyotrophic lateral sclerosis. Conversely, antioxidants, which may include carotenoids, can alleviate the harmful effects of the ROS.

It has been reported that supplementation with carotenoids can restore the age-related decline in both cell-mediated and humoural immune responses, in some cases to the levels found in younger individuals. High β-carotene status was associated with decreased incidence of acute respiratory infection (incidence rate ratio = 0.71, 95% CI 0.54-0.92) in elderly individuals compared to those who had low β-carotene [65]. In that study, no similar improvement was observed with α-carotene, β-cryptoxanthin (**55**), lycopene, lutein, or zeaxanthin (**119**). Geriatric dogs had lower Th and B cell sub-populations, lower T cell proliferation, and lower DTH response than age-matched controls and young dogs [66]. However, supplementation with β-carotene and α-tocopherol restored these impaired immune functions in older dogs.

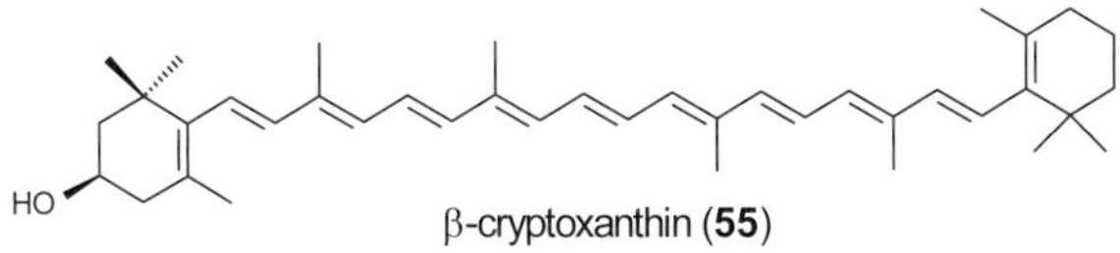

β-cryptoxanthin (**55**)

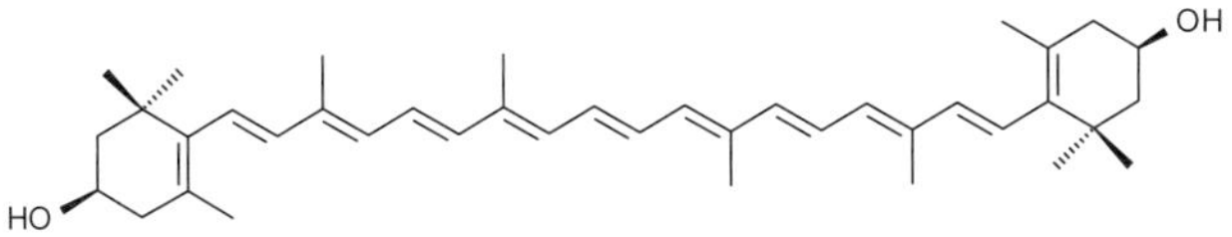

zeaxanthin (**119**)

b) Neurodegenerative conditions

A large body of evidence has emerged implicating impaired energy metabolism and oxidative damage in Alzheimer's disease [67]. In fact, oxidative damage occurs before the deposition of β-amyloid. In this disease, the inflammatory response is atypical in that there is an absence of an overt leukocyte infiltration [68]. Instead, the major factors include the resident cellular elements such as the microglia and astrocytes. The possible neuroprotective role of dietary antioxidants has been studied. Both β-carotene and vitamin E protected rat neurons against oxidative stress from ethanol exposure [69]. β-Carotene had a greater protective effect than vitamins E or C against neuro-vascular dysfunction [70].

c) Rheumatoid arthritis

Oxidative damage to the synovium can lead to the pathogenesis of rheumatoid arthritis [71]. Rheumatoid arthritis is an immune disorder in which lymphocytes accumulate and organize into lymphoid structures on the synovial surface of the cavities of small joints. CD4+ cells, activated B cells, and plasma cells are found in the inflamed synovium. In a large prospective population-based study with women aged 55-69 years [72], high intakes of β-cryptoxanthin were associated with protection against rheumatoid arthritis. A similar study recently reported that a modest increase in β-cryptoxanthin (equivalent to one glass of freshly squeezed orange juice per day) in human subjects was associated with a lower incidence of developing rheumatoid arthritis.

2. Cancer

Studies both *in vitro* and *in vivo* have reported effects of carotenoids in stimulating immunity against tumour growth. A specific action of β-carotene was reported [73] in augmenting immunity against syngeneic fibrosarcoma cells in mice. *In vitro*, β-carotene and lycopene inhibited the growth of human breast cancer cells; their action was related to the presence of oestrogen receptors [74]. Astaxanthin, canthaxanthin or β-carotene fed to mice injected with a transplantable mammary tumour cell line inhibited tumour growth, with astaxanthin having the highest inhibitory activity. Astaxanthin also inhibited the growth of fibrosarcoma cells and concomitantly increased Tc cell activity and IFNγ production by splenocytes and tumour-draining lymph node [75]. A transplantable mammary tumour model in BALB/c mice has been used to demonstrate the antitumour activity of dietary lutein and to study the mechanism involved. In several studies, dietary lutein consistently inhibited the growth of mammary tumours in mice [76-78]. When a lower tumour load was used, lutein decreased the incidence of tumour development [77]. The presence of a mammary tumour suppressed the populations of total T, Th, and Tc cells, but increased the populations of IL-2Rα+ T cells and B cells compared to those in mice not carrying tumours [78]. However, lutein prevented these tumour-associated lymphocyte sub-population changes. In addition, lutein increased IFN-γ mRNA expression but decreased IL-10 expression in splenocytes of tumour-bearing mice; these changes were associated with the inhibitory action of lutein against tumour growth [78].

Tumour growth is highly dependent on angiogenesis, *i.e.* formation of small blood vessels to increase blood flow [79]. Without proper neovascularization, tumour cells will lack growth factors and therefore undergo apoptosis. Mice fed lutein had fewer blood vessels associated with their tumours than did unsupplemented mice [80]. Other studies have also demonstrated the ability of carotenoids to reduce tumour blood flow [81].

3. Human immunodeficiency: HIV and AIDS

The human immunodeficiency virus, HIV, is a retrovirus that circulates in the bloodstrean but will specifically infect only CD4+ cells. It has a surface glycoprotein that precisely fits the cell surface receptor protein CD4 on the surface of macrophages. It replicates in the macrophage and is released from the cell by budding. After some years of HIV infection, the surface glycoprotein may undergo mutation to a form that binds the surface receptor of CD4+ T cells. These T cells are destroyed and the immune response is blocked. The consequence of this is the onset of AIDS (Acquired Immunodeficiency Syndrome). The body's defences against infection, cancer, *etc.* are compromised, usually with fatal consequences.

Low levels of carotene and other carotenoids are common in HIV patients and are more marked than any other micronutrient deficiencies [82-84]. In HIV-infected subjects, there is a significant depletion of all carotenoids analysed (lutein, cryptoxanthin, lycopene, β-carotene, α-carotene) but not of vitamin A or E [82]. Low plasma carotenoid concentrations were associated with increased risk of death during HIV infection among infants in Uganda [85]. Also, low serum carotene concentration is positively correlated with severity of the disease; children with AIDS had a greater magnitude decrease in serum carotene than children with HIV infection.

A correlation was demonstrated between serum carotene and both CD4+ cell counts and CD4+/CD8+ ratio in HIV-infected individuals [84]. Healthy individuals given 180 mg β-carotene daily for 14 days had higher CD4+ populations [29]. A transient increase of 60% in lymphocyte counts was reported in AIDS patients given 60 mg β-carotene per day for 4 weeks. Patients with HIV who were given 60 mg β-carotene daily showed a significant increase in leukocyte counts and CD4+:CD8+ ratio [86]. Also, patients administered 60 mg/day β-carotene had higher CD4+ counts and alleviated symptoms of the disease over 24-36 months [87]. Another study [30] showed a slight but not significant increase in CD4+ numbers with supplementation with 60 mg β-carotene daily for 4 weeks.

The potential ability of carotenoids to increase CD4+ lymphocytes has led to studies on the use of carotenoids as immuno-enhancing agents in the treatment of HIV infection. Giving natural mixed carotenoids to patients who had advanced AIDS and were on antiviral therapy improved survival rate [88].

In contrast, elderly women supplemented with 89 mg β-carotene daily for 21 days, or elderly men given 50 mg/day on alternate days for 10-12 years showed no significant changes in lymphocyte subsets [33]. Similarly, there was no significant change in CD4, CD8, or CD11 sub-populations in HIV-positive veterans given 60 mg β-carotene daily for 4 months [34]. Treatment of HIV seropositive subjects with 60-120 mg β-carotene daily for 3-7 months resulted in no improvement in infection or lymphocyte counts [89]. Several clinical trials with β-carotene supplementation failed to show significant or sustained improvement in immune response of patients with HIV infection or AIDS. A combination of β-carotene and vitamin A given daily to women during pregnancy and lactation increased the risk of mother-to-child transmission of HIV [90]. In a smaller study in South Africa where women were given β-carotene and vitamin A daily during pregnancy and at delivery, no beneficial effect on mother-to-child transmission of HIV was observed with the supplements [91]. Multivitamins (B complex, vitamins C and E), administered during pregnancy and lactation, were effective in improving postnatal growth; β-carotene + vitamin A reduced this beneficial effect [92].

Whilst a consistent negative relationship is reported between blood concentrations of carotenoids and HIV infection, carotenoid intervention studies to date have not produced a consistent beneficial effect on the clinical course of the disease. β-Carotene doses used have been very high; perhaps carotenoids can be included at a more optimal dose or other carotenoids can be studied. Studies on the possible action of other carotenoids are lacking.

D. Mechanism of Action

Carotenoids may regulate cell cycle progression, apoptosis and signalling pathways by modulating genes and transcription factors. Mechanisms and the general significance of these effects are discussed in detail in *Chapter 11*. Here only a brief outline will be given of those aspects that seem most relevant to modulation of the immune system. Central to the regulation of redox-sensitive molecular signalling pathways are the mitochondria, which are critical for processing and integrating the pro-apoptotic and anti-apoptotic signals. Mitochondrial dysfunction, therefore, can lead to pro-oxidative changes in redox homeostasis, resulting in an efflux of mitochondrial components, further increasing oxidative stress.

The Bcl-2 protein family, comprising the pro-apoptotic members Bax, Bak, Bad, and Bid, and the anti-apoptotic members Bcl-2 and Bcl-xL, are important in the regulation of apoptosis. The main target site for both groups is in the mitochondria where they facilitate or inhibit the release of cytochrome *c* that is located between the inner and outer mitochondrial membranes. Bcl-2 resides in the outer mitochondrial membrane and prevents the release of cytochrome *c*. On the other hand, the predominance of Bax over Bcl-2 accelerates apoptosis; Bax is inactive until it is translocated to the mitochondria where it binds to Bcl-2 to induce the release of cytochrome *c*, which then activates caspases to bring about apoptosis. Singlet oxygen [93] and nitric oxide [94] activate caspase-8.

Uncontrolled cell proliferation can lead to cancer and autoimmune diseases whereas excessive cell death can lead to neurodegenerative diseases and AIDS. In human leukaemia, colon adenocarcinoma and melanoma cells, β-carotene altered mitochondrial membrane potential ($\Delta\Psi m$) and induced the release of cytochrome *c* [95]. The first evidence for a gene regulatory role of lutein came when mice fed lutein, but not ones fed astaxanthin or β-carotene, showed increased *Pim-1* gene expression in lymphocytes [96]. Mice fed lutein and injected with a mammary tumour cell line had smaller tumours, higher p53 and Bax mRNA expression, lower Bcl-2 expression, and higher Bax:Bcl-2 ratio in tumours [80]. In contrast, lutein down-regulated the tumour-suppressive p53 and Bax mRNA and up-regulated Bcl-2 expression in circulating leukocytes; p53 can induce cell cycle arrest to allow DNA repair or apoptosis. The regulation of apoptotic genes by lutein parallels the observations of apoptosis rate in tumour tissues and leukocytes, lutein decreasing apoptosis in blood leukocytes but increasing apoptosis in tumour cells [80]. These results demonstrate a differential action of lutein on apoptosis in tumour cells and immune cells. Other work has shown similarly that lutein selectively induced apoptosis in transformed but not in normal human mammary cells *in vitro* [97]. In colon cancer cells, β-carotene also decreased the expression of Bcl-2 caused by ROS production [98].

Cyclins are essential for cell cycle progression from G_1 to S-phase (see *Chapter 11*). The D cyclins bind to and activate the cyclin-dependent kinases cdk4 and cdk6; this is promoted by the proteins p21 and p27. Activation of the cyclin-dependent kinases results in the phosphorylation of the Rb protein leading to the release of the E2F transcription factors, resulting in proper G_1/S transition. Lycopene inhibited cell cycle progression in breast and endometrial cancer cells by decreasing cyclin D and retaining $p27^{Kip1}$ in cyclin E-cdk2 complexes [99]. Damage to DNA elicits a complex response mediated by various intracellular and extracellular factors such as p53, abl, Rb, E2F, and growth factors, resulting in cell cycle arrest and apoptosis [100,101]. An abundance of intra-cellular or extra-cellular ROS can result in the over-stimulation of cell signalling mechanisms such as NFκB, resulting in the production of inflammatory cytokines and in inflammatory diseases. NFκB is a redox-sensitive transcription factor induced by TNF-α and IL-1, leading to the generation of ROS [102,103]. β-Carotene inhibited the growth of HL-60 and colon carcinoma cells through sustained NFκB expression and the induction of ROS production and glutathione content [95]. β-Carotene also modulates the activation of another redox-sensitive transcription factor, Ap-1, that is involved in cell growth regulation [104]. Similarly, lycopene inhibited NFκB p65 translocation in murine myeloid dendritic cells, thereby preventing the maturation of these cells [38].

As discussed in *Chapters 11* and *18*, carotenoids can modulate cancer cell growth by modulating the expression of Cox-2 which is involved in carcinogenesis and tumour promotion [105]. Cox-2 is an inducible enzyme [106] and it can act as an anti-apoptotic factor. Its expression is regulated by peroxisome proliferation-activated receptor PPARγ that in turn is regulated by carotenoids [107]. β-Carotene down-regulated Cox-2 expression in colon cancer cells [108] and this was accompanied by induction of apoptosis, decrease in intracellular ROS production, increase in the activation of ERK1/2, and decrease in production of the prostaglandin PGE2, which is a major prostaglandin synthesized by monocytes and macrophages, and is immunosuppressive. Therefore, carotenoids may alter the arachidonic acid metabolism cascade to suppress PGE2 production.

The anticancer action of carotenoids can be mediated through the induction of phase II detoxication enzymes, expression of which is regulated by the antioxidant response element (ARE) and the transcription factor Nrf2 (Nuclear factor E2-related factor 2). Indeed, lycopene, and to a lesser extent β-carotene and astaxanthin, are potent activators of ARE [109]. Induction of phase II enzymes by carotenoids and their metabolites is discussed more fully in *Chapters 11* and *18*.

Evidence has accumulated to show that carotenoids can exhibit pro-oxidant activity, especially at high concentrations and depending on the biological environment in which they act [110,111] (*Chapter 12*). The relevance of this to effects on signalling pathways and apoptosis, and to effects of carotenoids and their oxidative breakdown products on cancer, especially lung cancer in smokers, is discussed in *Chapters 11* and *18*.

Many mechanisms have been proposed by which carotenoids could modulate immune responses. The action of the provitamin A carotenoids could be mediated through their prior conversion to vitamin A and especially retinoic acid (*1*) (see *Chapter 8* and *Volume 4, Chapter 16*). The action of retinoic acid on the immune system is well studied; it can modulate immune cell differentiation and proliferation, apoptosis, and gene regulation. β-Carotene also can be cleaved excentrically to products such as 10′-apo-β-caroten-10′-al (**499**). The biological actions of the apocarotenoids are discussed in *Chapter 18*; it is not known if they have any effects on the immune system.

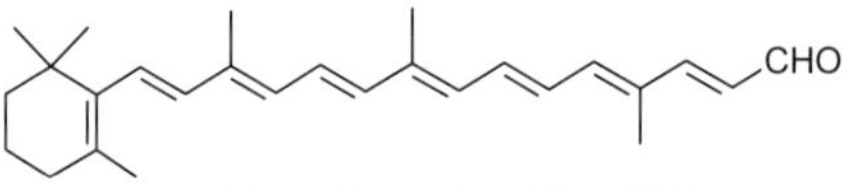

10'-apo-β-caroten-10'-al (**499**)

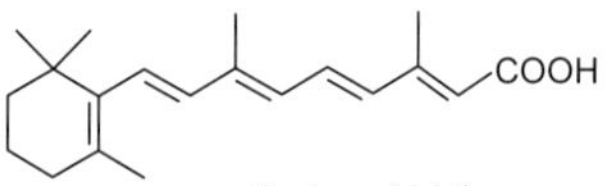

retinoic acid (*1*)

E. Summary and Conclusions

Carotenoids in general have been shown to improve cell-mediated and humoural immune response in healthy individuals. Improvements in immune responses following supplementation with carotenoids are observed more consistently when the immune system is compromised or is antigenically-challenged, conditions associated with age-related immune suppression, inflammation and disease states, and exposure to environmental pollutants. Carotenoids modulate many facets of the immune system, notably lymphocyte proliferation and cytotoxic activity, cytokine and Ig production, cutaneous DTH response, and phagocytic cell activity. The actions of carotenoids are mediated through their ability to regulate ROS in the immediate cellular environment, ultimately modulating shifts in immune cell sub-populations, and to regulate the expression of genes and gene products that are associated with cell signalling, cell-cycle progression and apoptosis. Interpretation of results is challenging; the interrelationships between pathways, and the effects of carotenoids, are complex (see *Chapter 18*, Fig. 2).

Animal studies have produced more consistent results than have human nutrition intervention studies, perhaps because of the greater ability to control dietary manipulations and the lower genetic variations in animals. Inconsistent results among studies have largely been due first to differences in the particular carotenoids used, with different carotenoids exerting somewhat different but overlapping immune modulation action and, second, to the carotenoid dose and duration of supplementation used, with high carotenoid amounts exerting effects opposite to those of an optimal dose. The source/matrix of the carotenoids (pure form *versus* whole food) which is related to the interaction of one carotenoid with another or with other food components, and the different uptake efficiency of a particular carotenoid into blood and tissue in different species are also significant factors.

Interpretation of an immune response must consider the physiological state of the individuals, *i.e.* healthy *versus* disease state, or young *versus* aged. Stimulation of certain aspects of immune function is generally considered desirable, but over-stimulation can be harmful. A strategy for alleviating immune suppression and inflammation, and slowing progression of the associated diseases, would involve limiting the over-production of ROS.

References

[1] S. N. Meydani, D. Wu, M. S. Santos and M. G. Hayek, *Am. J. Clin. Nutr.*, **62**, 1462S (1995).

[2] S. Gruner, H. D. Volk, P. Falck and R. V. Baehr, *Eur. J. Immunol.*, **16**, 212 (1986).

[3] M. K. Shigenaga, T. M. Hagen and B. N. Ames, *Proc. Natl. Acad. Sci. USA*, **91**, 10771 (1994).

[4] J. S. Park, B. P. Chew, T. S. Wong, B. C. Weng, M. G. Hayek and G. A. Reinhart, *FASEB J.*, **13**, A552 (1999).

[5] B. P. Chew, T. S. Wong, J. S. Park, B. C. Weng, N. Cha, H. W. Kim, M. G. Hayek and G. A. Reinhart, in *Recent Advances in Canine and Feline Nutrition. Vol. II*, (ed. G. A. Reinhart and D. P. Carey), p. 547, Orange Fraser Press, Wilmington, Ohio (1998).

[6] B. P. Chew, J. S. Park, B. C. Weng, T. S. Wong, M. G. Hayek and G. A. Reinhart, *J. Nutr.*, **130**, 1788 (2000).

[7] B. P. Chew, J. S. Park, B. C. Weng, T. S. Wong, M. G. Hayek and G. A. Reinhart, *J. Nutr.*, **130**, 2322 (2000).

[8] B. P. Chew, J. S. Park, M. G. Hayek, S. Massimino and G. A. Reinhart, *FASEB J.*, **18**, A158 (2004).

[9] M. M. Mathews-Roth, *Clin. Chem.*, **24**, 700 (1978).

[10] B. P. Chew, T. S. Wong and J. J. Michal, *J. Anim. Sci.*, **71**, 730 (1993).

[11] B. P. Chew, T. S. Wong, J. J. Michal, F. E. Standaert and L. R. Heirman, *J. Anim. Sci.*, **69**, 4892 (1991).

[12] B. P. Chew, T. S. Wong, J. J. Michal, F. E. Standaert and L. R. Heirman, *J. Anim. Sci.*, **68**, 393 (1990).

[13] B. P. Chew and J. S. Park, *J. Nutr.*, **134**, 257S (2004).
[14] C. Lingen, L. Ernster and O. Lindberg, *Exp. Cell Res.*, **16**, 384 (1959).
[15] E. Seifter, G. Rettura and S. M. Levenson, in *Chemistry and Technology, Vol. 2*, (ed. G. Charalambois and G. Inglett), p. 335, Academic Press, New York (1981).
[16] A. L. Garcia, R. Rihl, U. Herz, C. Koebnick, F. Schweigert and M. Worm, *Immunology*, **110**, 180 (2003).
[17] R. R. Watson, R. H. Prabhala, P. M. Plzia and D. S. Alberts, *Am. J. Clin. Nutr.*, **53**, 90 (1991).
[18] C. J. Fuller, H. Faulkner, A. Bendich, R. S. Parker and D. A. Roe, *Am. J. Clin. Nutr.*, **56**, 684 (1992).
[19] G. Van Poppel, S. Spanhaak and T. Ockhuizen, *Am. J. Clin. Nutr.*, **57**, 402 (1993).
[20] T. Murata, H. Tamai, T. Morinobu, M. Manago, H. Takenaka, K. Hayashi and M. Mino, *Am. J. Clin. Nutr.*, **60**, 597 (1994).
[21] M. S. Santos, S. N. Meydani, L. Leka, D. Wu, N. Fotouhi, M. Meydani, C. H. Hennekens and J. M. Gaziano, *Am. J. Clin. Nutr.*, **64**, 772 (1996).
[22] D. A. Hughes, *Nutrition*, **17**, 823 (2001).
[23] S. Moriguchi, N. Okishima, S. Sumida, K. Okamura, T. Doi and Y. Kishino, *Nutr. Res.*, **16**, 211 (1996).
[24] T. R. Kramer and B. J. Burri, *Am. J. Clin. Nutr.*, **65**, 871 (1997).
[25] A. Bendich and S. S. Shapiro, *J. Nutr.*, **116**, 2254 (1986).
[26] C. D. Hoskinson, B. P. Chew and T. S. Wong, *Biol. Neonate*, **62**, 325 (1992).
[27] L. R. Daniel, B. P. Chew, T. S. Tanaka and L. W. Tjoelker, *J. Dairy Sci.*, **74**, 124 (1990).
[28] B. P. Chew, J. S. Park, M. W. Wong and T. S. Wong, *Anticancer Res.*, **19**, 1849 (1999).
[29] M. Alexander, H. Newmark and R. G. Miller, *Immunol. Lett.*, **9**, 221 (1985).
[30] D. A. Fryburg, R. J. Mark, B. P. Griffith, P. W. Askenase and T. F. Patterson, *Yale J. Biol. Med.*, **68**, 19 (1995).
[31] M. Corridan, M. O'Donoughue, D. A. Hughes and P. A. Morrissey, *Eur. J. Clin. Nutr.*, **55**, 627 (2001).
[32] C. Gossage, M. Deyhim, P. B. Moser-Veillon, L. W. Douglas and T. R. Cramer, *Am. J. Clin. Nutr.*, **71**, 950 (2000).
[33] M. S. Santos, L. S. Leka, J. D. Ribaya-Mercado, R. M. Russell, M. Meydani, C. H. Hennekens, J. M. Gaziano and S. N. Meydani, *Am. J. Clin. Nutr.*, **66**, 917 (1997).
[34] H. S. Garewal, N. M. Ampel, R. R. Watson, R. H. Prabhala and C. L. Dols, *J. Nutr.*, **122**, 728 (1992).
[35] R. H. Prabhala, G. S. Harinder, M. J. Hicks, R. E Sampliner and R. R. Watson, *Cancer*, **67**, 1556 (1991).
[36] J. S. Park, J. H. Chyun, Y. K. Kim, L. L. Line, M. C. Maloney and B. P. Chew, *FASEB J.*, **18**, A479 (2004).
[37] H. Jyonouchi, L. Zhang, M. Gross and Y. Tomita, *Nutr. Cancer*, **21**, 47 (1994).
[38] G. Y. Kim, J. H. Kim, S. C. Ahn, H. J. Lee, D. O. Moon, C. M. Lee and Y. M. Park, *Immunology*, **113**, 203 (2004).
[39] W .C . Van Hoorhis, J. Valinski, E. Hoffman, J. Luban, L. S. Hair and R. M. Steinman, *J. Exp. Med.*, **158**, 174 (1983).
[40] P. J. Mosca, A. C. Hobeika, T. M. Clay, S. K. Nair, E. K. Thomas, M. A. Morse and H. K. Lyerly, *Blood*, **96**, 3499 (2000).
[41] R. Lapointe, J. F. Toso, C. Butts, H. A. Young and P. Hwu, *Eur. J. Immunol.*, **30**, 3291 (2000).
[42] J. L. Schwartz, E. Flynn and G. Shklar, *Micronutr. Immunol. Function*, **587**, 92 (1990).
[43] J. J. Michal, B. P. Chew, T. S. Wong, L. R. Heirman and F. E. Standaert, *J. Dairy Sci.*, **77**, 1408 (1994).
[44] L. W. Tjoelker, B. P. Chew, T. S. Tanaka and L. R. Daniel, *J. Dairy Sci.*, **71**, 3112 (1988).
[45] W. Zhao, Y. Han, B. Zhao, S. Hirota, J. Hou and W. Xin, *Biochim. Biophys. Acta*, **138**, 77 (1998).
[46] B. P. Chew, J. S. Park, T. S. Wong, H. W. Kim, B. C. Weng, K. M. Byrne, M. G. Hayek and G. A. Reinhart, *J. Nutr.*, **130**, 1910 (2000).
[47] H. W. Kim, B. P. Chew, T. S. Wong, J. S. Park, B. C. Weng, K. M. Byrne, M. G. Hayek, and G. A. Reinhart, *Vet. Immunol. Immunopath.*, **74**, 315 (2000).
[48] B. P. Chew, J. S. Park, M. G. Hayek, S. Massimino and G. A. Reinhart, *FASEB J.*, **18**, A533a (2004).

[49] B. P. Chew, J. S. Park, M. G. Hayek, S. Massimino and G. A. Reinhart, *FASEB J.*, **18**, A533b (2004).
[50] J. S. Park, B. P. Chew, M. G. Hayek, S. Massimino and G. A. Reinhart, *FASEB J.*, **18**, A533c (2004).
[51] H. W. Kim, B. P. Chew, T. S. Wong, J. S. Park, B. C. Weng, K. M. Byrne, M. G. Hayek and G. A. Reinhart, *Vet. Immunol. Immunopath.*, **73**, 331 (2000).
[52] W. Fawzi, M. G. Herrera and P. Nestel, *J. Nutr.*, **130**, 2537 (2000).
[53] B. Watzl, A. Bub, B. R. Brandstetter and G. Rechkemmer, *Brit. J. Nutr.*, **82**, 383 (1999).
[54] B. Watzl, A. Bub, K. Briviba and G. Rechkemmer, *Ann. Nutr. Metab.*, **47**, 255 (2003).
[55] B. Watzl, A. Bub, M. Blockhaus, B. M. Herbert, P. M. Luhrmann, M. Neuhauser-Berthold and G. Rechkemmer, *J. Nutr.*, **130**, 1719 (2000).
[56] K. Briviba, S. E. Kulling, J. Moseneder, B. Watzl, G. Rechkemmer and N. Abub, *Carcinogenesis*, **25**, 2373 (2004).
[57] E. Hagiwara, K. I. Takahashi, T. Okubo, S. Ohno, A. Ueda, A. Aoki, S. Odagiri and Y. Ishigatsubo, *Cytokine*, **14**, 121 (2001).
[58] B. P. Chew, J. S. Park, H. W. Kim, T. S. Wong, C. R. Baskin. K. W. Hinchcliff, R. A. Swenson, G. A. Reinhart, J. R. Burr and M. G. Hayek, in *Recent Advances in Canine and Feline Nutrition. Vol. III*, (ed. G. A. Reinhart and D. P. Carey), p. 531, Orange Fraser Press, Wilmington, Ohio (2000).
[59] C. R. Baskin, K. W. Hinchcliff, R. A. DiSilvestro, G. A. Reinhart, M. G. Hayek, B. P. Chew, J. R. Burr and R. A. Swensen, *Am. J. Vet. Res.*, **61**, 886 (2000).
[60] L. A. Herraiz, W. C. Hsieh, R. S. Parker, J. E. Swanson, A. Bendich and D. A. Roe, *J. Am. Coll. Nutr.*, **17**, 617 (1998).
[61] T. G. Blanchard and S. J. Czinn, *Int. J. Cancer*, **73**, 684 (1997).
[62] C. Lindholm, M. Quiding-Jarbink, H. Lonroth, A. Hamlet and A. M. Svernnerholm, *Infect. Immunol.*, **66**, 5964 (1998).
[63] M. Bennedsen, X. Wang, R. Willen, T. Wadstrom and L. P. Andersen, *Immunol. Lett.*, **70**, 185 (1999).
[64] H. Van Remmen and A. Richardson, *Exp. Gerontol.*, **36**, 957 (2001).
[65] J. M. van der Horst-Graat, F. J. Kok and E. G. Schouten, *Brit. J. Nutr.*, **92**, 113 (2004).
[66] S. Massimino, R. J. Kearns, K. M. Loos, J. Burr, J. S. Park, B. P. Chew, S. Adams and M. G. Hayek, *J. Vet. Intern. Med.*, **17**, 835 (2004).
[67] M. F. Beal, *Ann. Neurol.*, **58**, 495 (2005).
[68] V. H. Perry, C. Cunningham and D. Boche, *Curr. Opin. Neurol.*, **15**, 349 (2002).
[69] R. P. Copp, T. Wisniewski, F. Hentati, A. Larnaout, M. B. Hamida and H. J. Kayden, *Brain Res.*, **822**, 80 (1999).
[70] J. J. Mitchell, M. Paiva and M. B. Heaton, *Neurosci. Lett.*, **263**, 189 (1999).
[71] D. J. Pattison, D. P. M. Symmons, M. Lunt, A. Welch, S. A. Bingham, N. E. Day and A. J. Silman, *Am. J. Clin. Nutr.*, **82**, 451 (2005).
[72] J. R. Cerhan, K. J. Saag, L. A. Merlino, T. R. Mikuls and L. A. Criswell, *Am. J. Epidemiol.*, **157**, 345 (2003).
[73] Y. Tomita, K. Himeno, K. Nomoto, H. Endo and T. Hirohata, *J. Natl. Cancer Inst.*, **78**, 679 (1987).
[74] P. Prakash, R. M. Russell and N. I. Krinsky, *Nutr.*, **131**, 1574 (1991).
[75] H. Jyonouchi, S. Sun, K. Iijima and M. D. Gross, *Nutr. Cancer*, **36**, 59 (2000).
[76] B. P. Chew, M. W. Wong and T. S. Wong, *Anticancer Res.*, **17**, 3689 (1996).
[77] J. S. Park, B. P. Chew and T. S. Wong, *J. Nutr.*, **128**, 1650 (1998).
[78] C. G. Cerveny, B. P. Chew, J. S. Park and T. S. Wong, *FASEB J.*, **13**, A210 (1999).
[79] J. Folkman, *Perspect. Biol. Med.*, **29**, 10 (1985).
[80] B. P. Chew, C. M. Brown, J. S. Park and P. F. Mixter, *Anticancer Res.*, **23**, 3333 (2003).
[81] J. L. Schwartz and G. Shklar, *Nutr. Cancer*, **27**, 192 (1997).
[82] C. J. Lacey, M. E. Murphy, M. J. Sanderson, E. F. Monteiro, A. Vail and C. J. Schorah, *Int. J. STD AIDS*, **7**, 485 (1996).

[83] F. L. Tomaka, P. J. Cimoch, W. M. Rieter, R. J. Keller, D. S. Berger, J. Piperado, P. M. Nemechek, S. D. Loss and R. A. Houghton, *Int. Conf. AIDS*, **10**, 221 (1994).
[84] R. Ullrich, T. Schneider, W. Heise, W. Schmidt, R. Averdunk, E. O. Riecken and M. Zeitz, *AIDS*, **8**, 661 (1994).
[85] G. Melikian, F. Mmiro, C. Ndugwa, R. Pery, J. Brooks-Jackson, E. Garrett, J. Tielsch and R. D. Semba, *Appl. Nat. Nutr. Invest.*, **17**, 567 (2001).
[86] G. O. Coodley, H. D. Nelson, M. O. Loveless and C. Folk, *J. Acquir. Imm. Defic. Syndr.*, **6**, 272 (1993).
[87] A. Bianchi-Santamaria, S. Fedeli and L. Santamaria, *Med. Oncol. Med. Tumor Pharmacother.*, **9**, 151 (1992).
[88] J. Austin, N. Sibghal, R. Voigt, F. Smaill, M. J. Gill, S. Walmsley, I. Salit, J. Gilmour, W. F. Schlech 3rd, S. Choudhri, A. Rachlis, J. Cohen, S. Trottier, E. Toma, P. Phillips, P. M. Ford, R. Woods, J. Singer, D. P. Zarowny and D. W. Cameron, *Eur. J. Clin. Nutr.*, **60**, 1266 (2006).
[89] S. Silverman Jr., G. E. Kaugars, J. Gallo, J. S. Thompson, D. P. Stites, W. T. Riley and R. B. Brandt, *Oral Surg. Oral Med. Oral Path.*, **78**, 442 (1994).
[90] W. W. Fawzi, *AIDS*, **16**, 1935 (2002).
[91] A. Coutsoudis, K. Pillay, E. Spooner, L. Kuhn and H. M. Goovadia, *AIDS*, **13**, 1517 (1999).
[92] E. Villamor and W. W. Fawzi, *Clin. Microbiol. Rev.*, **18**, 446 (2005).
[93] S. Zhuang, M. C. Lynch and I. E. Kocheva, *Exp. Cell. Res.*, **250**, 203 (1999).
[94] K. Chlichlia, M. E. Peter, M. Rocha, C. Scaffidi, M. Bucur, P. H. Krammer, V. Schirrmacher and V. Umansky, *Blood*, **91**, 4311 (1998).
[95] P. Palozza, S. Serini, A. Torsello, F. Di Nicuolo, E. Piccioni, V. Ubaldi, C. Pioli, F. I. Wolf and G. Calviello, *J. Nutr.*, **133**, 381 (2003).
[96] J. S. Park, B. P. Chew, T. S. Wong, J. X. Zhang and N. S. Magnuson, *Nutr. Cancer*, **33**, 206 (1999).
[97] V. N. Sumatran, R. Zhang, D. S. Lee and M. S. Wicha, *Cancer Epidemiol. Biomarkers Prevent.*, **9**, 257 (2000).
[98] P. Palozza, S. Serini, N. Maggiano, M. Angelini, A. Boninsegna, F. Di Nicuolo, F. O. Ranelletti and G. Calviello, *Carcinogenesis*, **23**, 11 (2002).
[99] A. Nahum, K. Hirsch, M. Danilenko, C. K. W. Watts, O. W. J. Prall, J. Levy and Y. Sharoni, *Oncogene*, **20**, 3428 (2001).
[100] M. L. Smith and A. J. Fornace Jr., *Mutat. Res.*, **340**, 109 (1996).
[101] J. L. M. Gervais, P. Seth and H. Zhang, *J. Biol. Chem.*, **273**, 19207 (1998).
[102] K. Schulze-Osthoff, M. Los and P. A. Baeuerle, *Biochem. Pharmacol.*, **50**, 735 (1995).
[103] A. Bowie and L. A. O'Neill, *Biochem. Pharmacol.*, **59**, 13 (2000).
[104] E. C. Tibaduiza, J. C. Fleet, R. M. Russell and N. I. Krinsky, *Nutr. Cancer*, **132**, 1368 (2002).
[105] H. Sano, Y. Kawahito, R. Wilder, A. Hashiramoto, S. Mukai, K. Asai, S. Kimura, H. Kato, M. Kondo and T. Hla, *Cancer Res.*, **55**, 3785 (1995).
[106] K. Ortmann, T. Mayerhofer, N. Getoff and R. Kodym, *Radiation Res.*, **161**, 48 (2004).
[107] R. A. Kowluru and P. Koppolu, *Free Radic. Res.*, **36**, 993 (2002).
[108] P. Palozza, S. Serini, N. Maggiano, G. Tringali, P. Navarra, F. O. Ranelletti and G. Calviello, *J. Nutr.*, **135**, 129 (2005).
[109] A. Ben-Dor, M. Steiner, L. Gheber, M. Danilenko, N. Dubi, K. Linniwiel, A. Zick, Y. Sharoni and J. Levy, *Molec. Cancer Ther.*, **4**, 177 (2005).
[110] G. W. Burton and K. U. Ingold, *Science*, **224**, 569 (1984).
[111] P. Palozza, *Nutr. Rev.*, **56**, 257 (1998).

Carotenoids
Volume 5: Nutrition and Health

Chapter 18

Biological Activities of Carotenoid Metabolites

Xiang-Dong Wang

A. Introduction

Considerable research effort has been expended in an attempt to substantiate and understand the potential roles of carotenoids in human health and disease, as described in previous *Chapters* in this *Volume*. Early studies dealt with β-carotene (**3**) and other provitamin A carotenoids, but more recent research efforts have focused on the potential roles in health and disease of the non-provitamin A carotenoids, such as lycopene (**31**) and lutein (**133**).

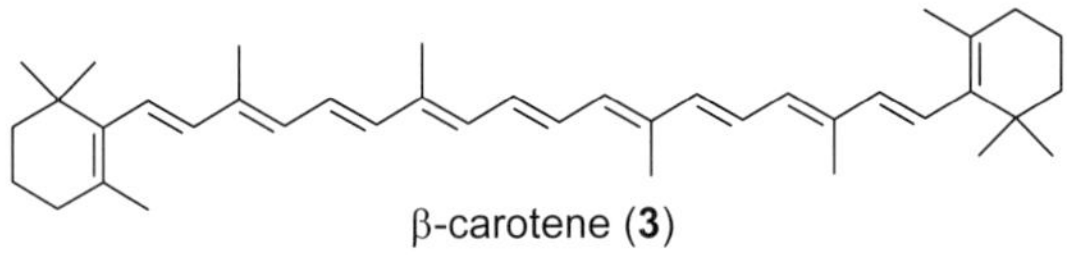

β-carotene (**3**)

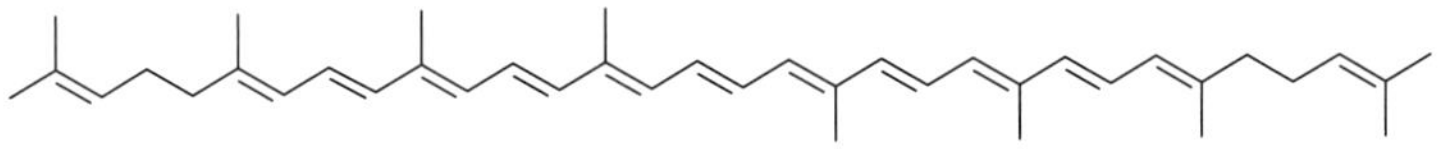

lycopene (**31**)

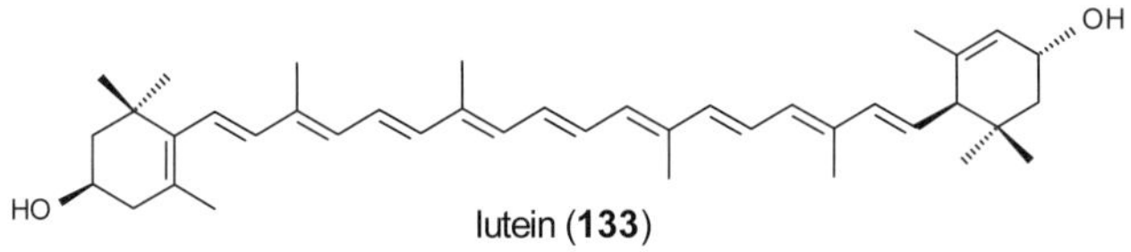

lutein (**133**)

Carotenoids are lipophilic and the series of conjugated double bonds in the central chain of the molecule makes them susceptible to oxidative cleavage [1], to isomerization between the *trans* (*E*) and *cis* (*Z*) forms [2], and to the formation of potentially bioactive metabolites [3]. The best known metabolite of carotenoids is vitamin A, as retinal (*1*), retinol (*2*) and retinoic acid (*3*). In recent years, considerable efforts have been made to identify biological properties of carotenoid metabolites other than vitamin A and related retinoids. Better understanding of the molecular details behind the actions of these carotenoid oxidative metabolites may yield insights into both physiological and pathophysiological processes in human health and disease.

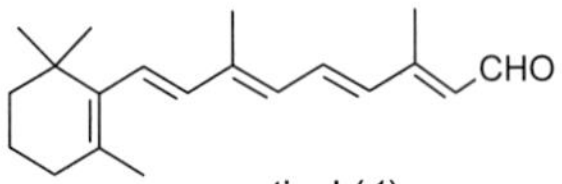

retinal (*1*)

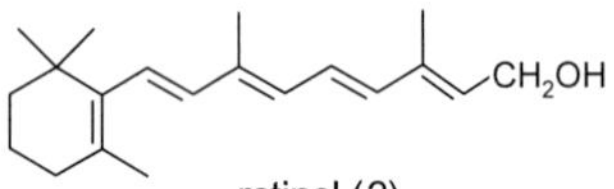

retinol (*2*)

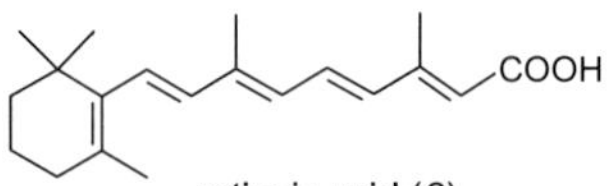

retinoic acid (*3*)

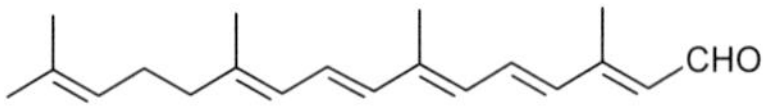

acycloretinal (*4*)

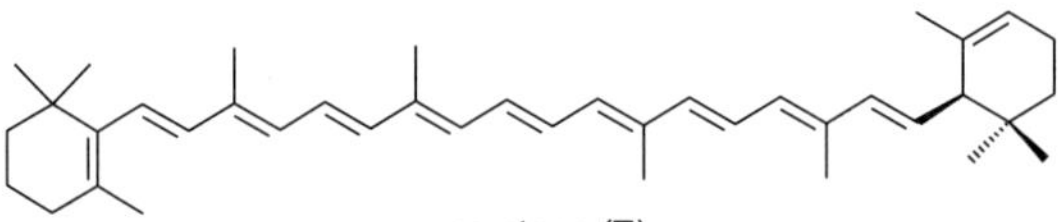

α-carotene (**7**)

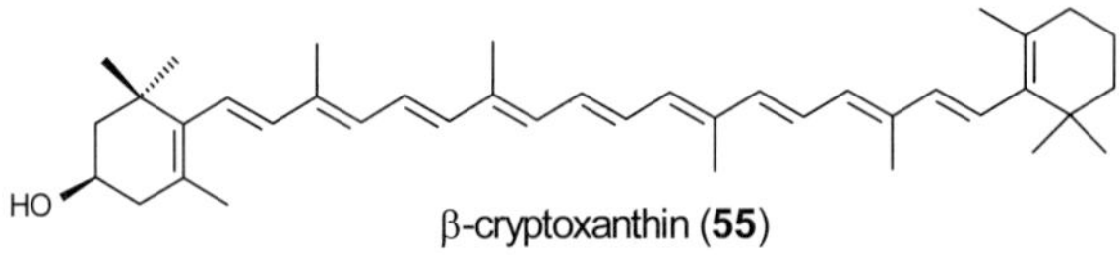

β-cryptoxanthin (**55**)

For provitamin A carotenoids, such as β-carotene, α-carotene (**7**), and β-cryptoxanthin (**55**), central cleavage is a major pathway leading to vitamin A and its derivatives [4,5] (see *Chapter 8* and *Volume 4, Chapter 16*). This pathway has been substantiated by the cloning of a central cleavage enzyme, β-carotene 15,15'-oxygenase (BCO1), which can cleave carotenoids at their C(15,15') double bond. It has been well demonstrated that retinoids, the most important oxidative products of provitamin A carotenoids, play an essential role in many critical biological processes, including vision, reproduction, metabolism, differentiation, haematopoiesis, bone development, and pattern formation during embryogenesis [6]. Considerable evidence demonstrates that the natural and synthetic retinoids may be effective in the prevention and treatment of a variety of human chronic diseases, including cancer [7]. Retinoids elicit these responses through their ability to regulate gene expression at specific target sites within the body [8,9].

An alternative pathway for carotenoid metabolism in mammals, the excentric cleavage pathway, was confirmed by the molecular identification of β-carotene 9,10-oxygenase (BCO2) in humans and animals. Recent biochemical characterization of BCO2 demonstrates that this enzyme catalyses the excentric cleavage not only of provitamin A carotenoids, but also of non-provitamin A carotenoids, such as lycopene. Recent experimental data suggest that carotenoid metabolites from the excentric cleavage pathway may have more important biological roles than their parent compounds. These metabolites may have specific actions on several important cellular signalling pathways and molecular targets, and may have both beneficial and detrimental effects in relation to cancer prevention [3,10,11]. The ability of carotenoids to modulate cell communication and signalling pathways, especially in relation to the cell cycle and apoptosis, is described in *Chapter 11*. This *Chapter* now discusses recent findings on the formation of metabolites of carotenoids, in particular β-carotene and lycopene, and addresses the question of whether the reported biological actions of carotenoids and their potential significance in chronic diseases such as cancer are in fact mediated by metabolites and not by the intact carotenoids themselves.

B. Carotenoid Metabolites

1. Enzymic central cleavage *in vitro*

a) β-Carotene 15,15′-oxygenase (BCO1)

As described in detail in *Chapter 8* and *Volume 4, Chapter 16*, carotenoids such as β-carotene, α-carotene, and β-cryptoxanthin are cleaved symmetrically at their central double bond by BCO1 [12,13]. This enzyme has been cloned in several species and its biochemical and enzymological characterization has been reported [14-18]. It has been detected in or isolated from several mouse and human tissues (*e.g.* liver, kidney, intestinal tract, and testis) which are important in carotenoid/retinoid metabolism. A purified recombinant BCO1, obtained *via* a human liver cDNA library, showed cleavage activity towards both β-carotene and β-cryptoxanthin, which has only one unsubstituted β ring [18], but with an approximately 4-fold lower affinity towards β-cryptoxanthin ($K_m = 30.0 \pm 3.8$ μM) than towards β-carotene ($K_m = 7.1 \pm 1.8$ μM) [18]. No cleavage of lycopene or zeaxanthin was detected. In other studies, no detectable activity of human retinal pigment epithelium BCO1 towards lycopene or lutein was observed [19]. No lycopene cleavage products were detected when lycopene was incubated with the *Drosophila* homologue of BCO1 [14] or with crude preparations of rat liver and intestine [20]. The presence of an unsubstituted β ring in the substrate appears to be a prerequisite for activity [21].

b) Central cleavage of lycopene

Indirect evidence for central cleavage of lycopene has been obtained, however. When a lycopene-accumulating strain of *Escherichia coli* was engineered to express also mouse BCO1, a distinct bleaching of colour was seen following induction, suggesting cleavage of lycopene [17]. In addition, purified recombinant mouse BCO1 was shown to display cleavage activity towards lycopene, but the expected central cleavage product acycloretinal (*4*) was only detected when the lycopene concentrations used were 2.5-3 times higher than the observed K_m (6 μM) for β-carotene. Taken together, these studies suggest that lycopene is, at best, a poor substrate for BCO1.

It is unclear if the lycopene substrate used in the BCO1 studies *in vitro* described above [17-20] was the pure all-*E* form or contained *Z* isomers. As reported in the following Section, *Z* isomers of lycopene were better substrates than (all-*E*)-lycopene for BCO2 [22]. This raises the important question of whether BCO1 might also cleave *Z* isomers of lycopene to acycloretinoids. Reports on the use of (all-*E*)-lycopene as a supplement revealed dramatic increases in the 5*Z*, 9*Z* and 13*Z* isomers in blood and tissues [23-26].

2. Excentric enzymic cleavage *in vitro*

a) β-Carotene 9,10-oxygenase (BCO2)

An alternative pathway for carotenoid metabolism in vertebrates is asymmetric cleavage at one of the other double bonds of the polyene chain, *i.e.* excentric cleavage [27-29]. The existence of this pathway was for a long time controversial [4,30,31], but has been substantiated by the identification of a series of homologous carbonyl cleavage products, including 14′-apo-β-caroten-14′-al (**513**), 12′-apo-β-caroten-12′-al (**507**), 10′-apo-β-caroten-10′-al (**499**), 8′-apo-β-caroten-8′-al (**482**), and 13-apo-β-caroten-13-one (*5*), along with retinoic acid, in tissue homogenates of humans, ferrets, and rats [32-35].

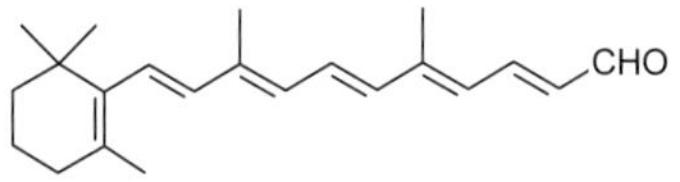

14'-apo-β-caroten-14'-al (**513**)

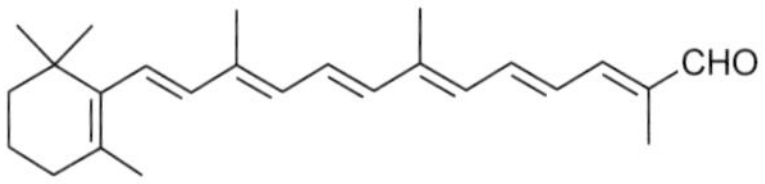

12'-apo-β-caroten-12'-al (**507**)

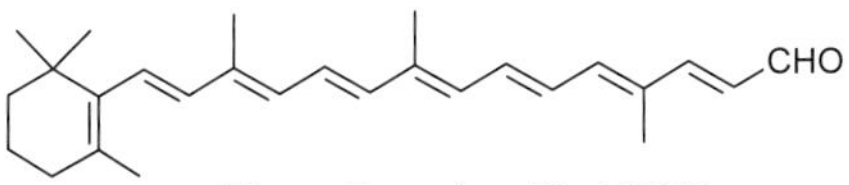

10'-apo-β-caroten-10'-al (**499**)

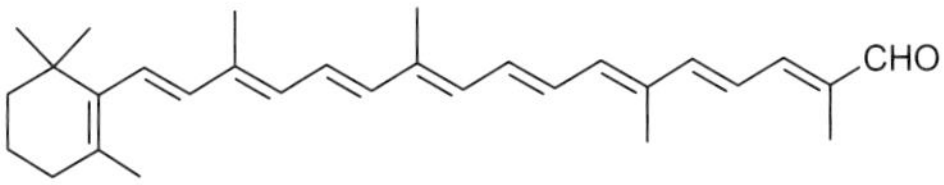

8'-apo-β-caroten-8'-al (**482**)

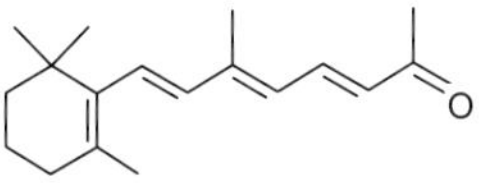

13-apo-β-caroten-13-one (*5*)

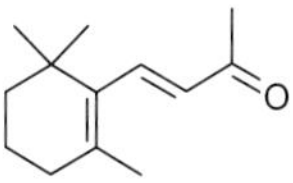

β-ionone (*6*)

A second cleavage enzyme, BCO2, has been cloned from mice, humans, and zebrafish [36]. BCO2 appears to be specific for the C(9,10) double bond; β-carotene, for example, gives rise to 10′-apo-β-caroten-10′-al (**499**) and β-ionone (*6*) [36]. Apo-β-carotenals can be precursors of vitamin A *in vitro* [28,37] and *in vivo* [38], by further cleavage. They can also be oxidized to their corresponding apo-β-carotenoic acids, which may then undergo a process similar to β-oxidation of fatty acids, to produce retinoic acid [35]. It is not known, however, whether other apo-β-carotenals with shorter carbon chain lengths are formed by further metabolism of the initial cleavage product, 10′-apo-β-caroten-10′-al, or are primary products of direct cleavage of other double bonds in the carotene molecule. Not much is known about the ability of BCO2 to cleave carotenoids other than β-carotene.

b) Excentric cleavage of lycopene

Ability to cleave lycopene was first demonstrated indirectly with strains of *Escherichia coli* engineered to synthesize and accumulate lycopene, and expressing the mouse BCO2 [36]. When BCO2 was induced, a distinct colour shift from red to white occurred, indicating cleavage. Following this, the ferret BCO2 gene has been cloned and characterized [22]; ferrets (*Mustela putorius furo*) and humans are similar in terms of carotenoid absorption, tissue distribution and concentrations, and metabolism [39,40]. The enzyme is expressed in the testis, liver, lung, prostate, intestine, stomach, and kidneys of ferrets, similar to the expression pattern of human BCO2 [41].

The recombinant ferret BCO2 catalysed the excentric cleavage of the C(9,10) double bond of (all-*E*)-β-carotene but not that of (all-*E*)-lycopene, though *Z* isomers of lycopene were cleaved effectively [22]. Based on the BCO2 expressed in Sf9 cells from the insect *Spodoptera frugiperda*, a K_m of 3.5 μM was estimated for (all-*E*)-β-carotene, but the kinetic constants for lycopene could not be calculated because of the difficulty in controlling auto-isomerization, so that mixed isomers of lycopene had to be used as the substrate. Because the lycopene substrate mixture contained only ~20% as *Z* isomers, and the ferret BCO2 would not cleave (all-*E*)-lycopene, it can be speculated that the K_m for (*Z*)-lycopene is actually much lower than that of the lycopene isomer mixture. This indicates that (*Z*)-lycopene might be a better substrate than (all-*E*)-β-carotene for the ferret BCO2. It is not known why ferret BCO2 preferentially cleaves the 5*Z* and 13*Z* isomers of lycopene into 10′-apolycopenal. It has been suggested that the structure of the *Z* isomers of lycopene could mimic the ring structure of the β-carotene molecule and fit into the substrate-enzyme binding pocket. The different solubility properties may be a key factor, however; the *Z* isomers are more readily solubilized and much less prone to aggregation and crystallization than is (all-*E*)-lycopene (see *Volume 4, Chapter 5*). The observation that supplementation with (all-*E*)-lycopene results in a significant increase in the tissue concentration of (*Z*)-lycopene in animals and humans supports this [23-26].

3. Non-enzymic oxidative breakdown

The non-enzymic formation of carotenoid oxidation products *in vitro* is well known (see *Chapter 12* and *Volume 4, Chapter 7*). Because of the susceptibility of carotenoids to cleavage by auto-oxidation, radical-mediated oxidation, and singlet oxygen, such breakdown products may be formed *in vivo* by non-enzymic processes if the tissues are exposed to oxidative stress such as smoking and drinking. The possible biological importance of such processes and products is poorly understood.

4. Detection of central and excentric cleavage products *in vivo*

a) Metabolites of β-carotene

Retinol, retinal, retinoic acid and retinyl ester can be detected in both plasma and tissues of animals and humans. Although the conversion of β-carotene by BCO2 to other apocarotenoids remains to be determined directly, a recent study [42], suggests that excentric cleavage of ingested β-carotene does occur in humans *in vivo*. Application of the highly sensitive technique accelerator mass spectrometry, that can measure attomole amounts (1 in 10^{-18} parts) of ^{14}C, enabled the detection in human plasma of [^{14}C]-apo-β-caroten-8′-al and several other, unidentified [^{14}C]-labelled metabolites from a true tracer oral dose of (all-*E*)-[10,11,10′,11′-$^{14}C_4$]-β-carotene (1.01 nmol; 543 ng; 100 nCi) in human plasma. Although further study is needed to identify and characterize the additional metabolites, this observation is in agreement with the previous identification of excentric cleavage metabolites in animal models. Significant amounts of 8′-apo-β-caroten-8′-al (**482**), 10′-apo-β-caroten-10′-al (**499**) and 12′-apo-β-caroten-12′-al (**507**) were isolated from the intestines of chickens given dietary β-carotene [28,29]. Also, 12′-apo-β-caroten-12′-al and 10′-apo-β-caroten-10′-al, as well as retinoids, were isolated from ferret intestinal mucosa after perfusion of β-carotene *in vivo* [43,44].

b) Metabolites of lycopene

Labelled 8′-apolycopen-8′-al (**491**) and 12′-apolycopen-12′-al (*7*) were detected in rat liver 24 hours after dosing with [^{14}C]-lycopene [45]. A large quantity of unidentified polar short-chain compounds was also detected. 10′-Apolycopen-10′-ol (**504.3**) has been detected, together with several unidentified compounds, in the HPLC profiles of lung tissue from ferrets supplemented with lycopene for 9 weeks [22]; this compound is the reduction product of the predicted aldehyde cleavage product 10′-apolycopen-10′-al (*8*). Neither the latter nor 10′-apolycopen-10′-oic acid (**504.4**) was detected, so it is likely that 10′-apolycopen-10′-al is a short-lived intermediate compound which, as soon as it is formed, is rapidly reduced to its alcohol form *in vivo*.

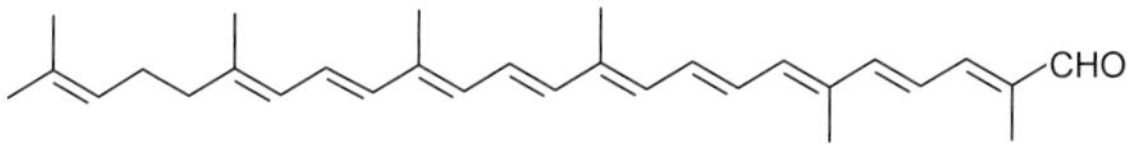

8'-apolycopen-8'-al (**491**)

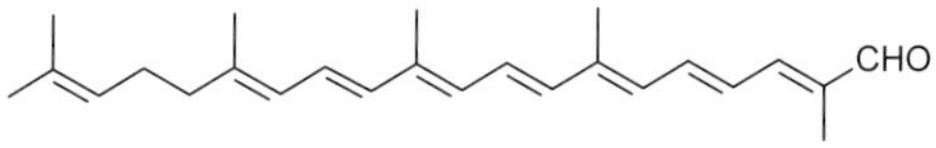

12'-apolycopen-12'-al (*7*)

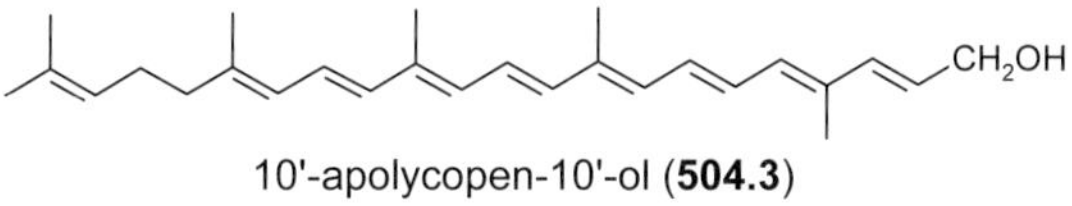

10'-apolycopen-10'-ol (**504.3**)

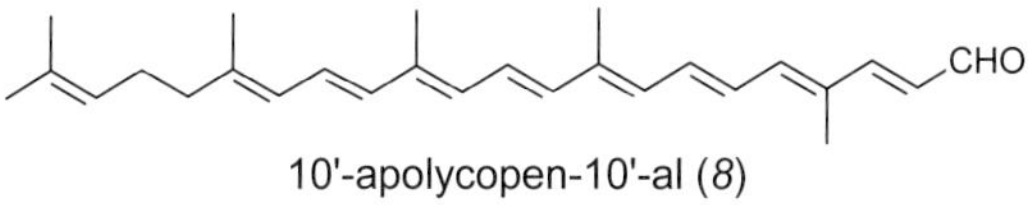

10'-apolycopen-10'-al (*8*)

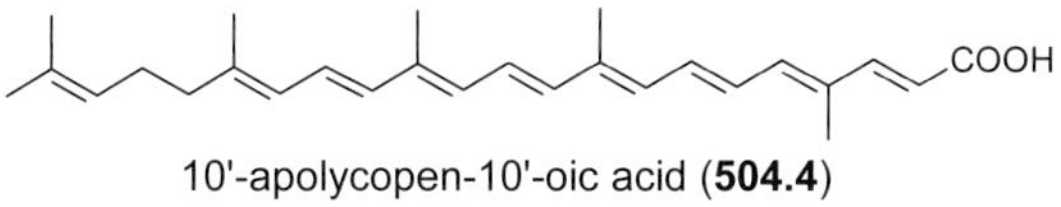

10'-apolycopen-10'-oic acid (**504.4**)

If 10′-apolycopen-10′-oic acid were present, its concentration was too low to be detected. It was demonstrated subsequently that incubation of 10′-apolycopen-10′-al with the post-nuclear fraction of hepatic tissue of ferrets resulted in the formation of either 10′-apolycopen-10′-ol or 10′-apolycopen-10′-oic acid, depending on the presence of either NAD^+ or NADH, respectively. Nonetheless, the presence of specific metabolites has not been consistent across different animal models.

C. Retinoids and the Retinoid Signalling Pathway

1. Retinoic acid and retinoic acid receptors

Provitamin A carotenoids, such as β-carotene and its excentric cleavage metabolites, can serve as direct precursors for (all-*trans*)-retinoic acid (*3*) and (9-*cis*)-retinoic acid (*9*) [35,46,47], which are ligands for retinoic acid receptors (RAR) and retinoid X receptors (RXR), respectively.

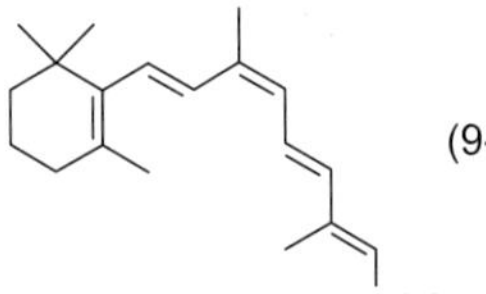

(9-*cis*)-retinoic acid (*9*)

Retinoid receptors function as ligand-dependent transcription factors and regulate gene expression by binding as dimeric complexes to the retinoic acid response element (RARE) and the retinoid X response element (RXRE), which are located in the 5′ promoter region of responsive genes. RXR can form dimeric complexes not only with RAR but also with other members of the nuclear hormone receptor superfamily, such as thyroid hormone receptors (TR), the vitamin D receptor (VDR), peroxisome proliferator-activated receptors (PPAR), and possibly other receptors with unknown ligands, designated orphan receptors. Recent results have shown that decreased expression of all RAR and RXR receptor subtypes is a frequent event in non-small cell lung cancer [48]. Particularly, studies *in vivo* and *in vitro* indicate that RARβ expression, which can be induced by retinoic acid, is frequently reduced in various cancer cells and tissues [49]. Recent evidence also suggests that the RARβ subtypes, RARβ2 and RARβ4, have contrasting biological effects, as tumour suppressor and tumour promoter, respectively, in human carcinogenesis [50]. The down-regulation of all retinoid subclasses suggests a fundamental disruption of the regulation of the retinoid pathway in lung cancer [48]. Conversely, restoration of RARβ2 in an RARβ-negative lung cancer cell line has been reported to inhibit tumorigenicity in nude mice [51]. Retinoic acid can reverse the suppression of RARβ protein caused by benzo(a)pyrene diol epoxide by increasing transcription of RARβ, in immortalized oesophageal epithelial cells [52] and lung cancer cells [53]. In a small-scale human trial, daily treatment with (9-*cis*)-retinoic acid for three months restored RARβ expression in the bronchial epithelium of former smokers [54]. Supplementing carcinogen-initiated AJ mice with (9-*cis*)-retinoic acid decreased lung tumour multiplicity and increased lung RARβ mRNA levels [55]. It has been shown that β-carotene supplementation prevents skin carcinoma formation by upregulating RARβ [56].

2. Effects of provitamin A carotenoids and their metabolites

a) β-Carotene and 14′-apo-β-caroten-14′-oic acid

Previously it was observed that the down-regulation of RARβ by smoke-borne carcinogens was completely reversed by treatment with either β-carotene or its oxidative metabolite, 14′-apo-β-caroten-14′-oic acid (*10*), in normal bronchial epithelium cells [57]. Further, transactivation of the RARβ2 promoter appeared to occur mainly as a result of the metabolism of 14′-apo-β-caroten-14′-oic acid to (all-*trans*)-retinoic acid [57]. Therefore, the molecular mode of action of provitamin A carotenoids can be mediated by retinoic acid, *via* transcriptional activation of a series of genes with distinct antiproliferative or proapoptotic activity, thereby eliminating cells with irreparable alterations in the genome, or killing neoplastic cells.

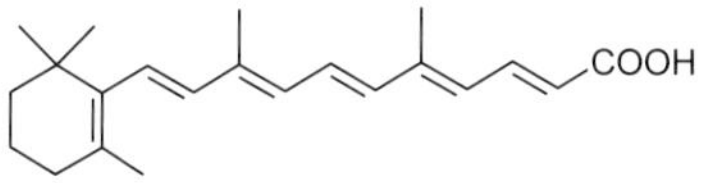

14′-apo-β-caroten-14′-oic acid (*10*)

It has been reported recently, however, that 14′-apo-β-caroten-14′-al, in contrast to 14′-apo-β-caroten-14′-oic acid, inhibited activation and responses of the nuclear receptors PPARγ, PPARα, or RXR, and promoted inflammation *in vivo* [58,59]. Although the question of whether this proinflammatory effect of 14′-apo-β-caroten-14′-al was related to dose was not addressed, this finding may help to explain the detrimental effect of β-carotene supplementation trials in smokers.

The basis of one explanation for this lies in the doses used and the free-radical-rich atmosphere in lungs of cigarette smokers [60-62]. This environment alters β-carotene metabolism and produces undesirable oxidative metabolites [62], which can affect many processes, *e.g.* they can facilitate the binding of metabolites of benzo(a)pyrene to DNA [63], down-regulate RARβ [61], up-regulate activator protein 1 (AP-1, c-Jun and c-Fos) activity [60], induce carcinogen-activating enzymes [64], enhance the induction of BALB/c 3T3 cell transformation by benzo(a)pyrene [65], inhibit gap junction communication in A549 lung cancer cells [66] or impair mitochondrial functions [67]. The doses of β-carotene used in the ATBC (Alpha-Tocopherol, Beta-Carotene Cancer Prevention Study) and CARET (Beta-Carotene and Retinol Efficacy Trial) studies were 20 to 30 mg per day, for 2-8 years, and these doses are 10-15 times the average daily dietary intake of β-carotene in the U. S. Such a pharmacological dose of β-carotene in humans could result in the accumulation of relatively high levels of β-carotene and its oxidative metabolites in the lung tissue, especially after long periods of supplementation. Potentially this could also lead to a decrease in lung retinoic acid concentration *via* induction of cytochrome P450 (CYP) enzymes [68]. It should be noted that excentric cleavage products, which may be formed in excess in cancerous lung tissue, have not been shown to bind competitively to RARβ at physiologically relevant levels [69]. It is possible, however, that the excentric cleavage products of carotenoids interfere with the binding of retinoic acid to its receptors when the retinoic acid level in tissues is low. This may be seen in the case of cigarette smoking and excessive alcohol drinking, which result in higher cytochrome P450 enzyme levels and breakdown of retinoic acid [68,70,71]. The loss of or low levels of retinoic acid, including both all-*trans* and 9-*cis* isomers, or the 'functional' down-regulation of retinoid receptors, because of the lack of retinoic acid, could interfere with retinoid signal transduction and result in enhanced cell proliferation and potentially malignant transformation. This is supported by previous studies with ferrets, showing that high dose β-carotene supplementation (equivalent to an intake of 30 mg of β-carotene/day/70 kg human, considered a pharmacological dose) and/or cigarette smoke exposure decreased levels of retinoic acid and RARβ protein, but increased levels of c-Jun and cyclin D1 proteins, and induced precancerous lesions in lung tissue [60,72].

Recently, further evidence was obtained to support the notion that the anti- or procarcinogenic response to β-carotene supplementation reported in human intervention trials and in animal studies may be related to the stability of β-carotene and its metabolites in different organ environments (such as high oxidative stress in the lung due to smoking or low antioxidants levels) as well as retinoic acid status in the lungs. A mixture of β-carotene (equivalent to 12 mg/day in human) together with the antioxidants α-tocopherol and ascorbic acid (which facilitates both recycling and stability of β-carotene and α-tocopherol, but was not used in the ATBC study and is expected to be low in this population of heavy smokers), provides protection against lung cancer risk by maintaining normal levels of retinoic acid [73]. This is in agreement with a previous study *in vitro* which showed that the addition of both ascorbic acid and α-tocopherol to an incubation mixture of β-carotene with ferret lung tissue can inhibit the smoke-enhanced production of excentric cleavage metabolites of β-carotene, increase the formation of retinal and retinoic acid [74] and decrease the smoke-induced catabolism of retinoic acid [68]. These studies and the known biochemical interactions of β-carotene, vitamin E and vitamin C (see *Chapter 12*) suggest that this combination of nutrients, rather than the individual agents, could be an effective chemopreventive strategy against lung cancer in smokers.

b) Other provitamin A carotenoids

Protective actions of other provitamin A carotenoids, namely β-cryptoxanthin and α-carotene, have been reported recently. Mechanisms for any effects of these carotenoids and their metabolites on the retinoid signalling pathway have not been elucidated, although their interactions with cleavage enzymes, depending on dose and the oxidative environment of the lungs, may be similar to those of β-carotene. In a recent cell culture study, it was observed that β-cryptoxanthin can inhibit lung cancer cell growth by increasing the expression of RARβ and transactivating RARE [75]. Another study [76] demonstrated that, in a yeast two-hybrid system, both β-cryptoxanthin and lutein exhibited RAR ligand activity but this was completely abolished by the RAR pan-antagonist LE540. Although their binding affinity was three orders of magnitude lower than that of (all-*trans*)-retinoic acid, β-cryptoxanthin and lutein were shown to bind to the RAR ligand-binding domain in the CoA-BAP system but not to the RXR ligand-binding domain, indicating that they can serve as ligands for RAR without being metabolized.

3. Effects of lycopene and its metabolites

Whereas up-regulation of retinoid receptor expression and function by provitamin A carotenoids may play a role in mediating the growth inhibitory effects of retinoids in cancer cells [57,75], it is not clear if non-provitamin A carotenoids and their metabolites may function in a similar fashion.

a) Acycloretinoic acid

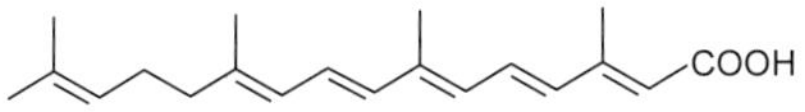

acycloretinoic acid (*11*)

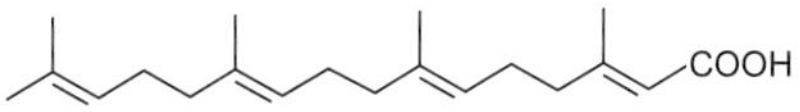

geranylgeranoic acid (*12*)

Several reports have evaluated the ability of the 'acycloretinoid' acycloretinoic acid (*11*), which would be a product of central cleavage of lycopene, to transactivate the RARE. It was demonstrated [77] that acycloretinoic acid can transactivate a RARE-reporter gene through an interaction with RARα. The potency of activation was approximately 100-fold lower than with retinoic acid, however. Binding affinity studies indicated that acycloretinoic acid had no appreciable binding affinity for RXRα, but bound RARα with an equilibrium dissociation constant in the range of 50-150 nM, two orders of magnitude lower than that of (all-*trans*)-retinoic acid. Intact lycopene did not show any significant binding to either receptor, but administration of lycopene led to a weak transactivation of the RARE-reporter gene [77]. Similar findings were reported for the RARβ2 promoter. Only when acycloretinoic acid was provided at concentrations 500-fold higher than retinoic acid was an effect on luciferase and β-galactosidase reporter activity observed [78]. At the concentrations used in this study there was no effect of intact lycopene on reporter transactivation. In another study, acycloretinoic acid was found to have no significant effect on transactivation of RAR and RXR reporter systems [79]. Whereas no effect of acycloretinoic acid on retinoid signalling *in vivo* has been substantiated, a synthetic acyclic retinoid, E-5166 (geranylgeranoic acid, *12*), has been shown to transactivate retinoid reporter systems, and to have potential benefits in treatment of hepatocellular carcinoma [80-82].

b) Other lycopene metabolites

The question arises of whether 10′-apolycopen-10′-oic acid could also be an activator of RARs. Three cell lines, which represent different stages of lung carcinogenesis, namely NHBE, a normal human bronchial epithelial cell line, BEAS-2B, an immortalized human bronchial epithelial cell line, and A549 cells, a non-small cell lung cancer cell, were incubated with increasing concentrations of 10′-apolycopen-10′-oic acid (3-5 μmol/L) [83]. After 48 hours, a dose-dependent increase in RARβ mRNA expression was observed in both NHBE and BEAS-2B cell lines. The effect of 10′-apolycopen-10′-oic acid was similar to that of (all-*trans*)-retinoic acid. Was the increased RARβ mRNA expression due to increased transactivation of the RARβ promoter region? To investigate the involvement of the RARE in the promoter, located between -53 and -37 bp, site-directed mutagenesis was utilized to abolish the RAR binding site. This mutation completely abolished induction of promoter activity by both retinoic acid and 10′-apolycopen-10′-oic acid. These results suggest that the growth inhibitory actions of 10′-apolycopen-10′-oic acid may be mediated through retinoid signalling. On the other hand, other studies show that 12′-apo-β-caroten-12′-oic acid (**510**) can inhibit the growth of HL-60 cells [84] and 14′-apo-β-caroten-14′-oic acid (*10*) can stimulate the differentiation of U937 leukaemia cells [85] and inhibit the growth of breast cancer cells [69].

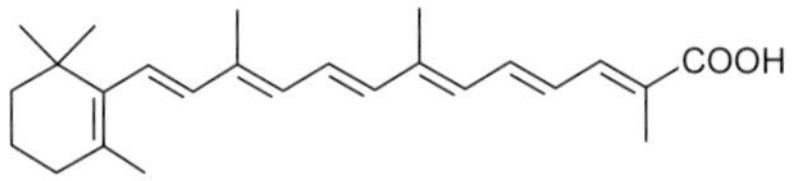

12′-apo-β-caroten-12′-oic acid (**510**)

These effects appear not to be due to cellular conversion of the apo-β-carotenoid to retinoic acid because no retinoids were detected in the cells after treatment with the apocarotenoids. It is possible, therefore, that breakdown products of β-carotene may play a role in regulating cell function which does not depend on their ability to be metabolized to retinoic acid. This is also supported by the finding that apocarotenoids have very low binding affinity to RAR [69]. Although 14′-apo-β-caroten-14′-oic acid can induce transcriptional activity of the RARβ2 promoter *via* its conversion into retinoic acid in the normal bronchial epithelial cells [57], it is possible that the conversion of β-carotene into retinoic acid is impaired in transformed cells.

This retinoid-independent activity of provitamin A carotenoid metabolites may be similar to the biological activity of non-provitamin A carotenoid metabolites. It has been shown that acycloretinoic acid, which is not a ligand for RAR and RXR [78], inhibited the growth of HL-60 human promyelocytic leukaemia cells [86], human mammary cancer cells [77], and human prostate cancer cells and this effect was significantly greater than those of (9-*cis*)-retinoic acid and (all-*trans*)-retinoic acid [87]. In addition, it has been shown that lycopene oxidation products enhance gap junctional communication [88] (see also Section **D**.4). A retinoic acid receptor antagonist did not suppress reporter activity induced by lycopene, indicating that gene activation by retinoids and by non-provitamin A carotenoids occurs by different mechanisms [89].

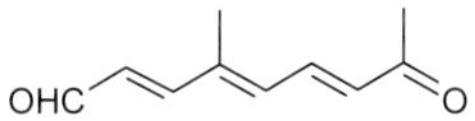

MON (*13*)

An oxidative product of lycopene, (*E*,*E*,*E*)-4-methyl-8-oxo-2,4,6-nonatrienal (MON, *13*), that induced apoptosis in HL-60 cells, was identified [90]. A dose-dependent decrease of cell viability was observed, with a concomitant increase in chromatin condensation and nuclear fragmentation, characteristic of apoptosis.

In spite of these observations with cell cultures, however, the physiological significance of these lycopene products remains unknown because none of them has been detected in biological systems.

c) Retinoid-dependent and retinoid-independent roles of carotenoid metabolites

Beyond participating in known retinoid signalling pathways, carotenoid metabolites appear to have retinoid-independent roles, signalling through other nuclear receptors (*e.g.* currently characterized orphan receptors with unknown ligands) or interacting with signalling pathways through transcriptional 'cross-talk'. Since RXRs function not only as heterodimeric partners of other nuclear receptors (*e.g.* VDR, PPAR), but also as active transducers of tumour suppressive signals [7], it will be interesting to investigate whether the biological activity of carotenoids or their metabolites is mediated through interaction with RARs, RXRs, PPAR, VDR or other orphan receptors. Recently, supplementation with (9-*cis*)-retinoic acid in combination with 1α,25-dihydroxyvitamin D3 was shown to reduce vitamin D-induced toxicity symptoms compared to those in mice that were supplemented with vitamin D alone, thereby suggesting an interaction between the two compounds [91]. In addition, carotenoids may be beneficial for bone formation by up-regulating vitamin D receptor levels [92]. It has been shown that both the PPARγ ligand ciglitazone and an RXR ligand cooperatively promoted transcriptional activity of RAREβ and induced RARβ expression in human lung cancer cells [57].

D. Effects of Carotenoid Metabolites on Other Signalling and Communication Pathways

1. Nuclear factor-E2 related factor 2 (Nrf2) signalling pathway

a) Phase II enzymes and antioxidant-response elements

In recent years, evidence has begun to accumulate indicating that some beneficial effects of carotenoids may be due to induction of the phase II enzymes that have important detoxifing and antioxidant properties in combating foreign substances (xenobiotics) including potential carcinogens [93]. Induction of phase II enzymes is mediated through *cis*-regulatory DNA sequences known as antioxidant response elements (ARE) that are located in the promoter or enhancer region of the gene [94]. The major ARE transcription factor Nrf2 (nuclear factor E2-related factor 2) is a primary agent in induction of antioxidant and detoxifying enzymes [95] and is essential for the induction of several phase II enzymes, including glutathione S-transferases (GSTs) and NAD(P)H:quinone oxidoreductase (NQO1) [96]. The induction of these and other phase II detoxifying/antioxidant enzymes, such as haem oxygenase-1 (HO-1), glutathione reductase (GR), glutamate-cysteine ligase (catalytic subunit, GCLC, and modifier subunit, GCLM), microsomal epoxide hydrolase 1 (mEH), and the UDP glucuronosyl-transferase 1 family polypeptide A6 (UGT1A6), results in the detoxication of carcinogens and the inactivation of reactive oxygen species (ROS), thus contributing to the protective effect of chemopreventive agents [95]. Under normal conditions, most of the Nrf2 is sequestered in the cytoplasm by 'Kelch-like erythroid Cap'n'Collar homologue-associated protein 1' (Keap 1) and only residual nuclear Nrf2 binds to the ARE to drive basal activities. Exposure to some chemopreventive agents leads to the dissociation of the Nrf2-Keap1 complex in the cytoplasm and the translocation of Nrf2 into the nucleus. The nuclear accumulation of Nrf2 subsequently activates target genes of phase II enzymes [95]. Because of its critical roles in detoxication and antioxidant processes in carcinogenesis, Nrf2 has been recognized as a potential molecular target for cancer prevention [95]. It has been shown that various dietary and synthetic compounds, *e.g.* sulforaphane [97], curcumin [98], and (-)-epigallocatechin 3-gallate [99], can induce gene expression mediated by Nrf2-ARE; this could be one mechanism for their reported chemopreventive effects.

b) Effects of carotenoids and their metabolites

Not only β-carotene but some non-provitamin A carotenoids, including lycopene, have been shown to induce several phase II enzymes both *in vivo* and *in vitro* [92,100,101]. An induction of UDP-glucuronosyltransferase and NQO1 was observed in rats fed various carotenoids [100]. Whilst canthaxanthin (**380**) and astaxanthin (**404-406**) induced phase II activity, lycopene and lutein had no effect after 15 days of feeding.

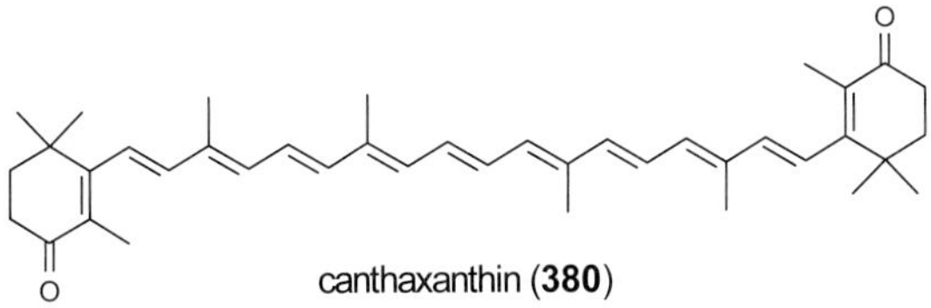

canthaxanthin (**380**)

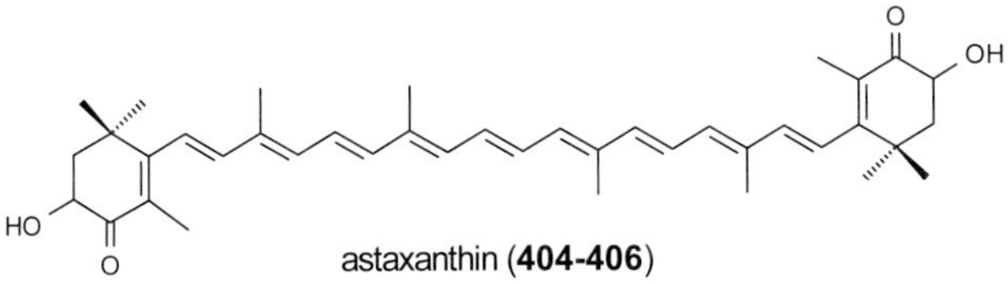

astaxanthin (**404-406**)

In another study, a dose-dependent induction of several phase I and II enzymes was demonstrated in female Wistar rats supplemented with lycopene at doses ranging from 0.001 to 0.1 g/kg body weight for 2 weeks [101]. Hepatic ethoxyresorufin O-dealkylase (EROD) and benzyloxyresorufin O-dealkylase (BROD) activity increased approximately 2-fold and 50%, respectively, suggesting activation of the cytochrome P450 enzyme CYP1A. In addition, several liver and red blood cell phase II enzyme activities, such as GR, GST and NQO1, were significantly increased by feeding lycopene. The induction of phase II enzymes by lycopene has been reported in other animal studies [102], but it was not determined whether this induction was due to the intact carotenoid or its metabolites. This has been addressed in other studies.

c) Lycopene metabolites

An ethanolic extract containing lycopene and unidentified hydrophilic oxidative derivatives was shown to induce phase II enzymes and activate ARE-driven reporter gene activity with a similar potency to lycopene [92], but those chemically produced oxidative derivatives have not been found in mammalian tissues. Evidence has been obtained recently to show that 10′-apolycopen-10′-oic acid, derived from cleavage of lycopene, induces phase II enzyme expression *in vitro* [103]. Work with BEAS-2B human bronchial epithelial cells has shown a dose-dependent and time-dependent increase in the accumulation of nuclear Nrf2 protein, following 10′-apolycopen-10′-oic acid treatment [103]. In addition, 10′-apolycopen-10′-oic acid significantly induced mRNA expression of several phase II enzymes, including HO-1, NQO1, GST, GR,GCLC and GCLM, mEH and UGT1A6, compared to treatment with THF alone [103]. Additionally, 10′-apolycopen-10′-al, 10′-apolycopen-10′-ol and 10′-apolycopen-10′-oic acid were all effective in activating the Nrf2-mediated induction of HO-1 [103], although the mechanisms of this remain unknown. The activation of Nrf2 is complex and is controlled through multiple regulatory mechanisms, including Keap1-mediated ubiquitination and degradation, subcellular distribution, and phosphorylation. 10′-Apolycopen-10′-al showed stronger induction of HO-1 than did 10′-apolycopen-10′-oic acid and 10′-apolycopen-10′-ol. Its aldehyde group is highly reactive, and is capable of forming Schiff bases with the amino groups of protein and of reacting with other cellular macromolecules, *e.g.* directly modifying the reactive cysteine residues in Keap1 and interrupting Keap1-mediated Nrf2 ubiquitination and degradation. It is also possible that these lycopenoids affect upstream signalling pathways, such as mitogen activated protein kinases (MAPKs), phosphoinositol 3-kinase (PI3K), epidermal growth factor receptor (EGFR) and protein kinase C (PKC), which all have been shown to play a role in the regulation of Nrf2-ARE in lung epithelial cells. Clearly, further investigation is needed.

It is known that intact lycopene can function as an antioxidant *in vitro* (see *Chapter 12*) and there is evidence that lycopene metabolites could also have antioxidant functions. Pre-treatment of BEAS-2B cells with 10′-apolycopen-10′-oic acid (3-10 μM) for 24 hours resulted in a dose-dependent inhibition of both endogenous ROS production and H_2O_2-induced oxidative damage, as measured by release of lactate dehydrogenase [103]. This decrease in ROS was comparable to that in control cells treated with the antioxidant *t*-butylhydroquinone.

Thus lycopene metabolites in general, and 10′-apolycopen-10′-oic acid in particular, may possess antioxidant activity by inducing antioxidant enzymes. It will be interesting to investigate whether metabolites of other carotenoids can induce phase II detoxifying/ antioxidant enzymes.

2. Carotenoid metabolites and the mitogen-activated protein kinase pathway

Among the members of the mitogen-activated protein kinase (MAPK) family are Jun N-terminal kinase (JNK), extracellular-signal-regulated protein kinase (ERK) and p38 mitogen-activated protein kinase. They are activated by phosphorylation in response to extracellular stimuli and environmental stress and may play an important role in carcinogenesis [104,105]. JNK was shown to phosphorylate the protein c-Jun on sites Ser-63 and Ser-73 and to increase activator protein 1 (AP-1) transcription activity, thereby mediating cell proliferation and apoptosis [104,105] (see *Chapter 11*). ERK also induces c-Jun through phosphorylation and activation of the AP-1 component ATF1 at Ser-63 [106]. On the other hand, MAPK phosphatases (MKPs), a family of dual-specificity protein phosphatases, can dephosphorylate both phosphothreonine and phosphotyrosine residues to inactivate JNK, ERK and p38 both *in vitro* and *in vivo* [107,108]. It has been shown that phosphorylated-JNK, phosphorylated-ERK, and phosphorylated-p38 are preferred substrates for the isomer MKP-1 *in vivo* [107,108].

a) β-Carotene and metabolites

Previously, expression of AP-1, c-Jun and c-Fos was shown to be up-regulated in the lungs of smoke-exposed ferrets supplemented with β-carotene [61], compared to the control animals. This overexpression of AP-1 was positively associated with increased levels of cyclin D1 protein and with squamous metaplasia in the lungs of animals exposed to smoke [61]. It is conceivable that chronic excess β-carotene intake may modulate MAPK signalling and cause abnormal cell cycle regulation, and promote carcinogenesis. This hypothesis is supported by the observation that smoke exposure and/or high dose β-carotene activated the phosphorylation of JNK and p38, but significantly reduced lung MKP-1 protein levels [109]. In contrast, low dose β-carotene attenuated smoke-induced JNK phosphorylation by preventing down-regulation of MKP-1 [109]. This inhibitory effect of low dose β-carotene supplementation could be due to increased lung retinoic acid levels in smoke-exposed animals; it is known that retinoic acid can inhibit phosphorylation of MAPKs, such as JNK and ERK, by upregulation of MKP-1 [110-112].

Relatively high β-carotene supplementation (equivalent to 12 mg/day in humans) in the presence of ascorbic acid and α-tocopherol blocked smoke-induced phosphorylation of JNK and ERK completely by preventing smoke-induced reductions in retinoic acid levels in the lungs of ferrets [113]. The combined antioxidants also inhibited smoke-induced oxidative stress, assessed by Comet analysis [113]. These data may help to explain the conflicting results of the negative human β-carotene intervention trials, which used high doses of β-carotene, *versus* the positive observational epidemiological studies which showed that diets high in fruits and vegetables containing β-carotene (but at much lower concentrations than in the intervention studies and with other antioxidants present) are associated with a decreased risk for lung cancer.

b) Lycopene and metabolites

Lycopene has also been shown to inhibit JNK, p38 and ERK, and the transcription factor, nuclear factor-κB (NF-κB) [114]. 10′-Apolycopen-10′-oic acid showed dose-dependent inhibition of cell growth and induced apoptosis in human THLE-2 liver cells by stimulating the cyclin-dependent kinase inhibitor p21 and by reducing activation of JNK and cyclin D1 gene expression [115]. It is possible, therefore, that the inhibition of JNK activation by combined antioxidants, including both provitamin A and non-provitamin A carotenoids, may help to 'rescue' the functions of RARs; it has been reported recently that activation of JNK contributes to RAR dysfunction by phosphorylation of RARα and by inducing its degradation through the ubiquitin-proteasomal pathway [116]. It has been shown that RARα can activate the RARE of RARβ, suggesting a possible accessory role for RARα in RARβ expression [117]. Further examination of effects of provitamin A and non-provitamin A carotenoids on the stability and degradation of RARs through JNK-mediated pathways should be considered.

3. Carotenoid metabolites and the insulin-like growth factor-1 (IGF-1) pathway

It has been suggested that the signalling system involving insulin-like growth factors (IGF) may play a role in the biological action of lycopene [118,119]. IGF-1 and IGF-2 are mitogens (mitosis inducers) that play a central role in regulation of cellular proliferation, differentiation, and apoptosis [120]. By binding to membrane IGF-1 receptors, IGFs activate intracellular phosphatidylinositol 3′-kinase (PI3K)/Akt/protein kinase B and Ras/Raf/MAPK pathways, which regulate various biological processes such as cell cycle progression, survival, and transformation [121]. IGFs are sequestered in circulation by a family of binding proteins (IGFBP1 – IGFBP6), which regulate the availability of IGFs to bind to IGF receptors [121]. Disruption of normal IGF signalling, leading to hyperproliferation and survival signal expression, has been implicated in the development of several tumour types [122]. Indeed, strong positive associations have been found between plasma IGF-1 levels and risk of prostate cancer [123], breast cancer [124], and colorectal cancer [125]. Recent epidemiological studies provide supportive evidence that lycopene may have a chemopreventive effect against a broad range of epithelial cancers, particularly prostate, breast, colorectal, and lung cancer [126-129].

A possible mechanism was indicated when it was shown [118-120] that IGF-1-stimulated cell growth and DNA-binding activity of the AP-1 transcription factor were reduced by physiological concentrations of lycopene in endometrial, mammary (MCF-7), and lung (NCI-H226) cancer cell lines. Lycopene has been shown to inhibit IGF-1-stimulated insulin receptor substrate 1 phosphorylation and cyclin D1 expression, block IGF-1-stimulated cell-cycle progression [118,130], and increase membrane-associated IGFBPs [118,131]. Consistent with previous findings from studies *in vitro*, recent epidemiological studies demonstrated that higher dietary intake of lycopene is associated with lower circulating levels of IGF-1 [132] and higher levels of IGFBPs [133,134].

The effect of lycopene on prevention of IGF signalling in cigarette smoke-related lung carcinogenesis has been examined in the ferret model [24]. Plasma IGF-1 levels were not affected by cigarette smoke exposure or lycopene supplementation, but IGFBP-3 levels were raised by lycopene supplementation and decreased by smoke exposure. Lycopene increased plasma IGFBP-3 regardless of smoke exposure status. Increased plasma IGFBP-3 was associated with inhibition of cigarette smoke-induced lung squamous metaplasia, and with decreased levels of proliferating cell nuclear antigen (PCNA), phosphorylated Bad, and cleaved caspase 3, suggesting inhibition of cell proliferation and induction of apoptosis [24]. These results, along with others, suggest that interference with IGF-1 signalling could be an important mechanism by which lycopene may exert an anticancer action. There is recent evidence that lycopene metabolites may be partly responsible for this effect. Treatment with 10′-apolycopen-10′-oic acid (5-20 μM) resulted in a dose-dependent increase in IGFBP-3 mRNA levels in THLE-2 human liver cells, whereas similar concentrations of retinoic acid, lycopene, and acycloretinoic acid showed no significant effect [135].

4. Carotenoid metabolites and gap-junction communication

Gap-junction communication has been implicated in the control of cell growth *via* differentiation, proliferation and apoptosis [136]. A large body of evidence now indicates loss of gap-junctional communication (GJC) to be a hallmark of carcinogenesis [137] and the targeting of the gap-junction proteins, connexins, has been suggested as a possible strategy for chemoprevention. Retinoids and carotenoids increase gap-junction communication between normal and transformed cells [90,138]. Both provitamin A and non-provitamin A carotenoids were shown to inhibit carcinogen-induced neoplastic transformation [139] and to upregulate connexin 43 (Cx43) mRNA expression [90,138]. Treatment with retinoic acid increased Cx43 expression within 6 hours, but carotenoid treatment required approximately three times longer to produce the same response [140,141]; this lag in activity is often attributed to the formation of active metabolites.

Several lines of evidence from experiments *in vitro* indicate that carotenoid oxidation products/metabolites may be responsible for increased GJC, especially in the case of lycopene. After oxidation of lycopene with hydrogen peroxide and osmium tetroxide, a product, 2,7,11-trimethyltetradecahexaene-1,14-dial (*14*), was isolated and this induced gap-junction communication as effectively as did retinoic acid. The oxidation product lycopene 5,6-epoxide (**222**), which is found in tomatoes, was shown to increase Cx43 expression in human keratinocytes [142].

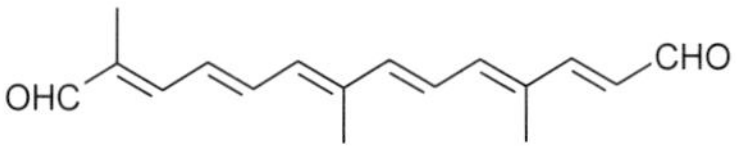

2,7,11-trimethyltetradecahexaene-1,14-dial (*14*)

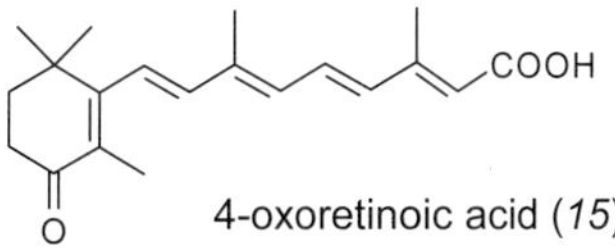

4-oxoretinoic acid (*15*)

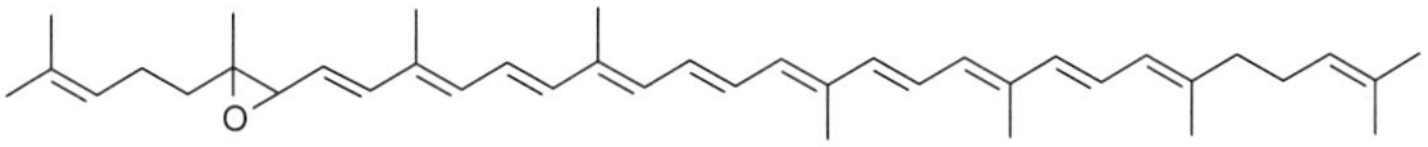

lycopene 5,6-epoxide (**222**)

Acycloretinoic acid was also shown to increase GJC [78], but this effect was achieved only at high concentrations, indicating that the contribution of acycloretinoic acid to the activity of lycopene on GJC appears to be minimal. While the Cx43 promoter does not contain a RARE, it has been reported that RAR antagonists inhibited upregulation by retinoids and had no influence on the effect of carotenoids [143]. The modulating effect of oxidation products and enzymic cleavage metabolites of lycopene on GJC could, therefore, provide two separate pathways for increasing GJC. Whether 10′-apolycopenoids contribute to the activity of lycopene on GJC warrants further study, however. In addition, two decomposition products of the non-provitamin A carotenoid canthaxanthin, namely the (all-*trans*) and (13-*cis*) isomers of 4-oxoretinoic acid (*15*) have the same activity as canthaxanthin on enhancing cell-cell gap-junctional communication in murine fibroblasts [144,145]. 4-Oxoretinoic acid has been shown to serve as a ligand of the nuclear receptor, RARβ [146]. Whether canthaxanthin can regulate gene expression *via* this metabolite remains to be determined.

E. Overview and Conclusions

An understanding of the impact of carotenoid oxidation products and bioactive metabolites is important in understanding the health effects of carotenoids. It appears that while small quantities of carotenoid metabolites can offer protection against chronic diseases and certain cancers, larger amounts may actually be harmful, especially when coupled with a highly oxidative environment (*e.g.* the lungs of a cigarette smoker or liver of an excessive alcohol drinker). The potential effects, beneficial and harmful, attributed to carotenoids and their metabolites are summarized in Fig. 1.

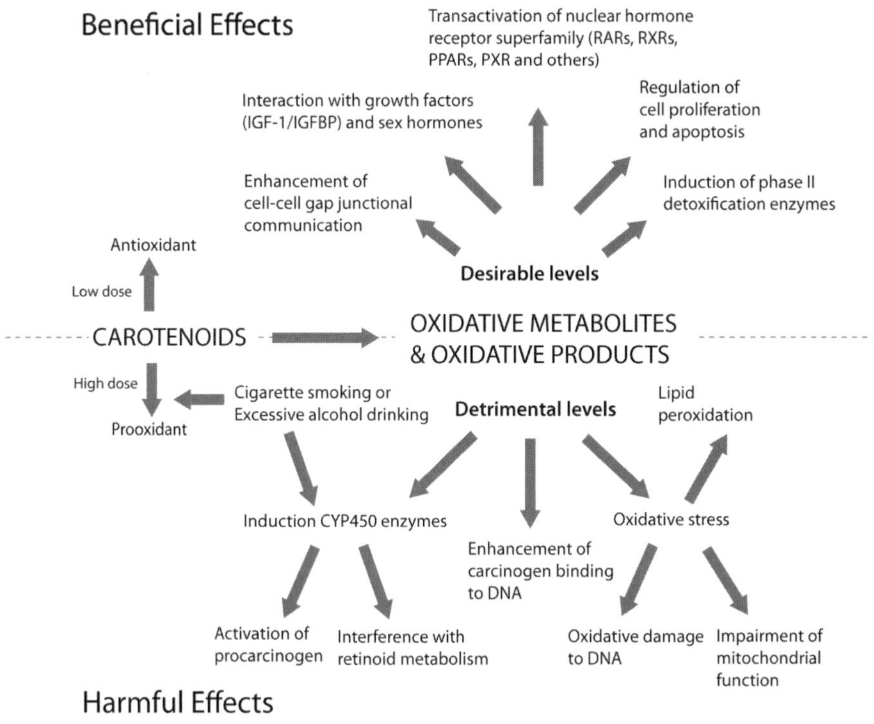

Fig. 1. Summary of biological effects of carotenoids and their metabolites and oxidation products. With low-dose treatment, carotenoids are likely to have antioxidant properties and produce small, desirable levels of metabolites, leading to beneficial effects. With high-dose treatment, carotenoids may have pro-oxidant properties, especially in smokers. The higher levels of oxidative products may be detrimental and lead to harmful effects. (Adapted from [4]).

Various effects of carotenoids on cellular functions and signalling pathways have been reported, as summarized in Fig. 2. An important question that remains unanswered is whether these effects are a result of the direct actions of intact carotenoids or of their derivatives, for example products of central or excentric cleavage of provitamin A and non-provitamin A carotenoids. Whilst evidence is presented in this *Chapter* to support the latter, more research is needed to identify and characterize additional carotenoid metabolites and breakdown products, and their biological activities; this could provide invaluable insights into the mechanisms underlying the actions of carotenoids.

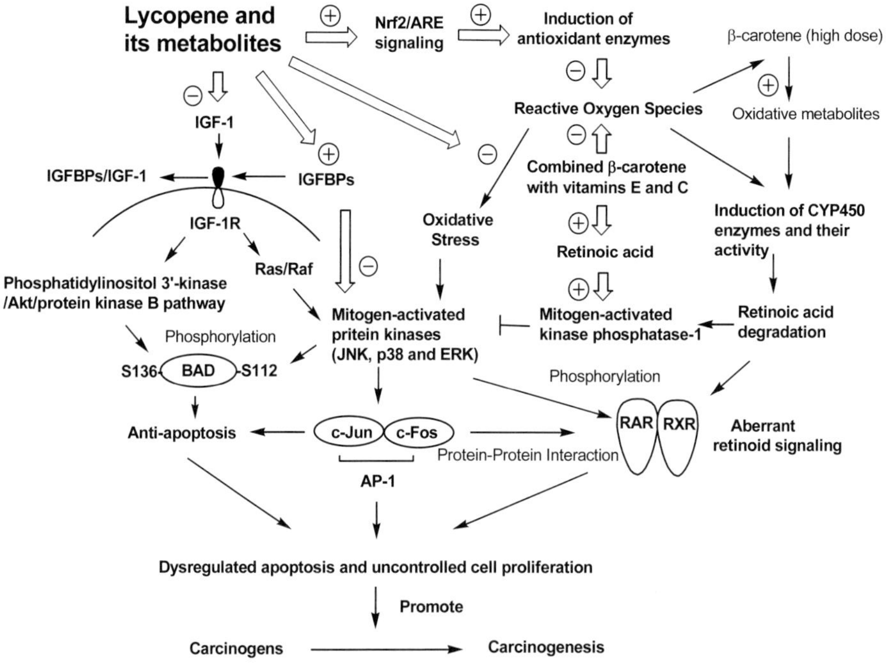

Fig. 2. Diagram illustrating the complex interactions between signalling pathways, especially those that result in impaired regulation of apoptosis and uncontrolled cell proliferation, leading to carcinogenesis. Interactions of carotenoids and their metabolites with these processes are indicated here and discussed in the text.

Finally, in considering the efficacy and complex biological functions of carotenoids in human chronic disease prevention, it appears that combining provitamin A carotenoids (*e.g.*, β-cryptoxanthin) with other antioxidants would be a particularly useful approach for chemoprevention. Antioxidants such as ascorbic acid and α-tocopherol limit the formation of excessive oxidative cleavage products of carotenoids in an oxidative environment. In addition, provitamin A carotenoids combined with non-provitamin A carotenoids (such as lycopene and lutein), which target different signalling pathways, could provide complementary or synergistic protective effects against chronic diseases including certain kind of cancers.

References

[1] F. Khachik, in *Carotenoids and Retinoids. Molecular Aspects and Health Issues* (ed. L. Packer, K. Kraemer, U. Obermüller-Jevic and H. Sies), p. 61, AOCS Press, Champaign, Illinois (2005).

[2] A. C. Boileau and J. W. Erdman Jr., in *Carotenoids in Health and Disease* (ed. N. I. Krinsky, S. T. Mayne and H. Sies), p. 209, Marcel Dekker, New York, NY (2004).

[3] X. D. Wang, in *Carotenoids in Health and Disease* (ed. N. I. Krinsky, S. T. Mayne and H. Sies), p. 313, Marcel Dekker, New York, NY (2004).

[4] X. D. Wang and N. I. Krinsky, *Subcell. Biochem.,* **30**, 159 (1998).

[5] H. Merintz and X.-D. Wang, in *Vitamin A: New Research* (ed. I. T. Loessing), p. 39, Nova Science Publisher, Columbia (2007).

[6] P. Chambon, *FASEB J.*, **10**, 940 (1996).

[7] L. Altucci and H. Gronemeyer, *Nat. Rev. Cancer*, **1**, 181 (2001).

[8] A. L. Fields, D. R. Soprano and K. J. Soprano, *J. Cell. Biochem.*, **102**, 886 (2007).

[9] G. Duester, *Cell*, **134**, 921 (2008).

[10] Y. Sharoni, M. Danilenko and J. Levy, in *Carotenoids in Health and Disease* (ed. N. I. Krinsky, S. T. Mayne and H. Sies), p. 165, Marcel Dekker, New York, NY (2004).

[11] Y. Sharoni, M. Danilenko, N. Dubi, A. Ben-Dor and J. Levy, *Arch. Biochem. Biophys*., **430**, 89, (2004).

[12] D. S. Goodman and H. S. Huang, *Science*, **149**, 879 (1965).

[13] J. A. Olson and O. Hayaishi, *Proc. Natl. Acad. Sci. USA*, **54**, 1364 (1965).

[14] J. von Lintig and K. Vogt, *J. Biol. Chem*., **275**, 11915 (2000).

[15] A. Wyss, G. Wirtz, W. Woggon, R. Brugger, M. Wyss, A. Friedlein, H. Bachmann and W. Hunziker, *Biochem. Biophys. Res. Commun*., **271**, 334 (2000).

[16] M. G. Leuenberger, C. Engeloch-Jarret and W. D. Woggon, *Angew. Chem. Int. Ed. Engl*., **40**, 2613 (2001).

[17] T. M. Redmond, S. Gentleman, T. Duncan, S. Yu, B. Wiggert, E. Gantt and F. X. Cunningham Jr., *J. Biol. Chem*., **276**, 6560 (2001).

[18] A. Lindqvist and S. Andersson, *J. Biol. Chem*., **277**, 23942 (2002).

[19] W. Yan, G. F. Jang, F. Haeseleer, N. Esumi, J. Chang, M. Kerrigan, M. Campochiaro, P. Campochiaro, K. Palczewski and D. J. Zack, *Genomics*, **72**, 193 (2001).

[20] A. Nagao and J. A. Olson, *FASEB J*., **8**, 968 (1994).

[21] E. Poliakov, S. Gentleman, F. X. Cunningham Jr., N. J. Miller-Ihli and T. M. Redmond, *J. Biol. Chem*., **280**, 29217 (2005).

[22] K. Q. Hu, C. Liu, H. Ernst, N. I. Krinsky, R. M. Russell and X. D. Wang, *J. Biol. Chem*., **281**, 19327 (2006).

[23] A. C. Boileau, N. R. Merchen, K. Wasson, C. A. Atkinson and J. W. Erdman Jr., *J. Nutr*., **129**, 1176 (1999).

[24] C. Liu, F. Lian, D. E. Smith, R. M. Russell and X. D. Wang, *Cancer Res*., **63**, 3138 (2003).

[25] C. Liu, R. M. Russell and X. D. Wang, *J. Nutr*., **136**, 106 (2006).

[26] K. Wu, S. J. Schwartz, E. A. Platz, S. K. Clinton, J. W. Erdman Jr., M. G. Ferruzzi, W. C. Willett and E. L. Giovannucci, *J. Nutr*., **133**, 1930 (2003).

[27] J. Glover, *Vitam. Horm*., **18**, 371 (1960).

[28] R. V. Sharma, S. N. Mathur and J. Ganguly, *Biochem. J*, **158**, 377 (1976).

[29] R. V. Sharma, S. N. Mathur, A. A. Dmitrovskii, R. C. Das and J. Ganguly, *Biochim. Biophys. Acta*, **486**, 183 (1976).

[30] J. Ganguly and P. S. Sastry, *World Rev. Nutr. Diet*., **45**, 199 (1985).

[31] G. Wolf, *Nutr. Rev*., **53**, 134 (1995).

[32] X. D. Wang, G. W. Tang, J. G. Fox, N. I. Krinsky and R. M. Russell, *Arch. Biochem. Biophys*., **285**, 8 (1991).
[33] G. W. Tang, X. D. Wang, R. M. Russell and N. I. Krinsky, *Biochemistry*, **30**, 9829 (1991).
[34] X. D. Wang, N. I. Krinsky, G. W. Tang and R. M. Russell, *Arch. Biochem. Biophys*., **293**, 298 (1992).
[35] X. D. Wang, R. M. Russell, C. Liu, F. Stickel, D. E. Smith and N. I. Krinsky, *J. Biol. Chem*., **271**, 26490 (1996).
[36] C. Kiefer, S. Hessel, J. M. Lampert, K. Vogt, M. O. Lederer, D. E. Breithaupt and J. von Lintig, *J. Biol. Chem*., **276**, 14110 (2001).
[37] M. R. Lakshmanan, J. L. Pope and J. A. Olson, *Biochem. Biophys. Res. Commun*., **33**, 347 (1968).
[38] C. Liu, X. D. Wang and R. M. Russell, *J. Nutr. Biochem*., **8**, 652 (1997).
[39] X. D. Wang, N. I. Krinsky, R. P. Marini, G. Tang, J. Yu, R. Hurley, J. G. Fox and R. M. Russell, *Am. J. Physiol*., **263**, G480 (1992).
[40] X. D. Wang, *J. Nutr*., **135**, 2053S (2005).
[41] A. Lindqvist, Y. G. He and S. Andersson, *J. Histochem. Cytochem*., **53**, 1403 (2005).
[42] C. C. Ho, F. F. de Moura, S. H. Kim and A. J. Clifford, *Am. J. Clin. Nutr*., **85**, 770 (2007).
[43] X. D. Wang, R. P. Marini, X. Hebuterne, J. G. Fox, N. I. Krinsky and R. M. Russell, *Gastroenterology*, **108**, 719 (1995).
[44] X. Hebuterne, X. D. Wang, D. E. Smith, G. Tang and R. M. Russell, *J. Lipid Res*., **37**, 482 (1996).
[45] M. Gajic, S. Zaripheh, F. Sun and J. W. Erdman Jr., *J. Nutr*., **136**, 1552 (2006).
[46] J. L. Napoli and K. R. Race, *J. Biol. Chem*., **263**, 17372 (1988).
[47] X. D. Wang, N. I. Krinsky, P. N. Benotti and R. M. Russell, *Arch. Biochem. Biophys*., **313**, 150 (1994).
[48] J. Brabender, R. Metzger, D. Salonga, K. D. Danenberg, P. V. Danenberg, A. H. Holscher and P. M. Schneider, *Carcinogenesis*, **26**, 525 (2005).
[49] S. M. Lippman and R. Lotan, *J. Nutr*., **130**, 479S (2000).
[50] X. C. Xu, J. J. Lee, T. T. Wu, A. Hoque, J. A. Ajani and S. M. Lippman, *Cancer Epidemiol. Biomarkers Prev*., **14**, 826 (2005).
[51] B. Houle, C. Rochette-Egly and W. E. Bradley, *Proc. Natl. Acad. Sci. USA*, **90**, 985 (1993).
[52] S. Song and X. C. Xu, *Biochem. Biophys. Res. Commun*., **281**, 872 (2001).
[53] G. Q. Chen, B. Lin, M. I. Dawson and X. K. Zhang, *Int. J. Cancer*, **99**, 171 (2002).
[54] J. M. Kurie, R. Lotan, J. J. Lee, J. S. Lee, R. C. Morice, D. D. Liu, X. C. Xu, F. R. Khuri, J. Y. Ro, W. N. Hittelman, G. L. Walsh, J. A. Roth, J. D. Minna and W. K. Hong, *J. Natl. Cancer Inst*., **95**, 206 (2003).
[55] H. Mernitz, D. E. Smith, A. X. Zhu and X. D. Wang, *Cancer Lett*., 101 (2006).
[56] R. M. Ponnamperuma, Y. Shimizu, S. M. Kirchhof and L. M. De Luca, *Nutr. Cancer*, **37**, 82 (2000).
[57] P. Prakash, C. Liu, K. Q. Hu, N. I. Krinsky, R. M. Russell and X. D. Wang, *J. Nutr*., **134**, 667 (2004).
[58] O. Ziouzenkova and J. Plutzky, *FEBS Lett*., **582**, 32 (2008).
[59] O. Ziouzenkova, G. Orasanu, G. Sukhova, E. Lau, J. P. Berger, G. Tang, N. I. Krinsky, G. G. Dolnikowski and J. Plutzky, *Mol. Endocrinol*., **21**, 77 (2007).
[60] C. Liu, X. D. Wang, R. T. Bronson, D. E. Smith, N. I. Krinsky and R. M. Russell, *Carcinogenesis*, **21**, 2245 (2000).
[61] X. D. Wang, C. Liu, R. T. Bronson, D. E. Smith, N. I. Krinsky and R. M. Russell, *J. Natl. Cancer Inst*., **91**, 60 (1999).
[62] X. D. Wang and R. M. Russell, *Nutr. Rev*., **57**, 263 (1999).
[63] M. G. Salgo, R. Cueto, G. W. Winston and W. A. Pryor, *Free Radic. Biol. Med*., **26**, 162 (1999).
[64] M. Paolini, A. Antelli, L. Pozzetti, D. Spetlova, P. Perocco, L. Valgimigli, G. F. Pedulli and G. Cantelli-Forti, *Carcinogenesis*, **22**, 1483 (2001).
[65] P. Perocco, M. Paolini, M. Mazzullo, G. L. Biagi and G. Cantelli-Forti, *Mutation Res*., **440**, 83 (1999).
[66] S. L. Yeh and M. L. Hu, *Food Chem. Toxicol*., **41**, 1677 (2003).

[67] W. Siems, I. Wiswedel, C. Salerno, C. Crifo, W. Augustin, L. Schild, C. D. Langhans and O. Sommerburg, *J. Nutr, Biochem.*, **16**, 385 (2005).
[68] C. Liu, R. M. Russell and X. D. Wang, *J. Nutr.*, **133**, 173 (2003).
[69] E. C. Tibaduiza, J. C. Fleet, R. M. Russell and N. I. Krinsky, *J. Nutr.*, **132**, 1368 (2002).
[70] J. Chung, C. Liu, D. E. Smith, H. K. Seitz, R. M. Russell and X. D. Wang, *Carcinogenesis*, **22**, 1213 (2001).
[71] C. Liu, R. M. Russell, H. K. Seitz and X. D. Wang, *Gastroenterology*, **120**, 179 (2001).
[72] X. D. Wang, C. Liu, R. T. Bronson, D. E. Smith, N. I. Krinsky and R. M. Russell, *J. Natl. Cancer Inst.*, **91**, 60 (1999).
[73] Y. Kim, N. Chongviriyaphan, C. Liu, R. M. Russell and X. D. Wang, *Carcinogenesis*, **27**, 1410 (2006).
[74] C. Liu, R. M. Russell and X. D. Wang, *J. Nutr.*, **134**, 426 (2004).
[75] F. Lian, K. Q. Hu, R. M. Russell and X. D. Wang, *Int. J. Cancer*, **119**, 2084 (2006).
[76] A. Matsumoto, H. Mizukami, S. Mizuno, K. Umegaki, J. Nishikawa, K. Shudo, H. Kagechika and M. Inoue, *Biochem. Pharmacol.*, **74**, 256 (2007).
[77] A. Ben-Dor, A. Nahum, M. Danilenko, Y. Giat, W. Stahl, H. D. Martin, T. Emmerich, N. Noy, J. Levy and Y. Sharoni, *Arch. Biochem. Biophys.*, **391**, 295 (2001).
[78] W. Stahl, J. von Laar, H. D. Martin, T. Emmerich and H. Sies, *Arch. Biochem. Biophys.*, **373**, 271 (2000).
[79] H. Araki, Y. Shidoji, Y. Yamada, H. Moriwaki and Y. Muto, *Biochem. Biophys. Res. Commun.*, **209**, 66 (1995).
[80] Y. Muto, H. Moriwaki and M. Omori, *Gann*, **72**, 974 (1981).
[81] M. Suzui, M. Masuda, J. T. Lim, C. Albanese, R. G. Pestell and I. B. Weinstein, *Cancer Res.*, **62**, 3997 (2002).
[82] Y. Muto, H. Moriwaki, M. Ninomiya, S. Adachi, A. Saito, K. T. Takasaki, T. Tanaka, K. Tsurumi, M. Okuno, E. Tomita, T. Nakamura and T. Kojima, *New Engl. J. Med.*, **334**, 1561 (1996).
[83] F. Lian, D. E. Smith, H. Ernst, R. M. Russell and X. D. Wang, *Carcinogenesis*, **28**, 1567 (2007).
[84] T. Suzuki, M. Matsui and A. Murayama, *J. Nutr. Sci. Vitaminol.*, **41**, 575 (1995).
[85] J. Y. Winum, M. Kamal, H. Defacque, T. Commes, C. Chavis, M. Lucas, J. Marti and J. L. Montero, *Farmaco*, **52**, 39 (1997).
[86] E. Nara, H. Hayashi, M. Kotake, K. Miyashita and A. Nagao, *Nutr. Cancer*, **39**, 273 (2001).
[87] E. Kotake-Nara, S. J. Kim, M. Kobori, K. Miyashita and A. Nagao, *Anticancer Res.*, **22**, 689 (2002).
[88] O. Aust, N. Ale-Agha, L. Zhang, H. Wollersen, H. Sies and W. Stahl, *Food Chem. Toxicol.*, **41**, 1399 (2003).
[89] A. L. Vine, Y. M. Leung and J. S. Bertram, *Mol. Carcinogenesis*, **43**, 75 (2005).
[90] H. Zhang, E. Kotake-Nara, H. Ono and A. Nagao, *Free Radic. Biol. Med.*, **35**, 1653 (2003).
[91] H. Mernitz, D. E. Smith, R. J. Wood, R. M. Russell and X. D. Wang, *Int. J. Cancer*, **120**, 1402 (2007).
[92] A. Ben-Dor, M. Steiner, L. Gheber, M. Danilenko, N. Dubi, K. Linnewiel, A. Zick, Y. Sharoni and J. Levy, *Mol. Cancer Ther.*, **4**, 177 (2005).
[93] P. Talalay, *Biofactors*, **12**, 5 (2000).
[94] P. Talalay, A. T. Dinkova-Kostova and W. D. Holtzclaw, *Adv. Enzyme Regul.*, **43**, 121 (2003).
[95] A. Giudice and M. Montella, *Bioessays*, **28**, 169 (2006).
[96] M. Ramos-Gomez, M. K. Kwak, P. M. Dolan, K. Itoh, M. Yamamoto, P Talalay and T. W. Kensler, *Proc. Natl. Acad. Sci. USA*, **98**, 3410 (2001).
[97] X. Gao and P. Talalay, *Proc. Natl. Acad. Sci. USA*, **101**, 10446 (2004).
[98] E. Balogun, M. Hoque, P. Gong, E. Killeen, C. J. Green, R. Foresti, J. Alam and R. Motterlini, *Biochem. J.*, **371**, 887 (2003).
[99] G. Shen, C. Xu, R. Hu, M. R. Jain, S. Nair, W. Lin, C. S. Yang, J. Y. Chan and A. N. Kong, *Pharm. Res.*, **22**, 1805 (2005).

[100] S. Gradelet, P. Astorg, J. Leclerc, J. Chevalier, M. F. Vernevaut and M. H. Siess, *Xenobiotica*, **26**, 49 (1996).

[101] V. Breinholt, S. T. Lauridsen, B. Daneshvar and J. Jakobsen, *Cancer Lett.*, **154**, 201 (2000).

[102] V. Bhuvaneswari, B. Velmurugan, S. Balasenthil, C. R. Ramachandran and S. Nagini, *Fitoterapia*, **72**, 865 (2001).

[103] F. Lian and X. D. Wang, *Int. J. Cancer*, **123**, 1262 (2008).

[104] R. J. Davis, *Cell*, **103**, 239 (2000).

[105] M. Karin, Z. Liu and E. Zandi, *Curr. Opin. Cell Biol.*, **9**, 240 (1997).

[106] P. Gupta and R. Prywes, *J. Biol. Chem.*, **277**, 50550 (2002).

[107] Y. Liu, M. Gorospe, C. Yang and N. J. Holbrook, *J. Biol. Chem.*, **270**, 8377 (1995).

[108] D. N. Slack, O. M. Seternes, M. Gabrielsen and S. M. Keyse, *J. Biol. Chem.*, **276**, 16491 (2001).

[109] C. Liu, R. M. Russell and X. D. Wang, *J. Nutr.*, **134**, 2705 (2004).

[110] D. D. Hirsch and P. J. Stork, *J. Biol. Chem.*, **272**, 4568 (1997).

[111] F. Furukawa, A. Nishikawa, K. Kasahara, I. S. Lee, K. Wakabayashi, M. Takahashi and M. Hirose, *Jpn. J. Cancer Res.*, **90**, 154 (1999).

[112] J. Chung, P. R. Chavez, R. M. Russell and X. D. Wang, *Oncogene*, **21**, 1539 (2002).

[113] Y. Kim, F. Lian, K. J. Yeum, N. Chongviriyaphan, S. W. Choi, R. M. Russell and X. D. Wang, *Int. J. Cancer*, **120**, 1847 (2007).

[114] G. Y. Kim, J. H. Kim, S. C. Ahn, H. J. Lee, D. O. Moon, C. M. Lee and Y. M. Park, *Immunology*, **113**, 203 (2004).

[115] K. Q. Hu, Y. Wang, R. M. Russell and X. D. Wang, *Carotenoid Sci.*, **12**, 180 (2008).

[116] H. Srinivas, D. M. Juroske, S. Kalyankrishna, D. D. Cody, R. E. Price, X. C. Xu, R. Narayanan, N. L. Weigel and J. M. Kurie, *Mol. Cell Biol.*, **25**, 1054 (2005).

[117] N. Inui, S. Sasaki, T. Suda, K. Chida and H. Nakamura, *Respirology*, **8**, 302 (2003).

[118] M. Karas, H. Amir, D. Fishman, M. Danilenko, S. Segal, A. Nahum, A. Koifmann, Y. Giat, J. Levy and Y. Sharoni, *Nutr. Cancer*, **36**, 101 (2000).

[119] J. Levy, E. Bosin, B. Feldman, Y. Giat, A. Miinster, M. Danilenko and Y. Sharoni, *Nutr. Cancer*, **24**, 257 (1995).

[120] H. Yu and T. Rohan, *J. Natl. Cancer Inst.*, **92**, 1472 (2000).

[121] D. R. Clemmons, W. H. Busby, T. Arai, T. J. Nam, J. B. Clarke, J. I. Jones and D. K. Ankrapp, *Prog. Growth Factor Res.*, **6**, 357 (1995).

[122] L. Jerome, L. Shiry and B. Leyland-Jones, *Endocr. Relat. Cancer*, **10**, 561 (2003).

[123] J. M. Chan, M. J. Stampfer, E. Giovannucci, P. H. Gann, J. Ma, P. Wilkinson, C. H. Hennekens and M. Pollak, *Science*, **279**, 563 (1998).

[124] S. E. Hankinson, W. C. Willett, G. A. Colditz, D. J. Hunter, D. S. Michaud, B. Deroo, B. Rosner, F. E. Speizer and M. Pollak, *Lancet*, **351**, 1393 (1998).

[125] J. Ma, M. N. Pollak, E. Giovannucci, J. M. Chan, T. Tao, C. H. Hennekens and M. J. Stampfer, *J. Natl. Cancer Inst.*, **91**, 620 (1999).

[126] E. Giovannucci, *J. Natl. Cancer Inst.*, **91**, 317 (1999).

[127] L. Arab, S. Steck-Scott and P. Bowen, *Epidemiol. Rev.*, **23**, 211 (2001).

[128] S. K. Clinton, C. Emenhiser, S. J. Schwartz, D. G. Bostwick, A. W. Williams, B. J. Moore and J. W. Erdman Jr., *Cancer Epidemiol. Biomarkers Prev.*, **5**, 823 (1996).

[129] E. Giovannucci, *Exp. Biol. Med.*, **227**, 852 (2002).

[130] A. Nahum, L. Zeller, M. Danilenko, O. W. Prall, C. K. Watts, R. L. Sutherland, J. Levy and Y. Sharoni, *Eur. J. Nutr.*, **45**, 275 (2006).

[131] M. Karas, M. Danilenko, D. Fishman, D. LeRoith, J. Levy and Y. Sharoni, *J. Biol. Chem.*, **272**, 16514 (1997).

[132] L. A. Mucci, R. Tamimi, P. Lagiou, A. Trichopoulou, B. Benetou, E. Spanos and D. Trichopoulos, *BJU Int.*, **87**, 814 (2001).

[133] M. D. Holmes, M. N. Pollak, W. C. Willett and S. E. Hankinson, *Cancer Epidemiol. Biomarkers Prev.*, **11**, 852 (2002).

[134] A. Vrieling, D. W. Voskuil, J. M. Bonfrer, C. M. Korse, J. van Doorn, A. Cats, A. C. Depla, R. Timmer, B. J. Witteman, F. E. van Leeuwen, L. J. Van't Veer, M. A. Rookus and E. Kampman, *Am. J. Clin. Nutr.*, **86**, 1456 (2007).

[135] K. Q. Hu and X. D. Wang, *unpublished results*.

[136] J. E. Trosko, C. C. Chang, B. Upham and M. Wilson, *Toxicol. Lett.*, **102-103**, 71 (1998).

[137] T. J. King and J. S. Bertram, *Biochim. Biophys. Acta*, **1719**, 146 (2005).

[138] M. Z. Hossain, L. R. Wilkens, P. P. Mehta, W. Loewenstein and J. S. Bertram, *Carcinogenesis*, **10**, 1743 (1989).

[139] J. S. Bertram, A. Pung, M. Churley, T. D. Kappock, L. R. Wilkins and R. V. Cooney, *Carcinogenesis*, **12**, 671 (1991).

[140] M. Rogers, J. M. Berestecky, M. Z. Hossain, H. M. Guo, R. Kadle, B. J. Nicholson and J. S. Bertram, *Mol. Carcinogenesis*, **3**, 335 (1990).

[141] L. X. Zhang, R. V. Cooney and J. S. Bertram, *Carcinogenesis*, **12**, 2109 (1991).

[142] F. Khachik, G. R. Beecher and J. C. Smith Jr., *J. Cell Biochem. Suppl.*, **22**, 236 (1995).

[143] L. M. Hix, A. L. Vine, S. F. Lockwood and J. S. Bertram, in *Carotenoids and Retinoids: Molecular Aspects and Health Issues* (ed. L. Packer, U. Obermüller-Jevic, K. Kraemer and H. Sies), p. 182, AOCS Press, Champaign, Illinois (2005).

[144] M. Hanusch, W. Stahl, W. A. Schulz and H. Sies, *Arch. Biochem. Biophys.*, **317**, 423 (1995).

[145] T. Nikawa, W. A. Schulz, C. E. van den Brink, M. Hanusch, P. van der Saag, W. Stahl and H. Sies, *Arch. Biochem. Biophys.*, **316**, 665 (1995).

[146] W. W. Pijnappel, H. F. Hendriks, G. E. Folkers, C. E. van den Brink, E. J. Dekker, C. Edelenbosch, P. T. van der Saag and A. J. Durston, *Nature*, **366**, 340 (1993).

Carotenoids
Volume 5: Nutrition and Health

Chapter 19

Editors' Assessment

George Britton, Synnøve Liaaen-Jensen and Hanspeter Pfander

A. Introduction

Since the question: "Can dietary β-carotene materially affect cancer rates?" first surfaced in 1981 [1], a large research effort has been directed to trying to determine if this is indeed the case. Much of the work has been based on the premise that any effect is likely to involve antioxidant action (*Chapter 12*) or effects on cellular and molecular processes (*Chapter 11*). The participation of the immune response system (*Chapter 17*) and suggestions that effects attributed to carotenoids may be mediated *via* retinoids or other metabolites/breakdown products (*Chapter 18*) have also attracted much attention. A variety of experimental approaches have been used to investigate the relationship between carotenoids and the incidence of cancer (*Chapter 13*) or coronary heart disease (CHD) (*Chapter 14*), particularly human, animal and cell studies. With the eye (*Chapter 15*) and skin (*Chapter 16*), the situation is different. These tissues are exposed to high intensity light, that can lead to photodamage. Do carotenoids have any protective roles against this damage?

Some 20 years ago it was recommended that biological properties and effects of carotenoids should be divided into functions, actions and associations [2]. In the present context, a **function** is an essential role played by the carotenoid, under normal physiological conditions, in growth, development and maturation, and maintaining life. An effect that can be demonstrated after administration of a carotenoid is considered as an **action**. It may or may not have general physiological significance. The term '**association**' is used when a demonstrated effect is associated with the presence of a carotenoid but cannot be directly attributed to that carotenoid. This perceptive insight has helped to shape thinking about the subject since then.

There is no doubt that carotenoids are the main source of vitamin A for most people, particularly those most at risk from vitamin A deficiency (VAD) (*Chapter 9*). Whatever other effects of carotenoids may or may not be substantiated, carotenoids will always be of vital importance for their role as provitamin A. But what of the other actions? What do we really know about carotenoids and human health? Where is there still uncertainty? There is much information about the major topics such as cancer and coronary heart disease (CHD) but rarely any definite proof. There are tantalizing glimpses of other interesting observations that would merit further investigation. In all this we must be guided by knowledge of the properties of the carotenoids as they exist *in vivo*.

So much is published. We can read so much in the popular press, publicity literature and articles on the internet. We read and hear extravagant claims 'supported by scientific research'. It is possible to find scientific literature that will contain some selected material which, taken out of context, will appear to give such support to almost any claim, though this may be based on uncritical experimental design and/or uncritical interpretation of results. This can be very confusing and lead the inexperienced reader to take it all at face value. It is therefore the duty of the 'carotenoid world' to plot a way through this and give informed judgement and guidance. This is what the authors in this *Volume* were asked to do and have done so well. They have given reasoned judgement and address some frequently asked questions. To a large extent, however, each chapter is a detailed account of one particular topic. The editors now attempt to put all this together to build a picture of where the subject stands today and what the future may hold.

We do not judge if studies and interpretations are good or bad. Our authors have done this when researching their chapters. Rather, we ask questions and raise points and recurring themes which readers should take into account when forming their own judgement and evaluation of published material.

B. From Food to Tissues

1. Sources, bioavailability and conversion

Apart from supplements (*Chapters 4* and *5*), we obtain our carotenoids from our food, primarily vegetables and fruit (*Chapter 3*). We therefore need to know what carotenoids are present and how much. Following the complex processes of digestion, absorption, transport and deposition, ingested carotenoids can be found in blood and other tissues and organs. Powerful methods are now available for analysing carotenoids in food, and in blood and body tissues (*Chapter 2*). For routine analysis, the use of HPLC is widespread but, without proper knowledge of the principles of separation, and without rigorous identification and peak assignment, this can lead to misleading information. Though upwards of 100 different carotenoids are present in all food sources in a varied diet when food is in good supply, few carotenoids are present in blood and tissues and those for which associations with health and biological activity have been studied are fewer still. Generally, these are the only ones that are included in the food composition tables. Quantitative analysis by HPLC is extremely precise but requires careful attention to sample preparation and critical appraisal of the level of precision that is realistic and acceptable. Also, analytical results recorded are for a particular sample grown in a particular place under particular, often optimized conditions, so the quantitative figures often recorded in food composition tables at high levels of precision cannot be expected to apply universally, when there is such variability between samples. If this is not recognized, food composition tables can be misleading. But, as general guides to what foods may be good sources of total or particular carotenoids, they are very valuable. There is also a need for a simple, inexpensive and reliable method for rapid basic analysis of, for example, the food that is actually being eaten, even in remote areas where facilities for laboratory analysis do not exist.

It is once the carotenoids are in the human body that the major uncertainty begins. Many factors affect bioavailability (*Chapter 7*). Apart from the great variability between individuals, a major factor is release from the food structural matrix. Bioavailability from oils or supplements is much better. We can make generalizations but these do not necessarily relate to any particular individual or food source/product.

A wide range of precise numerical factors for the conversion of provitamin A carotenoids to vitamin A have been reported in different studies (*Chapter 8*). Again there is much variability among individuals, so concentrating on the numbers can divert attention from the important questions about what factors strongly regulate or influence the conversion, such as the vitamin A status of the subject, the dose of provitamin given, the form and formulation presented, the food structure, and methods of cooking and processing. The numbers do, however, give us a picture and information on which to base guidance on important points and take steps to optimize the conversion. The accelerator mass spectrometry (AMS) technique [3] provides an extremely sensitive means of analysing isotopic tracers at a very low level and may prove invaluable for bioavailabilty/conversion studies.

2. Variability between individuals

In all questions about bioavailability and conversion, the great variability between individuals is a large and uncontrollable factor. There is great variability between individuals, even between members of the same population, community or family. It is recognized that there are 'responders' and 'non-responders' or 'low-responders' in terms of uptake, deposition and conversion of carotenoids, leading to great variability in carotenoid concentrations in blood and tissues following similar intake. Mean values for a population sample are likely to be derived from a wide span of values and may be of limited value. It may be the personal parameters of each individual that are important and there may be an optimal beneficial level for each individual. Below this, the risk of serious disease increases; above it, harmful effects become a possibility. Working on the basis of mean values could, therefore, be risky. The same value could correspond to a low carotenoid status in one individual but a high status in another. With the mapping of the human genome, new technologies of molecular biology and molecular genetics hold the key to solving these mysteries. An important example is understanding the basis of 'responders' and 'non-responders' by identifying the genetic and other factors that determine how efficiently an individual absorbs and stores carotenoids and converts them into vitamin A. This would open the door to real progress in defining the needs of individuals and populations. Recent work has revealed that genetic variations (single nucleotide polymorphisms, SNPs) can have a profound influence on the efficiency of the β-carotene-cleaving enzymes [4].

C. Carotenoids and Major Diseases: Practical Concerns and General Points

Chemistry and physics generally give definite answers. With biology this is often not so, and certainly not in the context of human health. Progress relies on building up a body of information, often based on statistical analysis and probability. The large variability between individuals is problematic. The biological cell is a complex system, multicellular organisms and their organs and tissues even more so. The complexity of the human body is almost incomprehensible. When looking at one factor we must always be aware that something completely different may be happening somewhere else or in another functional system. It is important not to think of one disease or effect in isolation. For example, when considering what may be recommended for skin health, it is necessary to consider what the consequences may be for cancer, CHD *etc.*

There are many reports of associations between higher carotenoid intake and reduced risk of major diseases such as cancer and CHD, but generally there is no proof of direct involvement and many questions remain. None of the experimental approaches on its own provides conclusive proof about whether carotenoids do have any effects and benefits for a real person under normal physiological conditions. But when they are taken together, indications begin to emerge. Here we draw attention to some aspects of experimental procedures that readers should bear in mind when evaluating results.

1. Human studies

Epidemiological surveys (*Chapter 10*), based upon reported normal food intake, may identify associations between high daily intake of total carotenoids or a particular carotenoid and reduced risk of serious degenerative diseases such as cancer and coronary heart disease. But there are particular areas of uncertainty. It is very difficult to distinguish whether any effect seen is due to the carotenoids or to the foods in which the carotenoids are concentrated. When carotenoids are given in pure form as supplements in intervention trials (*Chapter 10*), they introduce other uncertainties about the formulations and especially the doses administered, which are usually much higher than those obtained from the diet. Even if the doses given are comparable to levels in food, they have greater bioavailability and thus provide larger amounts. On the other hand, the slow release and absorption of the carotenoids during the digestion of food could also be a factor.

a) Are effects due to carotenoids or to food?

There is no doubt that there are associations between high carotenoid intake from natural food and hence higher concentration of carotenoids in blood and tissues, and reduced risk of serious diseases and conditions, especially cancer and CHD. But this does not prove that any protective effect is due to the carotenoid. The presence of a high level of carotenoid may simply be an indicator of a healthy diet rich in fruit and vegetables. Some other factor could be the biologically active principle; possibilities include fibre, other phytochemicals and antioxidants.

Advances in plant breeding and GM (*Chapter 6*) provide a possible approach to this problem. Red or yellow strains of carrots have been produced that accumulate lycopene (**31**) or lutein (**133**) instead of the β-carotene (**3**) and α-carotene (**7**) in familiar orange carrots.

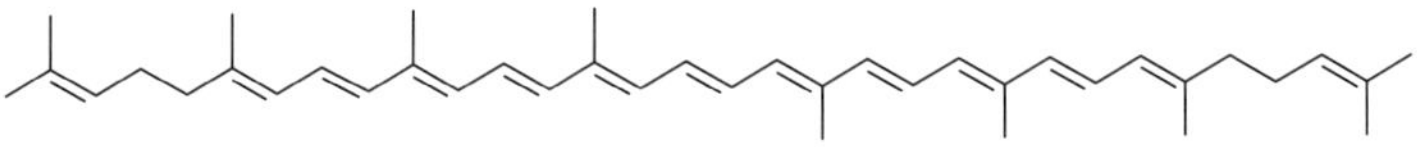

lycopene (**31**)

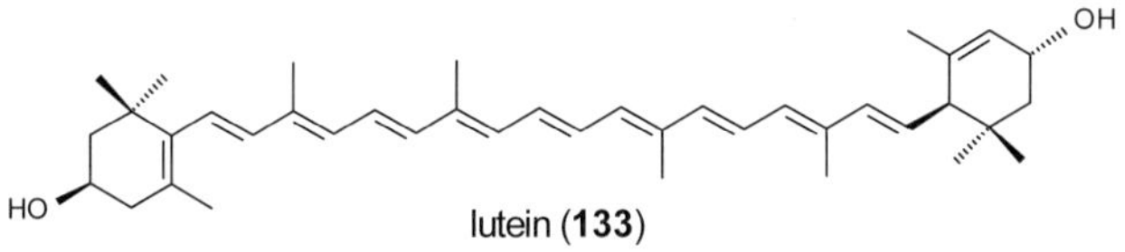

lutein (**133**)

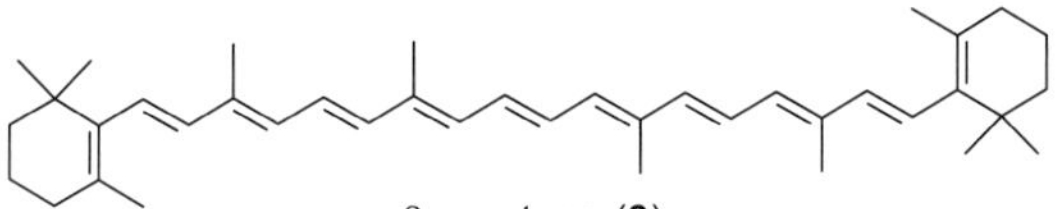

β-carotene (**3**)

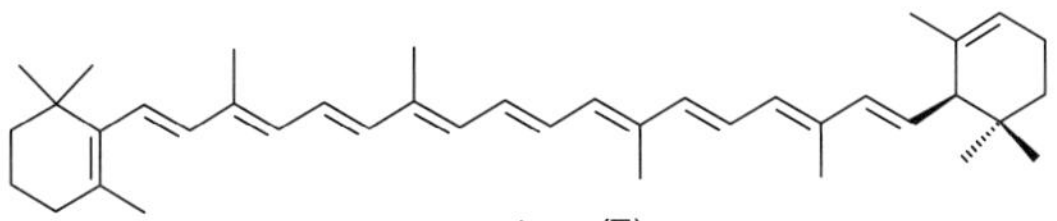

α-carotene (**7**)

Similarly, tomato strains that accumulate high concentrations of β-carotene or δ-carotene (**21**) in place of the usual lycopene, or even no carotenoid at all, are available. Comparison of feeding trials with carrots or tomatoes that provide different carotenoids, or between strains of the different sources that provide the same carotenoid, *e.g.* lycopene or β-carotene, may allow effects of carotenoids and the food material itself to be distinguished.

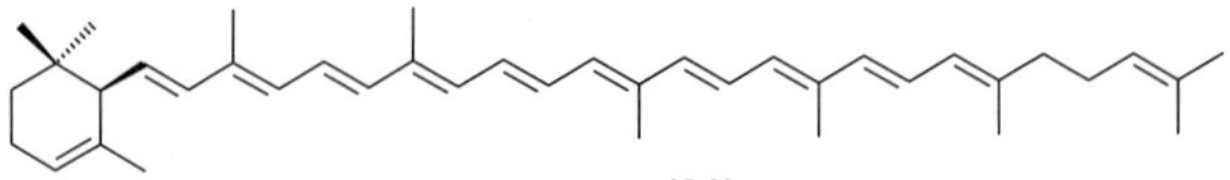

δ-carotene (**21**)

b) Biomarkers

It is important to have reliable biomarkers of carotenoid status. Reported, often retrospective, food intake patterns carry a degree of uncertainty and cannot take into account factors such as variation in bioavailability among individuals. An analytical biomarker such as blood or tissue carotenoid concentration should be more reliable, but a non-invasive method that does not require the taking of blood samples or tissue biopsies would be ideal. Some recent developments are worthy of note. Non-invasive methods, *e.g.* resonance Raman spectroscopic or reflectance photometric methods, for rapid analysis of carotenoids in the skin *etc.* as a biomarker of carotenoid status look promising, but further rigorous validation is necessary before results can be accepted with complete confidence and the methods can be applied extensively.

2. Cell cultures

Carotenoids can be shown to influence cell signalling at both protein and transcriptional levels in cell cultures *in vitro* (*Chapter 11*), but can the findings be related to the complex intact biological system that is the human body?

At high concentration, carotenoids and almost any other substance can be shown to have some effect. But that does not mean that this is relevant *in vivo* and at concentrations thay may be present in the cell under normal conditions.

3. Animal models

Research on effects of carotenoids in laboratory animals, mostly with rats, mice and ferrets, has generated much information. But these animals differ from humans in important biological features, so the data obtained may well not be applicable to the human. The most scientifically supportable solution would be to use our closest relatives, other primates, as models for experimental studies, but the cost of extensive trials would be prohibitive and many people have moral concerns about using primates in such a way.

4. High dose, low dose and balance

Much of the experimental work to study effects of carotenoids has used non-physiological doses or has applied carotenoids to cell cultures *etc.* in higher than physiological concentrations. The natural amounts of carotenoids in food are around 5 mg per day, whereas human trials typically use 20-100 mg/day, and equivalent levels are used in animal studies. It is common that beneficial associations are seen with carotenoids provided at the normal levels in food or with low levels as supplements, but adverse effects may be seen with the high doses. The unnaturally high pharmacological doses may be treated by the body as foreign substances, triggering detoxication mechanisms, including cytochrome P450 enzymes.

Prolonged, repeated intake of small amounts may be more effective than a single large dose. Good intake throughout life, starting at an early age, may be better than taking large doses as supplements later in life.

It is important to maintain a balance in cells and within the body, *e.g.* among carotenoids and with other factors, *e.g.* antioxidant vitamins C and E. High-dose supplements disturb the balance and may lead to unexpected and totally different results and consequences.

5. Safety and toxicity

The results of the ATBC and CARET trials from which it was concluded that daily high-dose β-carotene supplements increase the risk of lung cancer in heavy smokers have received much publicity. In non-smokers there is no evidence that high intake of carotene, either as supplements or from food, such as large amounts of carrots, has any serious adverse effect. The carotenodermia sometimes seen appears not to be harmful and the orange colour soon disappears when the excessive carotene consumption is discontinued. Despite much discussion, no recommended safe upper limit for carotene intake has been agreed. The taking of high-dose supplements of β-carotene by the general population is not recommended, though there is no recommendation that the use of such supplements for treatment of the photosensitivity disorder erythropoietic protoporphyria has dangerous consequences or should be discontinued.

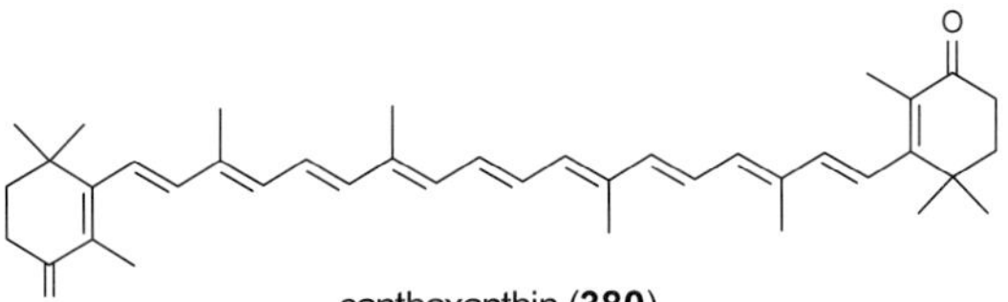

canthaxanthin (**380**)

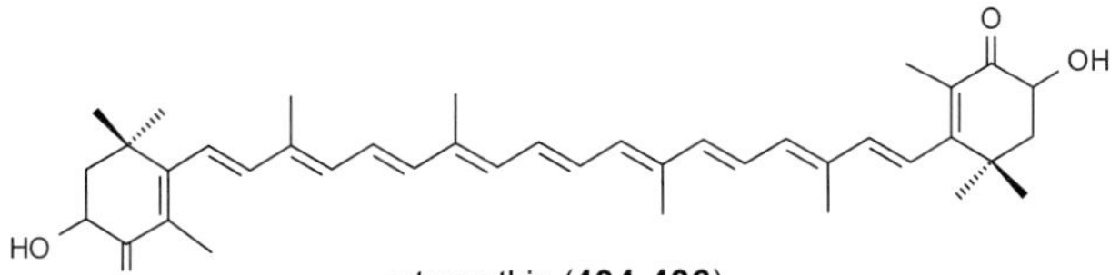

astaxanthin (**404-406**)

The practice of using oral canthaxanthin (**380**) capsules to give a tanned appearance of the skin has been discontinued because of cases in which microcrystalline deposits of canthaxanthin were seen in the eye. This appeared to have no long-term consequences and was rapidly reversed when canthaxanthin was no longer given.

Astaxanthin (**404-406**) supplements are now being promoted and are reported to have many beneficial effects. At normal dietary levels, astaxanthin is not detected in blood. With high dose supplements it may be, and it has been found to be taken up directly into erythrocytes. The consequences, good or bad, of this supplementation are unclear and must be evaluated.

No safety issues have been raised about other carotenoids, notably those that are now available and taken by some as dietary supplements, especially lutein, lycopene and astaxanthin. Equally, however, there is no direct evidence to prove that these compounds are completely safe. Knowledge of the properties of lycopene as a carotenoid that is easily oxidized and can have strong pro-oxidant properties suggests that, if trials with lycopene similar to the ATBC trials were undertaken, the adverse effects seen in smokers could be worse than with β-carotene.

Carotenoid supplements are subject to food legislation (*Chapter 4*). This seems reasonable if they do not lead to a total intake much greater than the normal dietary level that can be obtained from food. The high supplementary levels represent an unnatural situation, and the long-term effects are generally not known. It is not unreasonable that the use of carotenoids at these high pharmacological doses should be subject to the same stringent testing that is applied to pharmaceutical products.

6. Geometrical isomers

Now that modern analytical methods, especially HPLC, can readily detect geometrical isomers of carotenoids, and *Z* isomers are routinely found in food, blood and tissues, the question of whether they have any biological significance arises frequently. Some of the findings seem strange. For example, why should there be such great differences in the proportion of *Z* isomers in blood and tissues compared to the food sources? Various possibilities, such as enzymic conversion in the tissues have been suggested to explain this. Usually the only factor considered is the interconversion of isomers. This is always an equilibrium and should always lead to the same equilibrium mixture. Conditions will determine how long it takes for the equilibrium to be reached. But the *E/Z* isomerization is not the only equilibrium to consider. It is linked to other equilibria and mass action effects. All-*E* and *Z* isomers have different shapes and solubilities. All-*E* isomers aggregate and crystallize easily. Only the relatively small proportion that is in the monomeric form is available to participate in the isomerization equilibrium. Once formed, the *Z* isomers remain in solution, do not aggregate to any appreciable extent, and can be taken up and transported. The all-*E* isomer is more likely to re-aggregate, so the amount of this available to be taken up is small.

So, overall, the proportion of Z isomers that can be transported and deposited in tissues increases.

7. Natural *versus* synthetic

For some years there has been a public or at least media perception that 'natural is preferable to synthetic'. But a debate of this kind should be based on facts and not driven by emotion and deeply entrenched positions. If a carotenoid or other compound is pure, then natural and synthetic samples are chemically identical. Their biological actions must also be identical, provided they are in the same physical state and formulation. A natural extract rich in a particular carotenoid is not comparable to the pure form, natural or synthetic, because of the presence in the extract of many other components, any of which could have biological activity. There are, for example, reports that describe some biological effect of tomato-based products or tomato oleoresin and then attribute this effect to lycopene. The reader should look out for this in publications.

Production from natural sources is also generally considered to be more 'environmentally friendly' than industrial chemical synthesis, but is this now necessarily so? In industrial chemical synthesis there are strict controls on emissions, waste recycling or disposal *etc.*, so damage to the environment is minimized. Similar controls should be in place for industrial-scale extraction from natural materials, with its large requirement for solvents. The environmental cost of production of natural sources must also be considered. Production of a single crop in large areas of monoculture may involve a high cost in terms of the destruction of natural ecosystems or may use land and resources that would otherwise be used for food production.

The arguments are not simple, however. It is often said that the spread of oil palm plantations has come at the cost of large scale destruction of natural forests. But the main objective of the vast oil palm industry is to produce chemicals for soaps and detergents, and especially biodiesel. Production of the nutritionally rich carotene products is only a very minor part of the palm oil operation, and provides a useful and valuable product from what would otherwise be waste material from the major operations.

The use of other waste products would also seem to be an ideal goal, *e.g.* tomato skins for lycopene production, and would reduce the pressure for demand to be met from natural wild resources or additional extensive cultivation.

It is likely that new potentially valuable sources of carotenoids will be discovered, especially in remote places and these may be part of specialized ecosystems. As with so many natural resources, their uncontrolled exploitation could lead to serious environmental damage.

It is right that this debate should continue openly but should be based on the facts. Each case should be considered on its merits, and all aspects should be taken into account.

D. How Might the Effects be Mediated?

1. *Via* antioxidant action

Harmful reactive oxygen species and free radicals are continuously produced in the body during normal cellular functioning and are introduced from exogenous sources. Damage caused by these is associated with aging and with the incidence and progression of serious diseases including cancer, CHD, age-related macular degeneration and neurodegenerative conditions.

Although the body has a battery of defences against this (enzymes and endogenous antioxidants), it is widely believed that dietary supplementation with antioxidants can be a part of a protective strategy to minimize the oxidative damage, especially in the elderly and other vulnerable populations. But can carotenoids be counted among the effective antioxidants *in vivo*? Since the concept was first raised in 1984, much of the research on carotenoids and human health has been driven by the prospect that carotenoids may be among the group of antioxidants in fruits and vegetables that help to prevent damage caused by oxidizing free radicals. It is enlightening to compare the conclusions given in that paper [5] with the many other interpretations published by other reviewers and commentators.

Carotenoids have been shown to have antioxidant activity *in vitro* at physiological oxygen tensions. Most work has been done with model systems with oxidizing free radicals generated artificially from azo-initiators. Assays have usually addressed parameters of lipid peroxidation, most commonly the TBARS reaction, the reliability of which in the presence of carotenoids has been questioned; carotenoid oxidation products can give a positive reaction. Experimental design and control must be rigorous. Purity of the carotenoid tested is essential; if it is not free from peroxides, completely different results can be obtained. Comparison between different studies and different systems is, therefore, fraught with difficulty.

Model and non-biological studies of antioxidant effects are usually undertaken under conditions that do not resemble those *in vivo*. To demonstrate antioxidant activity in a natural system *in vivo* is extremely difficult; the system is so complex, conditions are difficult to control and are not uniform within the system, natural carotenoid concentrations are low and many interactions with other substances are possible.

There is, however, evidence that a synergistic/cooperative interaction between carotenoids and antioxidants such as tocopherols, ascorbic acid, and flavonoids, may play a role in the biological antioxidant network. This does not prove that carotenoids are antioxidants *in vivo*, however. It may be that the antioxidants protect the carotenoid and prevent the formation of pro-oxidant carotenoid peroxides and peroxy radicals, or regulate the formation of oxidative breakdown products that could have biological actions?

2. *Via* metabolites

A question that frequently arises is whether any effect attributed to a carotenoid is in fact due to a metabolite/breakdown product rather than to the intact carotenoid itself (*Chapter 18*). For any carotenoid, there are many products, as complex mixtures. Any one, or combination, could have some biological activity. The roles of vitamin A and retinoic acid are well known but evidence is accumulating to suggest that other non-retinoid breakdown products may be biologically active. Some possibilities that have been suggested are:

(i) β-Carotene cleavage products of different chain length or cleavage products of other carotenoids, bearing structural features (end group or chain) in common with retinoids, could act as retinoid agonists or antagonists.

(ii) Many carotenoid oxidation products have reactive α,β-unsaturated aldehyde structures. By binding to side-chain amino groups or, for dialdehydes, cross-linking, these could modify properties and activity of enzymes or other proteins.

(iii) The biological activity of many compounds is based on size, shape and position of functional groups. Although otherwise structurally unrelated, a carotenoid breakdown product could have the right topography to mimic some other molecule, *e.g.* vitamin D, hormones, and act as an agonist or antagonist.

3. *Via* the immune system

Carotenoid status is associated with immunocompetence (*Chapter 17*). Effects of several different carotenoids on various parameters of the human immune response system have been demonstrated, mostly in experiments *in vitro*. It is known that vitamin A strongly influences the immune system. When this is extended to provitamin A carotenoids, it is always possible that any action is *via* vitamin A. Effects are also seen, however, with some non-provitamin A carotenoids, especially lutein and astaxanthin. The question of whether antioxidant effects, or action *via* non-retinoid carotenoid metabolites may be involved in any such effects has been raised. This potentially interesting topic merits further careful investigation.

E. Reports of Other Health Effects

Possibilities of other effects of carotenoids on other aspects of health have been proposed. Most have not been pursued, but some do look interesting, though much more work is needed to substantiate any of these suggestions. The work is so far not extensive enough to warrant specific coverage in this *Volume*, but the topics are worth further consideration.

1. Water-soluble carotenoids

Carotenoids are usually not soluble in water and are therefore located in a hydrophobic environment *in vivo*. Carotenoid dicarboxylic acids, however, have appreciable solubility in water, allowing them to remain in an aqueous environment in cells or fluids.

Effects of the carotenoid dicarboxylic acid crocetin (**538**) and its disodium salt on haemorrhagic shock and wound trauma have been reported and attributed to increasing pulmonary oxygen flow to tissues [6]. It is proposed that the dipolar structure and its associated water shell impart special properties to the molecule. It would be interesting to know if similar effects could be found with carotenoid dicarboxylic acids of other chain lengths.

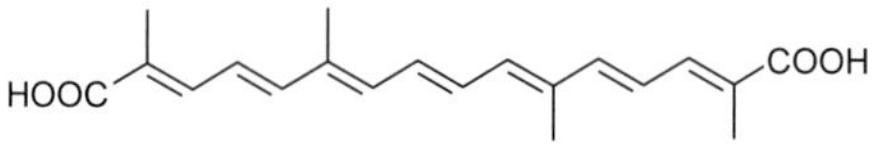

crocetin (**538**)

2. Bone health

Bone health is the result of two opposing processes, bone formation and bone resorption. These are regulated by various factors, notably hormones and vitamin D. β-Carotene, lycopene and other carotenoids have been associated with a role in supporting bone formation *via* increasing levels of bone-forming proteins [7]. This would constitute a benefit of carotenoids against brittle bones in the elderly. In contrast, high intake of vitamin A is associated with loss of bone mineral density and risk of osteoporosis (*Chapter 9*).

3. Metabolism and mitochondria

In recent years there are increasing suggestions that carotenoids (most experimental work has been with astaxanthin) may help to offset the imbalance known as 'metabolic syndrome' by increasing mitochondrial efficiency, energy metabolism and especially fat metabolism, thereby helping to reduce obesity and enhance athletic performance [8]. In intense and prolonged exercise, the level of metabolism in mitochondria in the muscles is high and oxidative stress greatly increases. Also, according to the 'mitochondrial theory of aging', oxidative damage, to DNA, protein and lipids, accumulates in the mitochondria over the lifetime of the organism. Mitochondrial dysfunction leads to many consequences, including pro-oxidative changes in redox homeostasis and efflux of mitochondrial components, notably cytochrome *c*. This has been linked to effects on signalling mechanisms, impairment of the immune response and neurodegenerative conditions (*Chapters 11* and *17*). The indications are interesting enough to merit further rigorous study, which should be extended to other carotenoids.

F. Final Comments: The Big Questions

(i) "Apart from their clear function as dietary precursors of vitamin A, do carotenoids have other functions or actions that are beneficial to human health?"

A number of associations have been reported between higher concentrations of carotenoids in the blood and reduced risk of serious diseases such as cancer and CHD. Also actions of carotenoids on various cellular and molecular processes have been demonstrated in cells in culture. Intervention trials to try to show a direct relationship between administered pure carotenoid intake and disease risk have been less informative; conflicting results are common.

(ii) "Are carotenoids important antioxidants *in vivo*?"

Although it is clear that carotenoids can serve as antioxidants *in vitro*, there is no unequivocal evidence for their functioning in this way *in vivo*. Indeed, carotenoid concentrations are low compared to those of recognized antioxidants such as vitamins C and E, so activity as a general antioxidant seems unlikely, but some specialized action in a particular sub-cellular environment such as a membrane, or in particular tissues, cannot be ruled out. All aspects must be considered. It is suggested, for example, that protection of LDL by lycopene as an antioxidant may be a factor in reducing risk of CHD. If this is looked at carefully, calculations show that, on average, there is only about one carotenoid molecule per four LDL particles. This does not seem compatible with a major antioxidant role.

(iii) "Do carotenoids help to provide protection and reduce risks of serious degenerative diseases?"

The honest answer to this and to most of the questions about the roles of carotenoids in human health is that we simply do not know. After many years of extensive and painstaking research, definite answers and unequivocal evidence are hard to find, though there is much circumstantial and indicative evidence. Much time and many resources have been spent chasing one popular idea to the exclusion of others or simply repeating experiments with one carotenoid after another. Would we now be better informed and have a better understanding, for example, if some of the massive research effort devoted to elucidating the proposed role of carotenoids as antioxidants had been channelled in other directions?

(iv) "Might some other carotenoids be important for health?"

Few of the hundreds of natural carotenoids have been studied for effects on human health. It is always possible that some, even ones that are not components of a normal diet, could have biological activity, though the question of safety and toxicity must be observed. One interesting structure is 3,3′-dihydroxyisorenieratene (**161**) which combines structural elements of carotenoids and tocopherol.

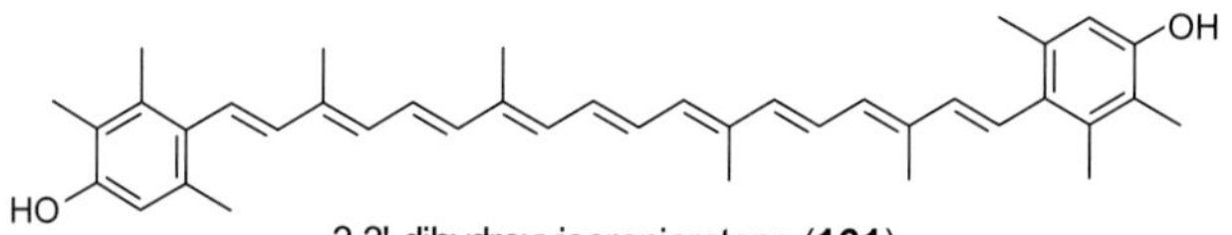

3,3'-dihydroxyisorenieratene (**161**)

(v) "Is it good for health to take carotenoid supplements?"

The bulk of evidence and the consensus of opinion supports the conclusion that intake of carotenoids at the levels obtained in a normal balanced diet is safe and may be beneficial, but the safety of larger intake, especially the high doses often taken as supplements is, at best, not proven. The reader must judge from the evidence. There is much uncertainty and it would not be wise or responsible of us to advocate the routine use of carotenoid supplements at levels above the normal dietary level of 5 mg/day.

We make no recommendation either way; personally, however, we are not convinced to take supplements but do try to maintain a good supply of carotenoids from our normal diet.

(vi) Finally, have we any advice for the researcher taking up the exciting challenges in this field?

Evaluate the literature carefully and critically. Look at experimental design and make your own interpretation. Don't rely only on interpretation by the author or by other reviewers or commentators. Make sure that any ideas and conclusions are compatible with the properties of the carotenoid as it exists in its natural state and surroundings. So enjoy it and:

Above all, be open-minded and expect the unexpected!!

References

[1] R. Peto, R. Doll, J. D. Buckley and M. B. Sporn, *Nature*, **290**, 201 (1981).

[2] J. A. Olson, in *Modern Nutrition in Health and Disease* (ed. M. E. Shils, J. A. Olson, A. C. Ross and M. Shike), p. 521, Saunders, Philadelphia (1998).

[3] R. S. Dueker, Y. Lin, A. B. Buchholz, P. D. Schneider, M. W. Lamé, H. J. Segall, J. S. Vogel and A. J. Clifford, *J. Lipid Res.*, **41**, 1790 (2000).

[4] F. Tourniaire, W. Leung, C. Méplan, A.-M. Minihane, S. Hessel, J. von Lintig, J. Flint, H. Gilbert, J. Hesketh and G. Lietz, *Carotenoid Sci.*, **12**, 57 (2008).

[5] G. W. Burton and K. U. Ingold, *Science*, **224**, 569 (1984).

[6] L. J. Giassi, A. K. Poynter and J. L. Gainer, *Shock*, **18**, 585 (2002).

[7] S. Sahni, M. T. Hannan, J. Blumberg, L. A. Cupples, D. P. Kiel and K. L. Tucker, *Am. J. Clin. Nutr.*, **89**, 416 (2009).

[8] M. Ishikara, *Carotenoid Sci.*, **12**, 3 (2008).

Index

A

B

C

D

E

F

G

H

I

K

L

M

N

O

P

R

S

T

U

V

W

X

Y

Postscript

This *Volume* brings to an end our work on this *Carotenoids* series, in which we have tried to tell the whole carotenoid story. It has been a demanding task that has taken over our lives for many years. But it has also been a 'labour of love'. We have learned a lot and hope this passes on to the reader. When the time comes for a new 'Carotenoids' project, it will be in new hands, seen with fresh eyes and bringing fresh ideas. Whoever undertakes this task, we wish them well.

In devoting ourselves to this project we have inevitably deprived others of our time and attention, and we thank our friends and colleagues and especially our families for their understanding and support during these long years. And we hope all who read and use these books will come to enjoy the challenges and beauty of carotenoids as much as we have.

George Britton

Synnøve Liaaen-Jensen

Hanspeter Pfander